Ultra-thin Chip Technology and Applications

Joachim N. Burghartz
Editor

Ultra-thin Chip Technology and Applications

Springer

Editor
Joachim N. Burghartz
Institut für Mikroelektronik Stuttgart
(IMS CHIPS)
Allmandring 30a
70569 Stuttgart
Germany
burghartz@ims-chips.de

ISBN 978-1-4419-7275-0 e-ISBN 978-1-4419-7276-7
DOI 10.1007/978-1-4419-7276-7
Springer New York Dordrecht Heidelberg London

Printed on acid-free paper

Springer is part of Springer Science+Business Media (www.springer.com)

To
Ulrich Gösele
25.1.1949 – 8.11.2009

Preface

After more than 40 years of semiconductor technology advancement we have accepted the technical framework that has enabled us to bring the minimum feature size down to 45 nm, an incredible 1,000-fold reduction from the integrated circuits of the early 1970s. Going further, we may face limits in device downscaling due to the physical silicon crystalline structure. Therefore, besides the coordinated miniaturization (More Moore), according to Gordon Moore's prediction from 1965, the International Semiconductor Roadmap for Semiconductors (ITRS) now also considers applications of silicon technology that depend less on pushing down the minimum feature size (More than Moore).

However, the basic concept of silicon wafer manufacturing applies to both More Moore and More than Moore applications and essentially is not put into question. But, to a novice to silicon technology, several questions may come up that would not likely be raised by the expert: Why do we accept considerable material loss by cutting rectangular chips from a circular-shaped silicon wafer? Or, why do we work with very thick silicon wafer substrates if we only need the upper 1% layer to integrate the circuit components? In view of the possibility that economics may become a show stopper for semiconductor manufacturing even before the physical or technological limits are in sight, these are indeed good questions. And, such questions may stimulate ideas on new applications of silicon technology.

As silicon is not only the best suited semiconductor for complex circuit and system integration but also a material of excellent mechanical properties, ultra-thin chip technology and applications are to emerge as a new paradigm of silicon technology. Extremely thin and thus flexible silicon chips are expected to greatly enhance the emerging thin-film and organic semiconductor technologies by combining the well-known high performance of silicon chip technology with the large area and system-in-foil (SiF) applications. Moreover, very thin chips with through-silicon-via (TSV) interconnects will provide a means of overcoming the interconnect bottleneck of planar chip integration by allowing a migration to three-dimensional integrated circuits (3D IC). A small chip thickness also leads to a reduced thermal resistance and thus helps the management of the power density issue in modern high-performance silicon chips. Finally, there are likely numerous new applications in microsystems that will emerge with the availability of ultra-thin chips.

However, ultra-thin chip technology features not only a new paradigm due to the thus emerging applications, but also because the fabrication of thin silicon chips will require certain adjustments and innovations in silicon process technology and circuit integration: Ultra-thin silicon chips need to have very high edge and surface qualities in order to retain the excellent inherent mechanical properties of silicon. Processing of thin wafers requires suitable handling techniques. Singulation of chips from thin wafers likely must be arranged by methods different from the conventional dicing process. The effect of stressors in increasing carrier mobility in the channel region of CMOS transistors will likely be different for extremely thin compared to thick silicon substrates. Also, ultra-thin chips will have physical properties that are different from those found on thick bulk substrates. Transistors on thin chips under variable bending stress will suffer from the piezoresistive effect shifting their operating point, which is thus an additional aspect to be considered in circuit design. The optical spectrum absorbed in a thin silicon film will be narrower compared to that absorbed on thick silicon. Also, the optical response of photo detectors on thin silicon might be affected by the properties of the back surface of the thin chip. The thermal boundary conditions of an integrated circuit on a thin silicon chip will to a large extent depend on the assembly of that chip into a rigid package or onto a foil substrate. Finally, for economic reasons, the techniques used to fabricate an ultra-thin chip will have to be optimized in terms of minimizing the density of defects to a tolerable level.

This book presents the general scope and the current status of the emerging ultra-thin chip technologies and applications. It was very fortunate that 75 leading international experts from academia, the semiconductor industry and equipment manufacturers could be brought together to contribute their expertise and vision to this new topic, and to do so in 34 chapters. Part I of the book provides an introduction, explaining the reasons why silicon technology has been based on thick wafers and chips for 40 years (Chap. 1) and why recently thin silicon chips became an issue considered by the ITRS (Chap. 2). Part II describes subtractive and additive thin chip fabrication based on bulk wafer thinning (Chaps. 3 through 5), wafer thinning that exploits etch stop techniques (Chaps. 6 and 7), and a new additive concept based on sintered porous silicon and epitaxial layer growth (Chap. 8). Part III relates to add-on process steps and modules that are applied to thin wafers in order to prepare them for specific applications, such as 3D IC and SiF (Chaps. 9 through 12). The assembly and embedding of ultra-thin chips is addressed in Part IV with SiF-specific chip embedding in foil (Chaps. 13 and 14) and general purpose chip-to-wafer alignment and micro-bump assembly (Chaps. 15 and 16). Part V is devoted to characterization and modelling of mechanical (Chaps. 17 and 18), electromechanical (Chaps. 19 and 20), thermal (Chap. 23) and optical (Chap. 24) properties. Also in this part, compact modelling of CMOS (Chap. 21) and bipolar (Chap. 22) are discussed, with a focus on flexible chips. Finally, in Part VI of the book, the possible improvements of known silicon applications and the emerging applications with thin chip technology are mentioned. Those include power applications (Chap. 25), back-illuminated imagers (Chap. 26), thin solar cells (Chap. 27), biomedical applications (Chap. 28), sensor applications (Chap. 29), RF-ID tags (Chap. 30), security chips (Chap. 31), drive chips for flexible displays (Chap. 32), microwave and millimeter wave (Chap. 33) and specific 3D IC (Chap. 34) applications.

I hope that this book will serve as a useful resource to researchers and engineers in industry as well as to scientists in academia in their effort to turn the new paradigm of ultra-thin chip technology and application into major new applications of silicon technology and to overcome bottlenecks in conventional silicon technology development.

I am grateful to all authors and co-authors who have contributed their time and energy to make this book a reality. In their name I also thank those people within their organisations who provided assistance to them. The compilation and editing of this book was, with great enthusiasm, supported by Astrid Hamala and Joachim Deh from IMS CHIPS. My own contribution would not have been possible without the patience and encouragement of my wife Susanne and my children Pia, Julia and Tamara, who found their husband and dad trapped behind the PC late at night for several weeks.

Stuttgart
November 1, 2010

Joachim N. Burghartz

Contents

Part III Add-on Processing

Part IV Assembly and Embedding of Ultra-Thin Chips

Part V Characterization and Modelling

About the Editor

Joachim N. Burghartz was born in Aachen, Germany, in 1956. He received the M.S. (Dipl. Ing.) from the RWTH Aachen, Germany, in 1982 and the Ph.D. (Dr.-Ing.) from the University of Stuttgart, Germany, in 1987; both the author's degrees are in electrical engineering. From 1982 to 1987 he was with the University of Stuttgart, where he developed sensors with integrated signal conversion, with a special focus on magnetic field sensors. From 1987 to 1998 he was with the IBM Thomas J. Watson Research Center in Yorktown Heights, New York. His earlier research work at IBM included device applications of selective epitaxial growth of silicon, Si and SiGe high speed transistor design and integration processes and deep submicrometer CMOS technology. In his last four years at IBM Dr. Burghartz was engaged in the development of circuit building blocks for SiGe RF front-ends, with a special interest in the integration of high quality passive components on silicon, which also included micromachining techniques. From 1998 to 2005 Dr. Burghartz was a full professor at Delft University of Technology, where he chaired the High-Frequency Technology and Components (HiTeC) group. His research interests there continued to be focused on silicon RF technology with a broader scope, ranging from investigations on materials to the design of RF circuit building blocks. From March 2001 until September 2005 the author was Scientific Director of the Delft Institute of Microelectronics and Submicron Technology (DIMES). Since October of 2005 he has been director of the Institute for Microelectronics Stuttgart (IMS CHIPS) and full professor at the University of Stuttgart, Germany, and from March 2006, he also has headed the Institute for Nano and Microelectronic Systems at the University of Stuttgart. Research and development activities at IMS CHIPS include CMOS and add-on process technologies, ASIC design, nanostructuring and advanced lithography, as well as high dynamic range CMOS imager technology. The institute also conducts a large research program on ultra-thin chip technology and applications, which is the motivation for this new book on the subject.

Joachim Burghartz is an IEEE Fellow and a Distinguished Lecturer of the IEEE Electron Devices Society (EDS). Currently, he is the EDS Vice President of Technical Activities and an EDS AdCom Member. He has served on the technical and executive committees of numerous IEEE and IEEE-sponsored conferences, including IEDM, BCTM and ESSDERC. He recently received the 2008 Jack Raper Award for outstanding technology directions paper at the internationally leading IEEE Integrated Circuits Conference, ISSCC. In 2009 he was awarded the State Research Award of Baden–Württemberg in Germany.

Dr. Burghartz has published 280 peer-reviewed articles and holds 34 patents and patent applications.

Contributors

Kurt Aigner
Infineon Technologies Austria, Villach, Austria

Evangelos A. Angelopoulos
Institute for Microelectronics Stuttgart (IMS CHIPS),
Allmandring 30a, Stuttgart 70569, Germany

Wolfgang Appel
Institute for Microelectronics Stuttgart (IMS CHIPS), Stuttgart, Germany

Ali Asif
Institut für Mikroelektronik Stuttgart (IMS CHIPS),
Allmandring 30a, 70569 Stuttgart, Germany

Joerg Bagdahn
Fraunhofer Institute for Mechanics of Materials, Halle, Germany;
Fraunhofer Center for Silicon Photovoltaics, Halle, Germany

Eric Beyne
Interuniversity Microelectronics Centre (IMEC),
Kapeldreef 75, Leuven, 3001, Belgium

Gernot Bock
Datacon Technology GmbH, Innstr. 16, 6241, Radfeld, Austria

Karlheinz Bock
Fraunhofer Research Institution for Modular Solid State
Technologies (EMFT), Hansastrasse 27d, 80686 Munich, Germany

Erwin Bosman
IMEC Ghent, Technologie Park 914, B-9052 Gent, Belgium

Jeroen van den Brand
Holst Centre, Eindhoven, KN 5605, The Netherlands

Joachim N. Burghartz
Institute for Microelectronics Stuttgart (IMS CHIPS), Stuttgart, Germany

Wim Christiaens
IMEC Ghent, Technologie Park 914, B-9052 Gent, Belgium

J. Fredrik Creemer
Delft Institute of Microsystems and Nanoelectronics (DIMES),
Delft University of Technology, P.O. Box 5053,
2600 GB Delft, The Netherlands

Vincenzo d'Alessandro
Department of Biomedical, Electronics, and Telecommunications
Engineering, University of Naples Federico II, Via Claudio 21,
80125 Naples, Italy

Johan De Baets
IMEC Ghent, Technologie Park 914, B-9052 Gent, Belgium

Piet De Moor
Interuniversity Microelectronics Centre (IMEC), Kapeldreef 75, 3001
Leuven, Belgium

Koen De Munck
Interuniversity Microelectronics Centre (IMEC), Kapeldreef 75, 3001
Leuven, Belgium

Andreas Dietzel
Holst Centre, Eindhoven, KN 5605, The Netherlands

Mark Doherty
IBM Semiconductor Research and Development Center,
1000 River Rd, Essex Junction, VT, USA

Jim Dunn
IBM Semiconductor Research and Development Center,
1000 River Rd, Essex Junction, VT, USA

Sebastian J. Eisele
Institut für Physikalische Elektronik (ipe), Universität Stuttgart,
Pfaffenwaldring 47, Stuttgart, Germany

Stefan Endler
Institute für Mikroelektronik Stuttgart (IMS CHIPS), Stuttgart, Germany

Stefan Engensteiner
Datacon Technology GmbH, Innstr. 16, 6241 Radfeld, Austria

Mete Erturk
IBM Semiconductor Research and Development Center,
1000 River Rd, Essex Junction, VT, USA

Patrick J. French
Delft Institute of Microsystems and Nanoelectronics (DIMES),
Delft University of Technology, P.O. Box 5053, 2600 GB Delft, The Netherlands

Heinz-Gerd Graf
Institut für Mikroelektronik Stuttgart (IMS CHIPS), Stuttgart, Germany

Oliver Häberlen
Infineon Technologies Austria, Villach, Austria

Christine Harendt
Institute for Microelectronics Stuttgart (IMS CHIPS), Stuttgart, Germany

Mahadi-Ul Hassan
Institute für Mikroelektronik Stuttgart (IMS CHIPS), Stuttgart, Germany

Tu Hoang
Institute für Mikroelektronik Stuttgart (IMS CHIPS), Stuttgart, Germany

Midoriko Honda
Sales Engineering Division Marketing Group, Marketing Team,
DISCO Corporation, Tokyo, Japan

Herbert Hopfgartner
Infineon Technologies Austria, Villach, Austria

Cheun-Wen Huang
IBM Semiconductor Research and Development Center,
1000 River Rd, Essex Junction, VT, USA

Alvin Joseph
IBM Semiconductor Research and Development Center,
1000 River Rd, Essex Junction, VT, USA

Alexander Kaiser
Reinhardt Microtech GmbH,
Wörthstrasse 85, 89077 UCM, Germany

Yoshikazu Kobayashi
Sales Engineering Division Marketing Group, Marketing Team,
DISCO HI-TEC EUROPE GmbH, Tokyo, Japan

Jan Kostelnik
Robert Bosch GmbH, Corporate Research and Development–Dept.
Plastics Engineering (CR/APP4), 71301 Waiblingen, Germany

Hannes Kostner
Datacon Technology GmbH, Innstr. 16, 6241 Radfeld, Austria

Mitsumasa Koyanagi
New Industry Creation Hatchery Center (NICHe), Tohoku University,
28 Kawauchi, Aza-Aoba, Aramaki, Aoba-ku, Sendai, 980-8579 Japan

Metin Koyuncu
Robert Bosch GmbH, Corporate Research and Development–Dept. Plastics Engineering (CR/APP4), 71301 Waiblingen, Germany

Andreas Kugler
Robert Bosch GmbH, Corporate Research and Development–Dept. Plastics Engineering (CR/APP4), 71301 Waiblingen, Germany

Christof Landesberger
Fraunhofer Research Institution for Modular Solid State Technologies (EMFT), Hansastrasse 27d, 80686 Munich, Germany

Walter Lang
Institute for Microsensors, -Actuators and -Systems (IMSAS), Microsystems Center Bremen (MCB), University of Bremen, Bremen, Germany

Franz Laermer
Robert Bosch GmbH, Corporate Research and Advance Engineering Microsystems (CR/ARY-MST), P.O. Box 106050, D-70049 Stuttgart, Germany

Thomas Laska
Infineon Technologies Austria, Villach, Austria

Paresh Limaye
Interuniversity Microelectronics Centre (IMEC), Kapeldreef 75, 3001 Leuven, Belgium

Mike McPartlin
IBM Semiconductor Research and Development Center, 1000 River Rd, Essex Junction, VT, USA

Slobodan Mijalković
Silvaco Europe, Compass Point, St Ives, Cambridge, PE27 5JL, UK

Kyriaki Minoglou
Interuniversity Microelectronics Centre (IMEC), Kapeldreef 75, 3001 Leuven, Belgium

Mattias Petzold
Fraunhofer Institute for Mechanics of Materials, Halle, Germany

Martin Plankensteiner
Sales Engineering Division Marketing Group, Marketing Team, DISCO HI-TEC EUROPE GmbH, Tokyo, Japan

Horst Rempp
Institute für Mikroelektronik Stuttgart (IMS CHIPS), Stuttgart, Germany

Michael Reuter
Institut für Physikalische Elektronik (ipe), Universität Stuttgart, Pfaffenwaldring 47, Stuttgart, Germany

Behzad Rejaei
Department of Electrical Engineering, Sharif University of Technology,
11365 Tehran, Iran

Harald Richter
Institute für Mikroelektronik Stuttgart (IMS CHIPS), Stuttgart, Germany;
Institut für Mikroelektronik Stuttgart (IMS CHIPS), Allmandring 30a,
70569 Stuttgart, Germany

Niccolò Rinaldi
Department of Biomedical, Electronics and Telecommunications Engineering,
University of Naples Federico II, Via Claudio 21, 80125 Naples, Italy

Salvatore Russo
Department of Biomedical, Electronics and Telecommunications Engineering,
University of Naples Federico II, Via Claudio 21, 80125 Naples, Italy

Sabine Scherbaum
Fraunhofer Research Institution for Modular Solid State Technologies (EMFT),
Hansastrasse 27d, 80686 Munich, Germany

Florian Schmitt
NXP Semiconductors Germany GmbH, Stresemannallee 101,
22529 Hamburg, Germany

Fabian Schnegg
Datacon Technology GmbH, Innstr. 16, 6241 Radfeld, Austria

Stephan Schoenfelder
Fraunhofer Institute for Mechanics of Materials, Halle, Germany;
Fraunhofer Center for Silicon Photovoltaics, Halle, Germany

Cor Scherjon
Institut für Mikroelektronik Stuttgart (IMS CHIPS), Stuttgart, Germany

Peter Stallhofer
Siltronic AG, Burghausen, Germany

Anthony Stamper
IBM Semiconductor Research and Development Center,
1000 River Rd, Essex Junction, VT, USA

Hannes Sturm
Institute for Microsensors, -Actuators and -Systems (IMSAS), Microsystems
Center Bremen (MCB), University of Bremen, Bremen, Germany

Thomas Suwald
NXP Semiconductors Germany GmbH, Business Unit Identification,
Georg-Heyken-Str. 1, 21147 Hamburg, Germany

Andrea Urban
Robert Bosch GmbH, Engineering Sensor Process Technology (EPT), P.O. Box 1342, D-72703 Reutlinger, Germany

Jan Vanfleteren
IMEC Ghent, Technologie Park 914, B-9052 Gent, Belgium

Dan Vanslette
IBM Semiconductor Research and Development Center, 1000 River Rd, Essex Junction, VT, USA

Wayne Woods
IBM Semiconductor Research and Development Center, 1000 River Rd, Essex Junction, VT, USA

Nicoleta Wacker
Institut für Mikroelektronik Stuttgart, Allmandring 30a, 70569 Stuttgart, Germany;
Institute für Mikroelektronik Stuttgart (IMS CHIPS), Stuttgart, Germany

Michael Zernack
NXP Semiconductors Germany GmbH, Stresemannallee 101, 22529, Hamburg, Germany

Wenqi Zhang
Interuniversity Microelectronics Centre (IMEC), Kapeldreef 75, 3001, Leuven, Belgium

André Zimmermann
Robert Bosch GmbH, Corporate Research and Development–Dept. Plastics Engineering (CR/APP4), 71301 Waiblingen, Germany

Martin Zimmermann
Institute for Microelectronics Stuttgart (IMS CHIPS), Stuttgart, Germany

Part I
From Thick Wafers to Ultra-Thin Silicon Chips

For more than 40 years we have been used to silicon technology based on thick and thus mechanically stiff circular-shaped wafers. We have accepted the fact that we sacrifice valuable silicon wafer real estate due to the rectangular shape of the chips that are cut from the round wafers. We have also disregarded the fact that the electronic structures only occupy the upper 1% of the wafer and, that the other 99% of the bulk wafer is apparently useless. However, this major part of the silicon wafer provides mechanical stiffness required for reliability in wafer processing and automated wafer handling. On the other hand, there are good reasons to look for possibilities to separate the thin electronic layer of a wafer from its bulk part. The interconnect dilemma in CMOS technology scaling has clearly pointed to three-dimensional (3D) circuit integration, pushing toward ultra-thin chips.

Chapter 1 explains the reasons why silicon wafers are as thick as they are. Chapter 2 highlights the need for thin chips, as projected on the International Roadmap for Semiconductors (ITRS). This first part of the book will, therefore, provide a motivation for the in-depth discussion on thin chip technology, add-on processes to thin chips, assembly and packaging, characterisation and modelling as well as ultra-thin chip applications discussed throughout the book.

Chapter 1
Why Are Silicon Wafers as Thick as They Are?

Peter Stallhofer

Abstract Silicon wafers have been the building blocks of the electronic industry for more than 40 years. Wafer size increased from 2 in. in diameter in 1970 the wafer size increased to 300 mm in 2000, enhancing the productivity of the chip manufacturing significantly. With growing diameters, wafer thickness of wafers increased steadily, reaching 775 μm for 300-mm diameter wafers. The primary reason for this increase was the need to ensure safe wafer manufacturing without breakage and to provide sufficient mechanical and thermal stability of the wafers in IC fabrication during processing steps of lithography and heat treatments.

Beyond this, silicon wafers also must meet certain defect kinetic properties in device processing, which depend on the wafer thickness as well and are crucial for device yield and economic feasibility.

1.1 Moore's Law and Trend in Wafer Diameter

Silicon is the foundation of semiconductor electronics and forms the material basis for the technology of modern communication society. Silicon has permeated all areas of our life, doing so to such a large extent that we do not appreciate its role in our daily routines: We commonly work with computers, entrust complex operations to electronic units, store or evaluate huge mounds of data, disseminate information worldwide via the Internet in a second and expect to be reachable everywhere by mobile devices. Silicon is the material that enabled the electronic revolution by providing a suitable platform for all these electronic devices. These devices have become affordable for everybody and are available in almost unlimited quantities.

The enormous potential of integrated circuits was recognized early by Intel cofounder Gordon Moore in his visionary and famous paper of 1965 [1]. In this paper he described the paramount economic and technological importance of

P. Stallhofer (✉)
Siltronic AG, Burghausen, Germany
e-mail: peter.stallhofer@siltronic.com

J.N. Burghartz (ed.), *Ultra-thin Chip Technology and Applications*,
DOI 10.1007/978-1-4419-7276-7_1, © Springer Science+Business Media, LLC 2011

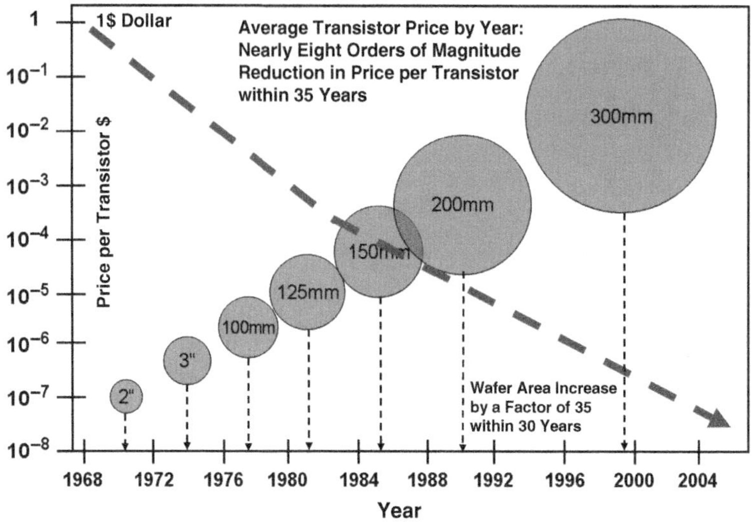

Fig. 1.1 Consequences of Moore´s law: price decay trend of electronic components and increase of wafer diameter

integrated semiconductor circuits. He recognized that permanent cost reduction and steady development of wafer technology together play the key role.

Moore predicted a progressive increase in integration density and simultaneous reduction of unit costs per circuit component. Moore's statement – known as Moore's law today – hypothesised that the density of components would be doubled every 2 years, a statement that has been valid for more than 40 years. The significant enlargement of wafer size by a factor of 35 from 2-in. to 300-mm diameter contributed an additional important element to enhancement of productivity in device fabrication and lower total costs (Fig. 1.1).

The validity of Moore's law has resulted in chips with a complexity and performance that had previously been inconceivable and with prices similar to common consumer goods. Advanced microprocessors contain more than one billion transistors on a silicon chip with an area of about 1 cm^2. For decades Moore's law has been established as a realistic guideline for development in the semiconductor industry. It forms the basis for the widely accepted technology roadmap, the International Technology Roadmap for Semiconductors (ITRS) [2] for the semiconductor community. The ITRS defines a set of technological requirements per device generation considered necessary to enable further miniaturization.

1.2 Wafer Size and Wafer Thickness

Larger wafer diameters are accompanied by a trend to increased wafer thicknesses. The thickness value is chosen to ensure mechanical and thermal stability during wafer manufacturing and device processing.

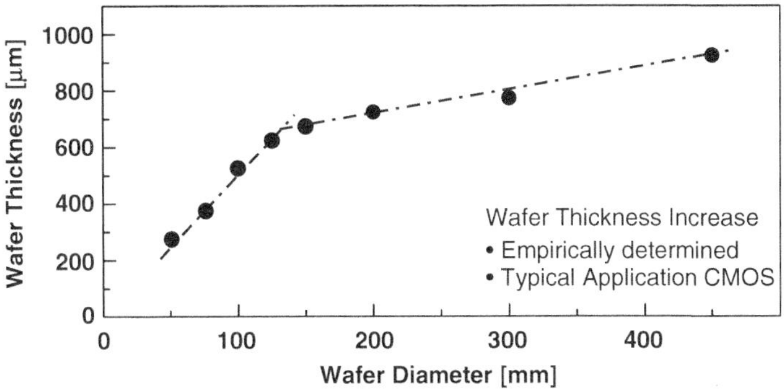

Fig. 1.2 Historical trend of wafer thickness and wafer diameter

The historical trend of wafer thickness as a function of wafer diameter is shown in Fig. 1.2. The wafer thickness was determined to be 275 μm for 2-in. wafers, 375 μm for 3-in. wafers, 525 μm for 100-mm wafers and 625 and 675 μm for 125- and 150-mm wafers, respectively. The current high volume wafers, 200- and 300-mm diameter, are produced with thicknesses of 725 and 775 μm, respectively.

Up to the 200-mm wafer generation, wafer thickness had been almost exclusively defined empirically, without much in-depth investigation. Wafer as well as device manufacturers carried out application-oriented experiments to determine optimal thickness, and, based on these results and their experience with previous wafer diameters, agreed upon a new wafer thickness value. At each wafer generation the Semiconductor Equipment and Materials International (SEMI), an international intertrade organization of leading semiconductor manufacturers, assigned a standard value to wafer diameter.

The most critical characteristics of a silicon wafer are fracture behavior and robustness in handling and transport operations. Silicon crystals are brittle and hence exposed to the risk of breakage during wafer manufacturing. Breakage is a major issue during the processes of ingot cutting, mechanical planarization and lapping. Furthermore, defects generated in these processes can cause wafer fracture in later wafering steps. The earlier inner diameter sawing technique used for smaller wafer diameters was accompanied by a higher breakage rate.

Wire sawing was introduced in the 200-mm wafer generation. In principle this allowed a transition to thinner wafers, which is considered in determining standardized wafer thickness for a given diameter. The wire sawing technique employs a metal wire for cutting with slurry, based on a silicon carbide (SiC) cutting grain immersed in a glycol carrier medium. This process is similar to a lapping process. In comparison, the previous ingot cutting technology – the inner-diameter saw – used diamond-coated sawing blades and exposed the silicon crystal to much higher mechanical forces during cutting. The forces generated during inner-diameter sawing inherently prohibited the transition to lower wafer thicknesses due to excessive crystal damage.

The key process for device miniaturization is optical lithography. It is used several times during device manufacturing and maps the fine geometric structures of the device onto the wafer surface. Lithography requires that the wafer maintain its mechanical stability throughout all process steps of device fabrication, including thermal processes such as diffusion or CVD deposition at elevated temperatures. Hence lithography requirements render it necessary to reproducibly place and accurately align the wafer on process chucks. Issues related to wafer bending, warpage and overlay problems between adjacent layers can lead to critical failures during lithography.

The standardized parameter describing wafer warpage is bow and is shown in Fig. 1.3. The degree of wafer warpage or twisting also depends on wafer thickness.

The bow of an unprocessed wafer has two sources: (1) every wafer has an intrinsic bow value resulting from the shape generated during wafer cutting; and (2) a bow can be generated by the weight of the wafer due to sagging. The latter is called gravitational bow.

The gravitational bow has become very relevant for the 300-mm wafer generation with a standardized thickness of 775 μm. The 300-mm wafer sags by about 200-μm when placed in a horizontal position and supported by three points around the wafer periphery.

Gravitational bow generates stress in the wafer (Fig. 1.3), which – amplified in thermal processes due to thermal gradients – can lead to slip and dislocation loops in the silicon crystal. Consequences of higher bow or warp are local crystalline defects, which cause functional failures in semiconductor devices and result in severe yield losses. In modern device production it has become necessary to develop appropriate support mechanisms for 300-mm wafers that allow an acceptable suspension of the wafer. Such support mechanisms are necessary to minimize the sagging effect in vertical furnaces and rapid thermal anneal (RTA) systems.

The gravity-induced bow is currently a very important issue. Extensive investigations have been performed to determine the standard thickness of the next proposed wafer generation with a diameter of 450 mm. If it were based on a three-point peripheral support mechanism, a 450-mm wafer with a thickness of

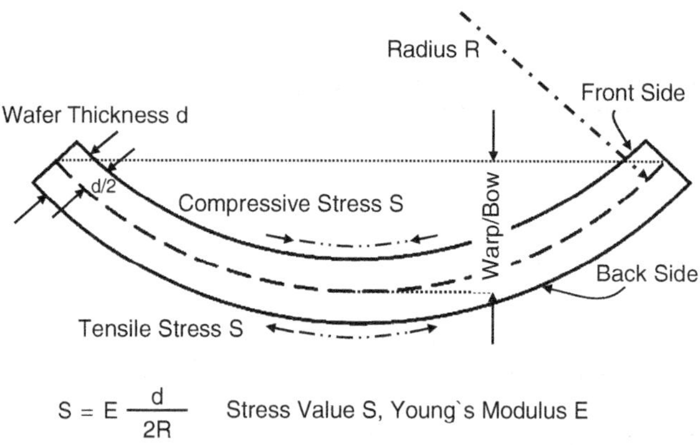

$$S = E\frac{d}{2R}$$ Stress Value S, Young`s Modulus E

Fig. 1.3 Schematic sketch of wafer with bow and related stress

775 μm would sag by more than 3 mm! To obtain the same sag as seen on the standardized 300-mm wafer, a thickness of 1,800 μm for 450 mm diameter wafer would be necessary [3]. This would require a factor of 2.3 times the silicon per processed chip area, which is unacceptable for the device community. The bulk crystal cost is the most significant cost factor in wafer manufacturing, one which has led to the latest compromise in wafer thickness of 925 μm. Present findings indicate that a thickness of 925 μm for a 450-mm diameter would be sufficient to effectively control the bow degradation during device processing [4].

Wafer thickness must not only result in acceptable material properties and affordable wafer bulk crystal costs, it must also be standardized. Thickness is a critical parameter for the set of sophisticated equipment used in an IC line and a standard is necessary for equipment compatibility. Once thickness is standardized, changes are only possible with considerably higher effort. Process chambers, pocket depths of susceptors, pitch heights, slit widths of process and transport carriers and handling by robots are only a few examples of equipment interaction with wafer thickness. Most tools used in the device manufacturing facility require standardized thickness specifications to enable fully automated device processing.

1.3 Silicon Wafer and Its Defect Kinetic Properties

The silicon wafer not only serves as a substrate to allow lithography and map the silicon surface with the device active structures, but it must also fulfill three additional requirements [5, 6]. These requirements are closely related to the silicon crystal and its defect kinetic properties. No highly integrated circuit would work and could be produced with economic yield without these properties (Fig. 1.4):

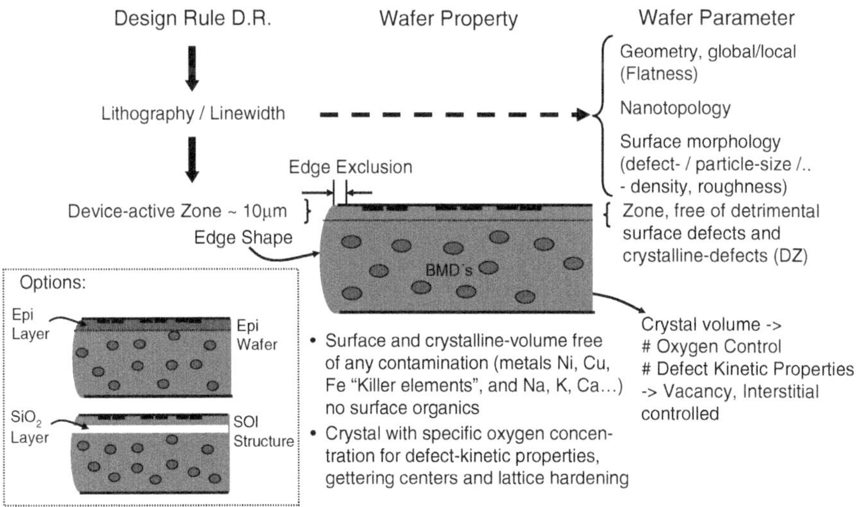

Fig. 1.4 Silicon wafer properties and related wafer parameters for a given design rule

1. The formation of a defect-free layer (denuded zone, DZ) below the surface of the silicon wafer, in which the active device islands are located.
2. The generation of gettering capable crystalline defects within the bulk of the wafer, which attract all detrimental impurities, in particular the fast diffusing metals like Ni, Cu, and Fe. This gettering keeps the impurities away from the active device zones (internal gettering [IG] is the technical term).
3. The silicon crystal has to contain a certain specified oxygen concentration, which is matched to the thermal budget of the device process. In addition to playing an important role in items (1) and (2), the oxygen concentration is critical in setting the crystal hardening behavior and resistance to slip and dislocations.

If, and only if, conditions (1) to (3) are fulfilled can the functionality of highly integrated devices be guaranteed. None of today's highly integrated complementary metal oxide semiconductor (CMOS) processes would be able to reach the economic yield necessary for low manufacturing costs without the defect kinetic properties of the silicon crystal.

The defect kinetic properties of silicon are closely linked to the oxygen content of the silicon crystal grown by the Czochralski method. The oxygen atoms (labeled Oi in figure 1.5) are incorporated into the growing crystal in a well-controlled manner. During crystal growth oxygen atoms are transferred from the crucible into the molten silicon and are segregated into the growing crystal. At 1,400°C the oxygen atoms occupy interstitial lattice sites with concentrations below 10^{18} cm^{-3} (DIN). During device processing, high temperature steps (~1,000–1,100°C, 1 h) cause oxygen in the vicinity of the wafer surface to out-diffuse, leaving a residual concentration below $(2–3) \times 10^{17}$ cm^{-3}, which corresponds to oxygen solubility at ~1,000°C. In the wafer volume, a certain amount of oxygen atoms conglomerate to form defect clusters and generate precipitates, the so-called bulk micro defects (BMDs). The near surface region remains cluster free and forms the so-called denuded zone (DZ). In the crystal volume the BMDs act as internal gettering sites due to the strain field associated with them. The remaining oxygen content on the interstitial sites enhances the hardening effect of the silicon lattice, showing higher resistance against slip formation.

The temperature steps necessary for the out-diffusion of the oxygen are inherent to device fabrication, e.g., in CMOS-like processes during the p-well drive in and in bipolar processes during the emitter drive-in step. Both process steps provide a suitable temperature-time window. The wafer has the following sandwich structure after device processing (Fig. 1.5).

Close to the wafer surface a defect-free layer exists having a depth of up to 20 μm, with the crystal volume containing the bulk micro defect region.

No other material combines all the exceptional advantages for integrated circuits as silicon does. It is readily available, allows for manufacturability of dislocation-free crystals, forms wafers with the highest possible perfection and purity and can prefectly match the defect kinetic properties to the needs of device manufacturing. All these attributes make silicon indeed it a special gift of nature.

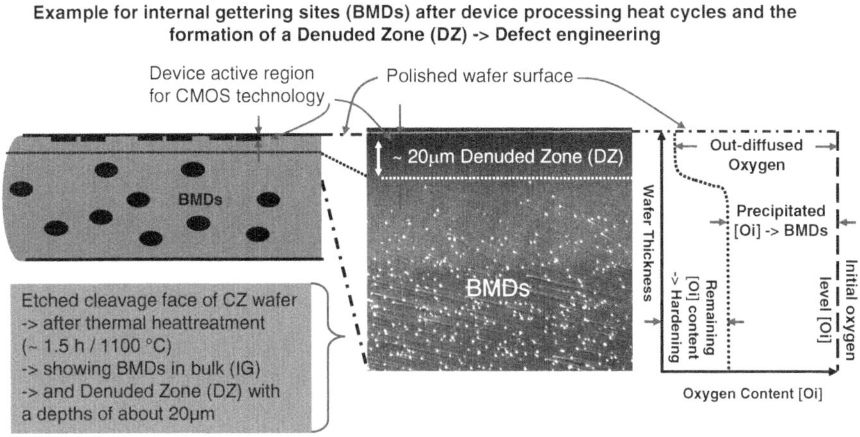

Fig. 1.5 Silicon wafer and the formation of oxygen-related bulk micro defects (*BMD*) with a denuded zone (*DZ*) formed during thermal heat treatment

All of the wafer characteristics described in the previous lines are driven by applications requiring highest integrated circuit density. These applications benefit strongly from Moore's law and are currently in the sub 100-nm regime. CMOS technology provides the platform for making the smallest possible feature size for electronic devices such as microprocessors, memory and logic circuits. Approximately 70–80% of silicon wafers are used in feature size critical CMOS applications. They are almost exclusively manufactured on 200-mm and 300-mm diameter wafers.

1.4 More than Moore Applications

In addition to the wafers used in CMOS-based applications a further 20% of silicon wafers are used for discrete (as opposed to integrated) devices, which find applications in power electronic sensors and micro-electro-mechanical systems (MEMS). These applications are sometimes referred to as "More than Moore" because they are not driven by the leading edge lithography and miniaturization trends but from other factors.

These discrete devices are not only dependent on the near surface region properties, as in the case of CMOS, but are strongly dependent on bulk crystal properties. The bulk material is actually an active part of the device structure. An extreme example of such a device is a 5-kV Thyristor which has lateral dimensions the size of one 100-mm wafer and a space charge region depth of 100 μm extending deep into the wafer.

PowerMOS transistors represent another product group that depends on the bulk electrical properties. PowerMOS transistors can switch currents of some hundred amperes and some thousand volts between terminals. The current is conducted through the substrate to a backside wafer contact. The substrate needs to have a very low resistivity of some mOhm-cm in order to reduce voltage loss in the device. The resistivity values and tolerances are critical for these devices. Consequently, power-MOS is an example of a device type that requires minimum substrate thickness to reduce voltage drop in the device.

Characteristics of more than Moore applications are:

- Relaxed lithography requirements, which can be as high as 3.5 μm minimum feature size.
- Active device area extending deep into the substrate.
- Very tight resistivity target and tolerance requirements. High resistivity of several hundred Ohm-cm or low resistivity in the mOhm-cm regime.
- Crystal volume free of critical crystal defects, i.e., very few oxygen conglomerations and prevention of BMD growth.
- Wafer diameters typically $\leq$150 mm, including both polished and nonpolished surfaces (lapped, etched or as-cut).
- The thickness spectrum of wafers used for these applications ranges from 200 to 1,000 μm.

Crystals grown for applications driven by Moore's law require interstitial oxygen Oi in the bulk material and are grown by the Czochralski method (CZ). The high defect requirements and resistivity tolerances for More than Moore applications prefer a different crystal growth process known as the float-zone (FZ) growth method. FZ crystals are free of Oi conglomerates and hence prevent the formation of BMDs. These wafers have no inherent internal gettering mechanism and other gettering means are required; e.g., mechanical gettering on the wafer backside. FZ grown wafers are also more susceptible to slip during high temperature processing, as the Oi hardening effect is not present. Wafer slip is avoided by reducing thermal gradients in furnace processes. This can be achieved by reducing push and pull rates into and out of the furnace. Increasing the distance between adjacent wafers in the furnace also helps to reduce thermal gradients.

In summary three factors are responsible for determining the wafer thickness:

1. Adequate mechanical and thermal stability of the wafer during the wafer manufacturing process and the device process to avoid wafer breakage. This stability is a prerequisite for processes that the lithography steps expose the wafer.
2. The silicon wafer must develop adequate gettering.
3. Device types requirements on the bulk material need to be considered.

For the highly integrated CMOS-based devices it is absolutely necessary to form a denuded zone and form internal gettering centers. These devices are manufactured in highly automated production lines with standardized equipment and hence forbid changes in wafer thickness. The situation changes if the wafer is further processed after completing the critical lithography steps. An example of further device processing after the critical lithography steps is device stacking to form three dimensional structures. The processed wafers may be thinned from the backside by grinding and/or etching to a thickness below 100 μm. A rest thickness of 25 μm has been demonstrated using this technology without causing damage to the CMOS-based devices. These extremely thinned wafers are very flexible and are temporarily bonded to support substrates, normally glass or ceramic, should additional processing be necessary. The support substrate may later be detached. The bonded wafer composite structure can further be processed using lithography on the wafer backside. The most frequent backside processing steps are void etching and contact metallization, which allow processed chips to be stacked on top of one another. This stacking of chips increases packaging density and reduces parasitic components by replacing lead-frame wires with very short metal contact lines. It also allows different types of chips to be stacked into one common package.

Gettering issues need careful consideration for the further processing of thin device wafers. Gettering centers that have formed during device processing (BMDs) can be removed during the thinning process. During post thinning metallization steps no getterer exists, and it is conceivable that metallic impurities can diffuse to the sensitive front-side device regions and cause contamination related failure. This failure mechanism is particularly dangerous for metals such as Cu, which diffuse quickly at relatively low temperatures ($<600^\circ$C). If the BMD gettering centers have been removed during the back thinning process then new centers can only be regenerated at high process temperatures (1,000°C). The engineering of the denuded zone and the BMD generation also needs to consider the post device processing including wafer thinning. It is also possible to bond a thinned wafer to a suitable substrate prior to device processing. In this case the denuded zone and BMD formation also need to be carefully engineered.

The details of gettering depend on the specific temperature-time profile of the device process. Two extreme cases help to understand the issues; in the first case the temperature budget is too high and interstitial oxygen diffuses out of the crystal surfaces resulting in two overlapping denuded zones. This leads to a situation where no gettering centers are formed and the Oi hardening effect is not present. In this case metallic impurities can diffuse to the critical device regions and cause failure. In the second case, the temperature budget is too low for out-diffusion of Oi atoms, and small defect centers form in the active device region resulting in either poor device characteristics or even complete functional failure.

The occurrence of gettering problems depends on the specific temperature-time profile of the specific device process. Consequently, the success of a suitable defect engineering measure depends on the specifics of the case at hand.

References

1. Moore GE (1965) Cramming more components onto integrated circuits. Electronics 38(8):114–117
2. http://www.itrs.net/links/2009ITRS/Home2009.htm

3. Kanda T, Fujiwara T, Takaishi K (12 Jan 2008) SUMCO. Semiconductor International
4. http://ismi.sematech.org/meetings/archives/ngf450/index.htm
5. Huff HR, Fabry L, Kishino S (eds) (2002) Semiconductor silicon 2002, vol 1 and 2. The Electrochemical Society, Inc., Pennington, NJ
6. Scheel HJ, Fukuda T (eds) (2003) Crystal growth technology. Wiley, New York

Chapter 2
Thin Chips on the ITRS Roadmap

Joachim N. Burghartz

Abstract The International Technology Roadmap for Semiconductors (ITRS) projects decreasing chip thickness in support of three-dimensional integrated circuit (3D IC) solutions. The 3D IC technology potential is not yet fully leveraged due to insufficient scaling of the through-silicon via (TSV) pitch. Since technological limits restrict the TSV ratio at about 10:1, minimum chip thickness enables us to take full advantage of the 3D IC concept in support of continued miniaturisation (More Moore). Moreover, the excellent mechanical properties of silicon qualify ultra-thin chips for applications of silicon technology for added functionality (More than Moore).

2.1 The ITRS Roadmap

In 1965 Gordon E. Moore presented his vision about the future exponential growth of the semiconductor industry, known today as 'Moore's Law' [1]. Following Moore's prediction the industry has kept a steady pace of miniaturisation, doubling device density every 18–24 months, doing so for 45 years now. It was not only about forecasting, but about commanding the silicon revolution. The idea has evolved into a more specific definition, a roadmap that is defined by semiconductor manufacturers and equipment providers.

Since the 1992 publication of the National Technology Roadmap for Semiconductors in the United States by the Semiconductor Industry Association (SIA), with new versions in 1994, 1997, and 1999, this treatise has been widely quoted throughout the industry [2]. Since 1998 the International Roadmap for Semiconductors (ITRS) has represented an international effort toward improving semiconductor device scaling by combining various national or regional roadmap initiatives worldwide, such as the European Electronic Component Manufacturers'

J.N. Burghartz (✉)
Institute for Microelectronics Stuttgart (IMS CHIPS), Stuttgart, Germany
e-mail: burghartz@ims-chips.de

J.N. Burghartz (ed.), *Ultra-thin Chip Technology and Applications*,
DOI 10.1007/978-1-4419-7276-7_2, © Springer Science+Business Media, LLC 2011

Association (EECA), the Electronic Industries Association of Japan (EIAJ), the Korea Semiconductor Industry Association (KSIA) and the Taiwan Semiconductor Industry Association (TSIA), besides SIA. Over the years ITRS has become the global industry's metronome, setting the pace for the development of new semiconductor technologies. Due to the fast-paced nature of the industry the roadmap needs to be reviewed annually. The international ITRS consortium with experts from over 900 companies organised in working groups projects the technological needs for the next 15 years on an annual basis.

2.2 Thin Chips for More Moore

The 2001 edition of the ITRS mentions for the first time about a need for thin dies to allow for three-dimensional (3D) chip stacking in system-in-package (SiP) solutions. In the 2003 edition very thin dies are mentioned, though not yet with a defined thickness target; only a forecast on the number of stacked dies was made. The 2005 ITRS edition put a strongly increased focus on wafer thinning and handling, small and thin die assembly and packaging of thin chips. A need for chips thinner than 20 μm was mentioned. Wafer thinning was the only technique considered for achieving thin chips. It was projected that at thicknesses below 10 μm a sequential combination of mechanical grinding, chemical–mechanical polishing (CMP), wet etching and plasma treatment, and dry chemical etching would be required to allow for control of such small chip thickness and to produce a die free of stress. The development of new pick and place techniques was viewed as a key issue for handling and assembling ultra-thin dies. The 2007 edition placed a stronger focus on the formation of TSV.

It is interesting that the projection of chip thickness requirements on the 2005 and 2007 Roadmap editions, as well as on the 2008 update, were identical

Table 2.1 Minimum wafer thickness projections on ITRS 2005, 2007, and 2008 update

	2005	2006	2007	2008	2009	2010	2011	2012	2013
ITRS-2005	50[1]	25[1]	20[1]	20[1]	15[1]	15[1]	10[1]	10[1]	10[1]
ITRS-2007		50[1]	20[1]	15[2]	15[2]	10[2]	10[2]	10[2]	10[2]
ITRS-2008			20[2]	15[2]	15[2]	10[2]	10[2]	10[2]	10[2]

2014	2015	2016	2017	2018	2019	2020	2021	2022	2023
10[1]	8[2]	8[2]	8[2]	8[2]	8[2]	8[2]			
10[2]	8[3]	8[3]	8[3]	8[3]	8[3]	8[3]	8[3]	8[3]	
10[2]	8[3]	8[3]	8[3]	8[3]	8[3]	8[3]	8[3]	8[3]	8[3]

[1] Manufacturing solutions exist
[2] Manufacturing solutions are known
[3] Interim solutions are known
[4] Manufacturing solutions are NOT known

(Table 2.1). However, confidence in technology's capacity to provide manufacturing solutions sank after the first projections were made in 2005.

This change is an indication that the challenges in wafer thinning and thin die fabrication had been underestimated and, that today manufacturing solutions are known but not in place for manufacturing.

The need for ultra-thin chips today comes primarily from 3D system integration, where multiple active dies having active and lateral interconnects are vertically connected through TSVs. Such a 3D interconnect scheme allows for effectively shorter lengths of intermediate and global wires compared to the conventional planar IC.

As shown in Fig. 2.1, the effective interconnect length of the active wires will continue to increase strongly. This relates to the ever-growing complexity of interconnects. However, investment into more interconnect levels does not alleviate that problem (Fig. 2.1). The only remaining solution to overcome the interconnect bottleneck is thus to consider true 3D ICs, in which the interconnect routing can exploit both the lateral and the vertical dimensions [3]. As Fig. 2.1 shows this concept is clearly more effective than the conventional multilevel interconnects even, if only one metal level per stratum is provided. When one exploits both the maximum number of interconnect layers on chip and the maximum number of strata, the increase in active wire length over time becomes subtle. Certainly, this concept may not be feasible both from a technological and an economic point of view. The trend in wire length increase with technology advancement remains in spite of the best technological effort; this may relate to the fact that the projections

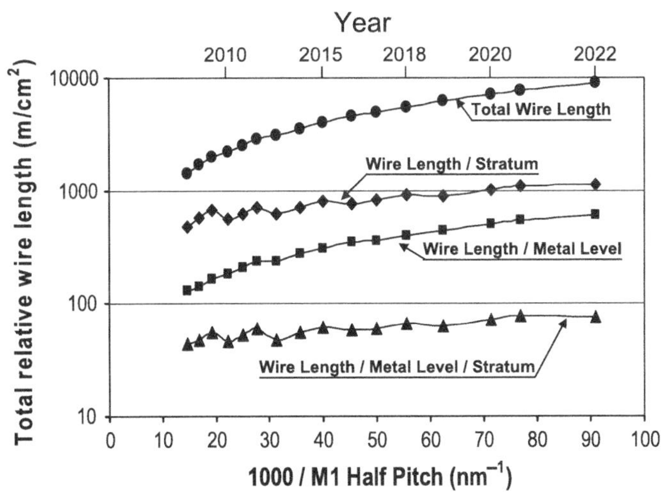

Fig. 2.1 Evolvement of the projection by the International Technology Roadmap for Semiconductors (*ITRS*) on total active wire length (M1 and intermediate interconnects) on chip. The wire length per metal layer, the wire length per stratum based on stratum dies having one metal layer only and the wire length per metal layer and stratum are calculated based on ITRS projections on the number of interconnect layers on chip and strata in 3D chips [2]

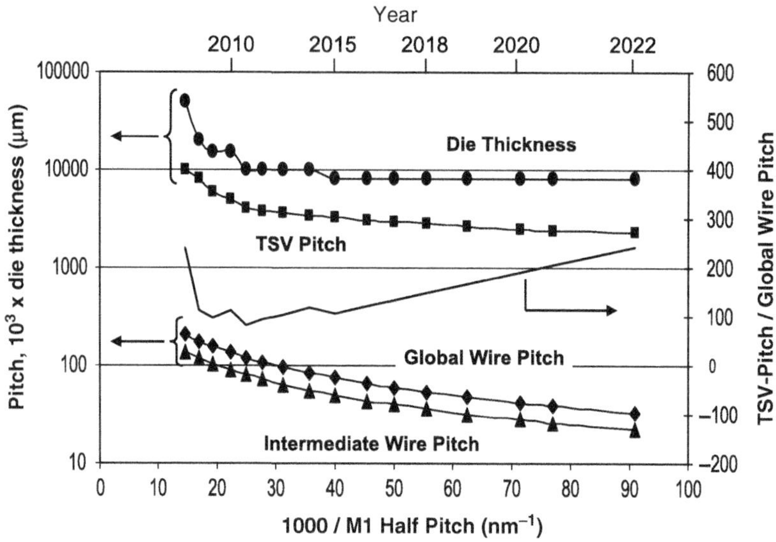

Fig. 2.2 Projection of the International Technology Roadmap for Semiconductors (*ITRS*) on global and intermediate on-chip pitch in comparison to the through-silicon via (*TSV*) pitch depending on die thickness. The calculated ratio of TSV and global wire pitch is large and will be increasing from 2010, thus indicating that utmost attention must be given to thin chip technology

on chip thickness assumptions, and thus on TSV pitch, have been too conservative. Figure 2.2 shows that from 2010 onward the advancement of global and intermediate on-chip interconnect pitch is stronger than that of TSV pitch. Unless the projections on chip thickness are revised in the coming editions of the ITRS, the effectiveness of TSV interconnects, indicated by the TSV/global wire pitch ratio in Fig. 2.2, will worsen over time. Note that the TSV pitch is directly related to the die thickness since, with current process solutions for via metal refill limit, the TSV depth/width ratio is 10:1 [2]. Clearly, considerably more attention must be put on ultra-thin chip fabrication techniques that can be made available for manufacturing [4].

2.3 Thin Chips for More than Moore

The 2005 edition of the ITRS first projected the need for focusing not only on device integration that relies on improvements in minimum feature size (More Moore) but also on applications leveraging silicon technology to provide added functionality (More than Moore). A need for thin chips was foreseen, e.g., in flip-chip packaging and chip assembly on flexible substrates and on textiles. Such applications rely less on electronic properties and technological advantages of silicon than on its excellent mechanical properties, which have been known for a long time [5]. Silicon features high stiffness, quite comparable to that of stainless steel and cast iron and about three times the stiffness of aluminium (Table 2.2). Silicon is known to be brittle, but its ultimate strength is eight times that of stainless steel and 35 and 15 times better than the values for cast iron and aluminium, respectively. The overall better mechanical properties of silicon when compared to stainless steel and cast iron come with a more than double higher thermal conductivity, which is only somewhat lower than that of aluminium. Silicon is also considerably lighter than the other materials listed in Table 2.2.

Table 2.2 Properties of silicon in comparison selected metals

	Young's modulus (GPa)	Ultimate strength (MPa)	Thermal conductivity (W/mK)	Density (g/cm^3)
Silicon	185	7000	150	2.33
Stainless steel	200	860	43	8.19
Cast iron	210	200	80	7.87
Aluminum	70	480	235	2.70

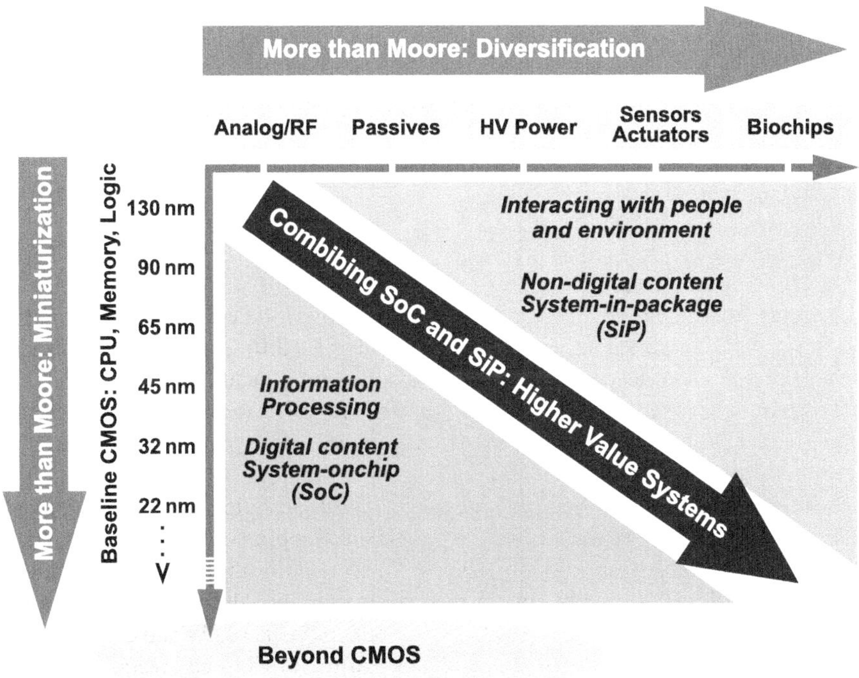

Fig. 2.3 Illustration of applications driven by strict miniaturisation according to Moore's law (More Moore), which may ultimately be superseded by novel device structures and physics (Beyond CMOS) and of applications of silicon technology providing added functionality and diversification (More than Moore)

References

1. Moore GE (1965) Cramming more components onto integrated circuits. Electronics 38 (8):114–116
2. http://www.itrs.net/reports.html
3. Joyner JW, Zarkesh-Ha P, Meindl JD (2001) A global interconnect design window for a three-dimensional system-on-a-chip. In: Proceedings of the IITC, Burlingame, CA, pp 154–156
4. Burghartz JN, Appel W, Harendt C, Rempp H, Richter H, Zimmermann M (2009) Ultra-thin chips and related applications, a new paradigm in silicon technology. In: Proceedings of the ESSDERC, Athens, Greece, pp 29–36
5. Peterson K (1982) Silicon as a mechanical material. In: Proceedings of the IEEE, vol 70, no. 5, pp 420–457

Part II
Thin Chip Fabrication Technologies

As we have explained, the past forty years microelectronic manufacturing has relied on mechanically stiff and stable wafer substrates (Chap. 1). However, the need for reducing the thermal resistance to the package and for achieving a suitable form factor for very small chips has driven techniques that allow for thinning wafers to smaller thicknesses afterwards. Such thinned wafers (100–300 μm) are still mechanically stiff but tend to fracture and thus have to be handled with great care. Below 100 μm they become even more fragile and in addition start to bend under their own weight. In the thickness range 50–100 μm, fracture of wafers can be initiated even by small deflections. Below 50 μm wafers and chips become flexible, i.e., they remain intact also with a considerable degree of deflection applied. At 10 μm and smaller thicknesses they could tolerate bending radii of the order of 1 mm and less, which is out of the range of practical application. Such ultra-thin chips are – in principle – unconditionally stable, flexible and reliable. The ranges of wafer thickness in summary:

- >300 μm Unconditional mechanical stability
- 100–300 μm Stiffness but limited mechanical stability
- 50–100 μm Limited stiffness and limited mechanical stability
- 10–50 μm Good flexibility and good mechanical stability
- <10 μm Excellent flexibility and unconditional mechanical stability

Until now wafer thinning has been a task of the assembly and packaging community. Thinning through grinding and subsequent auxiliary processes has matured to make wafers and chips available at thicknesses in the range of 50–100 μm in cost-effective manufacturing (Chap. 3). The recently increasing drive toward smaller thickness is motivated by new applications of silicon chip technology, which rely on mechanical flexibility, assembly on nonflat surfaces and chip stacking (Chaps. 9–12 and Chaps. 26–37). The excellent properties of chips having thicknesses below 50 μm can only be leveraged if they are not noticeably degraded by the thinning process. Conventional wafer grinding and chip dicing, however, introduces crystalline defects and micro cracks at a considerable level. Therefore, new techniques such as dicing-by-grinding are adopted to prevent edge chipping effects during chip dicing (Chap. 4).

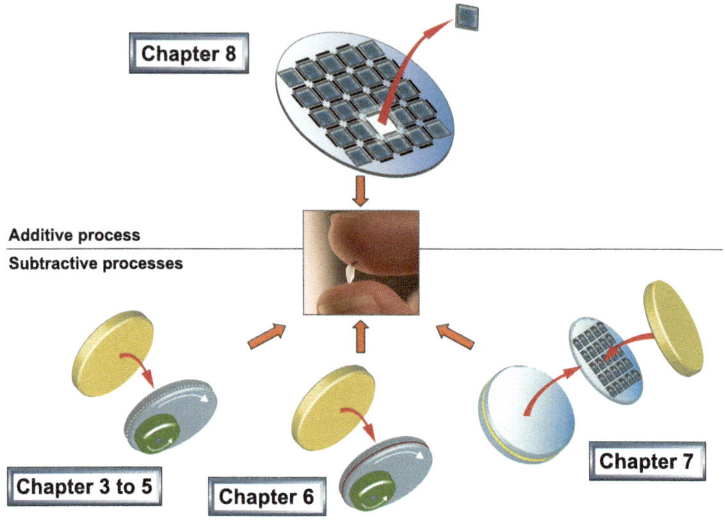

Fig. II.1 Illustrative comparison of generic subtractive technologies and one additive process technology targeted at fabrication of ultra-thin chips

Since silicon wafers at thickness below 50 μm lack mechanical stiffness they need to be supported by auxiliary carrier substrates during grinding. Bonding to – and debonding from – carriers considerably adds to process cost and may lead to reduced overall process yield, thus increasing the production expenditure even more. Therefore, thinning techniques that do not require any handle substrates hold great interest for development (Chap. 5).

The thinning process itself may be prone to excessive thickness nonuniformity across the wafer, particularly for an extremely small thickness target. Considerable progress has been made by using *in situ* thickness monitoring during wafer grinding, though it is uncertain if chip thickness far below 50 μm will be feasible in - cost-effective manufacturing. Better thickness uniformity requires the implementation of etch barriers into the wafer substrate. Such an etch barrier can be based on an epitaxially grown electrochemical etch stop layer (Chap. 6); alternatively, SOI wafer substrates can be employed (Chap. 7). Those technologies, particularly SOI substrates, will lead to much higher cost and are thus preferable for high-end applications. Evidently, all such subtractive thin chip fabrication processes lead to considerably higher cost and process effort as chip thickness decreases. A consequent step toward realising ultra-thin chips at reasonable cost is to consider additive thin chip fabrication, which offers hope for excellent chip thickness control. The recently introduced Chipfilm™ technology, e.g., features excellent thickness control by epitaxial growth and even reuse of the bulk silicon substrate (Chap. 8).

In general, different fabrication methods for thin wafers and chips (Fig. II.1) offer prospects for the new applications of silicon technology, exploiting the unique features of ultra-thin chips under the economic constraints of the semiconductor industry.

Chapter 3
Thin Wafer Manufacturing and Handling Using Low Cost Carriers

Florian Schmitt and Michael Zernack

This chapter focuses on thin wafer fabrication and processing that uses low cost sub carriers. It describes the state of the art and the technical boundaries of the application of foils for wafer support and protection. First, the thinning technology and the applied materials are described in terms of process capability and maturity. Methods of thin wafer characterization are presented. Subsequently, the impact of front end design on ultra-thin wafer manufacturing is highlighted. Finally, processes after wafer thinning that enable the "perfect" die are described.

3.1 Wafer Thinning Technologies for Different Thickness Ranges

During the standard industrial process flow, semiconductor wafers are thinned on wafer level after front end manufacturing and before integration of silicon chips into packages. Based on one dedicated circuit design, different chip thicknesses enable a broad range of different applications. Final wafer and chip thicknesses depend on application requirements such as package size (form factor), heat dissipation, electronic properties of the substrate, mechanical and thermal stability and flexibility. Wafer thickness of 150 μm is common and widespread and can be manufactured using standard grinding methods and equipment. Below 150 μm according wafer support is inevitable for a safe handling. Depending on wafer properties such as bow and warp, the structure and composition of active layers including topology and edge profile, wafers can be moved and processed on low cost grinding foils down to a thickness of 30 μm. Beyond that size rigid wafer carrier systems have to be applied in order to allow adequate wafer handling and thinning processes. However the step required for mounting and removing rigid carriers adds a significant contribution to the overall manufacturing costs.

F. Schmitt (✉)
NXP Semiconductors Germany GmbH, Stresemannallee 101, 22529 Hamburg, Germany
e-mail: florian.schmitt@nxp.com

J.N. Burghartz (ed.), *Ultra-thin Chip Technology and Applications*,
DOI 10.1007/978-1-4419-7276-7_3, © Springer Science+Business Media, LLC 2011

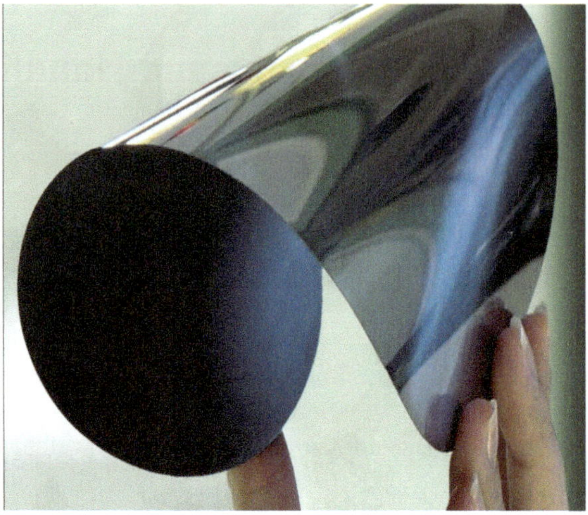

Fig. 3.1 Flexible 50 μm wafer

Fig. 3.2 Fifty micrometers wafer on film frame carrier

Below 150 μm thickness silicon wafers become increasingly flexible (Fig. 3.1). On the one hand flexibility of chips reduces the fracture strength when they are used in flexible card or board applications; on the other hand it complicates handling and subsequent manufacturing processes. The default delivery form today for wafers thinner than 150 μm is on dicing foil and film frame carrier (Fig. 3.2).

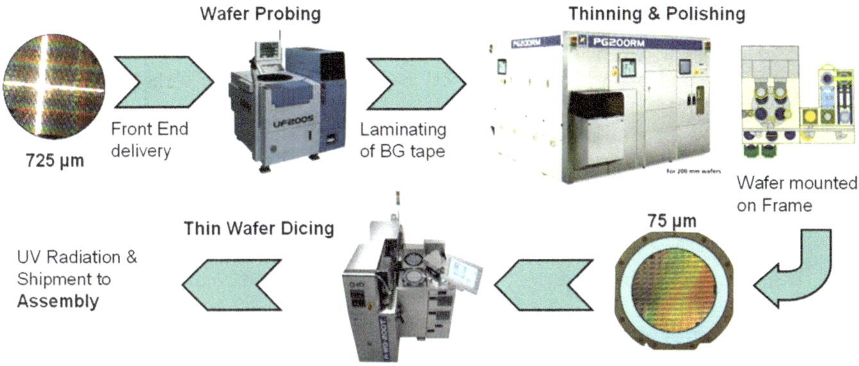

Fig. 3.3 Thin wafer process from wafer test to wafer separation

However, the wafer does not only have to be stabilised for transport after thinning, but also during the complete wafer thinning process. Wafer thinning encompasses several steps: lamination of a grinding tape on the active side of the wafer (protection from process chemistry and mechanical carrier), multigrinding step (coarse and fine grinding), stress relief (chemical-mechanical polishing or, alternatively, wet or dry etching), wafer mounting onto dicing foil and film frame carrier and grinding tape removal. The process sequence requires continuous handling of the thin wafer on a substrate. The thinning process module is fully integrated into a general purpose wafer treatment flow that enables the delivery of known good (tested) ultra-thin dies (Fig. 3.3). The process flow concept tolerates variations like wafer test after wafer thinning, separation on specialised probing equipment as well as laser separation of wafers.

3.2 Thinned Wafer Properties

Subsequent assembly processes and product specifications are highly demanding on the thinning process, and with respect to the applied materials. Principal requirements are:

- Thinning after bumping, i.e., embedding of topographic features
- Low target thickness and low total thickness variations
- Defect-free ultra-thin wafers
- High mechanical stability of chips

Figure 3.4 shows the impact of grinding tape inhomogeneities (upper left plot) on the thickness of the wafer (upper right plot). This effect can exceed the total thickness variation contributed by the grinding/polishing equipment itself.

Grinding tapes are composed of at least two layers, a harder polymeric backbone and a softer oligomeric or monomeric (UV) adhesion film. Although greater

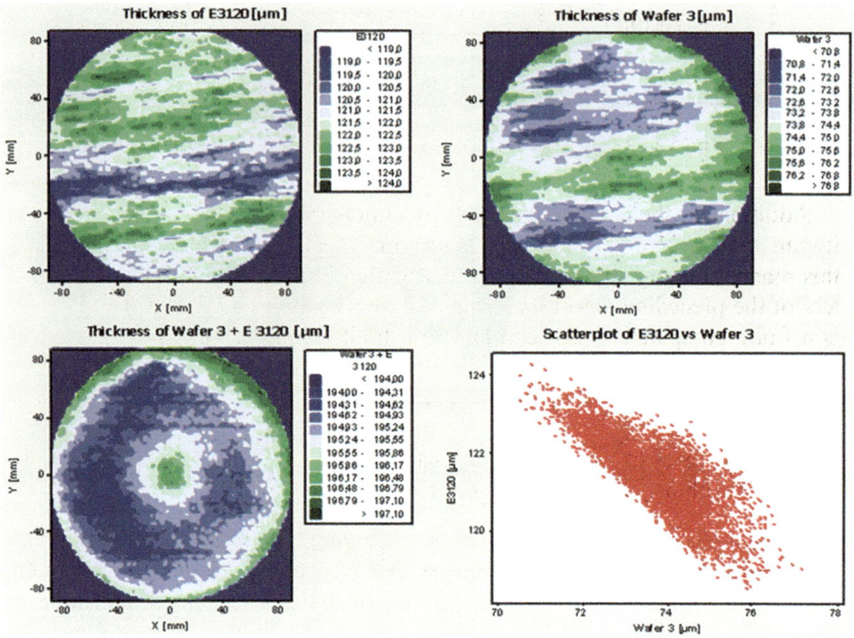

Fig. 3.4 Influence of grinding foil thickness on wafer thickness

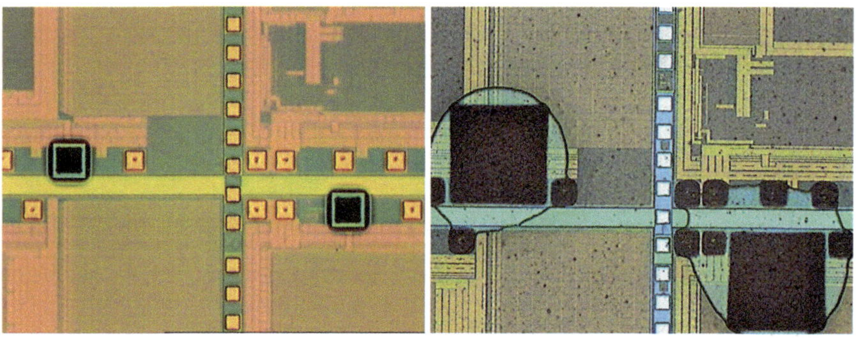

Fig. 3.5 Incomplete embedding of 400 µm bumps (*left* photo) and complete embedding of 100 µm bumps into the adhesion layer of the grinding tape (*right* photo)

stiffness of the grinding tape supports its carrier function, the thickness and viscosity of the adhesive film also determines the capability to embed elevated structures (Fig. 3.5), which in turn will affect the thinning process performance and the tape peel-off behaviour.

For process qualification, statistical process control and outgoing visual inspection the following method will be applied once the wafer thickness decreases to below

100 μm: After thinning, wafers are measured by optical coherence interferometry in the IR range of light. This contactless method enables the measurement of wafers on chip scale and of dicing foil thicknesses at micrometre resolution. It allows the monitoring and studying of thickness variations in critical areas like the wafer edge and in the vicinity of contact structures (due to topology).

Figure 3.6 depicts the wide thickness range (including the adhesive layer) of state of the art grinding tapes. The right graph shows that selected grinding tapes offer a thickness uniformity of less than 2 μm (upper right graph, box plots correspond to 80th percentile). Over a radius of more than 85 mm a wafer thickness variation of approximately 2 μm can be achieved (lower right graph). The mechanical tolerances of the thinning equipment are in the range of 1 μm.

The mechanical robustness of thinned wafers and chips is a not just a default requirement but a key differentiator for their application in flexible modules and chip cards. For the qualification of new thinning processes, ball-on-ring fracture tests are applied that provide information about the surface quality after grinding and polishing [1]. For the qualification of the complete preassembly process including wafer separation, three-point bending tests are applied (Fig. 3.7). The effective bending modulus puts stress on the chip surface and edges and is similar to the applied load of a silicon chip inside a card under field conditions.

The measured fracture forces are transformed into fracture stress data. By default, a defined number of chips well distributed over the wafer are broken. A force sensor measures fracture strength of the sample front- or back-side. These data can be transformed into fracture stress values and plotted as Weibull distribution graph [2]. Weibull modulus and characteristic stress are intrinsic values that depend on the pre-processing, but not on the thickness of the samples. However, the analytical equations that calculate maximum stress before fracture are valid up to a deflection roughly on the order of the chip thickness. For chips that are thinner than 100 μm only relative comparison between differently processed samples can be drawn. Absolute characteristic stresses cannot be deduced from this. FEM simulations result in significantly lower stress values for the bending of ultra-thin silicon chips (Fig. 3.8).

The thinner silicon wafers and chips get, the higher is the risk of microscopic defects, e.g., hair cracks due to particles in the thinning process. Automatic optical inspection of 100% at chip level is indispensable if the process is to satisfy low ppm failure requirements and exclude any mechanical failure mechanisms that would lead to field losses.

3.3 Impact of Wafer Frontside Design

Integral wafer handling is only one stringent prerequisite for thin wafer processing below 150 μm thickness. Moreover, the impact of front end design on ultra-thin wafer manufacturing has to be taken into consideration. Contacts with large area and stand-up heights present an additional challenge for the grinding foil material, as, on the one hand, it needs to embed topology into a soft adhesion layer, but on the

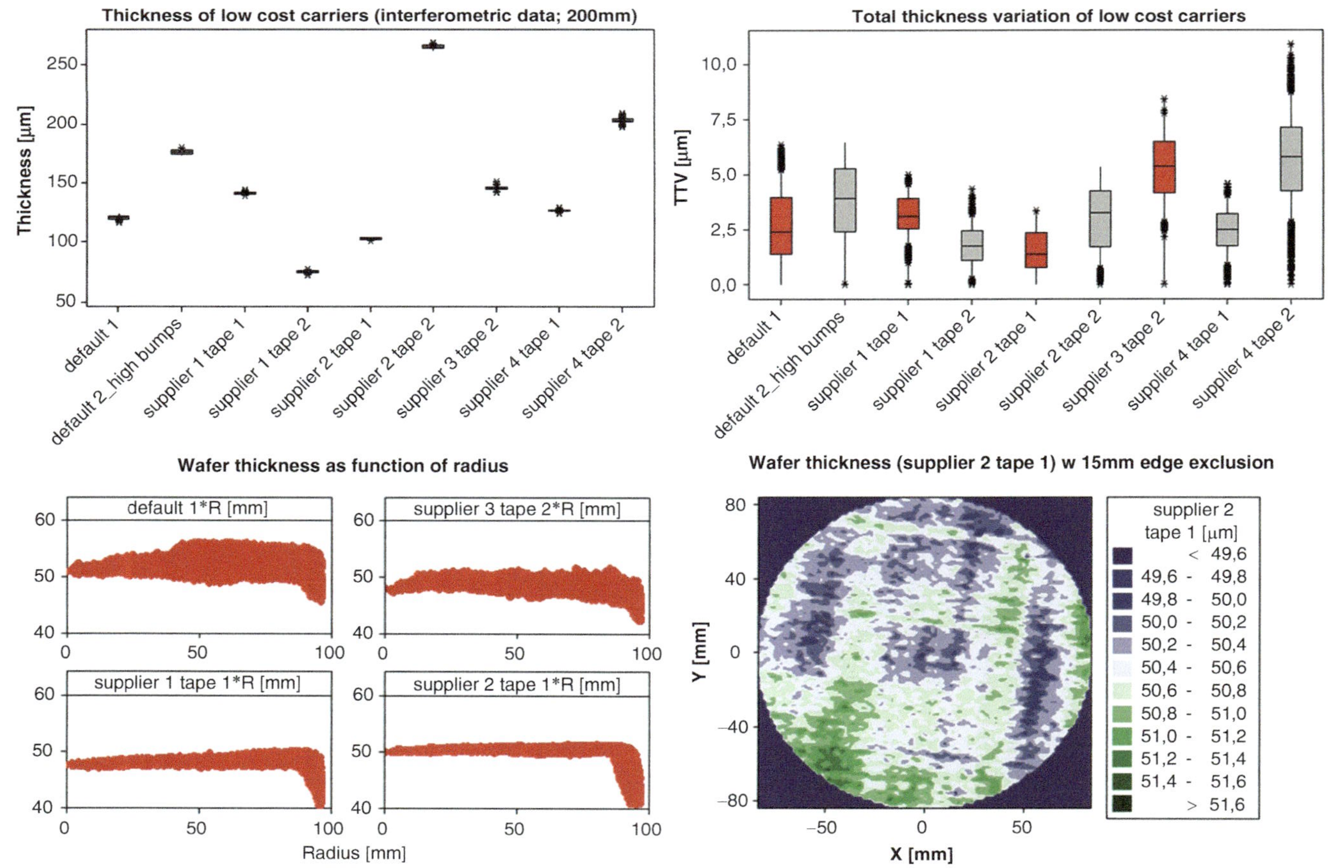

Fig. 3.6 Variation of measured thickness and thickness variation of different grinding tapes for ultra thin wafer processing (*upper row*) and resulting 50 μm wafer thickness uniformity (*lower row*)

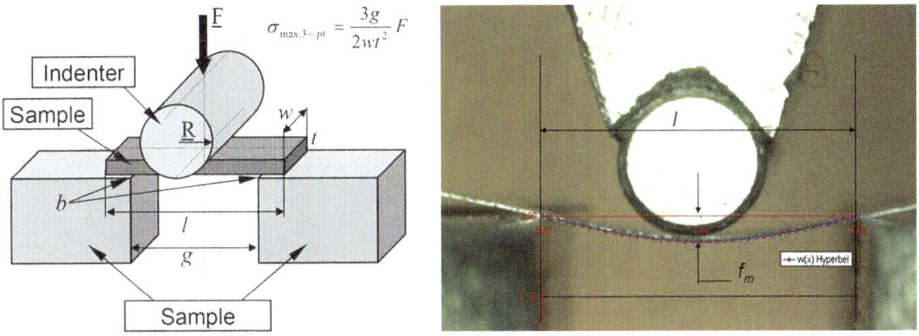

Fig. 3.7 Three-point bending test method with relation between fracture stress and force (*left*) and bending of 75 μm product sample (*right*)

other hand serve as a mechanical backbone of the ultra-thin wafer. The front end materials and design (metals, low and high k materials, multilayer metal stacks, etc.) do not only induce stress and cause wafer bow and warp, but also have a major influence on the mechanical robustness of ultra-thin wafers and chips. While the chip may be loaded with similar stresses on both sides in its final application, the ultra-thin wafer will be subjected to selective, high stress on either side (front/ back) depending on its position within the process flow.

Particles on a vacuum chuck table, for instance during the thinning process, may result in cross cracks, and put the backside of the wafer under tensile stress. Delamination of the grinding tape from the active wafer side also puts the wafer under a tensile stress that may cause edge breakage.

Particles on the wafer backside, however, that occur during subsequent process steps like wafer separation, put the wafer frontside under tensile stress. Consequently, the front end process needs to be optimised in order to achieve a similar robustness compared to the backside.

It was shown with ball-on-ring fracture tests that the mechanical robustness of the wafer frontside can be significantly increased by design for manufacturing (Fig. 3.9). Three-point bending tests of accordingly prepared 75-μm wafers demonstrate a similar mechanical robustness for wafer front- and backside.

After optimisation, the front end technology and the wafer thinning process, the engineering of the chip side is the last step on the way to yielding extremely robust ultra-thin chips. By default, product wafers are separated with a dual step cut process. The pre-cut removes nonsilicon materials from the saw lane. The second cut provides a well-defined separation of the remaining silicon. However, the thinner that products become, the higher is the thickness ratio of front end layer stacks consisting of metal and dielectric layers to substrate silicon. Below 75 μm the reduced process window of pre- and second cut is critical: Either metal structures cannot be properly removed out of the saw lane because of a too shallow pre-cut, or the remaining thin substrate is prone to self-breakage after a too deep pre-cut. As a consequence, frontside chipping or backside chipping can occur. One possible way to avoid those risks is the separation of wafers by laser.

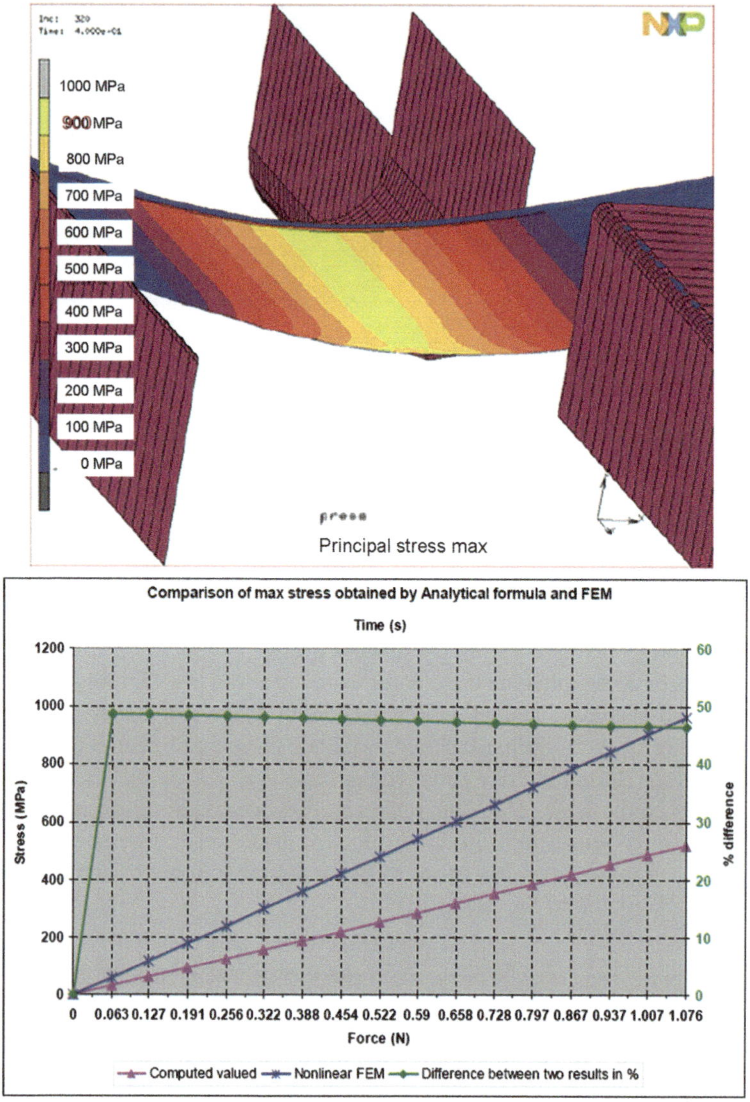

Fig. 3.8 FEM simulation of a three point bending test (*upper*) and comparison of maximum stress obtained by analytical formula and FEM simulation (*lower*)

An additional measure to further improve mechanical stability is the so-called chip side healing. Here plasma treatment will be used to remove additional silicon from the chip edge [3]. Not only can chip robustness be significantly optimised (Fig. 3.10), but the flexibility of chips under bending load is dramatically higher (Fig. 3.11).

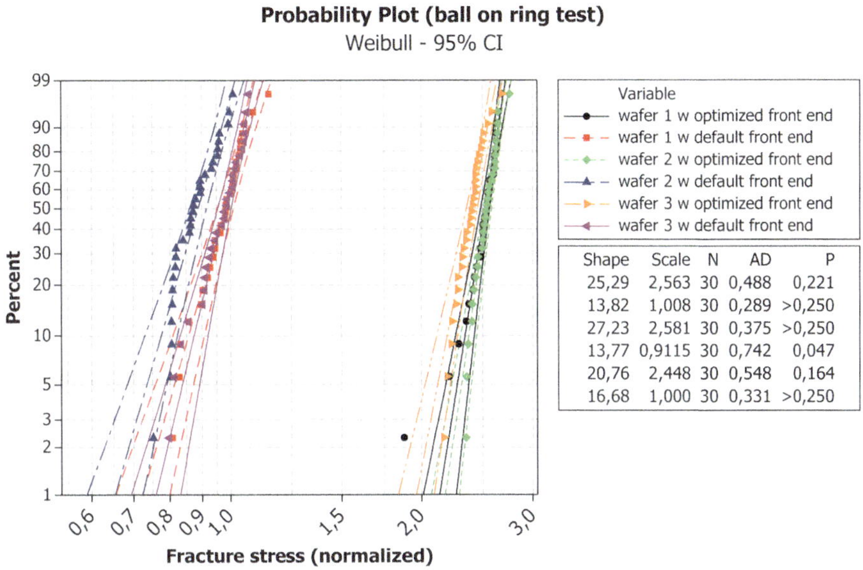

Fig. 3.9 Ball on ring test results for front side of standard and optimised CMOS wafers

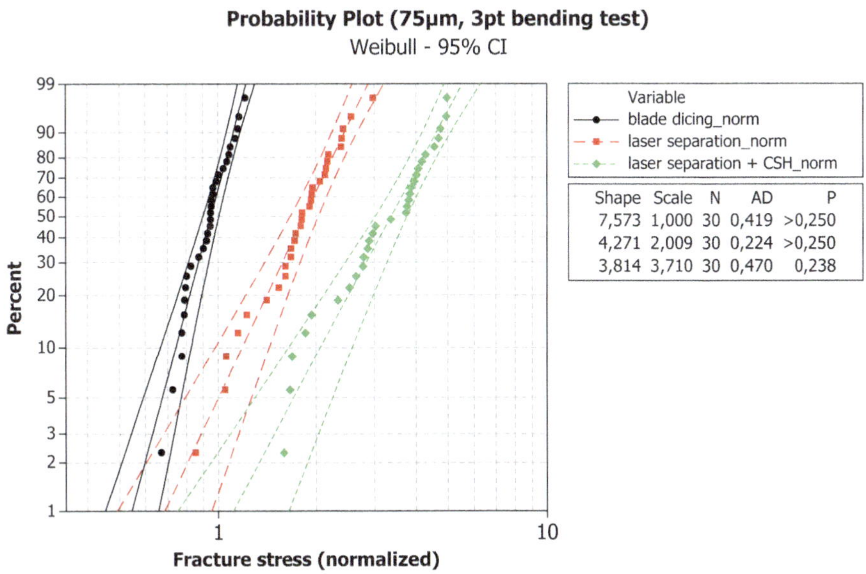

Fig. 3.10 Three point bending test results (chip back side) for blade diced, laser separated and laser separated + chip side healed samples

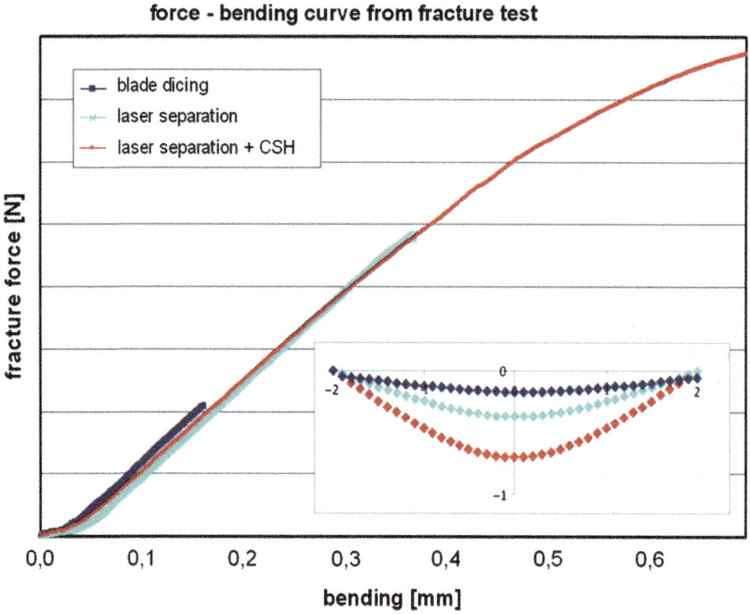

Fig. 3.11 Force bending curves with corresponding maximum chip deflection (inserted graphs show the sample bending depending on the chip side treatment)

The fracture test results show that the contribution of chip side quality to the overall mechanical robustness is significant. From the bending data at chip fracture, minimum bending radii can be calculated assuming hyperbolic deflection of the samples. For the case of laser-separated samples with additional chip side healing a bending radius of below 2.5 mm can be achieved.

3.4 Wafer Thinning Under Economic Constraints

How far can thin wafer fabrication and processing that uses low cost sub carriers be extended? The composition and layout of the circuit layers, including bumps (topology) and shape of wafer edge, limit the handling and processing of ultra-thin wafers. Currently available grinding tapes limit total thickness variations to some micrometres. Reduced polishing pressures limit the processing speed. The impact of particles during the handling or thinning process steps adds the additional risk of wafer cross cracks or other mechanical failures. Not taking into consideration specific product properties, a target thickness of 30 μm is considered to be controllable (Fig. 3.12, left column). Grinding tapes with better thickness uniformity (see Fig. 3.6) enables lower total thickness variation of the wafer.

The use of rigid carriers enables further options for stress relief without any mechanical force on the wafer, like wet etching. Technical challenges such as wafer

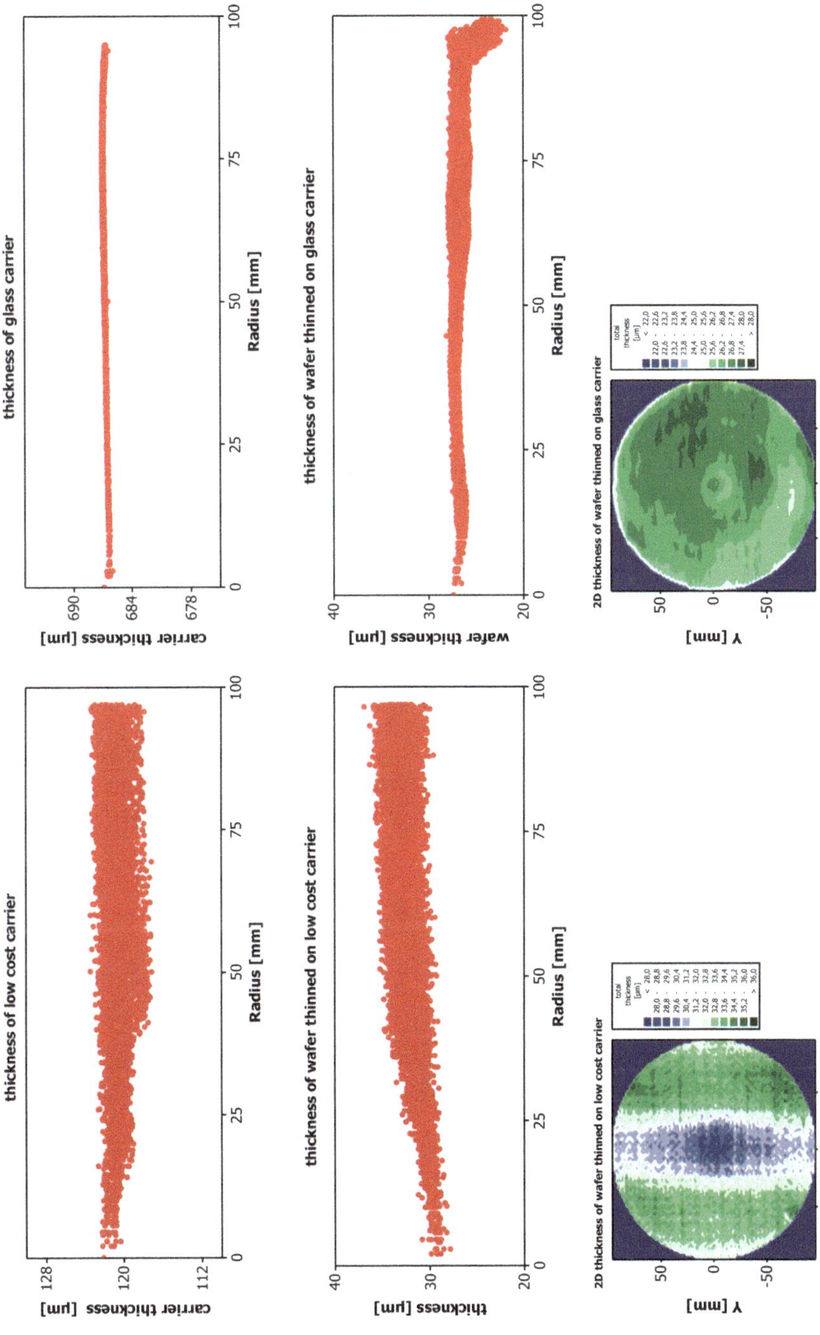

Fig. 3.12 Thinning of 200 mm wafers to 30 µm applying a low cost carrier (*left* column) and a glass carrier with transfer technology (*right* column)

edge thickness variations (Fig. 3.12, right column) can be addressed by wafer edge trimming and optimization of the adhesive film between wafer and carrier. However the costs for a two step (transfer bonding and de-bonding) process drastically exceeds the normal thinning process costs.

Coherent thin wafer handling and processing is the key technology for maximum process yield and optimised cost of ownership. Requirements of thicknesses of 50 μm and below combined with a high mechanical robustness and flexibility of chips are driven by security applications that are highlighted in Chap. 32. The interaction between circuit construction, composition and topology (front end), and preassembly processes must be understood and taken into consideration. The resulting ultra-thin wafer with separated devices on film frame carrier fulfils industrial standards of subsequent assembly processes. Due to the high mechanical robustness of the chips, state of the art high-speed pick and place processes can be applied.

Acknowledgements The findings and progress in the field of ultra-thin Silicon processing presented here are partially based on public funded innovation activities. The authors would like to thank the BMBF (Bundesministerium für Bildung und Forschung) for funding the projects SmartPack and SECUDIS.

References

1. Wu JD, Huang CY, Liao CC (2003) Fracture strength characterization and analysis of silicon dies. Microelectron Reliab 43:269–277
2. Weibull distributions are plotted corresponding to: http://www.weibull.de/WeibullHTML.htm
3. Heinze PM, Amberger M and Chabert T (2007) Five-side stress relief: the method to get the "perfect die," advanced packaging conference 10–11 October 2007, SEMICON Europa 2007, Stuttgart, Germany

Chapter 4
Ultra-thin Wafer Fabrication Through Dicing-by-Thinning

Christof Landesberger, Sabine Scherbaum, and Karlheinz Bock

The technological concept 'dicing-by-thinning' (DbyT) offers a new technique for die separation for ultra-thin wafers. Conventional sawing is replaced by preparation of frontside trenches and subsequent backside thinning. It will be shown that introducing plasma etched trenches allows for both preparation of ultra-thin dies of high fracture strength and for a significant increase in number of chips per wafer. It is concluded that dicing-by-thinning offers a new strategy for cost effective manufacture of very small and ultra-thin radio frequency identification (RFID) devices. Finally, a short outlook on the development of self-assembly processes for ultra-thin microelectronic components will be given.

4.1 Challenges in Thin Wafer Fabrication

Three main requirements must be fulfilled in order to establish a fabrication process for ultra-thin semiconductor substrates:

- Electronic functionality of ultra-thin semiconducting substrates must be retained or improved.
- Breakage of thin wafers must be prevented.
- Complete manufacture flow must include processes for die separation and chip assembly.

A perfect crystal lattice is a basic requirement to ensure optimum functionality of semiconductor devices. Crystalline distortions can be provoked by both backside grinding as well as dicing by means of a wafer saw. These processes result in micro cracks either at the back side or the sidewall of a wafer or chip, respectively. In the case of wafers of standard thickness micro defects are generally not an

C. Landesberger (✉)
Fraunhofer Research Institution for Modular Solid State Technologies (EMFT), Hansastrasse 27d, 80686 Munich, Germany
e-mail: christof.landesberger@emft.fraunhofer.de

J.N. Burghartz (ed.), *Ultra-thin Chip Technology and Applications*,
DOI 10.1007/978-1-4419-7276-7_4, © Springer Science+Business Media, LLC 2011

issue, because the rigid silicon substrate prevents the propagation of flaws towards the transistor regions. For ultra-thin wafers such a 'safety belt' against breakage no longer exists. Micro cracks from any side of a semiconductor chip will run through the crystal lattice as soon as the substrate is strongly bent.

A further challenge in thin wafer fabrication is related to the mechanical properties of the substrates. Ultra-thin wafers are neither robust nor flat like a wafer of standard thickness. They can be bent by several millimetres (sometimes even centimetres). State-of-the-art robot and cassette handling systems need to be modified in order to enable secure support of very thin wafers.

Finally, it should be mentioned that the economical product is not a thin wafer but a thin die, which needs to be assembled in the following process steps. Therefore, techniques for dicing and chip handling must become integral parts of thin wafer technology.

The following sections will explain the technological concept dicing-by-thinning, which enables both wafer handling during back thinning as well as a dicing technique that allows for preparation of separated ultra-thin dies of high crystalline perfection and high mechanical reliability.

4.2 Dicing-by-Thinning Concept

Wafer dicing is most commonly performed by means of sawing equipment running with diamond equipped rotating cutting wheels or high power lasers in the centre of a water beam. This die separation takes place as a back-end process, directly before assembly, and is also applicable for thin wafers. However, it has to be considered that sawing induces micro damage at the sidewall of silicon chips, and lasers also affect the surface with debris and induce micro cracks or degradation centers in the crystal structure.

As the wafers become thinner, apparent micro cracks may become a serious problem for the reliability of ICs. Mechanical stability of silicon chips is required for safe handling, mounting and packaging of IC products. Hence, micro cracks at the chip edge may lead to chip breakage during subsequent pick-and-place process or pressure-supported flip chip mounting. To overcome this problem, the dicing-by-thinning process was developed [1, 2], which is based on the following principles:

- Wafer-thinning: Various combinations of the processes grinding, wet- or dry-chemical etching and CMP polishing can be applied.
- Wafer-handling: In order to eliminate the risk for wafer breaking, a support substrate is temporarily bonded to the device wafer before thinning.
- Chip-separation: Instead of dicing an ultra-thin wafer, trenches are prepared at the frontside of the device wafer. Die separation takes place during back thinning.
- Complete process flow: Wafer thinning and dicing is done in one step. Separated thin dies can be transferred onto standard frame-foil holders.

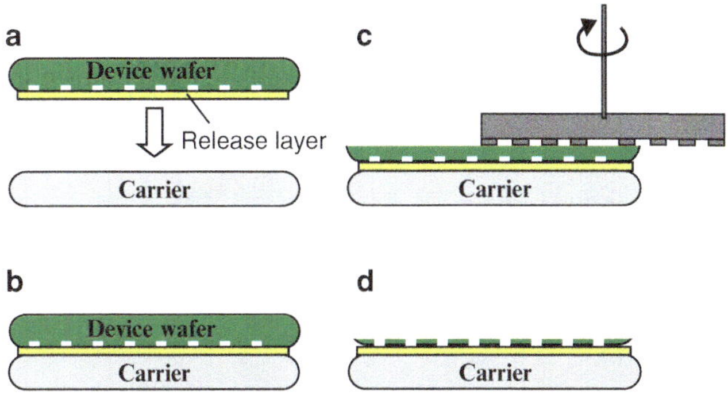

Fig. 4.1 Dicing-by-thinning process (*DbyT*): (**a**) Device wafer with frontside grooves and coated by a reversible bonding layer will be attached onto a carrier substrate; (**b**) temporary wafer bonding; (**c**) back-thinning by grinding and subsequent stress-relief process; (**d**) separated ultra-thin dies on carrier

The principle process flow according to the DbyT concept is shown in Fig. 4.1. It starts with a readily processed device wafer of standard thickness (500–800 μm). First, the trenches along the chip corners are prepared. The trench depth corresponds to the projected thickness of the device wafer. A carrier is attached to the device wafer by means of a temporary bonding technique. Subsequently, backside thinning is performed until the frontside grooves are opened. Then separated thin dies will be transferred onto a tape, which allows for a pick-and-place process for final assembly.

This basic concept allows for a variety of adaptations and variations. First, the frontside trenches may be prepared by half-cut sawing, wet etching or plasma etching. Reversible bonding of a rigid support substrate can be realised by releasable tapes, solvable glues or thermoplastic materials. A review on temporary bonding technique for thin wafer processing is given in Chap. 12 in this book.

The following section will focus on a dicing-by-thinning process flow that uses releasable tapes for temporary bonding. Such tapes use a base film (e.g. PET), which is coated by a combination of a releasable and a permanent adhesive layer. In order to enable easy handling a protective liner is applied on both sides. These liners will be removed directly before lamination or bonding. Widely used are, for instance, thermal release tapes. First, the tape is laminated onto the surface of a wafer by means of a roller. Bonding of two wafers, generally, is accomplished in a vacuum chamber. Thus, trapping of air is prevented. One idea experimentally evaluated was to laminate the thermal release side of such tape onto the frontside of device wafer. However, it turned out that some type of coatings sticks to the surface (especially onto aluminium pads of CMOS wafers) and thereby caused severe polymeric residues, which generally are not tolerable in subsequent assembly steps. Such difficulties can be circumvented by application of two tapes of different release mechanism. For instance, first a standard back grinding (BG) tape is laminated on the device wafers front and then applies a thermal release tape. This offers two advantages: BG tapes have been carefully developed to allow for

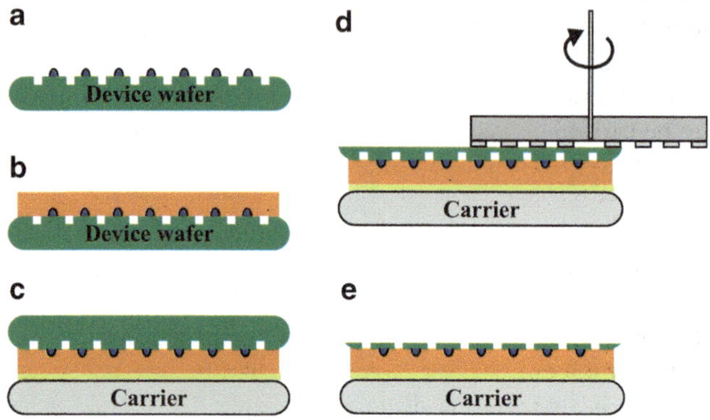

Fig. 4.2 Dicing-by-thinning concept for wafers with high topography: (**a**) Blue bumps indicate surface topography of device wafer; (**b**) lamination of a soft tape which embeds the topography; (**c**) bonding of carrier wafer; (**d**) back-thinning by grinding and subsequent stress-relief; (**e**) final state: separated ultra-thin dies

clean wafer surfaces after delamination and, second, specific BG tapes can be selected that enable embedding of surface topography (for instance bump metallization). The process flow is shown in Fig. 4.2.

Temporary bonding using a combination of releasable tapes enables a simple handling sequence for transfer of thin dies onto pick-up tape for final chip assembly. Such a transfer process is shown in Fig. 4.3.

Basically, it would be sufficient to mount a film-frame holder onto the back side of the separated thin chips (see Fig. 4.3a), apply some heat to release the thermal tape and, finally, delaminate the BG tape (Fig. 4.3b). However, some types of pick-up tape might get deteriorated when temperatures around 100°C are applied. Therefore, we propose to introduce an intermediate handling step, as follows. The carrier wafer is placed onto a hot plate. Thereby, the thermal tape releases its adhesion, which leads to a separation between BG tape with the thinned dies and the carrier wafer with the double side adhesion tape. All separated thin dies still stick to the BG tape and can be mounted onto a pick-up tape. Handling and transfer of such chip-on-tape arrangement (separated dies on flexible tape) can be done manually or automatically without introducing a risk for chip breakage.

4.3 Plasma Trench Etching for Die Separation

Etching of silicon substrates in a fluorine plasma gas is a well-known technique. Formation of trenches with perpendicular side walls can be achieved by the so-called 'Bosch process' (see Chap. 9). Trench pattern on the wafer surface is defined by a photolithographic process. The structured photoresist layer is then used as an etching mask during the subsequent dry-etching process for trench formation. Minimum width is given by the aspect ratio for trench-etching, with typically

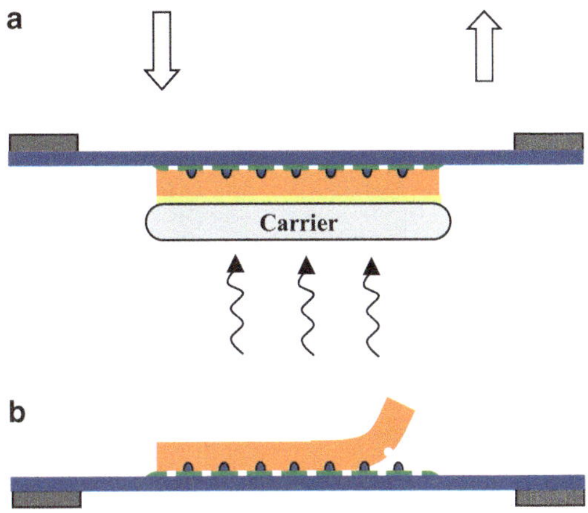

Fig. 4.3 Transfer of thinned dies from carrier wafer onto film-frame holder (blue) after DbyT process using thermal release tapes. (**a**) Apply heat to release thermal tape; (**b**) Delaminate surface tape after transfer on film-frame holder

allowed values of 5:1 (depth to width). This means a 25-μm deep trench may show a width of just 5 μm. The whole process sequence is done as an additional manufacture step in the wafer fabrication. The wafer is still of standard thickness and no restrictions in wafer diameter or process temperature will occur.

The dry etch approach for die separation offers several important advantages:

- Chip edges may have any form, for instance rounded corners or hexagonal shapes. Also, a combination of different chip sizes on one wafer design, see Fig. 4.4 ('multiproject wafers') can be easily realised.
- Sidewalls of chips show no mechanical damage. Crystalline perfection of the semiconductor substrate is not deteriorated. This results in strongly increased fracture strength of ultra-thin dies, which is also a basic requirement if the devices need to be mechanically flexible during their product life later on.
- Depth of trenches can be defined very precisely through control of etching time. After back thinning, chips will show very uniform thickness.
- In contrast to wafer sawing, the width of dry-etched chip trenches can be significantly reduced. Consequently, the number of product devices on one wafer substrate can be distinctly increased. In the case of very small chip products, for instance, RFID devices with a die length below 0.5 mm, chip yield may be drastically increased.
- Outer dimensions of dies are highly uniform and the chip size may have an accuracy of 1–3 μm. This feature strongly supports 3D integration techniques where precisely aligned chip stacking is a must.
- Etching of all trenches on a wafer surface is done in one step and takes less than 10 min for a 25-μm deep trench, independent of wafer diameter. In comparison, sequential dicing of very small devices by means of a wafer saw may take several hours for one substrate Fig. 4.4.

The advantages of plasma etched trenches for die separation for ultra-thin wafers are obvious. However, there is also an obstacle to be considered. As trench

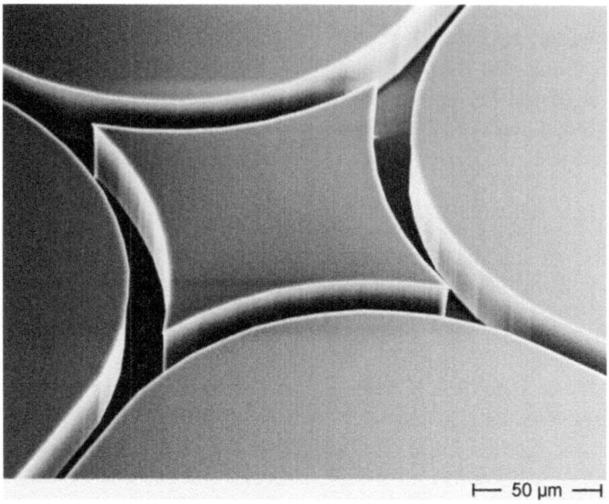

Fig. 4.4 Example of a plasma etched chip trench; the SEM picture shows the rounded corners of four neighbored dies

formation is performed at the frontside of device wafers, all thin film layers in the dicing lanes need to be etched as well, before the trench in the silicon substrate can be prepared. Semiconductor device wafers often show various test patterns in the dicing lanes. So, there might be dielectric layers (silicon oxides, silicon nitrides, low-k layers) and also metal structures (Al, Cu), which cannot be etched simultaneously. Etching of the silicon substrate by fluorine ions in a plasma chamber will not remove metal patterns. Therefore, below the metal patterns in the dicing lanes fine silicon bridges would remain after trench formation. The best way to circumvent such problems would be the redesign of the layout of test patterns between the active chip areas. So, application of the DbyT concept should be considered already within the design phase of a semiconductor product wafer.

In the case where already existing product wafers should be processed according to the DbyT concept, a simplified alternative could be applied, as explained in the following section.

4.4 Combination of Sawing and Plasma Etching for Die Separation

In the case where product wafers that show metal patterns in the layout of the dicing lanes need to be thinned, a combination of sawn grooves and subsequent plasma etching for trench formation could be of help [3]. According to such a concept, the device wafer's frontside is first coated by a photoresist layer. Then a wafer saw is used to prepare a frontside trench, which extends into the silicon substrate. Now the wafer can be brought into the plasma chamber for further etching of the pre-cut

Fig. 4.5 Comparison of sawn grooves (*left*) and a trench that was prepared by sawing and subsequent wet-chemical overetching (*right*). Micro damage that occurred after sawing and its removal after etching is clearly visible

trench. Thereby, the restriction of metal patterns between the chips is eliminated and the sidewalls of the chips undergo a dry etching process, which simultaneously removes the mechanical damage of the sawing step. Therefore, this might be a useful compromise that allows plasma trench etching, even in the case where metal patterns have not been redesigned. However, the benefit of narrower chip trenches cannot be achieved by this combination of sawing and etching.

In any case, overetching of sawn grooves by plasma dry etching or also by wet-chemical treatment enables the removal of crystalline damage from the sidewalls of sawn grooves and thereby results in ultra-thin silicon chips of higher fracture strength. The SEM pictures in Fig. 4.5 show the effect of wet-chemical overetching of sawn grooves. The surface defects along the saw line (also called 'chipping') visible in the left SEM picture of the figure are greatly reduced or disappear after overetching (SEM picture on the right side). Of course, the same effect can be achieved by plasma dry etching, performed either at the frontside or the backside of a thinned wafer.

The influence of dicing technology on the breaking behaviour of thin silicon dies has already been investigated by several working groups [1, 4–6]. It should be mentioned that defect free ultra-thin silicon material is a prerequisite for the realisation of microelectronic devices which need to be mechanically flexible during their product life. Micro damage at the sidewalls of thin chips would cause cracks after repeated bending of the flexible silicon product chip.

4.5 Ultra-thin RFID Devices Prepared Through Dicing-by-Thinning

Radio frequency identification devices (RFID) represent a type of semiconductor product that will benefit in manifold ways when dicing-by-thinning concept is introduced into its production process.

Passive RFID-tags (also called 'smart labels') are based on a thin silicon chip that is connected to an antenna via two contact pads. Depending on the chosen transmitter frequency this antenna is either a coil (for MHz range) or a dipole (for GHz range) which are made of a few microns thick copper lines on a flexible substrate (e.g. PET foils).

Smart labels enable simple detection and identification of single items that don't require a person to see or to grab the individual object. They are produced in large volume and are widely used in logistics, electronic passports, public transport, libraries, ski passes and many other applications.

Smart labels need to be very thin in order to allow for cost effective roll-to-roll processing during chip assembly and to enable a flexible foil package that can be laminated into chip cards, onto a sheet of paper or placed on curved surfaces.

Furthermore, there is another characteristic feature of RFID chips that makes them perfectly suited for the application of dicing-by-thinning technology: the very small size of each chip device. State of the art RFID chips show a side length of less than 0.5 mm and products of a side length of 200 μm have already been announced [7].

Figure 4.6 illustrates how RFID chip technology has evolved over the years. The larger chip is a 13.56-MHz device (Philips ICode) having a side length of some 2 mm. The smaller die in front of it is a 960 MHz chip (NXP UCode) having a size of 0.4 × 0.4 mm. The latter chip was thinned according to the DbyT concept to a final thickness of 25 μm. The picture shows both the impressive flatness of the thin die as well as the shrinkage of the active chip area required for a passive transponder chip.

According to DbyT concept, dicing lanes can be prepared by plasma trench etching. This allows for a reduction of dicing streets from some 80 μm to values down to 5 μm [8]. The width of the trench is related to the target thickness of the

Fig. 4.6 Two RFID chips of a thickness of 260 μm (Philips ICode) and 25 μm (NXP UCode, in front, size 0.4 × 0.4 mm). The shrinkage in size illustrates the ongoing miniaturisation of microelectronic devices

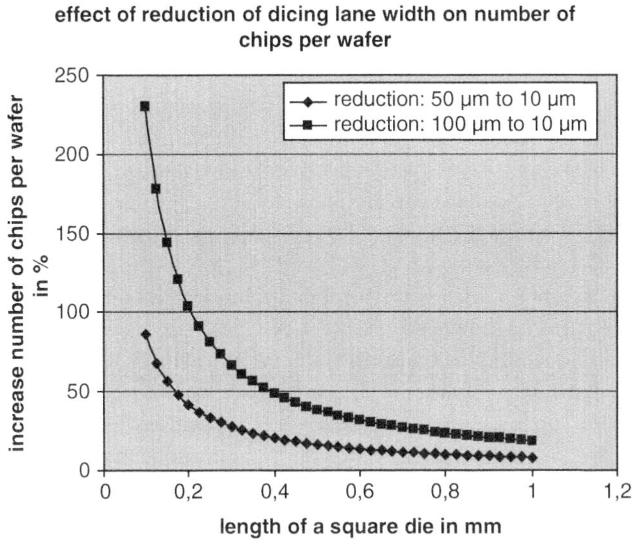

Fig. 4.7 Possible increase of number of chips per wafer when dicing lane width is reduced from 50 or 100 to 10 μm. Small trenches can be prepared by plasma etching. For dies of very small area the effect is of very high significance

chips. Anisotropic etching of the grooves generally allows an aspect ratio between 1:3 and 1:5 (width to depth). So, in order to prepare a flexible die of 25 μm in thickness a trench width of 5–10 μm can be achieved by a simple dry etch process.

In the case of small dies the smaller width of dicing lanes results in a valuable saving of wafer area and thereby in a much higher number of chip products per wafer. The correlation between increase of chip yield per wafer and die size according to a reduction of the trench width is shown Fig. 4.7. For instance, in the case of a die size of 0.4×0.4 mm, reducing the saw line from 100 to 10 μm leads to an increase in the number of chips per wafer of 50%.

It is concluded that the introduction of plasma dicing for RFID product wafers results in several advantages. First, there is the economical advantage of ‘more chips per wafer’. Second, the damage-free sidewalls of plasma-separated chips allow for reliable flexibility of thin dies. And finally, chips with rounded shapes can be manufactured, which again results in a further improvement of mechanical robustness during die handling.

4.6 Outlook: Self-assembly of Ultra-thin RFID Devices

As it was already mentioned, RFID chips are becoming increasingly smaller. Keeping in mind that chips of a size of several hundred micrometers have to be mounted in huge quantities for RFID labels, it becomes obvious, that robotic pick-and-place processes meet their physical limitations. New, highly parallel working assembly technologies and innovative concepts for electrical wiring are in demand to overcome this emerging assembly crisis.

One idea to enable high volume and low cost assembly of very small components is the introduction of self-assembly technology for die placement and interconnection. RFID chips are perfectly suited for self-assembly as they are small, thin and generally require just two contact pads.

Actually, the concept dicing-by-thinning fits very well to the further development of self-assembly processes. DbyT with plasma etched trenches allows the preparation of functional dies with specifically designed shape and without restriction concerning size or thickness.

In the case that fluidic self-assembly should be realised, DbyT offers a simple solution for preparation of single dies. Instead of tape-based support techniques during wafer thinning, soluble adhesive layers can be used for temporary bonding of the device wafer. After completion of the thinning sequence the wafer could be immersed into an appropriate solvent bath and all small and readily thinned dies would then be flushed away and may be introduced in the further fluidic self-assembly process.

A well-known example of a fluidic self-assembly process has already been industrialised by the US company Alien Technologies [9]. A new self-assembly concept was developed and demonstrated at Fraunhofer EMFT in Munich. It is based on the idea of that one supply a single small device in just one droplet of assembly medium for just one assembly site [10]. A maskless plasma treatment of a polymer foil with copper patterns enables a selective wetting behaviour of the assembly liquid on the foil substrate [11]. Surface tension forces of the liquid provoke self-alignment of the die at the target position. Furthermore, self-interconnection of the device can be realised by means of appropriate ACA coating at the front side of the RFID chip [12, 13].

The application example 'self-assembly of small dies' shows that the dicing-by-thinning concept is supposed to enable future applications of thin and small semiconductor devices.

References

1. Landesberger C, Klink G, Schwinn G, Aschenbrenner R (2001) New dicing and thinning concept improves mechanical reliability of ultra thin silicon. In: International symposium and exhibition on advanced packaging materials, Braselton, 92–97 March 2001
2. Landesberger C, Feil M, Klumpp A. Method of subdividing a wafer. US Patent 6,756,288 B1, European patent EP 1 192 657 B1
3. Landesberger C, Köthe O, Bleier M. Verfahren zum Bearbeiten eines Wafers. German Patent, DE 102 29 499 B4
4. Schönfelder S, Ebert M, Landesberger C, Bock K, Bagdahn J (2007) Investigations of the influence of dicing techniques on the strength properties of thin silicon. Microelectron Reliab 47:168–178

5. Heinze P, Amberger M, Chabert T (2008) Perfect chips: chip-side-wall stress relief boosts stability. Future Fab Int 25:111–117
6. Takyu S, Sagara J, Kurosawa T (2008) A study on chip thinning for ultra thin memory devices. In: Electronics components and technology conference (ECTC), Lake Buena Vista, 1511–1516.
7. Usami M (2004) An ultra-small RFID chip: μ-chip. In: 2004 IEEE Asia-Pacific conference on advanced system integrated circuits (AP-ASIC2004), 4–5 Aug 2004
8. Feil M, Adler C, Hemmetzberger D, Bock K (2004) The challenge of ultra-thin chip assembly. In: Electronics components and technology conference (ECTC), Las Vegas, NV, USA
9. Smith JS. Fluidic self-assembly of active antenna. US Patent 6,611,237 B2
10. Bock K. European Patent application EP 1 499 168 A2
11. Bock K, Scherbaum S, Yacoub-George E, Landesberger C (May 2008) Selective one-step plasma patterning process for fluidic self-assembly of silicon chips. In: Electronic components and technology conference (ECTC), Lake Buena Vista, 1099–1104.
12. Landesberger C, Hell W, Yacoub-George E, Scherbaum S, König M, Feil M, Bock K (June 2009) Assembly of thin and flexible silicon devices on large area foil substrates. In: International conference and exhibition for the organic and printed electronics industry, Frankfurt
13. Landesberger C (ed) (2010) Entwicklung eines Selbstmontage-Verfahrens (self-assembly) für die Systemintegration von sehr kleinen Silizium-Bausteinen. Abschlussbericht des BMBF Verbundvorhabens Assemble!, Fortschritt-Berichte VDI. 9(386)

Chapter 5
Thin Wafer Handling and Processing without Carrier Substrates

Yoshikazu Kobayashi, Martin Plankensteiner, and Midoriko Honda

SiP (system in package), IC cards and RFID (radio frequency identification) tags are more commonly used in digital and mobile devices, such as cell phones, and along with such recent trends, 30–50 μm thick product die are actually used. The thinning of silicon wafers to enhance performance is also in strong demand in power devices used for power conversion, such as IGBTs (insulated gate bipolar transistors) used in solar power generation and hybrid vehicles, which are attracting attention for energy conservation and global environmental protection. This chapter introduces the process technology of making wafers thinner, which also considers wafer handling and use in the subsequent steps, in anticipation of even further thinning of wafers.

5.1 Issues with Wafer Thinning

Previously, the backgrinding process for a thin wafer, finishing to a specified thickness and surface roughness was conventionally required. At present, the above mentioned need for thinner packages and higher functionality of devices requires that wafers be finished to an even thinner state with higher die strength. The thinner the wafer the higher the risk of wafer breakage, due, mainly, to the following:

5.1.1 Reduction in Wafer Rigidity (Warpage, Deflection)

A thick wafer can stand compression caused by residual stress of the surface film or tensile stress. A thinner wafer, however, has less rigidity, so that it cannot stand the residual stress itself, producing large warpage or deflection (Fig. 5.1).

Y. Kobayashi (✉)
Sales Engineering Division Marketing Group, Marketing Team,
DISCO HI-TEC EUROPE GmbH
e-mail: contactsales@discoeurope.com

J.N. Burghartz (ed.), *Ultra-thin Chip Technology and Applications*,
DOI 10.1007/978-1-4419-7276-7_5,

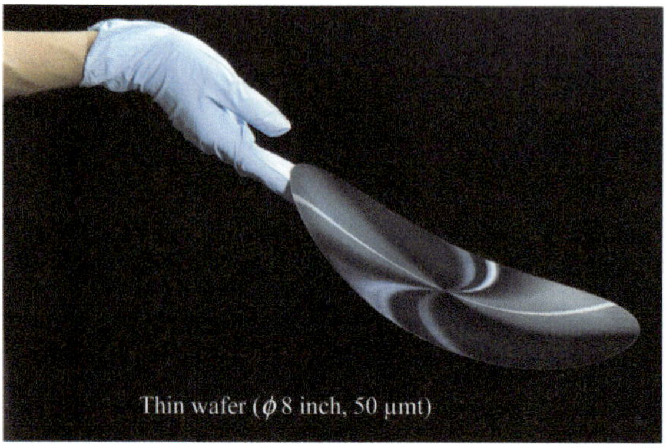

Fig. 5.1 Thinned wafer showing lack of mechanical stiffness

This kind of wafer condition presents a problem especially with handling between machines. With existing transport cassettes or during vacuuming for wafer handling, wafer damage easily occurs due to dropping or impact. Therefore, thin wafers are normally handled by affixing tape materials or attaching device wafers to glass or silicon substrates using adhesive, which creates other headaches: heat resistance of the tape and adhesive materials; generation of outgas in the vacuum process; and increased cost of consumables.

5.1.2 Grinding Damage

The purpose of backgrinding is to thin a wafer while processing silicon that is in a brittle mode; it is done with a grinding wheel using diamond abrasive grains. This method generally consists of two stages of grinding: rough grinding and fine grinding. In rough grinding a wafer is thinned efficiently with a wheel having diamond abrasive grains whose size is about several tens of micrometres, in order to improve the processing ability of the grinder. After that, in fine grinding a wheel having abrasive grains of several micrometers is used to finish the wafer to the designated thickness while removing the deep damaged layer produced in the rough grinding process. When the finished thickness was 200 micrometers or more, this damaged layer did not cause a problem, but since wafer thinning has progressed, even a very shallow damage of less than 1 μm that occurring in the fine grinding may lead to wafer damage.

5.1.3 Wafer Edge Chipping

At the start of the process the wafer edge cross-section is rounded, so thinning a wafer makes its edge sharp and greatly reduces its mechanical strength (Fig. 5.2).

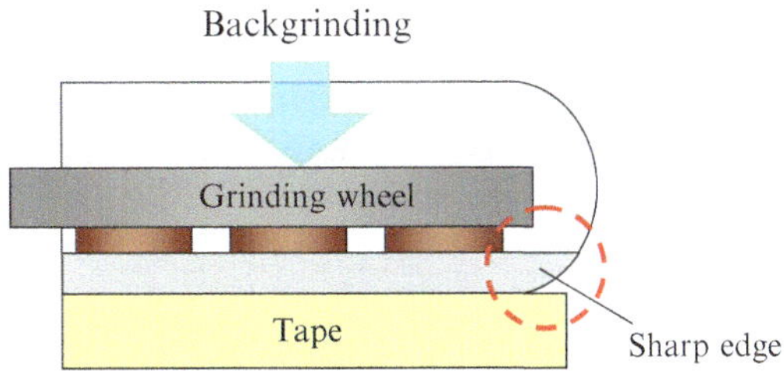

Fig. 5.2 Illustration of edge sharpening during wafer thinning

Thus, minute cracking called 'chipping' occurs at the edge during grinding, and a wafer tends to be broken starting from here.

5.1.4 Wafer Dicing

The dicing process, which cuts a wafer into die, generally uses a dicing blade with diamond abrasives. Compared with a thick wafer, dicing of a thin wafer is extremely difficult. First of all, the thin wafer has less strength and therefore larger backside chipping tends to occur from dicing. Even with the similar chipping size, since the rate of backside chipping in the wafer thickness increases in the thin wafer, its die strength is reduced. Furthermore, it is necessary to dice the TEG (test element group) on the street or DAF (die attach film) for wafer bonding together with the silicon, but since the amount of silicon that can be processed is small in the thin wafer, it causes another problem, one in which loading of the dicing blade tends to occur.

5.2 DBG (Dicing-Before-Grinding) Process

In the conventional thinning process, full-cut dicing is conducted after a wafer is thinned through backgrinding. This poses breakage risk when a thin wafer is handled or problems related to edge chipping and full-cut dicing. On the other hand, DBG reverses the existing process: After conducting half-cut dicing from the circuit side of a wafer first, backgrinding (BG) tape is affixed, and during the subsequent backgrinding, die are separated once the wafer thickness reaches the groove formed by dicing (Fig. 5.3).

This DBG process can provide various advantages. First of all, since die are separated when the wafer is already thinned, there is no need handling of the large and thin wafer, and wafer warpage or deflection does not have an impact on the handling. Therefore, DBG can realise easy handling of the wafer without using

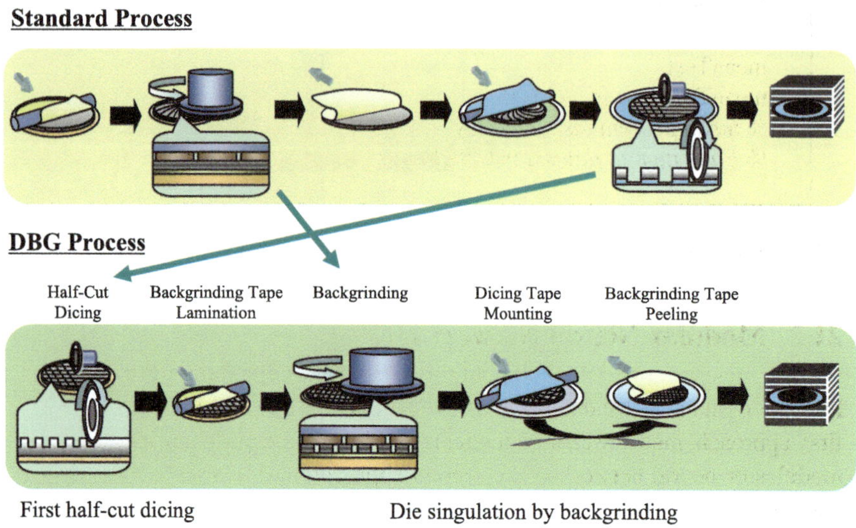

Fig. 5.3 Comparison of wafer thinning and chip dicing to the DBG (dicing-before-grinding) process

a substrate. Second, DBG can keep the influence of the sharp edge within the edge area, so it enables a production with high yield. Third, in normal full-cut dicing it is necessary to reduce the processing speed because backside chipping increases in the thin wafer; but DBG, which does not conduct full-cut dicing – and so can reduce backside chipping greatly – enhances die strength and increases dicing speed.

Generally, in backgrinding, because a wafer is thinned while breaking silicon, backgrinding damage remains on the processed surface side. When a wafer is thin, however, even shallow damage may cause wafer breakage due to various stresses produced during the assembly/packaging process or after it has actually been mounted on products. Because of this, stress relief, such as wet etching, dry etching, CMP (chemical mechanical polishing) or dry polishing, is proposed for the purpose of eliminating the damage produced in backgrinding.

In DBG, stress relief by dry polishing is usually used. Since this dry process does not require chemicals, such as an acid mixed liquid or polishing slurry, or water, different from wet etching and CMP, it can reduce costs for consumables and is more eco-friendly. Furthermore, the dry polishing has the following advantages: grinding damage can be eliminated adequately; die strength can be improved by more than six times than the grinding-only process, on average; processing rate is the same level achieved by the CMP which uses slurry; and a processing surface with high glaze equal to other processes can be obtained.

Stress relief using dry etching is also effective for the DBG. That is because it is possible to eliminate not only damage on the wafer backside but also dicing damage at the side of die. Thanks to this, further increase of die strength can be expected: the DBG and dry etching can provide an almost ideal die having no mechanical

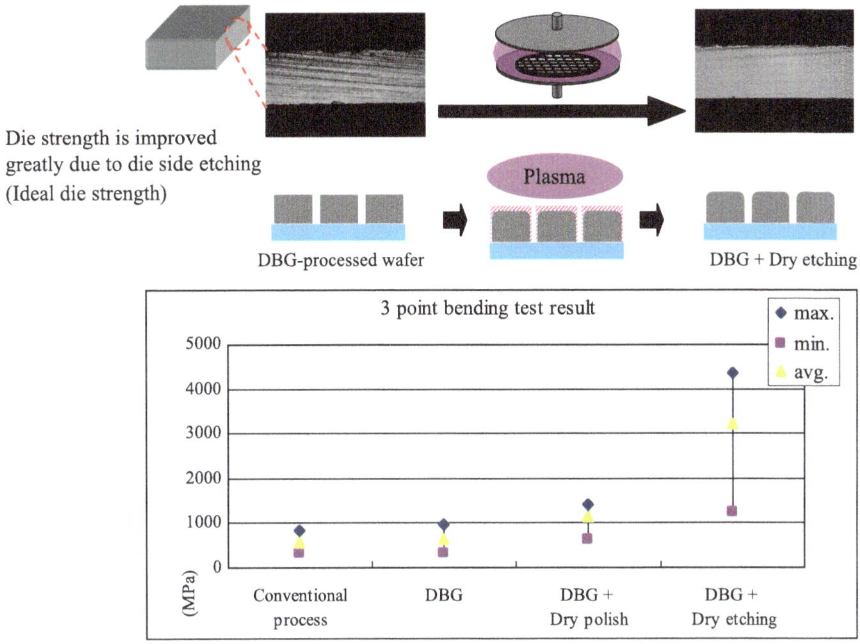

Fig. 5.4 Increased die strength through stress relief

damage (Fig. 5.4). Thus, the DBG process is increasingly adopted, especially for large diameter wafers of 300-mm diameter, because it makes possible the avoidance of the worst risk of wafer damage in the handling of a thin wafer.

5.3 TAIKO Process

In power devices, such as IGBT, collector electrodes are formed on the wafer backside after thinning, and it is necessary to transport the thin wafer to another machine. The process of forming electrodes on the wafer backside consists of many steps and varies depending on the device type, so an inline system cannot be realised between the machines. Usually, the wafer is transported using tape or a substrate, but this causes various problems, such as the heat resistance of the tape and adhesive materials, generation of outgas in the vacuum process, and increased cost of consumables, as explained before. The TAIKO process leaves the outermost area of the wafer backside as a ring shape and backgrinds only the inside. The name 'TAIKO' is the word for a Japanese drum, i.e. *taiko*; the processed wafer has this shape (Fig. 5.5).

- Retaining the outermost edge as a ring shape and thinning only the inside of it.

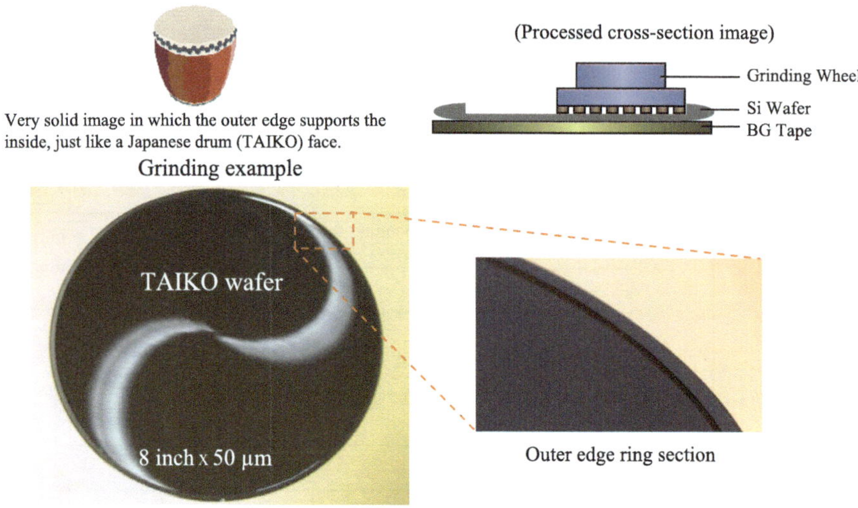

Fig. 5.5 The TAIKO process

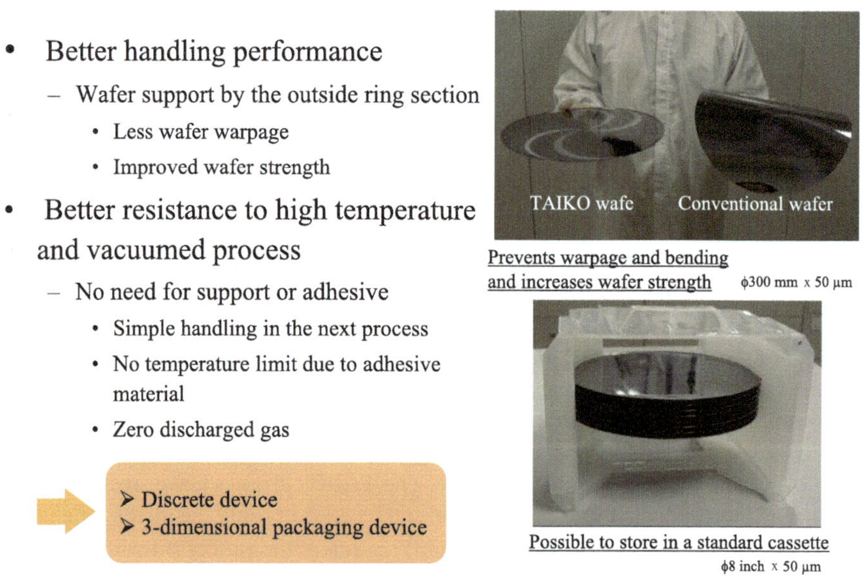

Fig. 5.6 Process stability of TAIKO wafers

Since the ring section at the outer edge works as a support, the wafer processed by the TAIKO process has better strength and less warpage and deflection; in addition, it can be unloaded to the standard wafer cassette and handled without a substrate (Fig. 5.6).

Furthermore, since the ring section is integrated with the wafer, consumable costs can also be reduced. Since the TAIKO process does not require adhesive materials, high-temperature and vacuum processes can be applied. Because it does not process the wafer edge, there is no need to worry about edge chipping. The TAIKO process is effective for devices on which high temperature and vacuum processes are used, as on thin wafers, and it is suitable for discrete devices and 3D packaging devices that use TSV (through-silicon via).

5.4 Conclusions

Recently, wafer thinning has attracted attention as a key technology not only for SiP applications but also power devices, such as IGBT and TSV devices. In order to realise thinning needs, it is important to consider solutions for various problems that are inherent to the thinning of wafers, including those related to peripheral processes, such as thin wafer handling and the dicing process, as well as final packaging formats.

Chapter 6
Epitaxial Growth and Selective Etching Techniques

Evangelos A. Angelopoulos and Alexander Kaiser

In this chapter a wafer-level thinning technique is presented, where, similar to what is discussed in Chapter 8, epitaxial growth of silicon (Si) determines the final chip thickness. A combination of backside grinding and selective wet chemical etching is used for thinning down the initial substrate. Since wafer bonding and grinding were thoroughly addressed in Chapters 4 and 5, this subsection deals only with the other key processing steps, namely the wet chemical etching of Si and the epitaxial growth of highly boron-doped Si required to trigger the etch selectivity. Finally, process parameters and significant results based on this technique are summarised.

6.1 Process Overview

The complete fabrication scheme is illustrated in Fig. 6.1. Relative to the point in time when device integration takes place, it can be divided into a pre-processing part, where the substrate is tailored to fit the process, and a post-processing part, where the substrate is back-thinned. On the front side of a standard Si wafer a highly boron- doped (p^+) film is formed, followed by a lightly doped (p^- or n^-) epitaxial Si layer, which serves as the active layer for the devices (Fig. 6.1a). Device integration is now carried out as usual (Fig. 6.1b). Next, the wafer is flip-bonded on a carrier substrate (Fig. 6.1c) and mechanical grinding is used to coarsely reduce the wafer thickness (Fig. 6.1d). The final wafer thickness, this being the combined thickness of the p^+ and p^- layers, is achieved by selective wet chemical etching and the p^+ film acting as an etch-stop (Fig. 6.1e). Further add-on processing may include through-silicon vias (TSVs) and three-dimensional (3D) system integration [1], as discussed further in Chapter. 3. For the process to be compatible with complementary metal

E.A. Angelopoulos (✉)
Institute for Microelectronics Stuttgart (IMS CHIPS), Allmandring 30a, Stuttgart 70569, Germany
e-mail: angelopoulos@ims-chips.de

J.N. Burghartz (ed.), *Ultra-thin Chip Technology and Applications*,
DOI 10.1007/978-1-4419-7276-7_6, © Springer Science+Business Media, LLC 2011

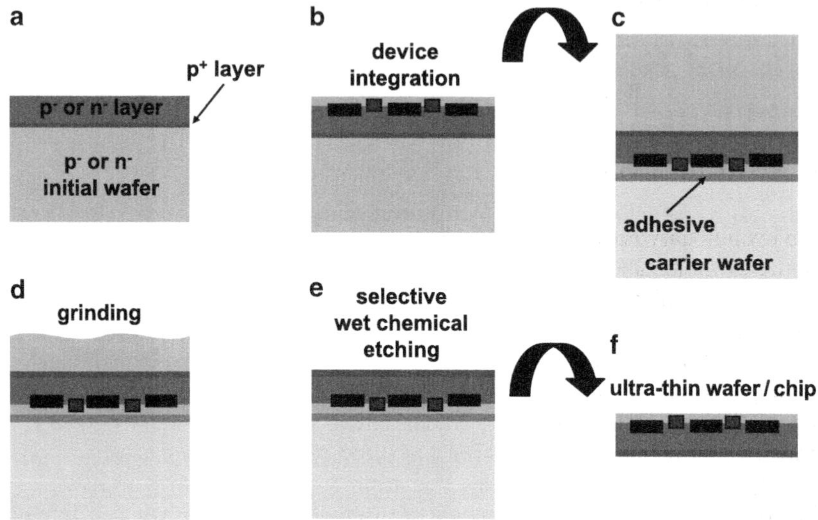

Fig. 6.1 Illustration of the complete step sequence of the back thinning technology described in the current subsection (**a**–**f**)

oxide semiconductor (CMOS) technology or applicable to temperature-sensitive carrier substrates, it is crucial that the overall thermal budget of the post-processing part be kept low.

The presented process offers an alternative to costly back-thinned silicon-on-insulator (SOI) wafers (see Chap. 7) and is compatible with conventional bulk Si technologies. An additional key advantage is the precise control of the final thickness and uniformity, a major challenge for the backgrinding and polishing techniques. The low resistivity of the p^+ backside of the thinned wafer is also favourable for applications requiring backside metal contacting, such as power electronics and backside illuminated image sensors [2].

6.1.1 Selective Wet Chemical Etching of Si

Apart from dry etching techniques, crystalline Si can be etched when in contact with certain chemical solutions. The dissolution of Si surface atoms is primarily an electrochemical process, as it involves charge transfer between the reacting species in the form of simultaneous oxidations (electron loss) and reductions (electron gain). HNA, a mixture of hydrofluoric and nitric acid ($HF + HNO_3$), is an example of an isotropic Si etchant, whereas basic solutions are commonly used in Si micromachining due to their anisotropic etch behaviour, offering up to 200 times higher dissolution reaction rate for the (100) planes compared to the (110) and (111)

Table 6.1 Reported values of etch rates and selectivity of Si planes for commonly used anisotropic etchants

Etchant	(100) Etch rate (μm/min)	(100)/(111) Etch ratio	Remarks
Potassium hydroxide (KOH)	0.5–3	Up to 200:1	Not compatible with FEOL
Tetramethyl ammonium hydroxide (TMAH)	0.2–1	Up to 35:1	High costs
Ethylene diamine pyrocatechol (EDP)	0.3–1.5	Up to 35:1	Toxic
Hydrazine (N_2H_4)	0.5–3	Up to 16:1	Toxic and explosive

planes. Alcohols, such as propanol and isopropanol, are usually included in the mixture to reduce surface roughness [3]. Since the reactions require a certain activation energy, elevated temperatures are needed for sufficient etch rates to be obtained [4]. Table 6.1 summarises rate and selectivity properties of the most important anisotropic etchants of Si along with some relevant remarks [5].

The dissolution reaction of Si in basic solutions can be split into two subreactions, namely an oxidation of Si atoms by hydroxide ions and a reduction of water molecules and release of molecular hydrogen gas. The chemical equations can be written as follows [4]:

Oxidation:

$$Si + 2OH^- \rightarrow Si(OH)_2^{++} + 4e^- \tag{6.1}$$

$$Si(OH)_2^{++} + 4OH^- \rightarrow SiO_2(OH)_2^{--} \tag{6.2}$$

Reduction:

$$4H_2O + 4e^- \rightarrow 4H_2O^-$$

$$4H_2O^- \rightarrow 4OH^- + 4H^+ + 4e^- \rightarrow 4OH^- + 2H_2\uparrow \tag{6.3}$$

And the complete dissolution reaction can be simplified to:

$$Si + 2OH^- + 2H_2O \rightarrow SiO_2(OH)_2^{--} + 2H_2\uparrow \tag{6.4}$$

From the above it becomes clear that Si can be etched as long as the reacting atoms on the surface have available electrons to offer for the oxidation reaction (6.1). Therefore, a significant reduction of the etching rate should be expected for p^+ Si, when the space charge region becomes thin enough for surface electrons to recombine with holes in the bulk. As depicted for different KOH concentrations in Fig. 6.2, hydroxides can etch p^- Si up to 100 times faster than p^+. Due to this effect, a p^+ Si film can function as an effective etch-stop layer using any of the anisotropic

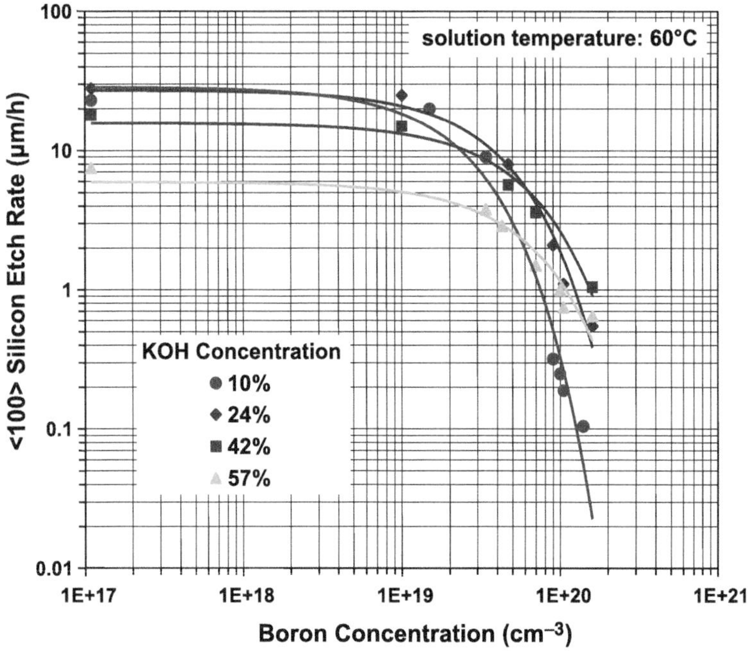

Fig. 6.2 Etch rate of the (100) Si planes for different KOH concentrations depending on the doping level [4]

etchants in Table 6.1. Several other chemical etch-stops have been reported, such as germanium(Ge)/Si alloys [6] and more recently gallium doping [7]. However, heavy boron doping remains the most investigated and effective option.

An alternative method is the so called electrochemical etch-stop, where a Si p-n junction is required and kept under reverse bias potential during the etching. The p-type Si layer is progressively dissolved and the process terminates upon reaching the n-type layer [8]. Nevertheless, this method can result in major complications when it is applied for wafer thinning.

6.1.2 *Highly Boron-Doped Si Epitaxy on Low-Doped Substrates*

Heavily doped p^+ layers can be obtained using several techniques, such as solid source or vapour phase diffusion, ion implantation and epitaxy. In any case, the introduction of boron atoms into the Si lattice is known to reduce the lattice constant of the crystal, doing so in proportion to the logarithm of their concentration. It has been measured, for example, that a boron concentration of 2.3×10^{19} cm^{-3} corresponds to a 0.014% smaller lattice parameter [9]. Thus, a mismatch exists when a p^+ layer is epitaxially grown on a lightly doped substrate,

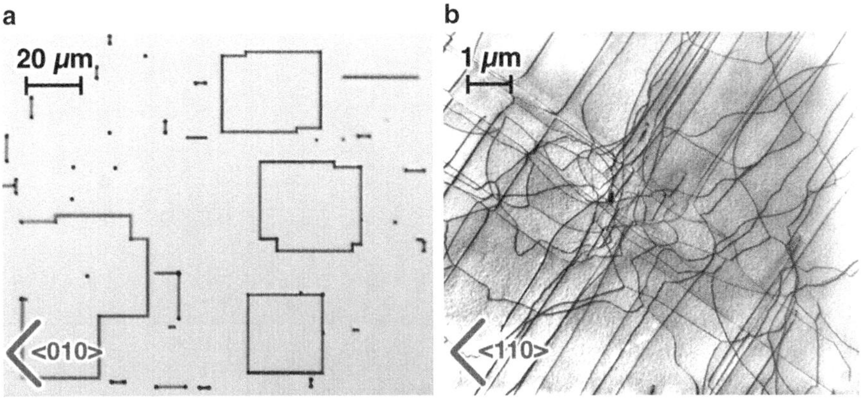

Fig. 6.3 (**a**) Optical microscope image of stacking faults on the (001) surface of a boron-doped Si layer ($N_A \sim 5 \times 10^{18}$) produced after ion implantation and thermal annealing, and (**b**) a transmission electron microscope (*TEM*) image of a dislocation network in a heavily doped p + epitaxial layer [11], both produced on lightly doped substrates

such as the one shown in Fig. 6.1a. Tensile strain builds up in the pseudomorphically growing layer and leads to the formation of misfit dislocations on the interface, part of which extend up to the surface [10] (see Fig. 6.3b). On the other hand, if the ion implantation approach is chosen to form the p^+ etch-stop, the high doses necessary can additionally induce high concentrations of interstitials, vacancies and finally stacking faults and point dislocations on the p^+ layer surface [9], as seen in Fig. 6.3a. The p^+ film in the process discussed here is capped by the active p^- layer and a high density of defects propagating from the p^+ surface up into the p^- layer would cause significant deterioration of device performance. It is therefore important to keep the layers as defect-free as possible. Another aspect, often overseen, is that a high defect density in a p^+ film weakens the selectivity of the etching process, which can drop to only a factor of 10 for densities $\sim 10^{10}$ cm^{-2} [12].

In order to suppress dislocations and defects, the p^+ film should be as thin as possible and the concentration of boron kept relatively low, something that creates an obvious trade-off with the selectivity during the wet chemical etching process. An alternative solution for the suppression of defects is strain compensation by germanium co-doping of the p^+ film. Ge has a larger covalent radius than Si and remains electrically inert. Thus, introduction of Ge atoms in the crystal can eliminate the lattice mismatch [13].

6.2 Process Results

Some of the results presented here were already introduced in [1]. The initial substrates were standard n^- Si wafers. Prior to the process steps in Fig. 6.1, allignment marks were etched on the wafer surface (see Fig. 6.4a), for through-Si via etching at a

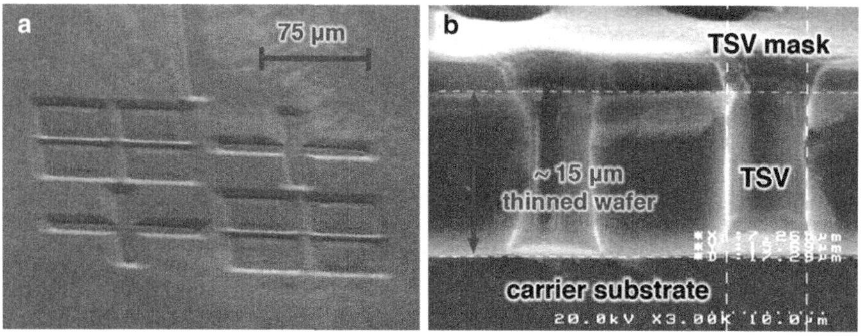

Fig. 6.4 Scanning electron microscope images (*SEM*) (**a**) of the alignment marks on the initial wafer surface and (**b**) of a thinned wafer cross-section including through-silicon vias (*TSVs*)

later stage. An epitaxial p^+ Si film with thickness of 750 nm and boron concentration of 8×10^{19} cm^{-3} was found to provide sufficient etch-stop function. The subsequently grown p^- active layer with boron concentration of $1 \times 10^{15} - 1 \times 10^{16}$ cm^{-3} had a thickness of ~10 µm. In this first experimental phase no device integration was performed. A combination of a polymethylglutarimide (PMGI) photoresist and benzocyclobutene (BCB) was used as adhesive to bond wafer pieces on a Si carrier substrate, and, with backside grinding, the thickness of the wafer was then reduced to 30–40 µm. In order to avoid direct contact of the PMGI with the KOH solution, which can result in the detachment of the wafer sample from the carrier, a spin-etcher was used for the selective wet chemical etching of the remaining 20–30 µm of n^- silicon. Highest selectivity (see Fig. 6.2) is achieved using a 10% KOH solution, and at a temperature of 60–62°C the etch rate is around 10–12 µm/h. Finally, wafer pieces with thickness down to ~15 µm were fabricated (Fig. 6.4b).

In the case device integration is included (Fig. 6.1b), the additional high temperature processes will affect the doping profile of the etch-stop layer. Boron atoms from the p^+ etch-stop film will diffuse into the lighter doped substrate and cap layer, and their maximum concentration within the p^+ film will be decreased. In Fig. 6.5, technology computer-aided design (TCAD) simulations using the Silvaco ATHENA software show the effect of different thermal budgets on the doping profile of the p^+ film. In this model, both epitaxial layers are grown using chemical vapour deposition (CVD) at 1,100°C. As grown, the p^+ film has a thickness of 750 nm and boron concentration of 8×10^{19} cm^{-3}, and the p^- layer has a thickness of 10 µm and boron concentration of 5×10^{15} cm^{-3}. Combining the simulation results with the etch rates diagram in Fig. 6.2, it becomes clear the p^+ film can lose its etch-stop capabilities after device integration. Therefore, additional care should be given when choosing the thickness and doping of the epitaxial layers and the total thermal budget of the processes that follow should be taken into account forehand.

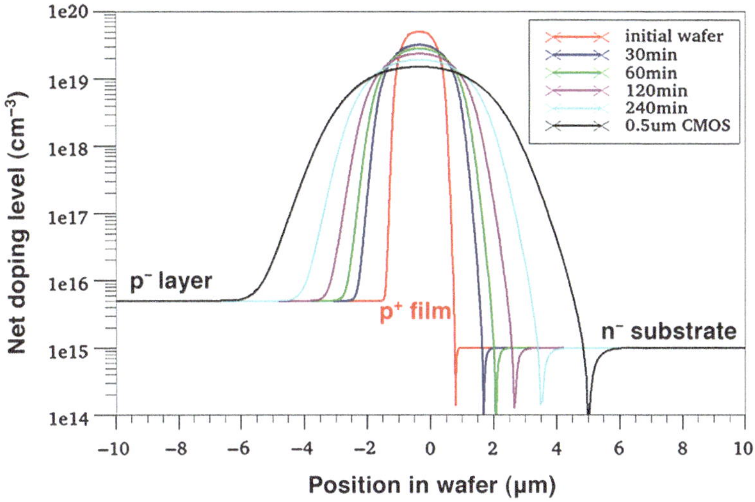

Fig. 6.5 Simulation results of the doping profile evolution for different thermal budgets (different annealing duration at 1,100°C) and for the realistic case of the thermal budget of a typical 0.5 μm CMOS process

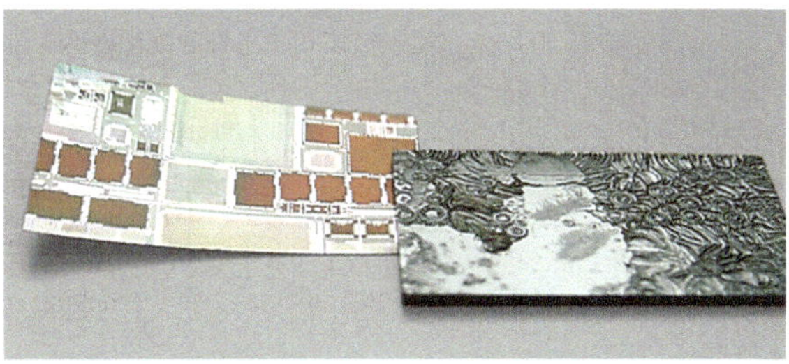

Fig. 6.6 Photograph of an ultra-thin (~30 μm) silicon wafer piece with full CMOS integration fabricated with the epitaxial growth and wet chemical etching technique and to the right a thick (d ~ 300 μm) piece from the Si substrate carrier for comparison [1]

In conclusion, the epitaxial growth and selective wet chemical etching technique provides precise thickness and uniformity control for the fabrication of ultra-thin silicon wafers/chips. It can be implemented using standard semiconductor processing equipment and the additional costs are relatively low. However, the efficiency of the etch-stop layer is limited by quality requirements for the active layer and by the total thermal budget of the integration processes. The latter limitation can be overcome by state-of-the-art CMOS integration technologies having sufficiently low thermal budgets. As demonstrated in [1], the technique can be most useful for investigating ultra-thin wafer/chip add-on technologies, such as 3D system integration.

References

1. Alexander Kaiser (2007) Dreidimensionale Systemintegration: technologische Entwicklung und Anwendung. Ph.D. dissertation, Cuvillier Verlag, Göttingen, Germany
2. Ninković J, Eckhart R, Hartmann R, Holl P, Koitsch C, Lutz G, Merck C, Mirzoyan R, Moser H-G, Otte A-N, Richter R, Schaller G, Schopper F, Soltau H, Teshima M, Vâlceanu G (2007) The avalanche drift diode a back illumination drift silicon photomultiplier. Nucl Instrum Methods Phys A 580:1013–1015
3. Campbell SA, Lewerenz HJ (1998) Semiconductor micromachining: techniques and industrial applications, vol 2. Wiley, West Sussex
4. Seidel H, Csepregi L, Heuberger A, Baumgärtel H (1990) Anisotropic etching of crystalline silicon in alkaline solutions. J Electrochem Soc 137:3626–3632
5. Cho D (Fall 2002) Bulk micromachining. Class presentation notes, for course Introduction to MEMS
6. Finne RM, Klein DL (1967) A water-amine-complexing agent system for etching silicon. J Electrochem Soc 114:965–968
7. Steckl AJ, Mogul HC, Mogren S (1992) Localized fabrication of Si nanostructures by focused ion beam implantation. Appl Phys Lett 60:1833–1835
8. Wallman L, Bengtsson J, Danielsen N, Laurell T (2002) Electrochemical etch-stop technique for silicon membranes with p- and n-type regions and its application to neural sieve electrodes. J Micromech Microeng 12:265–270
9. Kucytowski J, Wokulska K (2005) Lattice parameter measurements of boron doped Si single crystals. Cryst Res Technol 40:424–428
10. Lin W, Hill DW, Paulnack CL, Kelly MJ (1991) Misfit stress in p/p + epitaxial silicon wafers: effect and elimination. In: Defects in silicon II. The Electrochemical Society, Pennington, pp 163–171
11. Föll H. Defects in crystals, Hyperscript, Faculty of Engineering, University of Kiel, http://www.tf.unikiel.de/matwis/amat/def_en/index.html
12. Desmond C (1993) Thin-film silicon-on-insulator by bond and etch back (BESOI). Ph.D. dissertation, Department of Electrical and Computer Engineering, University of California, Davis
13. Maszara WP, Thompson T (1992) Strain compensation by Ge in B-doped silicon epitaxial films. J Appl Phys 9:4477–4479

Chapter 7
Silicon-on-Insulator (SOI) Wafer-Based Thin-Chip Fabrication

Joachim N. Burghartz

Abstract Silicon-on-insulator (SOI) is a wafer substrate technology with potential to fabricate ultra-thin silicon layers and thus ultra-thin chips. The high cost of SOI wafers and technical difficulties to derive ultra-thin chips from SOI substrates so far have hindered the industrial exploitation of SOI technology for thin chip manufacturing. This chapter provides an overview of the present and past SOI technologies, a discussion about the technical difficulties in thin-chip fabrication based on SOI, and a former industrial approach as a related example.

7.1 SOI Technologies

Silicon-on-insulator (SOI) was confined to niche markets for many years. During the 1980s and early 1990s SOI technology was mainly used for radiation-hard and high temperature applications. It took until the late 1990s for SOI wafers to be used for manufacturing VLSI circuit products by IBM, replacing bulk silicon substrates [1]. The advantages of SOI substrates in comparison to bulk silicon were primarily in the elimination of latch-up, the limitation of the source/drain junction's depth, and the reduction of junction capacitances. Those aspects, however were traded against new challenges, such as floating body effects and device self-heating. Today, the fabrication of SOI materials is a multibillion dollar industry and already accounts for a large part of the total wafer production. The potential of SOI to support future CMOS device scaling has been shown by the successful demonstration of CMOS transistors with channel length <5 nm [2]. The main reason SOI has not yet replaced the bulk technologies is the high cost of SOI substrates, which amount to about 25% of the total wafer cost.

J.N. Burghartz (✉)
Institute for Microelectronics Stuttgart (IMS CHIPS), Stuttgart, Germany
e-mail: burghartz@ims-chips.de

J.N. Burghartz (ed.), *Ultra-thin Chip Technology and Applications*,
DOI 10.1007/978-1-4419-7276-7_7,

7.1.1 History of SOI Wafer Technology

SOI technology has its roots in the 1960s, when the so-called silicon-on-sapphire (SOS) technology was first introduced [3]. From a conceptual point of view SOS substrates have clear advantages, because the thin device layer can be epitaxially grown onto the alumina substrate, which features excellent electrical insulation and good thermal conductivity [3]. However, SOS wafers are fragile and the lattice mismatch between sapphire and silicon leads to high defect density in the electrical device layer. Nevertheless, SOS technology is still used today for special and defence applications.

SOI materials developed later involve only silicon and silicon dioxide, the two materials most commonly used in silicon technology. SOI technology has been used for a long time in niche applications such as spacecraft electronics, device operation in a high temperature or radiation-contaminated environment and for the fabrication of high speed integrated circuits (ICs). However, more recently, much attention has been paid to SOI technology because it is well suited to the fabrication of high frequency and low voltage circuits based on complementary metal oxide semiconductor (CMOS) devices.

A variety of SOI materials has been introduced to dielectrically separate the active device region from the silicon substrate by use of a buried oxide (BOX) [4]. The basic issue addressed is that in a bulk silicon MOS transistor only the superficial layer (<0.2 μm thick) supports electron transport and device operation, whereas the underlying substrate parts are prone to undesirable effects. Those parasitic effects can – in principle – be eliminated by inserting a thin and high quality amorphous silicon dioxide layer between the silicon device layer and the silicon substrate. Various techniques have been developed to produce this class of material, often with mixed results but always with a lot of ingenuity. Among these techniques, three stood out and became very successful and commercially available; they were Separation by IMplantation of OXygen (SIMOX), Bonded and Etch-back SOI (BESOI) and the SmartCut® process.

7.1.2 Classification of SOI Wafer Fabrication

In this part we briefly describe the most relevant techniques that have been developed for SOI wafer synthesis. The challenge of SOI material fabrication is in the production of a thin film of single-crystalline silicon on top of an insulator, usually an oxide, placed onto a silicon wafer. Ideally, both the silicon layer and the oxide layer should be defect-free, stress-free and uniform in thickness, and should display good interface properties. SOI materials produced in the early 1980s had silicon layers that were full of defects and stress and were highly nonuniform in thickness. Early materials were produced by melting and recrystallisation of a polycrystalline silicon layer using a laser or a focused halogen lamp (zone melting and re-crystallisation (ZMR) process). Silicon-on-sapphire (SOS) and silicon-on-zirconia (SOZ) technologies aimed at deposition of an epitaxial device layer onto an insulating substrate having a crystalline structure with a similar, but not equal lattice constant [3]. Since also with this heteroepitaxy the defect density in the device layer was rather high, homoepitaxy in the form of epitaxial lateral overgrowth (ELO) from silicon seed regions was proposed. This method provided high quality, but local SOI regions and thus did not draw any major attention for a long time. Consequently, the next class of approaches aimed at insertion of a buried insulating layer into high quality silicon substrates. Here, oxidation of porous silicon (FIPOS) and high-dose ion implantation of oxygen (SIMOX) was proposed. SIMOX was the leading SOI wafer technology in the late 1980s and early 1990s [4]. More recently, silicon-on-nothing (SON) came up as a technique for producing local SOI with a buried air gap insulator. Wafer bonding was an important enabling technique presented around 1985 that allowed for joining two wafer substrates, keeping them firmly together by exploiting van der Waals forces [5]. If those two wafers were bonded together with an insulating oxide layer at the interface and one of the wafers etched back to a remaining thin silicon film, a high quality bonding-and-etch-back SOI wafer substrate (BESOI) was achieved. However, this technology is costly since two wafers and a considerable amount of processing are required to produce an SOI wafer. The invention of SmartCut® addressed that cost issue to some extent by splitting off the major part of one of the substrates instead of etching it back [6]. The splitting becomes possible by implantation of a high dose of hydrogen and successive annealing at a high temperature. The thickness of the resulting thin SOI layer relates directly to the implantation depth and is thus precisely controllable. Another benefit of SmartCut® is that the remainder of that substrate can be recycled. However, defects resulting from the hydrogen implantation and surface roughness remain serious challenges, requiring considerable additional processing, therefore keeping the cost level high. Another technique, one which exploits layer splitting and substrate reuse in a different way, is epitaxial layer transformation (ELTRAN®) introduced by Cannon in the early 1990s. There, the splitting force is not a high temperature step but a mechanical impact in form of a water beam that is applied parallel to the wafer surface in order to shear the two substrate parts apart. Table 7.1 shows a classification of those mentioned SOI wafer technologies and a comparison according to their main characteristics.

Almost any of the SOI wafers manufactured today is based on BESOI or SmartCut®. In the BESOI process two silicon wafers, which each have one surface covered by a high quality oxide layer, are joined together by the exploitation of van der Waals forces, provided that the surfaces can be brought into very close proximity everywhere (Fig. 7.1a). This wafer bonding process is thus prone to particles and air bubble inclusion. After the wafers are joined together the bond is strengthened by high temperature annealing so that the formerly two oxide films become merged into one effective buried oxide layer. The next task is to thin one of the two silicon substrates down to a thickness that is suitable for device integration, usually in the range of 50 nm to 2 μm (Fig. 7.1b).

Table 7.1 Main SOI technologies

Generic concepts	Technologies	Main characteristics
Recrystallization	• Laser Recrystallization • Zone Melting Recrystallization (ZMR)	• Random crystal orientation • High defect density
Heteroepitaxy	• Silicon-on-Sapphire (SOS) • Silicon-on-Zirconia (SOZ)	• Fragile substrate material • Excellent insulating properties
Homoepitaxy	• Epitaxial Lateral Overgrowth (ELO)	• Only suitable for local SOI • Excellent crystal quality
Buried Insulator Formation	• Full Isolation by Porous Silicon (FIPOS) • Separation by Implanted Oxygen (SIMOX) • Silicon-on-Nothing (SON)	• FIPOS, SON: only suitable for local SOI • SIMOX: Medium crystal quality
Wafer Bonding	• Wafer Bonding and Etch Back (BESOI)	• Good crystal quality • High fabrication cost • Limited thickness uniformity • BESOI is preferred for thick SOI wafer fabrication today
Layer Splitting	• Epitaxial Layer Transformation (ELTRAN®) • H+-Induced Layer Splitting (Smart Cut®)	• Excellent crystal quality • High/moderate fabrication cost • Combination of Wafer Bonding and Smart Cut® is preferred for thin SOI wafer fabrication today

This layer thinning is a challenge since the thickness of the remaining silicon film residing over the buried oxide needs to be well controlled across the whole wafer surface. It is therefore beneficial to have an inserted etch-stop layer that will allow for etching selectively to safely terminate the thinning process at the desired level (Fig. 7.1a). Such etch-stop layers or layer transitions, however, work well for wet or dry etching techniques but not for the combined grinding and polishing processes that are favourably used today for economic reasons. More recently such mechanical thinning techniques have been combined with in situ thickness measurement capabilities so that they can be applied with reasonable thickness control without any etch-stop feature.

BESOI has a disadvantage in that the major part of one of the two wafer substrates used is wasted during the thinning process. Both a well-defined separation layer and a means of preserving the redundant part of that wafer is accomplished with the SmartCut® process (Fig. 7.2). There, a high dose of hydrogen is positioned by implantation near the bottom part of the intended thin silicon layer that is to be separated from the thick substrate (Fig. 7.2a). After bonding of the hydrogen-implanted wafer to a second wafer face-to-face (Figs. 7.2b and c), cleavage occurs near that heavily implanted region during a subsequent high

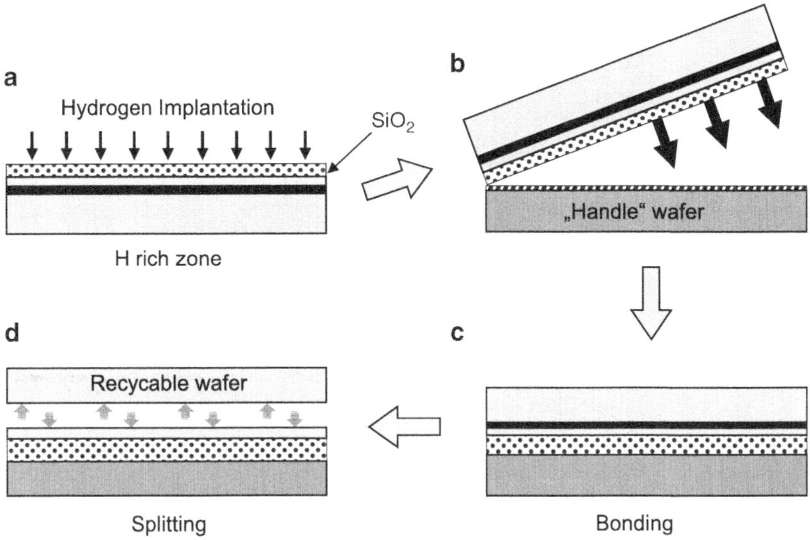

Fig. 7.1 Sequence of steps required to make SOI wafers by BESOI: (**a**) hydrogen implantation into surface of a wafer, (**b**) attaching and (**c**) face-down bonding of that SOI wafer to a handle carrier, and (**d**) splitting at the hydrogen-rich interface to form a thin SOI wafer while leaving a recyclable wafer substrate behind

temperature annealing step, and the residual substrate can be lifted off and be reused (Fig. 7.2.d).

7.2 SOI-Based Thin-Chip Fabrication

SOI wafer technology provides a class of wafer substrates for which a thin layer of single-crystalline silicon is a priori defined in thickness and is separated from the bulk silicon substrate by a buried oxide layer. Obviously, thin silicon chips having a well-defined thickness could be derived from that kind of substrate material. Both BESOI and SmartCut® wafers are successfully used for the integration of complex CMOS circuits and systems, thus demonstrating the excellent device quality of that material. The challenge left is in separating the thin chips from such SOI substrates. Conceptually, two techniques are envisioned: (1) trenching at the chip's edges down to the buried oxide and selective removal of the oxide to free the thin chips; and (2) temporary attachment of the SOI wafer face down to a carrier wafer, followed by back-thinning of the silicon substrate while using the buried oxide as an etch-stop layer, trenching of the chips and chip removal from the carrier [7]. Both concepts present considerable challenges and drawbacks: the first in the difficulty they pose to fix and transfer the thin chips during the removal process as well as in the cost; the second in the complexity of the processing, respectively. So

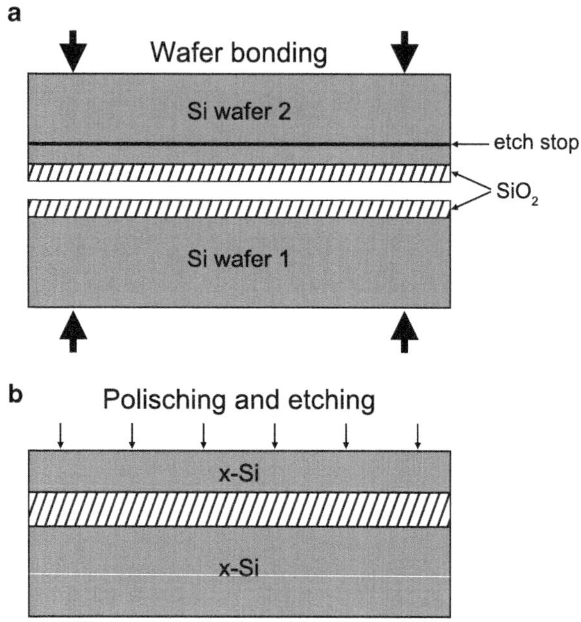

Fig. 7.2 Sequence of steps required to make SOI wafers by SmartCut® process

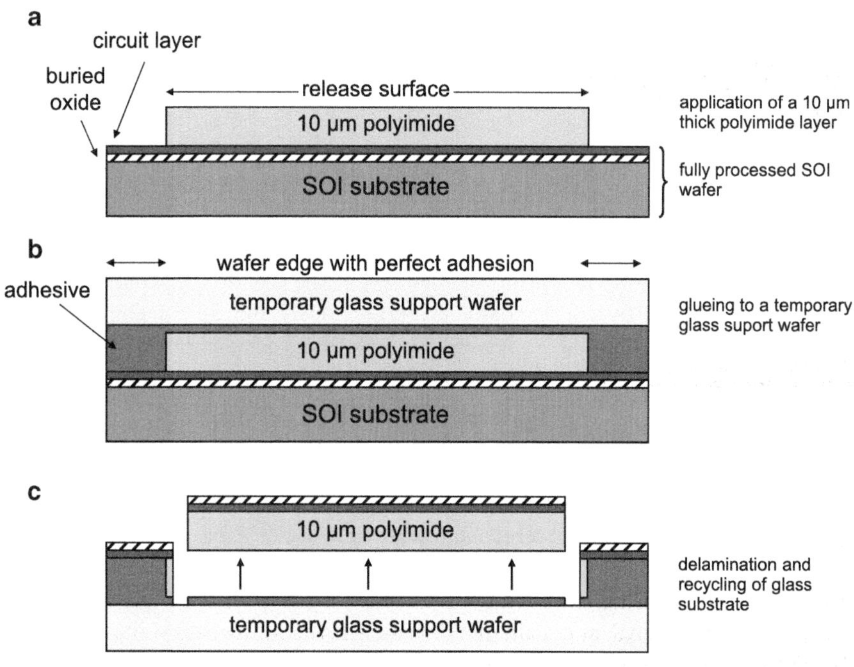

Fig. 7.3 Thin wafer and chip fabrication using Circonflex technology [8]: (**a**) polyimide coating of an SOI wafer, (**b**) attachment of a support wafer using an adhesive, and (**c**) delamination of the thin SOI film on polyimide after having removed the bulk silicon

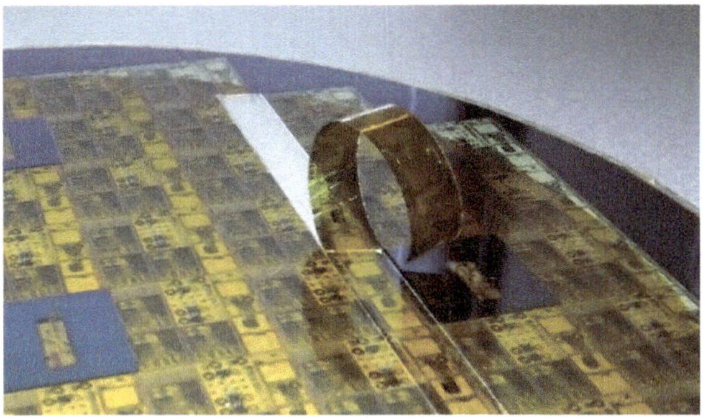

Fig. 7.4 Delamination of thin silicon on polyimide in Circonflex technology (Courtesy R. Dekker) [8]

far, only the second mentioned technique has been demonstrated and promoted as 'Circonflex' by Philips Research [8].

In the Circonflex process an SOI wafer, which includes already a circuit layer, is covered with a polyimide film, while the region near the wafer edge is excluded (Fig. 7.3a). By means of an adhesive a temporary glass wafer is then attached (Fig. 7.3b). That wafer stack is primarily kept together at the wafer edge where the adhesive is in direct contact with the SOI wafer. In the center part of the wafer the inserted polyimide only provides a weak attachment. The bulk silicon part of the SOI wafer is removed selectively, using the buried oxide as a stop layer (Fig. 7.3c). Trenching near the wafer edges finally allows for removing the thin silicon layer together with the attached polyimide film (Fig. 7.4).

The applications of Circonflex are found in integration of RF circuits and in flexible and stretchable electronics [7, 9].

References

1. Shahidi GG (2006) SOI technology for the GHz era. IBM J Res Dev 46:121–131
2. Doris B et al. (1992) Extreme scaling with ultra-thin silicon channel MOSFET's (XFET). In: International electron device meeting (IEDM), Dig. Techn. P., San Francisco, pp 267–270
3. Manasevit HM, Simpson WJ (1964) Single-crystal silicon on a sapphire substrate. J Appl Phys 35:1349–1351
4. Collinge JP (1991) Silicon-on-insulator technology: materials to VLSI. Springer, New York
5. Tong Q-Y, Goesele U (1998) Semiconductor wafer bonding: science and technology. Wiley-Interscience, New York
6. Soitec (2010) http://www.soitec.com/en/technology/smart-cut-smart-choice.php
7. Dekker R et al. (2003) Substrate transfer: enabling technology for RF applications. International electron device meeting (IEDM) Dig. Techn. P., San Francisco, pp 371–374, 2003
8. Dekker R et al. (2005) Circonflex: an ultrathin and flexible technology for RF-ID tags. In: Proceedings of the 15th European microelectronics and packaging conference, Bruges
9. Zoumpoulidis T (2010) Electronics on flexible and stretchable silicon arrays. Ph.D. thesis, TU Delft

Chapter 8
Fabrication of Ultra-thin Chips Using Silicon Wafers with Buried Cavities

Martin Zimmermann, Joachim N. Burghartz, Wolfgang Appel, and Christine Harendt

Abstract In this chapter the recently introduced Chipfilm™ technology is presented. In contrast to subtractive wafer thinning techniques this technology is inherently additive, allowing for excellent control of extremely small chip thickness through epitaxial growth and for reuse of the wafer substrate. The thickness of the chips is essentially identical to the thickness of the epitaxy layer. Therefore, process cost decreases withsmaller chip thickness, while the opposite trend applies to the conventional thinning concepts. The technology consists of a pre-process module Chipfilm™, in which wafer substrates with buried cavities in the chip areas are prepared, and a minimum complexity post-process module Pick, Crack&Place™ for singulation and assembly of the ultra-thin chips. The feasibility of Chipfilm™ technology is demonstrated for 20-μm thin chips mounted and interconnected on foil and exposed to tensile stress through bending to radii down to 20 mm. The device parameters and parameter statistical variations are well comparable to those achieved on bulk wafers for the given 0.8-μm CMOS technology.

8.1 The Chipfilm™ Concept

In contrast to subtractive techniques, i.e. wafer thinning from the wafer backside after CMOS integration, the Chipfilm™ concept considers an a priori definition of the chip thickness from the wafer frontside prior to the CMOS processing [1, 2]. This is achieved by the replacement of the starting substrates by pre-processed wafers that have buried cavities within the dedicated chip areas (Fig. 8.1a). Those pre-processed Chipfilm™ wafers are introduced to CMOS device integration like any conventional bulk substrate with the sole difference that a coarse alignment of the CMOS features to the dedicated chip areas needs to be arranged. This can be achieved either by a global alignment in the lithographic stepper tool or by using a

M. Zimmermann (✉)
Institute for Microelectronics Stuttgart (IMS CHIPS), Stuttgart, Germany
e-mail: zimmermann@ims-chips.de

J.N. Burghartz (ed.), *Ultra-thin Chip Technology and Applications*,
DOI 10.1007/978-1-4419-7276-7_8, © Springer Science+Business Media, LLC 2011

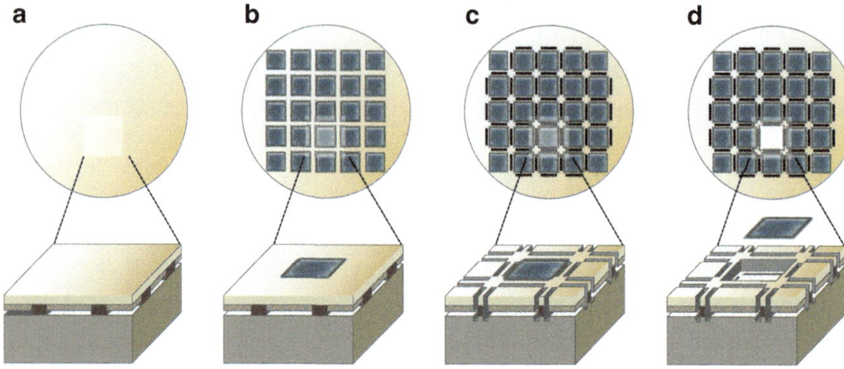

Fig. 8.1 Schematic illustration of the Chipfilm™ technology, involving (**a**) buried cavity formation in the pre-process module Chipfilm™, (**b**) CMOS circuit integration, (**c**) transformation from strong to weak chip-to-wafer attachment by trench etching at the chip edges into the buried cavity leaving only anchor connections at the chip corners, and (**d**) singulation, transfer and assembly of the ultra-thin chips in a Pick, Crack&Place™ post-process module

zero-level alignment mark on the Chipfilm™ wafers. After the CMOS fabrication is completed (Fig. 8.1b) trenches are etched at the periphery of the chip areas down into the buried cavities, while small anchors are left to the bulk substrate at selected places, such as the chip corners (Fig. 8.1c). The trench etching thus leads to a transition from strong connections between chip and bulk wafer along the entire perimeter to a weak attachment only at those anchor points. The anchors are designed to be sufficiently strong for keeping the chips reliably in place during wafer handling but weak enough to allow for breaking off the chips by reasonable mechanical force. Therefore, a conventional pick and place assembly tool can be employed to attach a vacuum chuck to the chips, break off the anchors and transfer the chips to their packaging destination. For this reason, this post-process is called Pick, Crack&Place™ (Fig. 8.1d).

8.2 The Pre-process Module Chipfilm™

The pre-process uses commercially available p-type wafer substrates with a resistivity in the range of 10–30 Ω-cm. Those wafers first receive a zero-level lithographic alignment mark to provide a means for accurate positioning of the chip circuitry over the buried cavities, which would otherwise not be detectable by the step-and-repeat lithography tool. Such proper alignment is required for reliable etching of trenches down to the buried cavity with minimum chip real estate wasted. Also, only when this is done are breakable anchors formed at the chip corners (or, optionally, anywhere at the chip edge) that allow for chip singulation in the Pick, Crack&Place™ post-process module.

Next, to prepare for the buried cavity formation, an anodic etching step is carried out to form a dual-layer porous silicon, consisting of a 1.5-μm fine porous layer above a 0.3-μm coarse porous layer. Good control of the porosity and thickness of those layers is a pre-requisite for a reliable formation of a continuous buried cavity within the chip area. An inherent feature of porous silicon is its extremely large surface area compared to its volume. During a sintering process in hydrogen atmosphere at high temperature the fine porous silicon therefore rearranges into single-crystalline silicon with enclosed nano cavities, while the coarse porous layer transforms into the desired continuous buried cavity. If the porosity in that layer were too small the cavity would be discontinuous, having randomly distributed vertical strings. If the porosity were too large the fine porous layer above could collapse during the sintering, or irreversible bonding to the substrate could take place, possibly preventing singulation of the thin chips or causing damage to the chips during Pick, Crack&Place™.

Controlled anodic etching of silicon requires a well-defined hole concentration and, thus, is sensitive to the boron-doping level in the silicon [3]. The boron concentration at the wafer surface is therefore adjusted by a boron implantation step, followed by thermal annealing. High uniformity of the etch current density across the wafer also relates to a low ohmic contact at the wafer backside [3]. Similar to a metal-silicon contact, which behaves like a Schottky diode in reverse biasing at the wafer backside, a quasi low-ohmic contact is achieved by providing a boron-doping concentration of at least $5 \times 10^{18}\ \text{cm}^{-3}$. Without sufficiently high backside-doping, local 'hot spots', in which the current density is increased considerably above average, will appear. The higher the doping concentration the less

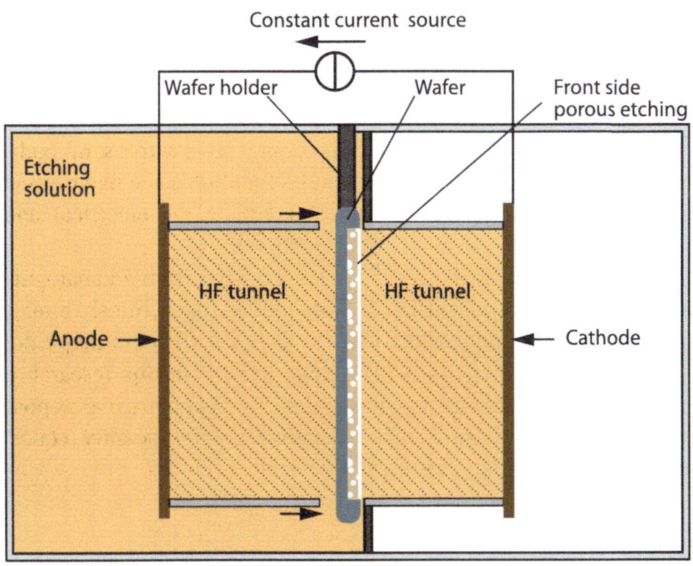

Fig. 8.2 Anodic etch cell schematic showing the wafer being clamped at half distance between anode and cathode, through which a constant current is supplied. The etch process is monitored by measuring etch current, the potential difference between anode and cathode and the time [3]

this undesirable effect becomes. At a doping concentration of $5 \times 10^{19}\ cm^{-3}$ the bias drop at that backside contact becomes a negligible fraction of the potential difference between the anode and cathode of the anodic etch cell (Fig. 8.2).

The buried cavities only extend over the target chip area. Between chip areas a spacer region is defined. This spacer features a massive connection to the silicon substrate and therefore anchors the silicon chip membrane firmly to the substrate during the CMOS integration process. Since anodic etching requires the presence of holes the spacer region can be protected from etching by introduction of an n-type doping. The phosphor implantation dose is chosen compensate the p-type background doping and thus eliminate mobile holes from that region [4]. All three implantation steps mentioned above are annealed in one high temperature step in order to minimise process cost.

Following that process step the anodic etching takes place using the etch cell in Fig. 8.2. The etching cell is filled with a mixture of hydrofluoric acid (HF) and isopropanol. The wafer is clamped at half distance between anode and cathode, through which a constant current is supplied. The etch process is monitored by measuring etch current, the potential difference between anode and cathode, and the etch time [3]. Holes injected from the wafer backside will polarise the hydrogen-silicon bonds of silicon surface atoms. Fluor ions can attack the polarised hydrogen bonds, leading to an instable SiF_2 interim molecule. Formation of SiF_4 or SiO_2 leads to a removal of surface silicon atoms from the substrate [5]. However, silicon is not removed layer by layer during anodic etching but instead pores are etched, which propagate into the bulk material while maintaining the single-crystalline structure [3]. The density of pores, i.e. the porosity, is increased for higher current density, lower HF concentration, and lower boron concentration [3]. The synthesis of the dual-layer porous structure results from varying the etch current density. The

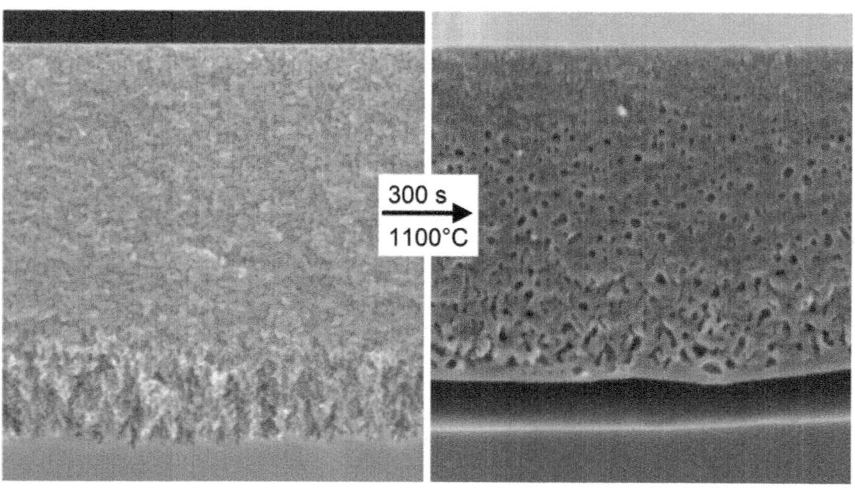

Fig. 8.3 Cross-sectional scanning electron micrograph (*SEM*) of the as-etched dual-layer porous silicon (*left*) and the structure after sintering at 1,100°C for 300 s (*right*)

initial fine porous 1.5-μm thick layer is formed with a lower etch current for a comparably longer time, followed by a higher current for a short time to form the 0.3-μm coarse porous layer (Fig. 8.3). The formation of this structure is supported by a boron concentration profile that is higher at the wafer surface and drops to a lower level near the transition from fine to coarse porous silicon. Therefore, the implanted boron concentration profile provides a means of self-adjustment.

Annealing of the dual-layer porous silicon in a non-oxidizing atmosphere at a temperature higher than 900°C dramatically alters the structure by sintering [6]. Sintering means restructuring with the aim of reaching a minimum free surface energy of the material. In the Chipfilm™ pre-process, sintering is carried out at 1,100°C. The result is a continuous cavity with above a ~1.5-μm thick single-crystalline layer having a high density of nano cavities and closed silicon surfaces (Fig. 8.3). The thus restructured single-crystalline and defect-free wafer surface serves as a seed for a subsequent epitaxial layer growth, which is used to accurately define the final chip thickness. The porosity and thickness of the fine porous layer have to be tailored for a proper formation of the underlying narrow cavity and for a restructuring of a closed defect-free wafer surface, thus allowing for epitaxial growth of a high quality silicon layer. Epitaxy is carried out in trichlorosilane ($SiHCl_3$) at atmospheric pressure and 1,100°C.

Besides the crystalline quality of the epitaxial layer, outdiffusion of boron from the p-type doped porous layer during sintering and epitaxial growth is of concern. Outdiffusion causes an unintended vertical autodoping of the epitaxial layer in which the electronic devices are to be integrated. Even more of concern is the lateral autodoping that results from boron autodoping from the higher doped backside of the wafer. In the Chipfilm™ pre-process this backside p-type layer, which was necessary for the anodic etching, is therefore removed prior to epitaxy.

During sintering the buried cavity becomes sealed. The remaining hydrogen, which has filled that cavity during the epitaxy pre-bake and the epitaxial growth, escapes during a second high temperature process step in a non-hydrogen

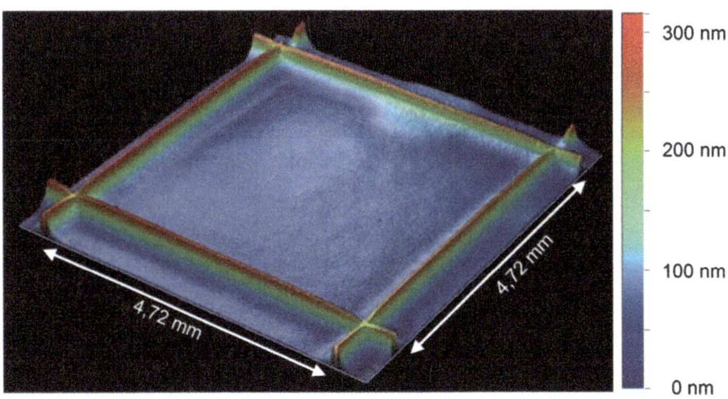

Fig. 8.4 Surface topography profile of a 4.72 × 4.72-mm^2 wide and 20-μm thick chip silicon membrane region

atmosphere. As a consequence the silicon membrane settles down against the bottom of the cavity, thus causing a ~300 nm depression within the chip areas (Fig. 8.4).

8.3 CMOS Circuit Integration on Chipfilm™ Wafers

Chipfilm™ wafers are intended to be used for CMOS process integration like any conventional bulk substrate with the restrictions mentioned in Sect. 8.2. The Chipfilm™ membranes wafers should be mechanically stable and in good thermal contact with the substrate. Otherwise the silicon membranes may warp and break, making the wafers unprocessable. Warpage of the membranes results from stress supplied by deposited layers or from a temperature difference between silicon membrane and substrate. The latter may occur during rapid thermal processing (RTP). The fact, that the chip membrane is in contact with the silicon substrate during the CMOS process as a result of the evacuated buried cavity, however, provides both mechanical stability and a good thermal contact. Owing to the long range roughness of the cavity ceiling and bottom planes the probability of van der Waals bonding is very small. The slight depression of chip areas may require an adjustment of the focus plane at lithography nodes of 0.1 μm and below, which can be accomplished. However, the ~300-nm surface topography hampers shallow trench formation for advanced technology nodes (<0.1 μm) because chemical-mechanical polishing (CMP) is used and the shallow trench depth is of the order of that wafer topography. Employment of Chipfilm™ technology for advanced CMOS will thus require shallower cavity formation or other means of levelling the wafer surface. In the initial demonstration of Chipfilm™, in which a 0.8 μm CMOS process was used, the slight depression was not an issue [1, 2].

8.4 The Pick, Crack&Place™ Post-process

After CMOS circuit integration has been finished the chips are still firmly attached to the substrate at the perimeter. In such a state they cannot be detached from the substrate without being mechanically destroyed. The lateral connection to the substrate must therefore be weakened. This is achieved by reactive ion etching (RIE) deep trenches at the chip edges with use of the 'Bosch process', described in Chap. 9. The trenches need to extend down to the buried cavities. Only at the corners, or any other selected small region at the perimeter, trench formation is avoided so that lateral anchors remain after trenching. The size and number of those anchors is to be tailored in a way that the connection of the chips to the substrate is still sufficiently reliable, but weak enough to be breakable by attachment of a vacuum chuck to the chip surface and application of a vertical force. This allows

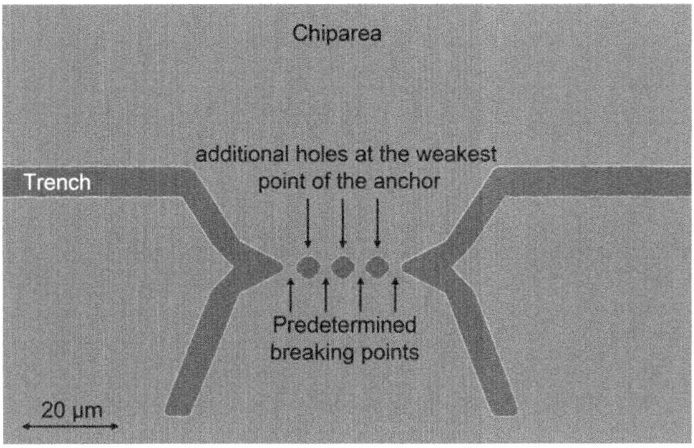

Fig. 8.5 Zipper-style designed anchors for localised maximum stress near the perforation

for using conventional pick-and-pace assembly equipment. Since, in addition, breakage of the anchors is involved, the process is called Pick, Crack&Place™. The maximum force to be applied is proportional to the chip area. Therefore, the number and the size of the anchors need to be adjusted to the size, shape and thickness of the chips. The following requirements are defined:

- Breakage of the anchors should reproducible at a yield >99%.
- Anchors should de designed to break at well-defined locations.
- Particle generation should be minimum.
- Anchor remains must not degrade the chip stability (Chaps. 19 and 20).

The anchor type and optimum design were explored through finite element (FEM) simulations. A maximum distance up to which the chips could move freely without breaking off the anchors was set as a constraint [7]. Beyond that distance the anchors were expected to break reliably at a location restricted by design. In that way extension of secondary micro cracks into the chip area could be avoided. The best-suited anchor design is shown in Fig. 8.5.

This type of anchor emerges from a chip area with a rather steep angle, leading to the intended breaking points (Fig. 8.5). Notably, the trench outside of the anchors does not extend over the full distance between chip and spacer region but is restricted to minimum size in order to ensure similar etch depths of trench and holes [7].

8.5 Characterisation of Chipfilm™ Dies

A mixed-signal (38,000 digital and 2,700 analogue transistors) circuit was fabricated on both bulk and 20 μm Chipfilm™ wafers to demonstrate the feasibility of the Chipfilm™ and Pick, Crack&Place™ process modules (Figs. 8.1a, d). For both

Table 8.1 Carrier mobilities μ_n and μ_p, effective channel lengths $L_{eff,\,n}$ and $L_{eff,\,p}$ and threshold voltages $V_{th,\,n}$ and $V_{th,\,p}$ of NMOS and PMOS transistors and delay TRO of CMOS ring oscillators on Chipfilm™ and on bulk substrates for comparison

	Chipfilm™	Bulk silicon	Difference
μ_n	(630 ± 20) cm²/Vs	(610 ± 10) cm²/Vs	4.28
μ_p	(178 ± 5) cm²/Vs	(174 ± 3) cm²/Vs	2.30
$L_{eff,n}$	(0.79 ± 0.02) μm	(0.80 ± 0.1) μm	−1.25
$L_{eff,p}$	(0.82 ± 0.03) μm	(0.83 ± 0.1) μm	−0.61
$V_{th,n}$	(0.87 ± 0.006) V	(0.87 ± 0.005) V	0.00
$V_{th,n}$	(0.917 ± 0.005) V	(0.915 ± 0.005) V	0.22
T_{RO}	(310 ± 6) ps	(300 ± 6) ps	3.33

bulk and Chipfilm™ samples, on-wafer and in-package tests were carried out, providing information about the device parameters and parameter variations as well as the mixed-signal circuit parameters (Table 8.1). The functional and parametric yields of bulk and Chipfilm™ wafers were very similar, indicating that no considerable increase in defect density results from the Chipfilm™ process. Parameters control monitors (PCM) were fabricated on the same wafers as the mixed-signal circuit chips. Average values and standard deviations of carrier mobilities, effective channel lengths, threshold voltages of NMOS and PMOS transistors and CMOS ring oscillator delays, are listed in Table 8.1 for Chipfilm™ and bulk wafers, in comparison. The minor differences are attributed to the different type and concentration level in the epitaxial device layer of Chipfilm™ and bulk wafers [2].

Chipfilm™ samples were also mounted onto 50-μm thin Kapton® tape containing copper wiring, to which the chip's pad connections were wire-bonded. Those tapes were attached to an Agilent 83,000 parameter analyzer. An in-house built stress apparatus was used to apply various levels of tensile stress by means of pulling the tape around aluminium cylinders having different radii (Chap. 21; [8]). A maximum stress up to 110 MPa, which has been applied through a bending radius of 20 mm, could reliably be applied to the mixed-signal product chips [8]. As one signature of the evident piezoresistive effect in the channels of the CMOS transistors, the operating and standby currents of the mixed-signal circuit exhibited a distinct dependence on the bending radius. It was found that the stress dependence of those supply currents can effectively be tailored by changing the physical orientation of the NMOS and PMOS transistors mounting the thin chips on tape in different orientations [8].

Acknowledgements The authors wish to acknowledge W. Appel, C. Burwick, S. Ferwana, C. Harendt, T. Hoang, F. Hutter, E. Penteker, N. Remmers, C. Reuter, H. Richter, S. Schmiel, I. Schindler, P. Vöhringer and all other IMS staff members involved in device and circuit fabrication and characterisation. Moreover, O. Tobail, M. Schubert and J. Werner from the Institute for Physical Electronics (ipe) at the University of Stuttgart are acknowledged for helpful discussions and support with the topic of porous silicon etching.

References

1. Zimmermann M et al. (2006) A seamless ultra-thin chip fabrication and assembly process. In: Proceedings of the IEDM, San Francisco, pp 1010–1012
2. Burghartz JN, Appel W, Rempp HD, Zimmermann M (2009) A new fabrication and assembly process for ultrathin chips. IEEE Trans Electr Dev 26(2):321–327
3. Beale MIJ et al. (1985) An experimental and theoretical study of the formation and microstructure of porous silicon. J Cryst Growth 73:622–636
4. Watanabe Y, Arita Y, Yokoyama T, Igarashi Y (1975) Formation and properties of porous silicon and its application. J Electrochem Soc 122:1351–1355
5. Lehmann V, Gösele U (1991) Porous silicon formation: a quantum wire effect. Appl Phys Lett 58(856–858):1991
6. Labunov V et al. (1985) Heat treatment effect on porous silicon. Thin Solid Films 137:123–134
7. Burghartz JN, Harendt C, Hoang T, Kiss A, Zimmermann M (2009) Ultra-thin chip fabrication for next-generation silicon processes. In: Proceedings of the BCTM, Capri, pp 131–137
8. Rempp H, Burghartz JN, Harendt C, Pricopi, Pritschow M, Reuter C, Richter H, Schindler I, Zimmermann M (2008) Ultra-thin chips on foil for flexible electronics. In: Proceedings of the ISSCC, San Francisco, pp 334–335

Part III
Add-on Processing

The fabrication of ultra-thin wafers and chips may only be a first step toward the target industrial products. Often, additional processing will be required prior to assembly or embedding of the ultra-thin chips (Table III.1). The metal pads of the chips may have to be capped by a noble metal layer for protection purposes and to allow for low-ohmic bonding to interconnect the chips. Thin chips used for biomedical applications may have to be completely protected and isolated because they may be implanted into the human body without any packaging, and erosion of materials such as aluminium, silicon and silicon dioxide may occur otherwise. Thin chips are of high interest to power device technology, since they lead to a lower thermal resistance and allow for novel device structures. Here, backside doping by ion implantation and rapid thermal processing and backside metallisation requires the use of suitable handle carriers and thermal processes, such as laser annealing. Finally, but currently of highest interest, is the goal to provide dense through-silicon vias (TSVs) to the ultra-thin dies to allow for building true 3D ICs and stacked micro systems.

Table III.1 Classification of add-on processes

Add-on processes	Processing (examples)	Applications
Capping of metal pads	Electroless plating Au evaporation/shadow mask	Chips in PCB Chips in foil Chips in paper Biomedical
Chip isolation/protection	Thermal oxidation Atomic layer deposition (ALD)	Biomedical
Backside metallisation	Handle carrier attachment Wafer thinning Laser annealing Backside metal deposition	Power devices
Through-silicon via (TSV)	TSV first (during FEOL) TSV middle (during BEOL) TSV last (after BEOL)	3D ICs 3D microsystems

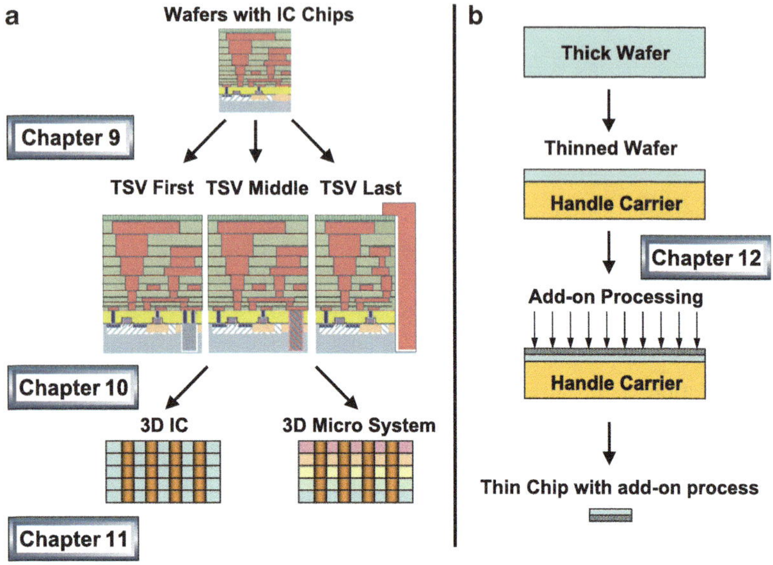

Fig. III.1 Illustration of (**a**) the three generic concepts for the implementation of through-silicon-vias (TSVs) to build 3D ICs or 3D micro systems; and of (**b**) wafer backside add-on processing by using handle carriers

Figure III.1 illustrates the three generic concepts to implement TSVs as vertical interconnects in stacked ultra-thin chips. TSVs can be inserted during the front-end-of line (FEOL), during the back-end-of-line (BEOL) or after BEOL (i.e., with chip assembly) in order to build three-dimensional (3D) ICs and 3D micro systems (Chap. 10). For TSVs to be very dense their aspect ratio (height/pitch) must be as high as possible. This requires high-aspect ratio holes to be etched through the thin silicon, an isolating liner material to be added and the remaining cavity to be refilled with a low-resistive metal. The Bosch process (Chap. 9) allows for etching holes into silicon with an aspect ratio as high as 100:1. Also, the liner formation in those high-aspect ratio holes has also become feasible through atomic layer deposition (ALD). However, as stated in Chap. 2, the International Roadmap for Semiconductors does not foresee any solution to metal refill of the vias at aspect ratios higher than 10:1. The technological issues and solutions of 3D IC technology and 3D micro systems are treated in Chaps. 10 and 11.

Chapter 9
Through-Silicon Vias Using Bosch DRIE Process Technology

Franz Laermer and Andrea Urban

Abstract Silicon deep reactive ion etching (DRIE) is having a great effect on micro-electro-mechanical systems technology (MEMS) and quite recently also on memory devices and through-silicon via (TSV) etch applications. Nowadays, it is an established fabrication process within the MEMS field. The basic technology was originally developed at Bosch in the early 1990s. At that time, classical wet etching in KOH was the state-of-the-art technology, with design options like the silicon vias and also many others remaining unthinkable. With the Bosch DRIE process it became possible to overcome the design restrictions and compatibility problems related to the old silicon wet etching technology. The etching performance of the Bosch process is not reached by any wet etchant or other microstructuring technology. Today, after more than a decade of Bosch process plasma etching technology shaping the MEMS field, it is emerging also as a new enabler in other areas, e.g., for extremely precise high aspect ratio TSVs for stacking memory chips and other purposes. The Bosch process has evolved throughout the years thanks to continuing progress in process and hardware development. Today a broad supplier base is supporting DRIE customer needs in many application areas, with the industry's high-density plasma equipment and process know-how all over the world.

9.1 Silicon Plasma Etching: Fundamentals on Etching Equipment and Processes

Plasma etching, or reactive ion etching (RIE), has been well established in the semiconductor area since the 1970s and 1980s [1–3]. The classical tool that was used in the early RIE work is the diode reactor, as shown in Fig. 9.1. Detailed descriptions of different plasma reactors and processes can be found in [4]. The

F. Laermer (✉)
Robert Bosch GmbH, Corporate Research and Advance Engineering Microsystems (CR/ARY-MST), P.O. Box 106050, D-70049 Stuttgart, Germany
e-mail: franz.laermer@de.bosch.com

J.N. Burghartz (ed.), *Ultra-thin Chip Technology and Applications*,
DOI 10.1007/978-1-4419-7276-7_9,

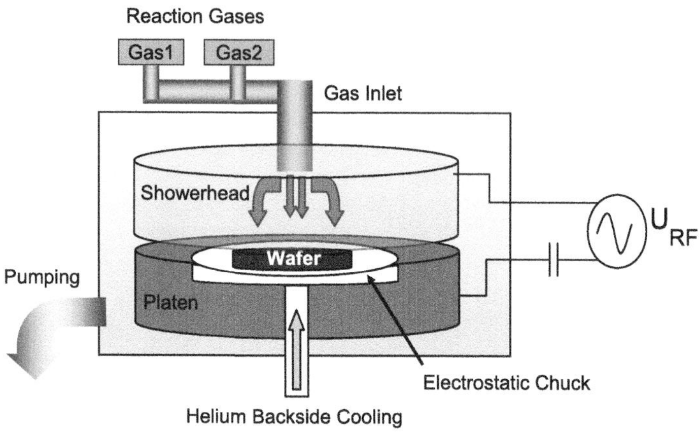

Fig. 9.1 RIE plasma reactor

scheme depicts the key elements of the typical RIE configuration: reaction gases provided to the process chamber by adjustable mass flow controllers, a grounded showerhead electrode as a gas inlet, a substrate electrode carrying the wafer and a radio frequency (RF) power supply coupled to the substrate electrode via a capacitor (which is mostly part of a more complex matching unit to adapt RF impedances). Radio frequencies generally used are in the low frequency (380 kHz), mid frequency (2 MHz) or high frequency (13.56 MHz) range; in most cases 13.56 MHz is used as a standard for plasma generation. Efficient pumping provides the low pressure conditions in the reactor chamber needed for the electrical discharge to form between the powered and the grounded electrode, which builds up the plasma in an atmosphere of one or more compound gases, at a pressure ranging from 0.1 to some tens of Pa. The coupling capacitor allows the powered electrode to float at a DC potential, which is generated from the applied RF voltage as a result of the different mobilities of low mass negative charges (electrons) and high mass positive charges (ions) in the plasma. This is the so-called "DC-bias voltage," which is accelerating the ions towards the powered substrate electrode.

In the classical RIE tool, RF power coupled to the substrate electrode generates both the plasma discharge and the ion acceleration voltage towards the wafer substrate (DC biasing). Isotropic, vertical and all different kinds of tapered profiles could be etched into silicon for many applications such as shallow trenches, trench isolations between transistors in IC manufacturing, contact hole formation and DRAM trench cells (trench capacitors for memory) during the manufacturing of electronic circuits. A general overview on reactive ion etching can be found in [2]; a review on silicon RIE is given in [3]. The capabilities of RIE, mainly its independence on crystal orientation and the potential to fabricate arbitrarily shaped geometries, made plasma etching a promising candidate for developing a new microstructuring technique for the MEMS field. Early work demonstrated that micromechanical elements containing shallower pattern features could well be realised using the classical RIE approaches, which were adapted from semiconductor manufacturing [5–8]. But in classical RIE, plasma density and ion energy are tightly linked to each other and cannot be adjusted independently. Therefore RIE is insufficient for micromachining deep structures, mainly for reasons of low mask selectivities, insufficient profile control and of the fact that it is limited to a low etching speed. However it was felt that it inherits the basic potential for etching deeper structures at high aspect ratios.

9.1.1 High-Density Plasma Sources

In order to overcome the tradeoffs of classical RIE technology and to reach both, high mask selectivities and high rates for anisotropic high aspect ratio silicon etches, a high-density high-pressure plasma tool as well as a so-called remote or decoupled plasma generation is necessary. In a remote plasma tool, the plasma density consisting of chemically active neutral species, the so-called uncharged radicals, can be provided independently from the ion impact onto the substrate. This hardware precondition has been the basis for the successful development of the Bosch DRIE process technology. The only high-density high-pressure plasma sources available in the early 1990s was the microwave surfatron source [9, 10] and later on the first inductively coupled plasma (ICP) source geometries. Both types of sources demonstrated astonishingly similar key performance parameters. The first choice for early development work at Bosch was the microwave surfatron source, mainly for the reasons of its virtually unlimited magnetron power levels at 2.45 GHz at very low cost. Even today, microwave power is still significantly cheaper than RF power, if compared on a cost-per-kilowatt base. In addition to the cost argument, the nearly unlimited reflected power tolerance of microwave sources makes it attractive for discontinuous process solutions like the switched Bosch process.

9.1.2 The Bosch Process: Repetitive Cycling of Passivation and Etching Steps

In 1992, the breakthrough in DRIE technology was achieved, using a surfatron plasma source at first. Later on, the Inductively Coupled Plasma (ICP) equipment as illustrated in Fig. 9.2 emerged more or less as an industry standard – and became the industry standard for the Bosch process technology as well.

In the Bosch process, a high-density plasma, which is generated remote from the wafer, meets the needs of both high-density of etching species and a moderately low ion energy impact onto the substrate. This is combined with a sophisticated strategy of alternating etching and passivation steps, which can be controlled independently from each other [11]. The process is based exclusively on fluorine chemistry. Sulphur hexafluoride (SF_6) as the etching gas readily delivers fluorine radicals

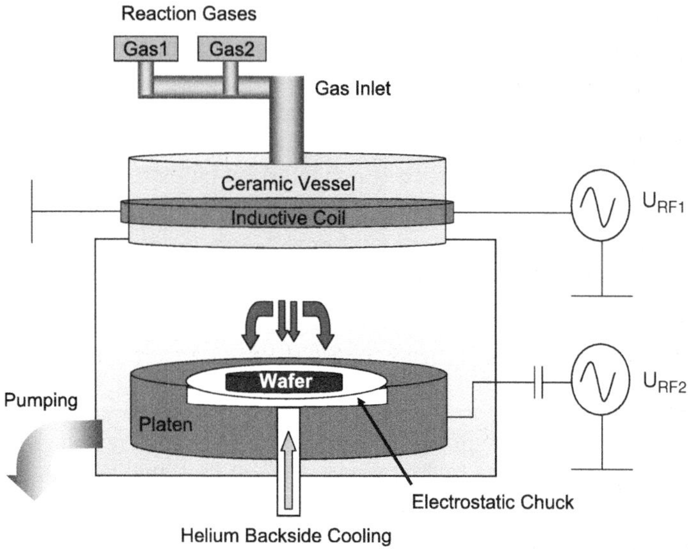

Fig. 9.2 Schematic of an inductively coupled plasma (ICP) tool. RF Generator 1 powers the inductive coil around the dielectric reactor vessel to generate a high density plasma within the reactor. RF Generator 2 powers the substrate electrode for biasing and accelerating ions to the wafer. Electrostatic clamping and helium backside pressurizing cool the substrate during processing

after excitation of the SF_6 gas molecules by electron impact from the plasma. These fluorine radicals have the capability to etch silicon spontaneously without a need for ion bombardment, forming volatile silicon fluorides like SiF_4 as the main reaction product.

$$SF_6 \rightarrow SF_n + (6-n)F^* \quad \text{with } n = 0, 1, 2, 3, 4, 5$$

$$Si + 4F^* \rightarrow SiF_4 \uparrow$$

Without any sidewall passivation, the silicon plasma etching in fluorine based chemistry is isotropic and thus leads to large mask underetching and imprecise features. The Bosch process enables an anisotropic silicon etch by inhibiting the lateral etch attack via a sidewall passivation film. (Hydro)Fluorocarbons are excited in the plasma to build up Teflon-like polymer films on the treated silicon wafer. During the early stage of process development, a number of hydrofluorocarbons like CHF_3, $C_2H_2F_2$, C_3F_6, C_4F_6, C_4F_8, and C_4F_{10} were compared for their availability and for their passivation efficiency and undesired particle formation in the plasma gas phase. As a result of these investigations, under ICP conditions typical for Bosch DRIE processing, octafluorocyclobutane (C_4F_8) was found to be the best choice. This somewhat exotic process gas at the time of establishing Bosch DRIE technology is nowadays widely available from a number of trustworthy gas suppliers.

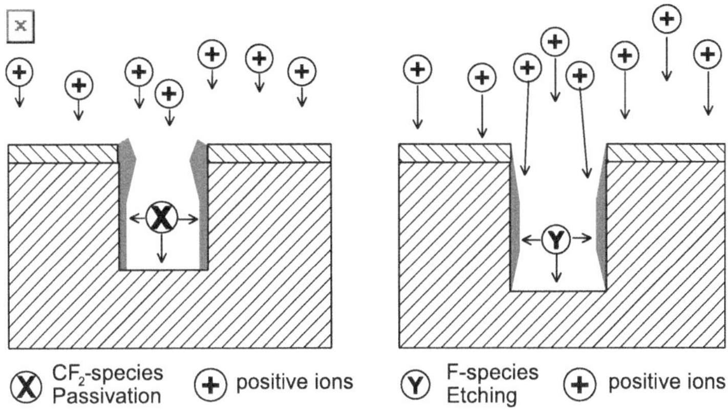

Fig. 9.3 Basic mechanism of the Bosch process. The protecting Teflon-like film deposited onto the structures during the passivation step is cleared from the trench floor and driven deeper into the trenches during the (intrinsically isotropic) fluorine-based etching step

The discontinuous nature of the process overcomes the restrictions caused by strong recombinations in the gas phase, which would happen in a mixture of etching fluorine radicals and passivating polymer-forming monomers. A mixture of etching and passivating species especially at the high concentration levels within the plasma source volume leads to their pairwise extinction and to the loss of etch rate and profile accuracy. Separating the etch and passivation species at least within the high-density source area is a process essential, which is from a process point of view preferably reached by switching etch and passivation cycles in the time domain. High etch rates and high mask selectivities combined with a high anisotropy at the same time is no longer a tradeoff within the Bosch process strategy, as the soft passivating Teflon-like polymer material requires only low energetic "soft" ion impact during the subsequent etch step for its complete removal from the bottom of the trenches. This basic mechanism of the Bosch process is shown in Fig. 9.3. Resist or oxide mask erosion is very small for this low ion energy flux, yielding in high resist selectivity. Selectivity values between 50:1 and 300:1 are typically achieved with the Bosch process, depending on the details of the etching equipment and process recipe chosen.

9.2 The Bosch Process: Characteristics and Effects Relevant for Through-Silicon Via Applications

The fast alternating sequence of etch and passivation cycles, the isotropic etching nature of the fluorine radicals and the partial erosion of the sidewall film during each etch step creates a small "undercut" during each cycle, leading to "ripples" or "scallops," which is a periodic nanostructure along the sidewall (see Fig. 9.4).

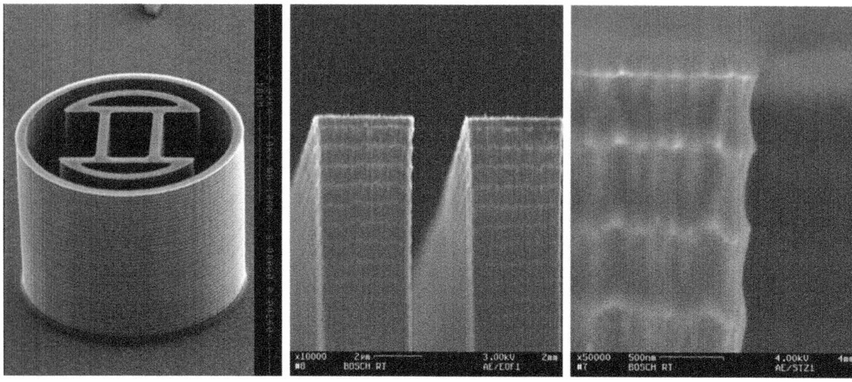

Fig. 9.4 Trenches and sidewall-scalloping resulting from a typical Bosch process recipe

The ripple size is affected from sealing capabilities and erosion behavior of the passivation layer under fluorine radicals attack, which is determined by passivation layer thickness and morphology. The typical ripple size is 10–100 nm depending on process details.

9.2.1 Solutions for Reduction of Sidewall Scalloping for TSVs

For through-silicon vias, especially for the metal-refill via approach, a positively tapered profile and smooth sidewalls are essential requirements for a void-free trench refill and high electric breakthrough strength. The trenches have to be coated conformally by LPCVD or PECVD oxide isolation layers and a subsequent metal seed layer prior to electroplating. The galvanodeposition step needs to refill the trenches conformally with the chosen contact metal, without formation of voids or other irregularities. Very fast switching times between etching and passivation steps of below 1 s and even down to 100 ms for each individual step, the so-called "ultrafast gas-switching conditions" [12], yield in very smooth sidewalls needed to meet the requirements of TSVs and their subsequent refill. Sidewalls of such trenches are etched with hardly any remaining scalloping at all, like illustrated in Fig. 9.5.

A different approach to minimise scalloping is a post-etch annealing step >1,000°C in hydrogen atmosphere, in case the total process flow can still tolerate such a temperature budget after the DRIE step. The mechanism for the post-etch sidewall planarization is an atomic migration of silicon atoms, which is enabled by their thermal activation at the high temperature, driven by the minimization of surface energy [13]. Besides this, the sidewall roughness of a trench can also be reduced post-etch by a thermal oxidation of the silicon surface and a subsequent removal of the grown thermal oxide layer [14] in liquid HF or HF vapour phase. This procedure can be repeated several times, until the desired surface roughness

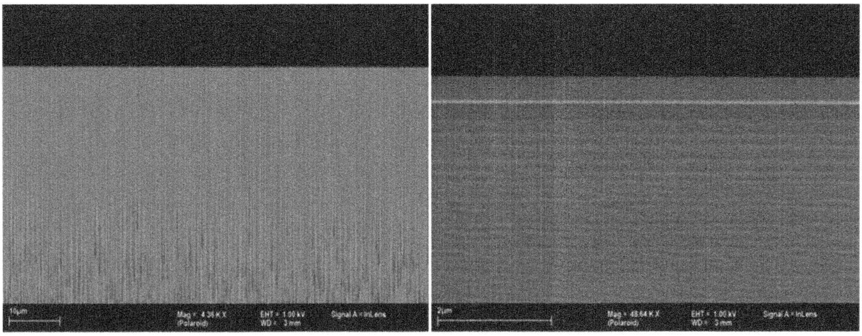

Fig. 9.5 SEM pictures of the sidewall resulting from a Bosch process using an ultrafast switching recipe. Hardly any scallopping is visible at the sidewall any more. Some vertical striations appear in the lower part of the left-hand side SEM picture, indicating the beginning of microroughness formation at the trench bottom, which also affects the sidewalls

reduction is achieved. Again, tolerance of the total process for temperature budgets like that even after the DRIE step is a prerequisite for applicability of this approach. Both post etch treatments yield in very smooth silicon trench sidewalls, as "scallops" and eventually also "spiking" get removed. Smoothing by thermal oxidation inherits a small but well controlled critical dimension (CD) loss, which may be a drawback, especially for MEMS inertial sensor applications, where the "ultrafast gas-switching" technique is often the preferred choice to achieve smoother trench sidewalls. A marginal CD loss is in most cases acceptable for the through-silicon via application, but it has to be taken into account that metallisation cannot withstand temperatures significantly above 400°C, which prevents high temperature annealing and thermal oxidation in the case of many "via-last" process approaches. Figure 9.6 illustrates void-free copper vias with smooth sidewalls, processed under optimised Bosch process trench conditions.

9.2.2 The Notching Phenomenon: Its Relevance to TSV Etching and Solutions to Overcome Notching

One characteristic of high-density plasma etching, as in the Bosch process, is the notching behaviour, when the etch meets a stop at dielectric interfaces. Silicon vias often terminate on top of a silicon oxide layer, which acts as a vertical isolation. High-aspect ratio Bosch process etching technology is used to define the lateral isolation of silicon vias. Notching occurs and is a well-understood phenomenon of electrical charging in high-aspect ratio trenches when the dielectric interface is reached and an overetch needs to take place for full clearing of the bottom area from remaining silicon. The notching effect is caused by a sidewall attack by redirected ions during the overetch phase [15–18]. Charge accumulation at the dielectric trench bottom and in the opposing conductive silicon sidewalls is responsible for the redirection of ions towards the interface between insulator and conductor. Notching is a highly undesirable effect, as notches eventually cut deep into the structure sidewalls and corrupt their mechanical properties; in addition, notching represents a serious particle source, with the risk of short circuits in smaller gaps. For a via-hole application, notches represent hard to refill structural deviations, involving risks of void-formation and later in-use insulation dielectric breakdown. Charging can be reduced by advanced pulsing schemes involving both the substrate bias power [19] and the plasma source power [20]. Pulsing gives the dielectrics the chance to discharge during bias off-periods, with discharge eventually supported by anions created in the afterglow phases of plasma shutdown. As a consequence, a wide range of overetch conditions is tolerated without notching (see Fig. 9.7 below).

Implementation of anti-notching solutions and their integration into the equipment hardware can be driven to an extent that notching is completely avoided even for long overetches, resulting in complete design freedom independent of etching depths and aspect ratios. This freedom is often needed for today's trench applications, like sensors and silicon TSVs shown in Fig. 9.8. Here, effective notch suppression is a key requirement.

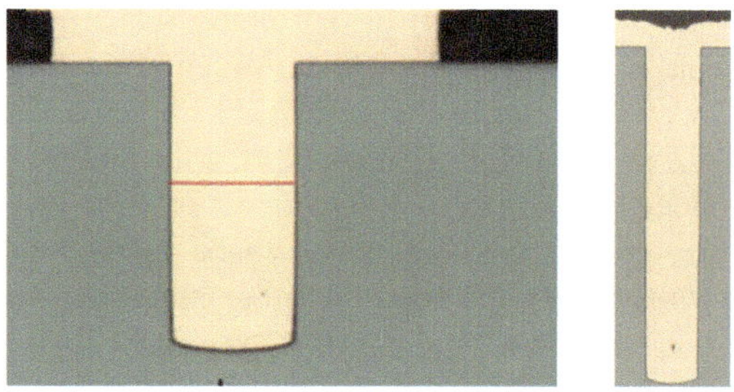

Fig. 9.6 Metal vias of 50-μm diameter. Void-free copper electroplated "Bosch-processed" trenches of aspect ratio of 2.4:1 (*left*) and 6:1 (*right*)

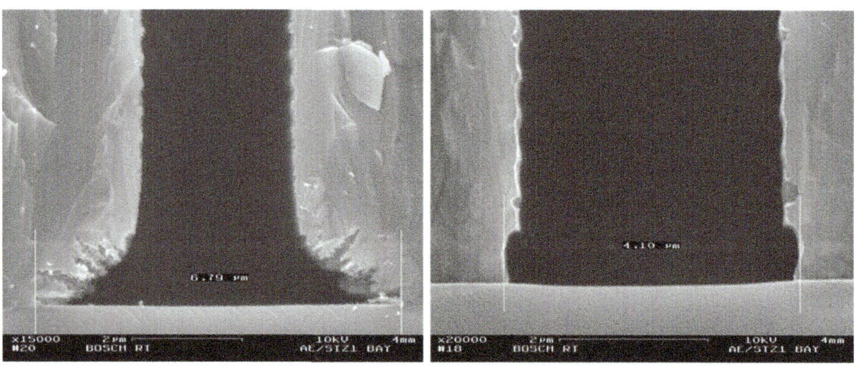

Fig. 9.7 Notching at dielectric SiO_2 interface without pulsing substrate bias power (*left*), and notch-free etch-stop on SiO_2 after a 100% overetch using pulsed bias power (*right*)

9.2.3 Parameter Ramping for High-Aspect Ratio TSV Etching

In order for CD loss to be minimised at the top of a via trench and for uncontrolled resist mask undercut to be avoided, it is often beneficial that the process parameters be adjusted or "ramped" during the course of the trench process. In general, the etch rate goes down with the progress of the etch and the increasing aspect ratio; the reason of this is the amount of ions (and neutrals) reaching the trench floor is slowly diminishing due to aperture effects at the trench opening in combination with the angular distribution of the ions. As a consequence, there is a trend towards more passivation with the increase in aspect ratio, and thus profiles tend to get positive slopes or end up tipping for aspect ratios of >20:1 if no adaptation of the process recipe is foreseen. A stepwise or continuous parameter adaptation or ramping [21–23] can reduce the amount of passivation during the progress of the etch. For example, passivation cycle time or passivation gas flow can be ramped down steadily, or etching cycle time or etching gas flow can be ramped up. Other options like source power, bias power or pressure ramping also exist. The most obvious parameter ramping strategy is to change cycle times, which will affect the etch-to-deposition

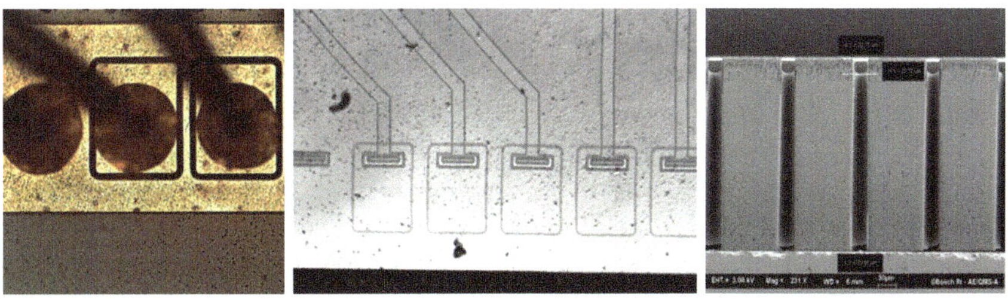

Fig. 9.8 Silicon via approach. Silicon vias: top view wire-bonded (*left*), thin film wiring (*center*), and via cross-section (*right*)

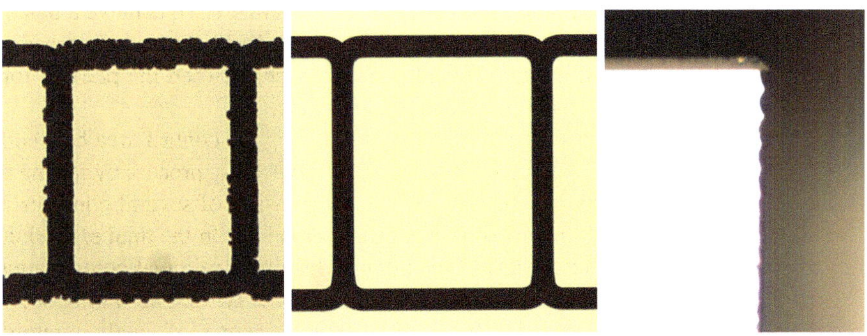

Fig. 9.9 Silicon vias after Bosch process etching using resist mask. Top view with large CD loss and uncontrolled mask undercut (*left*), top view with minimised CD loss and no mask undercut (*center*), cross-section of trench profile and resist mask after completion of an optimised via trench (*right*)

balance in a predictable and straightforward manner. For an optimised silicon via trench with low CD loss and resist mask undercut, the trench sidewall has to recede at about the same speed as the resist is eroded during the etch. Therefore, the trench process should start with a high etch-to-deposition ratio, leading to larger scallops at the top of the trench and a fast receding of the top sidewall directly below the resist mask. To maintain this faster recede of the silicon sidewall in relation to the erosion resist – and to avoid an excessively tapered profile – the via trench should proceed and finish with a slope with a slightly negative taper. With this strategy of "Bosch-processing," the demand for a minimised CD loss in combination with negligible mask undercut can be realised even for low cost resist masking in silicon via trenching (see Fig. 9.9).

9.3 Summary and Conclusions

Silicon via etch applications with their high-aspect ratios and crucial profile control requirements are demanding new challenges in trench processing. Aspect ratios of up to 30:1 have to be realised at reasonably high etch rates for an economical output under production conditions. Besides this, high mask selectivity and minimised mask undercut are essential issues. Low process pressure is, generally speaking, beneficial for high-aspect ratio etching, as exchange of species in narrow deep trenches is accelerated by higher mean-free path of the gas molecules and radicals. Low process pressure is, however, a trade-off with high etch rate requirements as, again speaking generally, etch rates increase with higher pressure levels. What is helpful in the case of via etching is that the exposed silicon area on the wafer be generally low as compared to other DRIE applications more typical for the MEMS field. This fact in principle enables high silicon etch rates in the mid pressure range from 3 to 5 Pa, which is the typical operating range for today's remote plasma equipments. For wire-bonded silicon vias the trench can be realised using top metallisation itself as a mask, which is somewhat beneficial from the selectivity point of view. Trench etching using this metal masking profits from the extremely high selectivities at no extra costs, in this case. A certain metal mask undercut by the trenches can be tolerated, as long as the metal overhang is not too large to risk break-off during subsequent wire-bonding which might short-circuit the via contact to substrate potential. For silicon vias with thin-film wiring, a low cost resist mask is the preferred option, although mask selectivity is lower for resist than for hard masks, and in particular for metal masks. CD loss at the top entry of the trench should not exceed certain limits during the etch, in order to refill more easily, and to allow for the thin film interconnection process steps to follow. With the different process strategies described in this chapter or combinations thereof, the various demands on mask selectivity and mask integrity, etch rate and profile control, including avoidance of scalloping and notching, can be met in the fabrication of through-silicon vias.

Acknowledgments We acknowledge support and assistance from our colleagues at Bosch Corporate Research Department CR/ARY2, Dr. Achim Trautmann, Dr. Ando Feyh and Dr. Ralf Reichenbach, together with our colleagues at Bosch Engineering Sensor Process Technology Department EPT, Eckhard Graf, Dr. Jens Frey and Dr. Barbara Will, for sharing the results of their development, our valuable discussions with them, their contributions and the illustrations on through-silicon vias.

References

1. Manos DM, Flamm DL (1989) Plasma etching: an introduction. Academic, Boston
2. Oehrlein GS (1990) Reactive ion etching. In: Rossnagel SM (ed) Reactive ion etching: handbook of plasma processing technology. Noyes, Park Ridge, pp 196–206
3. Schwartz GC, Schaible PM (1979) Reactive ion etching of silicon. J Vac Sci Technol 16:410–413
4. Chapman B (1980) Glow discharge processes. Wiley, New York
5. Li YX, French PJ, Sarro PM et al (1995) Fabrication of a single crystalline silicon capacitive lateral accelerometer using micromachining based on a single step plasma etching. Proc. IEEE Electro Systems Conference. IEEE, Amsterdam, pp 398–403
6. Diem B, Rey P, Reynard S et al (1995) SOI-Simox: from bulk to surface micromachining. A new age for silicon sensors and actuators. Sens Actuat A 4647:8–16
7. Sung KT, Pang SW (1993) J Vac Sci Technol A 11:1206–1210
8. Juan WH, Weigold JW, Pang SW (1996) Dry etching and boron diffusion of heavily doped, high aspect ratio Si trenches. Proceedings of SPIE, Austin, pp 45–55
9. Charlet B, Peccoud L, Dupeux T (1987) SFV-CIPG. Antibes France, 183
10. Charlet B, Peccoud L (1991) European Patent EP-0359777
11. Laermer F, Schilp A (2005) German Patent DE-4241045. US-Patent #5,501,893
12. Laermer F, Urban A (2003) European Patent EP-1554747
13. Lee JW, Lee JY (2000) Structural modification of a trench by hydrogen annealing. J Korean Phys Soc 37(6):1034–1039
14. Juan WH, Pang SW (1996) Controlling sidewall smoothness for micromachined Si mirrors and lenses. Vac Soc J Vacuum Sci Technol B 14:4080–4084
15. Nozawa T, Kinoshita T, Nishizuka T, Narai A, Inoue T, Nakaue A (1994) Dry process symposium I-8, Nagoya., pp 37–41
16. Kinoshita T, Hane M, McVittie JP (1996) J Vac Sci Technol B 14:560–565
17. Hwang GS, Giapis KP (1997) J Vac Sci Technol B15:70–87
18. Hwang GS, Giapis KP (1987) Jpn J Appl Phys 37:2291–2301
19. Laermer F, Schilp A (2003) US Patent #6,926,844
20. Laermer F (2005) US Patent #7,361,287
21. Hopkins J, Ashraf H, Bhardwaj JK, Hopkins J, Johnston I, Shepherd JN (1999) The benefits of process parameter ramping during plasma etching of high aspect ratio silicon structures. Mater Res Soc Symp Proc 546:63
22. Hynes AM, Ashraf H, Bhardwaj JK, Hopkins J, Johnston I, Shepherd JN (1999) Sens Actuat A 74:13
23. Laermer F, Schilp A (2008) US Patent #6,284,148

Chapter 10
Through-Silicon via Technology for 3D IC

Eric Beyne

Abstract Three-dimensional (3D) integration complements semiconductor scaling in enabling higher integration density as well as heterogeneous technology integration. Through 3D chip stacking it is possible to extend the number of functions per 3D chip well beyond the near-term capabilities of traditional scaling. The 3D strata may be realised using advanced CMOS technology nodes but may also exploit a wide variety of device technologies to optimise system performance. A key technology is the possibility for establishing electrical connections through the bulk of the silicon substrates on which semiconductor devices are realised. These connections are generally referred to as through-silicon-vias, or TSVs.

A wide variety of technologies has been and continue be being proposed to realise such TSV connections. In this chapter, a classification of 3D technologies is proposed based on the system-level interconnect hierarchy. Different technology options for realising TSV structures are put forth and the main technological challenges discussed.

10.1 Introduction

10.1.1 3D Interconnect Hierarchy

Electronic systems are characterised by a hierarchical interconnect structure. At the transistor level, interconnects are short and have small cross-sections (local interconnect). To interconnect blocks of transistors, longer lines with larger cross-sections are used (intermediate interconnect) and at the level of the integrated circuit, large circuit blocks (IP-blocks, cores, and so on) are integrated with the longest and fattest wires (global interconnect). In order to establish communication between ICs, signals are

E. Beyne (✉)
Interuniversity Microelectronics Centre (IMEC), Kapeldreef 75, Leuven 3001, Belgium
e-mail: Beyne@imec.be

J.N. Burghartz (ed.), *Ultra-thin Chip Technology and Applications*,
DOI 10.1007/978-1-4419-7276-7_10, © Springer Science+Business Media, LLC 2011

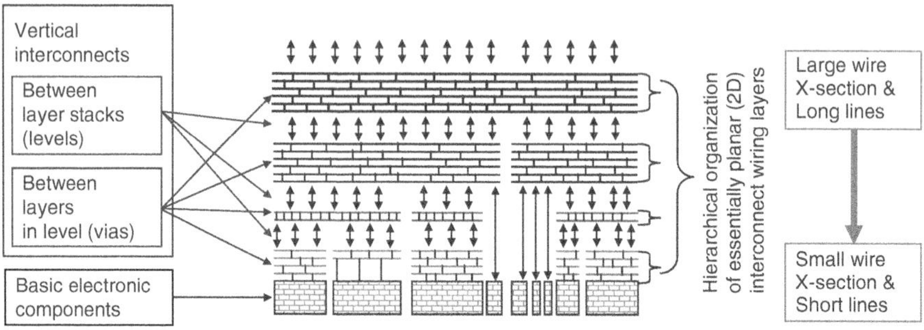

Fig. 10.1 Abstracted view of the 2D interconnect hierarchy of electronic systems

buffered and connected to a bond-pad level interconnect level, with the size (pitch) is larger to allow for compatibility with the package level interconnect capabilities. This type of hierarchy of interconnect levels is continued in the package and printed circuit board interconnect layers (see schematic in Fig. 10.1). The common feature is a scaling of interconnect geometries with interconnect line length. Each level of interconnect has essentially a two-dimensional topography: active devices are laid-out in a plane; long lines are required to connect distant devices. The number of interconnects in such a hierarchical interconnect exhibits an exponential behaviour. Crossing of lines is realised on adjacent interconnect planes; connections between planes are realised through features such as vias, plated through-holes, pins, solder balls and connectors. These 'via' interconnects allow for the 3D stacking of interconnect levels. The combination of basic circuit elements with multiple 2D-interconnect planes is considered a 2D-device, such as the integrated circuit or the printed circuit board.

Interconnects such as flexible printed circuit boards and moulded interconnect devices (MID) can be shaped or folded in three dimensions; however, their interconnect topology remains two-dimensional. In contrast, 3D system integration includes device capability to make direct connections in the third dimension, in which different layers of active devices are connected by a 2D interconnect fabric. These 3D interconnects can be viewed at different levels of the wiring hierarchy, as illustrated in Fig. 10.2. As 3D moves down in interconnect hierarchy – from long coarse to the short small interconnects – the density of 3D interconnects will need to increase exponentially. The choice of the 3D interconnect level(s) has a significant impact on the system design and the required 3D technologies, resulting in a strong interaction between system design and technology requirements.

10.1.2 Classification of 3D Interconnect Hierarchy

Based on the interconnect hierarchy concept, we can define the different layers and flavours of 3D technology as [1, 2]:

- Package-interconnect level: 3D packaging (3D-P):

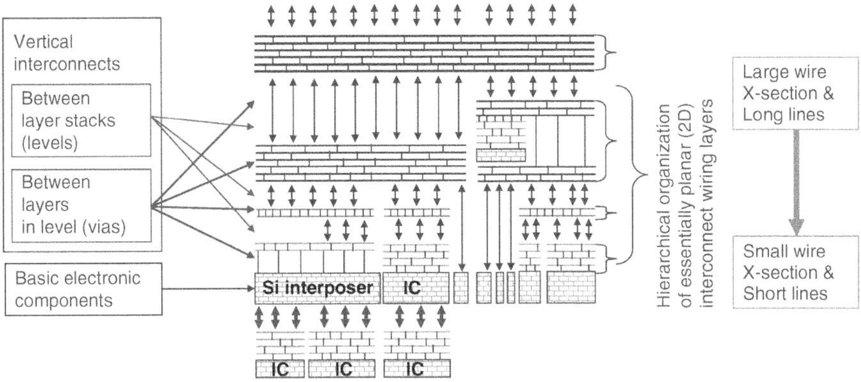

Fig. 10.2 Abstracted view of the 2D and 3D interconnect hierarchy of electronic systems

- Traditional packaging technologies such as wire-bonded die stacks and package-on-package stacks
- Includes die in PCB embedding
- Does not use Si TSVs
- Supply chain: OSAT, assembly, PCB

- Bond-pad interconnect level: 3D wafer level packaging (3D-WLP)
  - 3D interconnects are processed after the IC fabrication, ‘post IC-passivation’ (via last process)
  - Connections on bond-pad level
  - 3D connectivity density requirements follow bond-pad density roadmaps
  - Supply chain: WLP infrastructure, such as RDL and bumping
- Global on-chip interconnect level: 3D-stacked-integrated-circuit/system-on-chip (3D-SIC/3D-SOC)
  - Stacking of large circuit blocks (tiles, IP-blocks, memory banks), similar to a system-on-chip (SOC), approach but having circuits physically on different layers, 3D-system-on-chip (3D-SOC)
  - 3D-interconnects directly connect circuit tiles in different die levels. These interconnects are at the level of global on-chip interconnects. Allows for extensive use/reuse of IP-blocks.
  - Unbuffered I/O drivers (low C, little or no ESD protection on TSVs)
  - TSV density requirement significantly higher than 3D-WLP: pitch requirement down to 4–16 μm
  - Supply chain: wafer fab TSV technology
- Intermediate on-chip interconnect level: 3D-stacked-integrated-circuit (3D-SIC)
  - Stacking of smaller circuit blocks, parts of IP blocks stacked in vertical dimension
  - Mainly wafer-to-wafer stacking
  - TSV density requirements very high: pitch requirement down to 1–4 μm
  - Supply chain: wafer fab TSV technology
  - Local on-chip interconnect level: 3D integrated-circuit, 3D-IC

- Stacking at the transistor level
- Common BEOL interconnect stack on multiple layers of FEOL
- Requires 3D connections at the density levels of local interconnects.

10.1.3 3D Process Options Using Through-Si Via Technology

A wide variety of technologies can be used to realise 3D interconnect structures using Si TSV connections. A TSV is defined as a galvanic connection between both sides of a Si wafer that is electrically isolated from the substrate and from other TSV connections. The isolation layer surrounding the TSV conductor is called the TSV liner. The function of this layer is to electrically isolate the TSVs from the substrate and from each other. This layer also determines the TSV parasitic capacitance. In order to avoid diffusion of metal from the TSV into the Si substrate, a barrier layer is used between the liner and the TSV metal.

Numerous methods have been proposed for realizing these TSV-stacked 3D-SIC and 3D-WLP structures. Common to all these approaches are three basic technology modules:

1. The Through-Si Via process
2. Wafer thinning, thin wafer handling and backside processing
3. The actual 3D-stacking process

The 3D interconnects technology based on Through-Si Via interconnects basically consists of three main process modules: (1) The TSV module itself, (2) wafer thinning and backside processing, and (3) the die or wafer stacking process (permanent bonding and/or temporary bonding). Each of these steps requires rather specific equipment and process technologies and may be executed by different parts of the microelectronic supply chain. The discussion below on process modules is therefore organized along these three basic elements of any 3D interconnect technology.

The sequence of these process modules may vary, resulting in a large variation of proposed process flows, as shown in Fig. 10.3. The different process flows may however be characterized by four key differentiating characteristics:

- The order of the TSV process with respect to the device wafer fabrication process (see Fig. 10.4):
  - 'Via-first': fabrication of TSVs before the Si front-end-of-line (FEOL) device fabrication processing.
  - 'Via-middle': fabrication of TSVs after the Si front-end (FEOL) device fabrication processing but before the back-end (BEOL, back-end-of-line) interconnect process.
  - 'Via-last': Fabrication of TSVs after or in the middle of the Si back-end (BEOL) interconnect process.

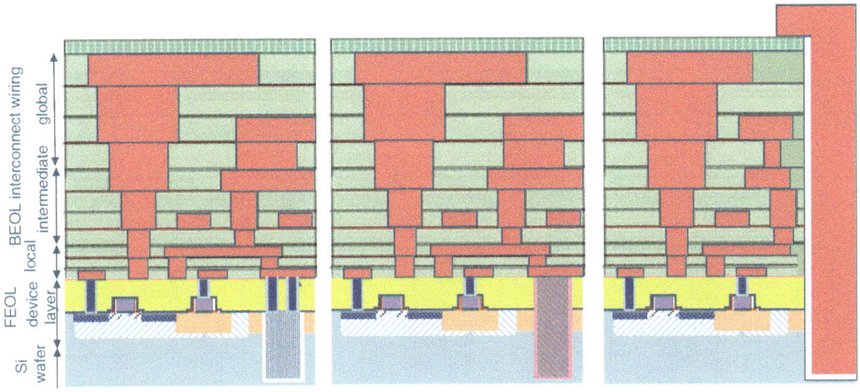

Fig. 10.3 Schematic representation of TSV first (*left*), middle (*middle*) and last (*right*) processes flows

- The order of TSV processing and 3D bonding: TSV processing before stacking the die or wafers or after 3D wafer-to-wafer bonding: before or after 3D bonding.
- The order of wafer thinning and 3D bonding: Wafer thinning before or after 3D bonding.
- The method of 3D bonding:
  - Wafer-to-wafer (W2W) bonding
  - Die-to-wafer (D2W) bonding
  - Die-to-die (D2D) bonding

In addition to these four main characteristics, three secondary characteristics are identified:

- Face-to-face (F2F) or back-to-face (B2F) bonding (the face or top surface of the wafer being the side with the active devices and back-end interconnect layers)
- In case of a 'via-last' approach, the TSV may be realised starting from the wafer frontside or from the wafer backside. Starting from the frontside of the wafer requires etching through the BEOL stack
- Removal of the carrier-wafer before or after bonding (i.e., temporary bonding and permanent bonding)

The generic flow characteristics defined above are applicable to 3D-WLP and global and intermediate interconnect level 3D-SIC process flows. For 3D-WLP TSV technology, the via-last route is the most important; it is realised before 3D bonding either as frontside or backside TSV.

The different approaches presented are not only applicable to regular semiconductor devices but could also be applied to passive redistribution or interposer substrate layers.

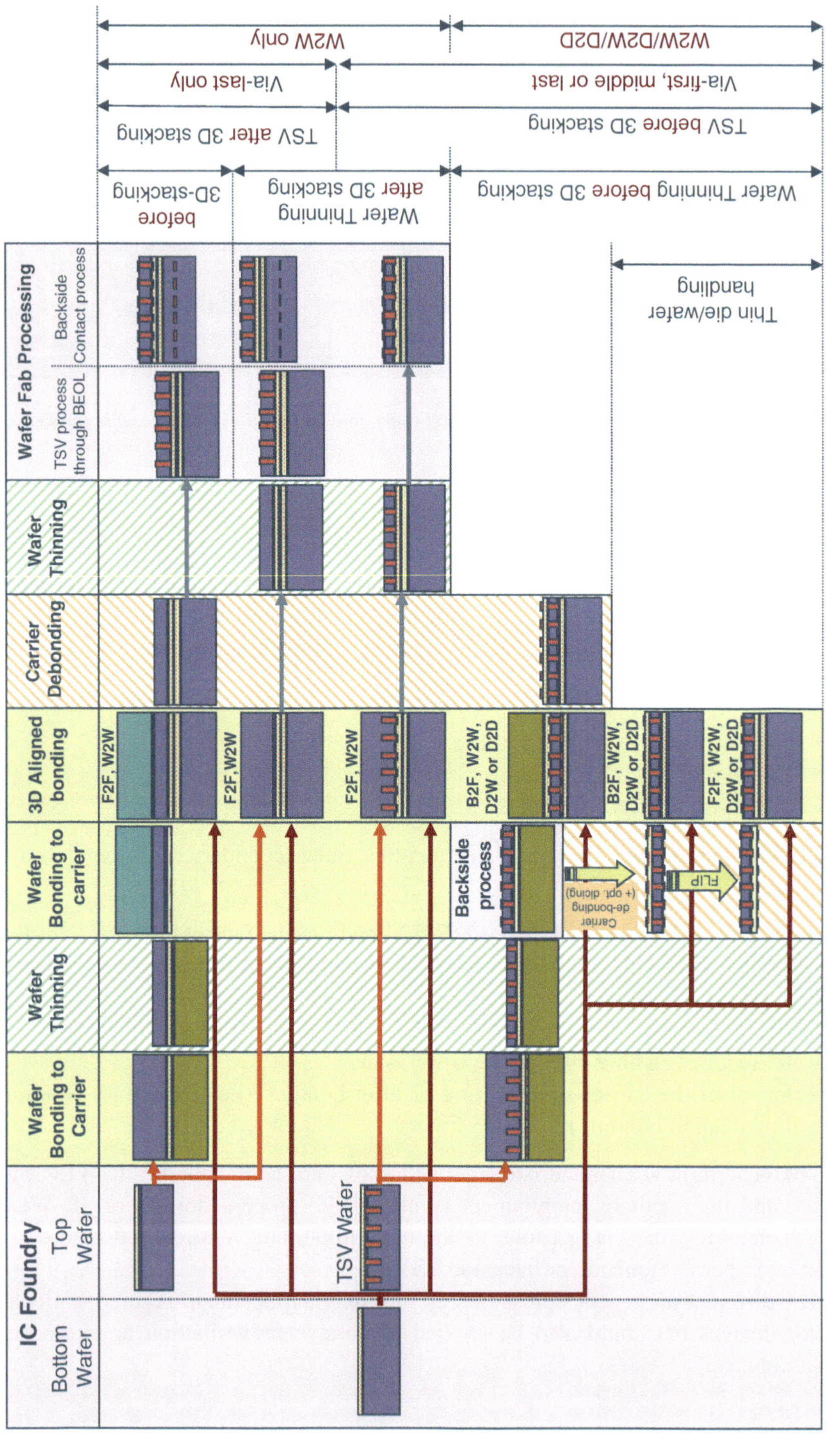

Fig. 10.4 Schematic representation of the various key process modules and 3D stacking options when using through-Si via 3D-SIC technologies

10.2 Through-Silicon via and 3D Stacking Technology

10.2.1 Introduction

Although a large variety of TSV approaches are proposed, they share many technology steps: A hole has to be etched in the Si substrate, an isolation layer has to be provided to isolate the TSV electrically from the Si substrate, a barrier layer has to be provided to prevent diffusion of metals into Si and the via must be filled with a conductive material.

The most common approaches to TSV technology provide for the TSV function before finalizing the wafer, which is prevalent for 3D-SIC technology, or realise the vias after finalizing the wafer, which is prevalent for 3D-WLP technology.

10.2.2 TSV Hole Formation

TSV holes are generally not etched through the entire wafer. Processing of wafers with actual through-Si holes is not compatible with standard semiconductor or wafer-level packaging processes and equipment. The prevalent technology is to use a 'blind' via approach: The TSV is etched to a certain depth or until an etch-stop layer is reached, as shown in Fig. 10.5.

Depending on the actual integration scheme used, etching a via hole in the Si substrate may require etching through resist, oxide or back-end-of-line (BEOL) layers like SiO, SiN, SiON, SiO(C) and, in certain cases, low-k materials. Before the TSV is etched in the Si substrate, such masking layers have to be etched. This may strongly complicate the etching processes, particularly if etching is required through metal containing layers or very thick masking layers. Common problems include undercutting of the Si right below the patterned passivating/masking layer when etching a via through a hard mask. Another during the etching down to a nonconductive stopping layer (e.g., via last from the backside) is so-called 'notching', an undercut of the Si etch at the bottom of the via (Fig. 10.5). This problem occurs when charged particles from the plasma accumulate on the nonconductive bottom of the via and cause a deflection of conductive particles from a vertical to a more horizontal

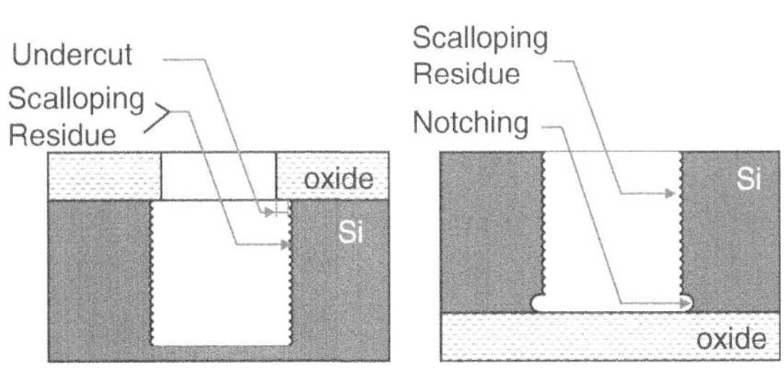

Fig. 10.5 Schematic representation of the etching challenges for Si TSV plasma etching

trajectory, causing preferential etching at the bottom of the via. This problem can be mitigated by employing a soft-landing approach.

The actual fabrication of the Si hole is commonly realised by plasma etching. A specific feature of TSV etching is the need for etching deep and often high aspect ratio holes in Si. This may require long processing times on expensive equipment, so fast etching processes are highly desirable.

Critical aspects of via hole etching include good control over sidewall tapering angle (both global and local), minimal sidewall roughness and scalloping, minimal residue/defect issues, minimal undercutting and notching issues, minimal local bowing effects right below masking layer, respectable etch rates and excellent repeatability and within-wafer center-to-edge depth and profile uniformity.

In order that isotropic etching of the Si may be avoided, the etching recipe balances sidewall passivation with bottom Si etch process chemistries. The prevalent technique used is the so-called 'Bosch' recipe, where passivation and etching steps are alternating in time. During the passivation step, a polymer is deposited on the Si surface. During the actual Si etching step, this polymer is easily removed from the bottom surface of the hole, while it remains on the via sidewall, protecting the previously etched Si sidewall. An undesired characteristic of this technique is the so-called 'scalloping' on the Si sidewall, as can be seen schematically in Fig. 10.5. The periodic circular ridges that are formed along the perimeter of the sidewall after each cycle of the etching process may make the following steps in the TSV processing flow more complex. An important advantage of this process is the high selectivity of the Si etch versus a masking photoresist layer, obviating the need for thick resist or hard mask layers.

Depending on the critical dimensions, aspect ratio and final depth of the TSV etch process, certain non-Bosch RIE process solutions may also be used. These solutions are more closely related to existing oxide or polysilicon CMOS plasma etch approaches (oxide or polysilicon) [3]. Modifications are required to address the main attributes that are unique to the nominal feature size of a TSV structure: high etch rates in the order of 5–15 μm/min, high anisotropy/ability to modulate the taper angle, and high selectivity to silicon etch. From a manufacturing perspective, some of the main advantages of the non-Bosch RIE process over the Bosch process include: smooth sidewalls with no scalloping, possible modulation of sidewall tapering angle, re-use of existing tools, minimal F-containing polymer residues and minimal undercutting. The main limitations are a low selectivity of resist masks and limitations in etch depth and aspect ratio.

After etching, cleaning of the Si via hole is a critical process. In particular the F-containing polymers deposited during the passivation cycle of a Bosch etch need to be fully removed before further processing.

Another inherent characteristic of deep Si etching processes is the aspect-ratio dependent etch rate. As vias are etching deeper in the Si wafer, or as vias have a smaller diameter, etch speed goes down. This typically causes a linear dependence between the average etch rate and the feature size aspect ratio. The consequence is that CD control for TSV patterning is critical toward obtaining a uniform processing speed across wafers.

10.2.3 TSV Isolation Liner Process

In order to electrically isolate the TSV connections from the Si substrate, an isolation layer is required. The key requirements for this layer are that it should exhibit low leakage current, sufficiently large breakdown voltage and low capacitance.

Deposition of the TSV liner layer must be compatible with the device process flow. For the deposition temperature this implies for via-middle a deposition temperature acceptable to the front-end process devices, and for via-last deposition a temperature acceptable for the back-end interconnect processes and for processing on carriers, compatible with the temporary bonding materials. In particular, for post-processing on DRAM memory devices, temperatures below 200°C may be required to prevent damage to the device wafers.

Ideally this layer should planarise the Si sidewall roughness (e.g., scallops from Bosch etching). Purely conformal deposition on sidewall scallops may cause a more difficult surface topology for the following processing steps.

The most popular liner materials are oxide or nitride layers, which are deposited by chemical vapour deposition (CVD) techniques, although physical vapour deposition (PVD) techniques are also being evaluated. Obtaining a conformal fill is more difficult at low processing temperatures. Nitride results in a higher capacitance but can also act as a barrier layer to prevent metal diffusion.

For 3D-WLP via-last TSVs, the use of polymer isolation layers is also possible. This allows for a significantly lower capacitance of this larger diameter structures and also allows absorption of some strain from the metal in the TSV structure. [4].

10.2.4 TSV Barrier Layer

In order that migration of TSV metal into the Si be avoided, a high quality pinhole free barrier layer is required. Prevalent barrier materials used are Ta and TiN. These barrier layers also fulfil the function of improving the adhesion between the TSV metal and the liner layer.

Prevalent technologies for barrier deposition are physical and chemical vapour deposition technologies. Different forms of CVD technology allow for barrier deposition on the most challenging, high aspect ratio TSV via holes. PVD technology has more limitations with respect to coating conformality and via aspect ratio, but it is often preferred because of superior adhesion and barrier properties of the deposited films and lower operational costs. Improvements in PVD equipment have extended the process window for PVD barrier deposition.

10.2.5 TSV Metal Fill Process

The main approaches for realising conductive TSV structures entail filling the via holes with copper (Cu) or tungsten (W) metal. For some via-first approaches, poly-Si via fills are also used.

10.2.5.1 Cu-TSV

The prevalent technology for Cu-TSV is derived from the commonly used single Damascene Cu plating in the BEOL process. The main difference is the high aspect ratio of the Cu-TSV features [5]. The main process steps include the deposition of a Cu plating seed layer, Cu-via fill by electrochemical deposition, ECD (plating) and CMP removal of Cu-overburden. An example of a Cu TSV process is illustrated in Figs. 10.6 and 10.7.

For the Cu-seed deposition process, the prevalent technology is physical vapour deposition. The main challenge here is to obtain a continuous Cu seed layer in high aspect ratio TSV structures. The highest TSV aspect ratios that can be successfully realised with Cu PVD is 10:1 for a 50-μm deep via hole. Alternative technologies for high aspect ratio TSVs are CVD Cu, CVD of metals such as W and Co as plating seed and direct-on-barrier plating and electrografting of Cu seed layers.

The main challenge for the ECD Cu filling process is to realise void-free Cu-filling of the Cu-TSV structures. This requires a 'superfilling' of the etched via structures, which is achieved by careful control of additives to the plating solution that accelerate the plating in the bottom of the via and suppress and smoothen the plating on the via sidewalls and wafer top surface. The resulting processes are slow and require equipment that can run multiple wafers in parallel on a single tool.

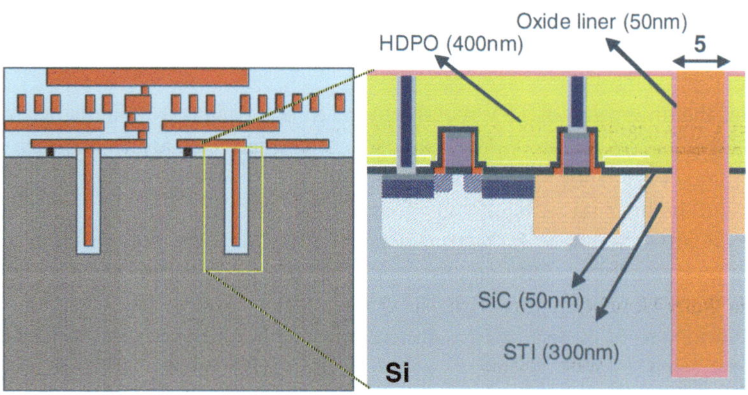

Fig. 10.6 Schematic representation of a 3D-SIC via-middle process (imec). The TSV is realised after the front-end-of-line of the CMOS process

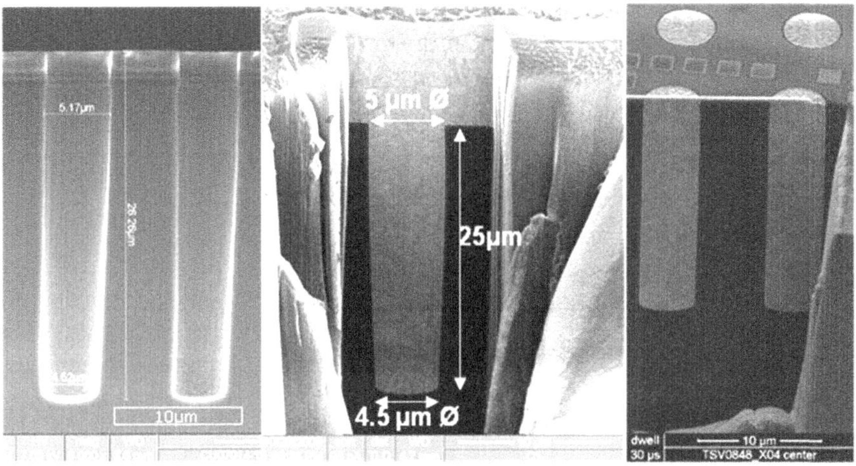

Fig. 10.7 Imec's 3D-SIC Cu-nail process for realising high density 3D-SIC TSV connections. *Left*: Aspect ratio 5:1 TSV Si etch after FEOL processing. *Middle*: Cu TSV after plating. *Right*: After Cu, barrier and liner CMP

After ECD Cu deposition, the Cu must be annealed in order that the so-called 'Cu-pumping' effect is avoided. If it is not properly annealed Cu will tend to expand during later processing steps, potentially causing fracture of the layers capping the TSV, as deposited Cu exhibits low stress. Upon annealing at higher temperatures, the Cu will try to expand, but it is constrained by the Si substrate. This causes high compressive stress in the Cu TSV, exceeding the compressive yield strength of the Cu. The Cu will show creep, causing plastic out-of plane extrusion and hilocking of Cu at the surface of the wafer. When the wafer is cooling down the Cu extrusion will not remain. When the Cu is properly annealed, out-of-plane Cu deformation can be avoided in future processing and annealing steps. This is illustrated in Fig. 10.8.

Except for some 3D-WLP process flow, where pattern plating is used during via filling, the Cu-TSV via formation is finalised using a CMP process to remove the Cu overplating. In addition to the Cu-CMP, the barrier and the liner layer also need to be removed from the wafer to allow further processing.

10.2.5.2 W-TSV

Tungsten is proposed as an alternative to Cu for TSV metal fill. W exhibits a much smaller mismatch in thermal expansion coefficient with Si than Cu. As a result less variation in stress with temperature and no metal extrusion is expected for W. The elastic modulus of W is, however, much higher than that of Cu and Si, and the deposition processes for W exhibit very large build-in stress, causing significant stress in the Si and limiting the thickness of W layers that can be deposited.

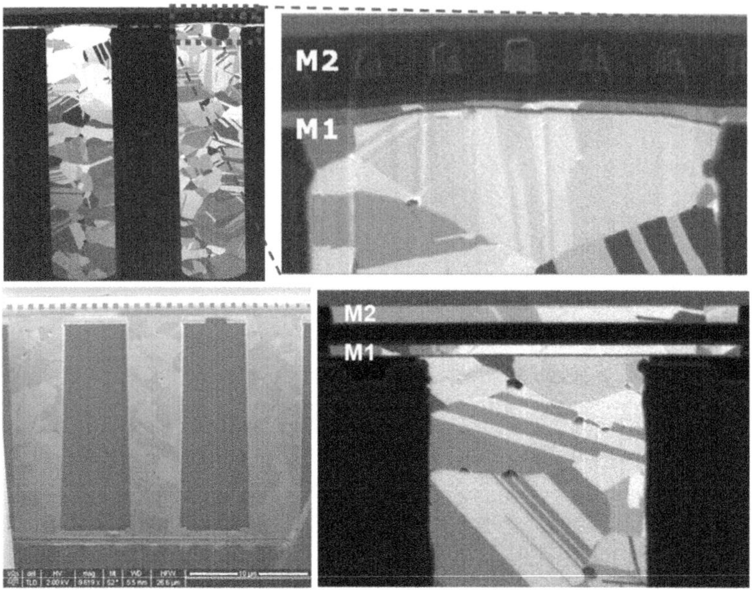

Fig. 10.8 Illustration of the 'Cu pumping' problem. *Top*: 3D-SIC TSV process with diameter 5 μm and depth 25 μm without pre-anneal before Cu CMP. *Bottom*: Same process with proper pre-anneal. No extrusion of Cu is observed when an appropriate temperature anneal is used before CMP

Chemical vapour deposition can be used to fill narrow TSV structures with large aspect ratios. W-filled TSVs with diameters up to 3 μm have been reported [6]. A thick W layer on the surface of the wafer causes large stress and wafer bow and may compromise film adhesion. Therefore, a partial etch-back of the tungsten on the surface of the wafer, without removing the metal, fills in the TSVs. Keeping the film thickness below about 500-nm prevents strong wafer bow and delamination (peeling) of the W layer. Stress inside the W-TSVs is, however, not reduced by this procedure. Larger TSV structures are realised by combining multiple TSVs in parallel, using narrow slits or using annular-ring type TSVs with a maximum width of 1–2 μm [3]. The W-CVD process is highly conformal. A typical W-TSV filled structure is characterised by a center-seam void.

After CVD-W fill, the typical process consists of a W-CMP step to remove the W on the wafer field. After this step, a barrier and liner layer CMP have to be performed to allow for further wafer processing.

As an alternative to the W-CMP step, a photoresist-structured W and barrier layer etch can be performed to pattern the contact pads to the W-TSV structures [7].

10.2.5.3 Poly-Si TSV

For via-first technologies, Cu- and W-TSV cannot be used because of compatibility problems with the FEOL process; but, poly-Si can be used as a TSV fill. In this case only a liner and no barrier layer is required. After poly-Si deposition, the wafers are polished and the standard process Si process flow can be performed. This requires a high quality of the pre-processing steps to prevent yield loss during device manufacturing. The higher resistivity of poly-silicon limits the use of this approach to applications that allow for high-impedance TSV interconnects.

10.3 Wafer Thinning and Backside Processing

Wafer thinning and backside processing are key technologies to complete the 3D TSV processes described in the previous sections and to allow for a 3D stackable solution. These process steps strongly deviate from classical IC processing technologies and novel equipment and materials. The actual process flows deviate depending on the chosen technology route, in particular, whether wafers are bonded before TSV processing or after TSV processing.

10.3.1 Wafer Thinning and Backside Processing for TSV Before 3D Stacking

This route allows for a parallel processing approach to 3D integration: Wafers are prepared for 3D stacking by performing TSV processing and contact pad formation in parallel. At the end of the process, the different die or wafers are combined to realise the 3D stack.

Realising TSVs before 3D bonding implies processing on thinned wafers. For a via-last process this can be the actual fabrication of the TSV connections. For via-first and via-middle processes this typically consists of processes to expose the TSVs on the wafer backside, provide a backside passivation and realise redistribution and bump structures on the wafer backside. These processes may be extensive and require relatively high temperatures.

For flows that use wafer thinning before bonding, a robust thin wafer carrier process is required. The requirements for 3D stacking are significantly more stringent than classical wafer thinning and singulation processes used for 3D-SIP applications and require dedicated solutions.

The key processing steps are as follows:

Wafer bonding to a Temporary thin wafer carrier

Thin wafer carrier systems should allow for extensive post-processing of the thinned wafers in standard semiconductor processing tools. The temporary glue layer between the thin device wafer and the carrier should be stable during all the (high temperature) TSV processing steps and able to detach without leaving residue or damaging the thin 3D die.

In this system, a major strategy uses glass substrates as carriers. Optical techniques may thus be used to cure (e.g., UV cure) or debond (e.g., laser ablation)

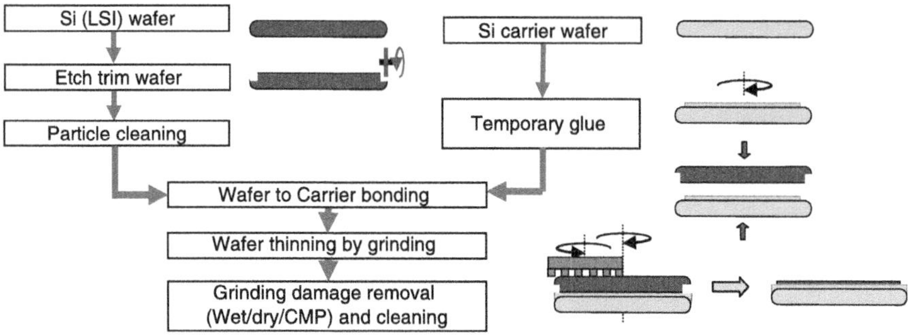

Fig. 10.9 Si temporary carrier strategy for thin wafer post-processing

the carrier wafer from the thin TSV wafers at the end of the process. It also allows for optical back-to-front alignment for backside processing. Disadvantages of glass carriers are the need for special Si CTE matched glass, the cost of the carrier wafers and the compatibility with standard semiconductor processing tools.

An alternative strategy is the use of silicon Si wafers as a temporary carrier substrate. A typical process flow is shown in Fig. 10.9. Wafer edge trimming is performed so that problems with the razor-sharp edges of silicon wafers after thinning may be avoided. As a result the thin wafer has a smaller diameter than the carrier wafer after thinning. This allows for a more robust handling of the wafer in standard semiconductor equipment.

The temporary glue layers for this process are very challenging to work with. They are critical to the success of 3D integration schemes. A complex combination of properties is required: there must be stability during processing but easy debonding must also be possible. A wide variety of debonding mechanisms is being studied, such as laser-assisted (glass carriers), melting and sliding (thermoplastic adhesives), dissolution in solvents and mechanical debonding (peeling).

Wafer thinning

Wafer thinning by grinding is a well-established processing step in semiconductor packaging. Critical for TSV technology is the control of Si thickness and Si surface quality. The total thickness variation of the thinned wafer is a combination of the thickness variation of the carrier wafer, the temporary glue layer thickness variation and the accuracy of the grinding tool.

After the Si wafer undergoes mechanical grinding, a thin, damaged Si layer is present on the wafer backside. CMP, dry etch and wet etch techniques are used to remove this damaged layer.

When Si wafers are grinded with already processed TSV structures, particular attention has to be paid to exposing the TSV structures from the wafer backside. This may require additional processing steps.

Wafer cleaning after thinning

Back grinding is a mechanical process that may leave particles on the wafer backside. In order to allow re-introduction of these wafers into a wafer-process line for backside processing, a thorough particle cleaning after wafer back grinding is essential.

Wafer backside process

The thin wafer-on-carrier must be compatible with standard semiconductor processing equipment to allow for processing such as:

- Via-last TSV processing (particularly typical for 3D-WLP)
- Backside wafer passivation
- Optional backside interconnect redistribution
- Backside interconnect 'bump'.

10.3.2 Wafer Thinning and Backside Processing for TSV after 3D Stacking

This is a sequential processing approach to 3D integration: Wafers are bonded together before 3D TSV processing. The process is repeated for multiple tier stacking. As a result, the bottom wafer will be going through all TSV processing steps:

- Wafer-to-wafer permanent bonding to bottom wafer or wafer stack
- Wafer thinning: total thickness variation and Si surface quality, impact on devices
- Wafer cleaning after thinning, allowing reintroduction into a wafer process line for further processing
- Wafer backside process requirements: TSV or pad metallization layer process.

10.4 Stacking Technology Module

Stacking technologies are discussed in more detail in Chap. 16. The main methods can be categorised as:

- Wafer-to-wafer bonding approaches
- Polymer or oxide W2W bonding
- Metal-to-metal W2W bonding
- Metal/oxide or metal/polymer W2W bonding
- Die-to-die or die-to-wafer bonding approaches
- Metal/metal thermo-compression bonding
- Cu/Sn and similar micro bump interconnect techniques.

References

1. International Technology Roadmap for Semiconductors, http://www.itrs.net
2. Beyne E (2004) IEEE ISSCC, 15–19 February 2004, San Francisco, pp 138–145

3. Teh WH, Caramto R, Arkalgud S, Saito T, Maruyama K, Maekawa K (2009) Proc. IEEE IITC, pp 53–55
4. Tezcan DS et al (2007) Proceedings of the 57th IEEE ECTC, Reno, 29 May–1 June 2007
5. Van Olmen J et al (2006) IEEE IEDM, 15–17 December 2008, San Francisco, pp 603–606
6. Klumpp A, Wieland R, Ecke R, Schulz SE (2008) Handbook of 3D integration. Wiley-VCH, Weinheim, pp 157–173
7. Ramm P, Bonfert D, Ecke R, Iberl F, Klumpp A, Riedel S, Schulz SE, Wieland R, Zacher M, Gessner T (2002) Proceedings of the advanced metallization conference AMC 2001, Montreal. Materials Research Society, Warrendale, pp 159–165

Chapter 11
3D-IC Technology Using Ultra-Thin Chips

Mitsumasa Koyanagi

Abstract Various generic methods for the three-dimensional (3D) integration of integrated circuits (ICs) are discussed. All these methods rely on ultra-thin chips. Wafer-to-wafer bonding, chip-to-wafer bonding, multichip-to-wafer bonding and reconfigured wafer-to-wafer bonding are described and compared. Several test chips fabricated by that use some of those concepts are briefly mentioned. Finally, specific concerns related to 3D-IC integration that use ultra-thin chips are indicated.

11.1 Introduction

The cross-sectional structure of a 3D-IC with stacking thin chips is illustrated in Fig. 11.1. Thin IC chips are stacked on a thick IC chip in the figure; the thick chip also acts as a supporting material. A thick chip is indispensable for a mechanical support of thin chips in a 3D-IC shown in Fig. 11.1. Three methods of wafer-to-wafer bonding, chip-to-wafer bonding and chip-to-chip bonding are used for the fabrication of such 3D-IC [1-16].

A wafer-to-wafer bonding method provides a high fabrication throughput, but production yield decreases as the number of stacking chips increases, since we cannot stack known good dies (KGDs). On the other hand, we can expect the high production yield in a chip-to-wafer bonding method and chip-to-chip bonding method since we can stack KGDs whereas a big problem in these methods is the low production throughput due to the sequential bonding of chips. To solve these problems new bonding methods of multichip-to-wafer bonding and reconfigured wafer-to-wafer bonding have been proposed [17-21]. These bonding methods are described below.

M. Koyanagi (✉)
New Industry Creation Hatchery Center (NICHe), Tohoku University, 28 Kawauchi, Aza-Aoba, Aramaki, Aoba-ku, Sendai 980-8579, Japan
e-mail: koyanagi@bmi.niche.tohoku.ac.jp

J.N. Burghartz (ed.), *Ultra-thin Chip Technology and Applications*,
DOI 10.1007/978-1-4419-7276-7_11, © Springer Science+Business Media, LLC 2011

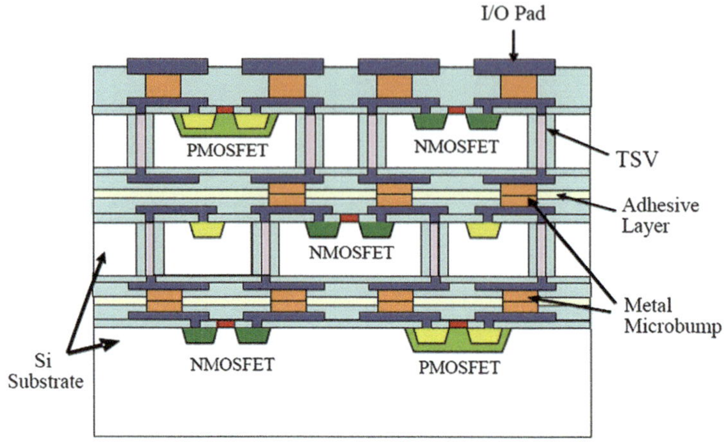

Fig. 11.1 Cross-sectional structure of 3D-IC

11.2 3D Integration by Wafer-to-Wafer Bonding

The thinned upper layers are stacked onto the thick IC wafer in a 3D integration technology based on a wafer-to-wafer bonding. In this 3D-IC, a relatively thick Si substrate remains after the fabrication process is complete. This remaining Si substrate is useful for reducing the damage caused to the devices during the 3D fabrication process. The electrical interconnection in the vertical direction is created by the through-silicon vias (TSVs) in such 3D-ICs. TSVs are fabricated before the transistor formation in a via-first process, whereas those are fabricated after the transistor formation in a via-middle process. Furthermore, TSVs are fabricated after completing the BEOL (back-end-of-line) of the CMOS process in a via-last process. The 3D-IC fabrication process flow for a front-via method is illustrated in Fig. 11.2. At first, a thick IC wafer with TSVs is glued to the supporting material as shown in Fig. 11.2a. The IC wafer glued to the supporting material is thinned from the back surface by mechanical grinding and chemical mechanical polishing (CMP) to expose the base of TSVs. This is followed by the formation of metal micro bumps as shown in Fig. 11.2b. The thinned IC wafer with the supporting material is then bonded to another thick IC wafer having TSVs, as shown in Fig. 11.2c. Then the bottom thick IC wafer is thinned from the back surface, and metal micro bumps are formed on the base of the TSVs. By repeating this sequence a 3D-IC can be easily fabricated. We can also use an IC wafer itself as a supporting material. In this case, the first IC wafer with TSVs and the supporting IC wafer are bonded face-to-face and the thick supporting IC wafer remains even after completing the 3D fabrication process. Therefore, a very wide range of thicknesses from several tens of nanometers to several tens of micrometers can be used for the thinned wafers bonded to the thick IC wafer.

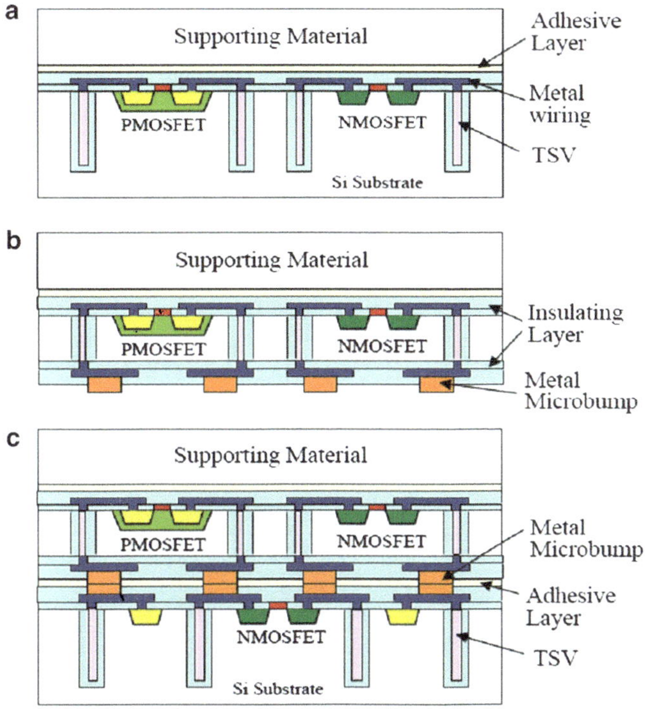

Fig. 11.2 Fabrication process flow for 3D-IC (W2W: front-via)

The 3-D IC fabrication process flow for a back-via method is illustrated in Fig. 11.3. In this 3D technology, the TSVs are formed after thinning the IC wafer. The thick IC wafer without the TSVs is glued to a supporting material as shown in Fig. 11.3a. A thick IC wafer can be employed as a supporting material. The IC wafer glued to the supporting material is thinned from the back surface by mechanical grinding and CMP. After that, deep trenches are formed in the thinned IC wafer from the back surface as shown in Fig. 11.3b. Then, an insulating film such as a silicon oxide film is formed on the trench surface and the trenches are filled with an electrically conducting material after the insulating film at the bottom of the trench has been selectively removed by reactive ion etching (RIE). After filling the trenches, metal micro bumps are formed on the tops of the TSVs. A 3D-IC can be fabricated by repetition of this process. This 3D integration technology is useful for the case when there are no spaces to form the TSVs on the front surface, as is the case for logic ICs having many metallization layers.

A wafer-to-wafer bonding method is the key for 3D integration technology. Three kinds of wafer bonding methods have been proposed: (a) adhesive bonding, (b) direct oxide bonding, and (c) direct metal bonding, as shown in Fig. 11.4. The basic idea is that mechanical pressure must be applied to wafers, raising the temperature in all three bonding methods. Thermal tolerance and shrinkage of adhesive are issues in adhesive bonding. Oxide surfaces have to be atomically flat

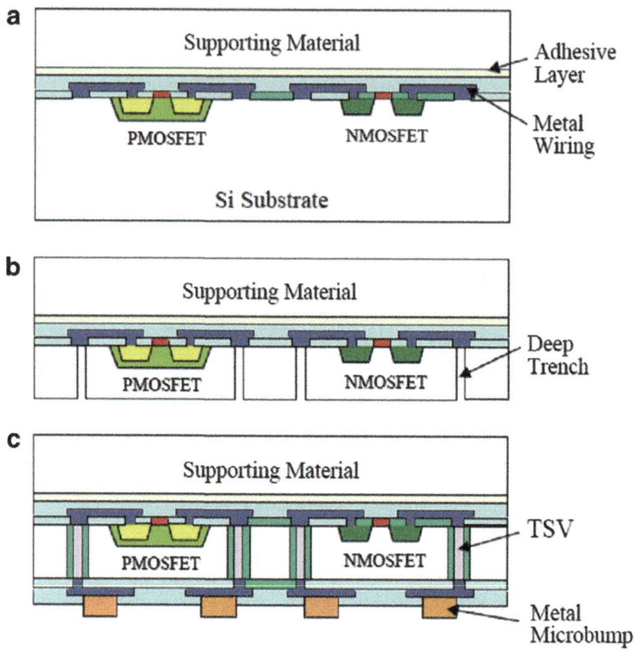

Fig. 11.3 Fabrication process flow for 3D-IC (W-to-W: back-via)

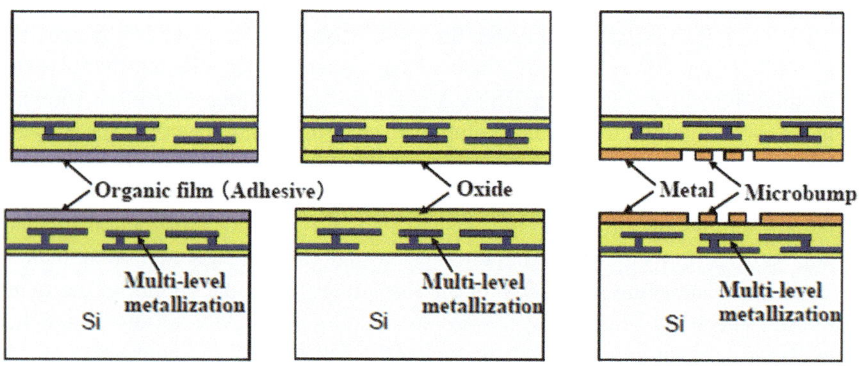

Fig. 11.4 Three kinds of wafer-to-wafer bonding

and special surface treatments are required in direct oxide bonding. In addition, a relatively higher temperature is required for bonding. Direct metal bonding is divided into the two categories of metal diffusion bonding and metal eutectic bonding. Metal surfaces also have to be atomically flat and special surface treatments are required in a metal diffusion bonding such as Cu–Cu direct bonding. The bonding temperature can be reduced in metal eutectic bonding, such as

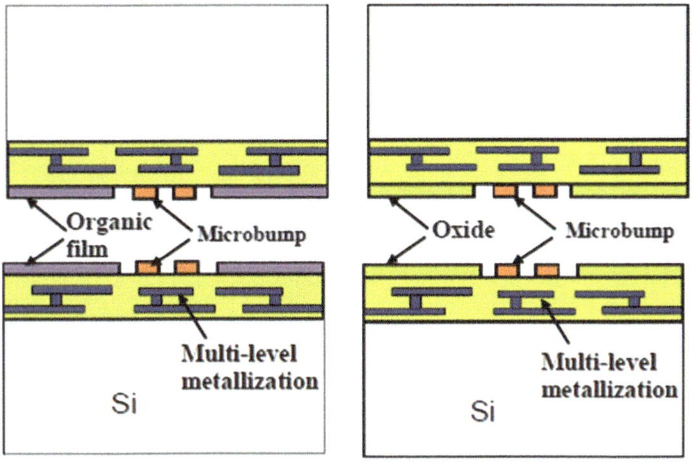

Fig. 11.5 Hybrid wafer-to-wafer bonding

Cu–Sn/Cu–Sn bonding. The thermal stability of the intermetallic compound formed after bonding is very important in this bonding method.

The difficulty of wafer-to-wafer bonding in 3D integration technology is that electrical connections have to be established between the upper wafer and lower wafer, even after the two wafers are bonded. Metal micro bumps such as Cu–Sn or Cu–Sn–Ag are used to establish electrical connections between two wafers. Hybrid bonding methods as shown in Fig. 11.5 have been proposed to simultaneously realize wafer-to-wafer bonding and electrical connections between two wafers. An adhesive organic material is used for wafer bonding and metal micro bumps are used for electrical connections between two wafers in the hybrid bonding of Fig. 11.5a, whereas direct oxide bonding is employed for the wafer bonding seen in Fig. 11.5b. We have developed our own hybrid bonding method where In-Au micro bumps are used for temporary bonding of two wafers and then a liquid adhesive is injected into a narrow gap between two wafers. By this method we have succeeded in bonding two wafers with a high density of metal micro bumps having the small size of $2 \times 2\ \mu m$, as shown in Fig. 11.6.

11.3 3D Integration by Chip-to-Wafer Bonding

A 3D integration technology based on chip-to-wafer bonding is preferable for stacking KGDs and different sizes of chips. Chips used for chip-to-wafer bonding are fabricated on a wafer as shown in Fig. 11.7. A thick IC wafer with TSVs is glued to a holding material as shown in Fig. 11.7a and then thinned from the back surface by mechanical grinding and CMP to expose the base of TSVs. After that, metal micro bumps are formed onto the bases of TSVs as shown in Fig. 11.7b. The thinned

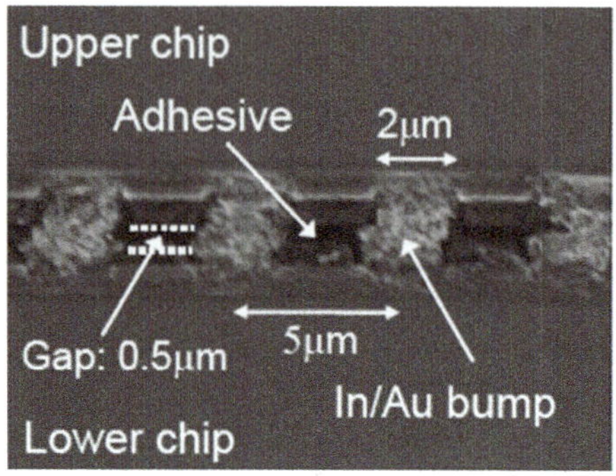

Fig. 11.6 Hybrid wafer-to-wafer bonding by adhesive injection method

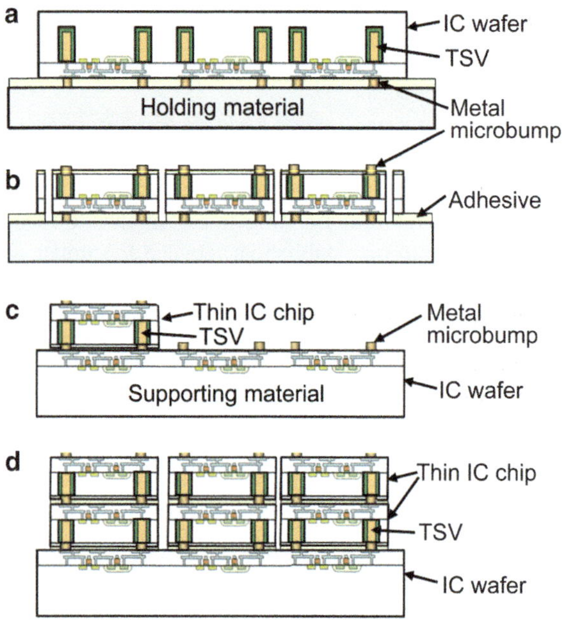

Fig. 11.7 Fabrication process flow for 3D-IC (C-to-W: thin chip)

IC wafer with TSVs and metal micro bumps is diced into thin chips. Then, these thin chips are picked up from the holding material and bonded to another thick IC wafer using a pick-and-place technique and bonding tools. Repeating this sequence a 3D-IC stacked with thin KGDs can be fabricated on a thick IC wafer, as shown in Fig. 11.7d. When KGDs are thinned to less than 20 μm, a handling substrate should be necessary for mechanical support of ultra-thin chips. In this case, an IC wafer with the handling substrate is glued to a holding material, as shown in Fig. 11.8a, and then thinned from the back surface by CMP down to less than 20 μm. A Si or

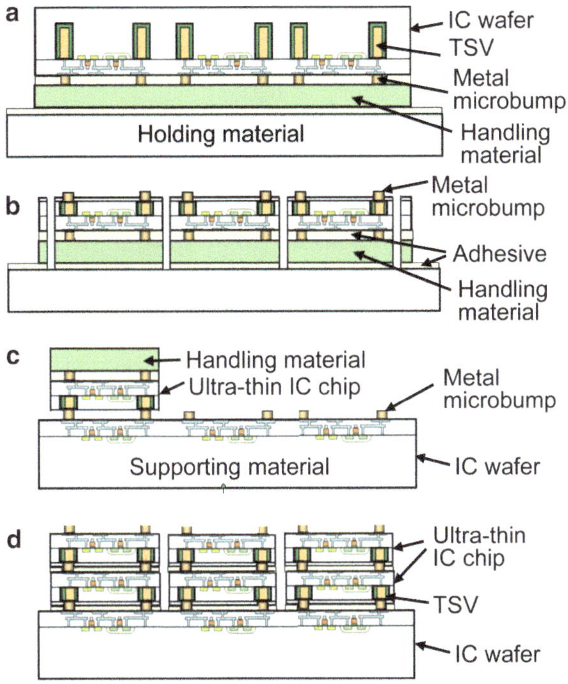

Fig. 11.8 Fabrication process flow for 3D-IC (C-to-W: ultra-thin chip)

glass substrate can be used as a handling substrate. The thinned IC wafer and handling substrate are diced to ultra-thin Si chips with handling substrates, as shown in Fig. 11.8b. These ultra-thin chips with handling substrates are picked up from the holding material and bonded to another thick IC wafer; then the handling substrates are removed from the ultra-thin chips.

11.4 3D Integration by Multichip-to-Wafer Bonding

The wafer-to-wafer 3D integration technology has a serious problem: the overall chip yield significantly decreases with an increase in the number of stacked layers. Therefore, this technology can be employed only when the chip yield of each IC wafer is very high. On the other hand, we can expect a higher overall chip yield in 3D integration technology using chip-to-wafer bonding since we can vertically stack known good dies. The inherent problem in the chip-to-wafer 3D integration technology, however, is the low production throughput. To solve these problems in the wafer-to-wafer 3D integration technology and the chip-to-wafer 3D integration

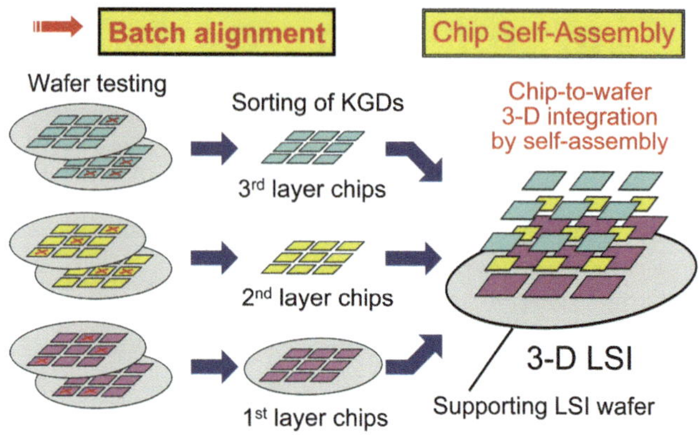

Fig. 11.9 3D-IC fabrication by multichip-to-wafer integration technology

technology, we have proposed a new 3D integration technology based on a multichip-to-wafer bonding; it is called a super-chip integration technology [17]. A number of KGDs are simultaneously aligned and bonded onto lower chips or wafers with high alignment accuracy using a self-assembly technique in a super-chip integration technology. We aim to form high density TSVs of more than one million vias/cm^2 in our super-chip integration.

Figure 11.9 describes the concept of super-chip integration. After wafer probing and dicing, many KGDs with TSVs and metal micro bumps are simultaneously aligned and bonded onto a thick IC wafer with or without stacked thin chips in a high alignment accuracy using a self-assembly technique. Then KGDs on the thick IC wafer are thinned from the backside by mechanical grinding and chemical mechanical polishing to expose the base of TSVs after coating witha high-viscosity resin. After that, the metal micro bumps are formed onto the base of TSVs. By repeating this sequence, we can obtain 3D-ICs with several layers of thin chips. We can fabricate a new 3D-IC called a super-chip as shown in Fig. 11.10 by this technology. The most striking feature of this super-chip is that various kinds of thin chips with different sizes – such as MEMS chips, sensor chips, CMOS RF-IC, MMIC, power IC, control IC, analog IC and logic IC – can be vertically stacked.

A chip self-assembly process is the key in our super-chip integration technology. The chip self-assembly process using a small volume of aqueous solution is schematically shown in Fig. 11.11, where Si wafers are used as the substrate. First, a thin silicon dioxide layer of thermally grown SiO_2 films is formed on a Si wafer. Then, hydrophilic areas are photolithographically patterned on the Si wafer, followed by the formation of hydrophobic areas surrounding the hydrophilic areas. After that, aqueous solutions are dropped onto the hydrophilic areas. Subsequently, many Si chips with the hydrophilic backside are roughly aligned on the hydrophilic bonding areas and then placed on them. Immediately after the chip placement the Si chips are simultaneously and precisely aligned on the hydrophilic areas in a short

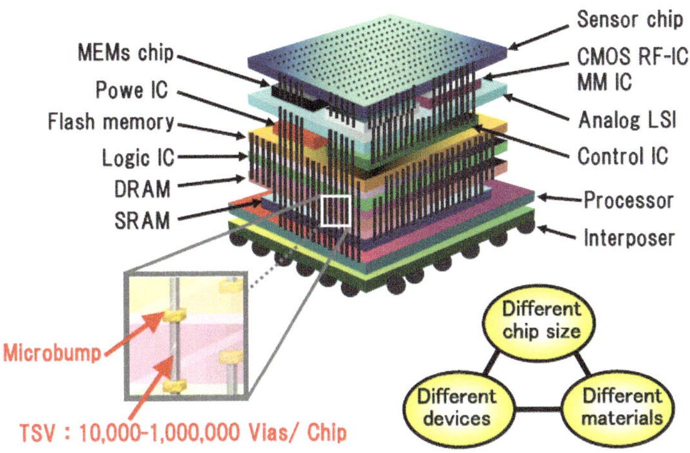

Fig. 11.10 Conceptual structure of 3D super-chip

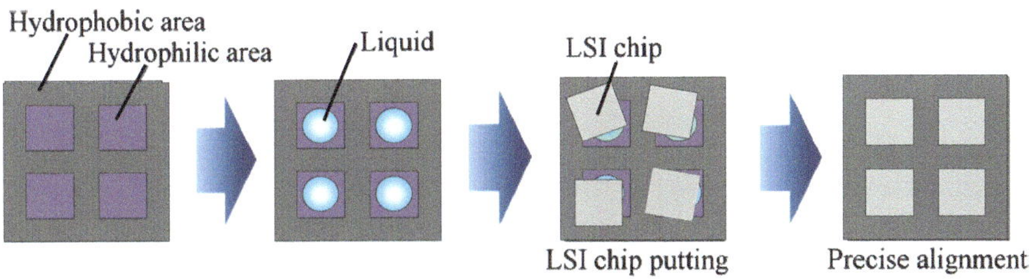

Fig. 11.11 Process sequence for self-assembly on Si substrate

time. Finally, these Si chips are tightly bonded on the hydrophilic areas after the liquids are evaporated at room temperature. We confirmed by self-assembly experiments that it takes less than 0.6 s to completely align the chips by surface tension of aqueous solution and obtain the average alignment accuracy of 0.43 μm. A self-assembly has been successfully performed even when Si chips with metal micro bumps are used.

To summarize, we have fabricated 3D test chips by the super-chip integration technology based on the multichip-to-wafer bonding using the self-assembly technique. Figure 11.12 shows photographs after stacking three layers of KGDs onto a thick IC wafer. The third layer KGD with the die size of 5 × 5 mm is stacked on the second layer KGD with the die size of 6 × 6 mm and on the first layer KGD with the die size 7 × 7 mm in the upper part of Fig. 11.12a. But, the third layer KGD with the die size of 7 × 7 mm is stacked on the second layer KGD with die size of 6 × 6 mm and the first layer KGD with die size 5 × 5 mm, as shown in the lower part of Fig. 11.12a. Figure 11.12b shows the cross-sectional view after stacking

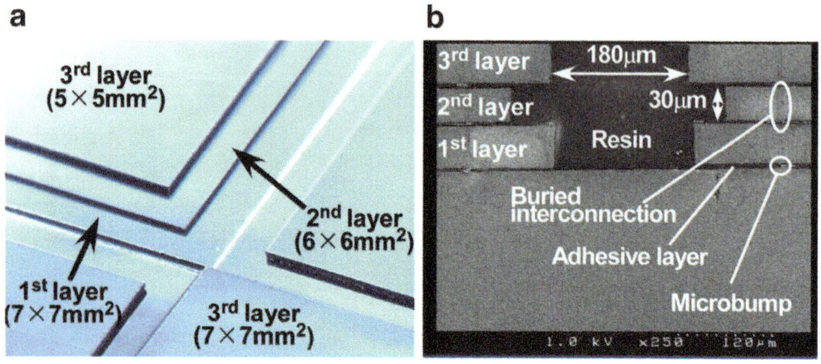

Fig. 11.12 Photomicrographs of plan view (**a**) and cross-sectional view (**b**) of three-layer stacked test chips with different chip sizes

three layers of KGDs with high alignment accuracy onto a thick IC wafer. As is clear in the figure, the die size of the second layer KGD is smaller than that of the third layer KGD and the first layer KGD.

11.5 3D Integration by Reconfigured Wafer-to-Wafer Bonding

Reconfigured wafers are used in place of conventional LSI wafers in the 3D integration technology based on a reconfigured wafer-to-wafer bonding. A reconfigured wafer is reconstructed using KGDs. Therefore, production yield of reconfigured wafers is very high (100%). Such reconfigured wafer with the yield of 100% is bonded to a thick IC wafer, which acts as supporting material. The reconfigured wafer is made by temporarily bonding a number of KGDs onto a thick holding wafer using a self-assembly technique. The 3D-IC fabrication process flow by a reconfigured wafer-to-wafer bonding method is illustrated in Fig. 11.13. At first, a number of KGDs with TSVs are simultaneously aligned and face-down-bonded to a thick holding wafer to fabricate a reconfigured wafer, as shown in Fig. 11.13a. Then KGDs on a reconfigured wafer are thinned from the backside by mechanical grinding and CMP to expose the base of TSVs after they have been coated with a high-viscosity resin, as shown in Fig. 11.13b. After that, the metal micro bumps are formed onto the base of TSVs. This reconfigured wafer with a number of thinned KGDs is firmly bonded to a thick IC wafer, which also acts as a supporting material, and then a holding wafer is removed as shown in Fig. 11.13c and d. Next, another reconfigured wafer with different kinds of KGDs is bonded to the thick IC wafer with more KGDs; finally, the wafer that holds another reconfigured wafer is removed. By repeating this sequence, we can obtain 3D-ICs with several chip layers. Thus, we can significantly improve both production yield and throughput of 3D-ICs by employing 3D integration technology based on a reconfigured wafer-to-wafer bonding. In addition, advantages of using a reconfigured wafer include the capability of adding extra processes, such as formation of multilevel metal redistribution wiring in.

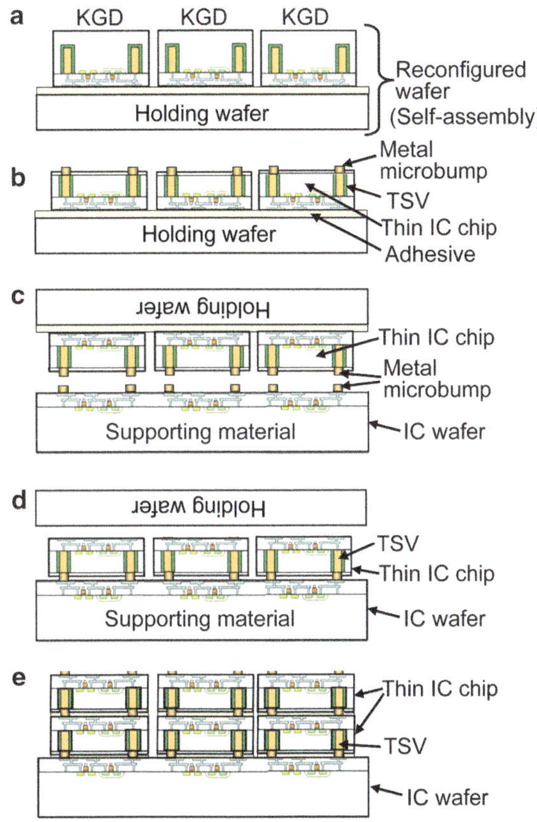

Fig. 11.13 Fabrication process flow for 3D-IC by reconfigured W-to-W bonding

11.6 3D-IC Test Chip Fabrication

3D-ICs are suitable for parallel processing and parallel data transferring because of the increased connectivity of their short vertical interconnections compared with conventional ICs. We can create various kinds of new ICs with parallel processing and parallel data transferring capabilities by employing 3D stacked structures having many TSVs.

So far we have fabricated several 3D-IC test chips, such as the 3D image sensor chip, 3D shared memory, 3D artificial retina chip and 3D microprocessor chip, using 3D integration technology based on wafer-to-wafer bonding [22–26]. Figure 11.14 shows a configuration of 3D microprocessor test chip and an SEM cross-sectional view of it. A processor and logic circuits and cache memory are formed in the first layer, second layer and third layer, respectively, in this 3D microprocessor chip. These three layers are connected by poly-Si TSVs with the size 2 by

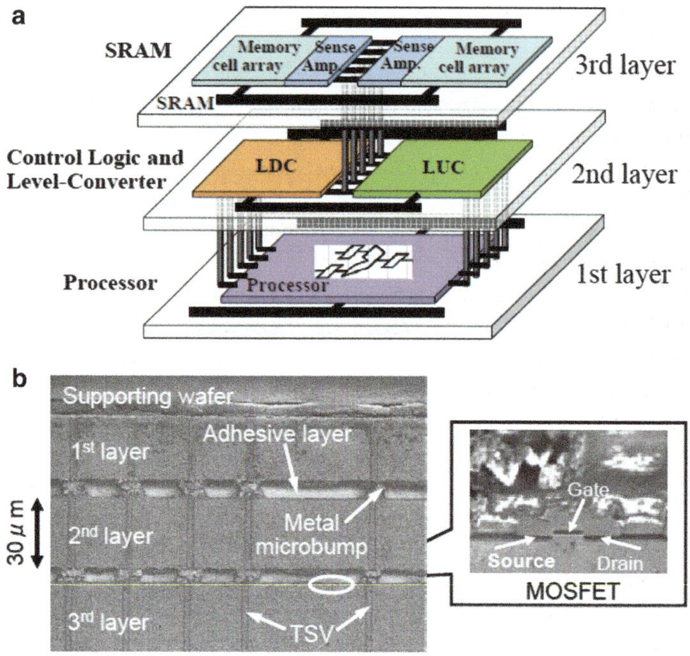

Fig. 11.14 A configuration of 3-D microprocessor test chip (**a**) and an SEM cross-sectional view of it (**b**)

12 μm and In-Au micro-bumps with a size of 5 × 5 μm as shown in Fig. 11.14b. Thec silicon layer thickness is approximately 30 μm. The epoxy adhesive was injected into the gap of approximately 4 μm between the upper and lower wafers. In this test chip, a processor in the first layer operates at the supply voltage of 2.5 V and SRAM cache memory in the third layer operates at 3.3 V. I/O pads of the test chip were formed in the third layer. It was confirmed in this test chip that data "1" (Vin1) and "0" (Vin2) are read from the SRAM cache memory in the third layer and then transferred to the processor in the first layer through the second layer; this allows the final device to successfully perform the arithmetic operation as shown in Fig. 11.15. A configuration of the 3D shared memory test chip and an SEM cross-sectional view of it are shown in Fig. 11.16. Several blocks of data in a memory layer are simultaneously transferred to other memory layers through a number of TSVs. CPUs are connected to the respective memory layers of this 3D shared memory. Therefore, many CPUs can share the identical data without any conflicts after the data transfer. The 3D shared memory test chip with ten memory layers was fabricated as shown in Fig. 11.16b. We confirmed in this test chip that block of data are simultaneously transferred in parallel among many memory layers.

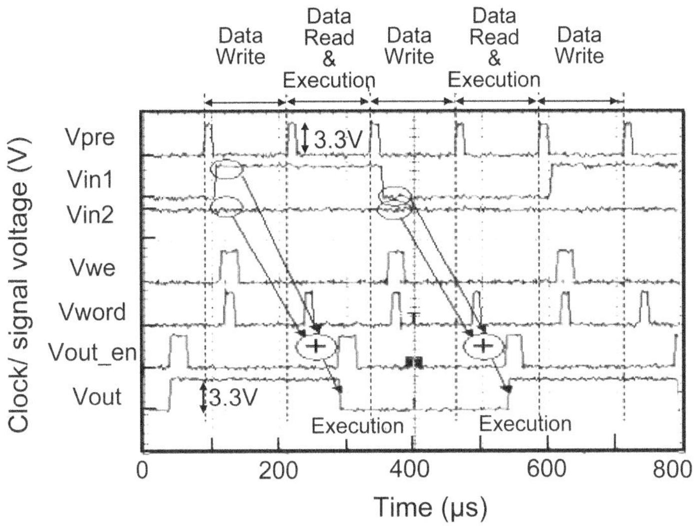

Fig. 11.15 Measured waveforms of 3D microprocessor test chip

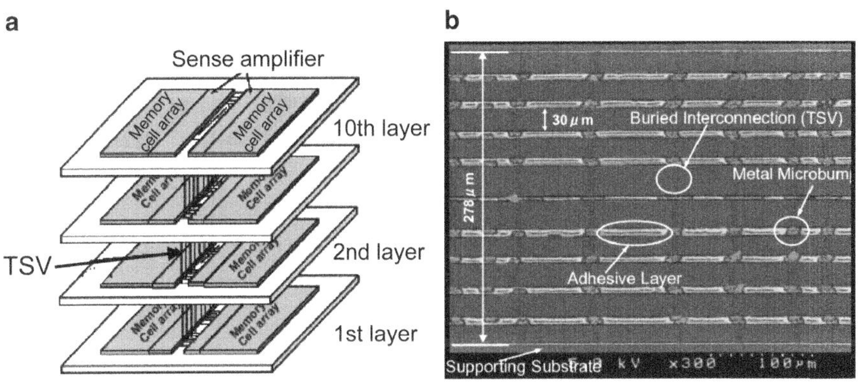

Fig. 11.16 A configuration of 3-D shared memory test chip (**a**) and an SEM cross-sectional view of it (**b**)

11.7 Concerns in 3D Integration Using Ultra-Thin Chips

3D-ICs using TSVs provide many advantages over conventional 2D ICs. However, there are several concerns involving 3D-ICs with TSVs that need to be solved before volume production may start. The most serious concern is heat accumulation in 3D stacked chips. TSVs and metal micro bumps act as effective heat conductors among many stacked chip layers. Therefore, heat generated at a hot layer is quickly transferred to cool layers, and consequently the average temperature of a 3D stacked chip increases. Influences of mechanical stress and strain introduced into

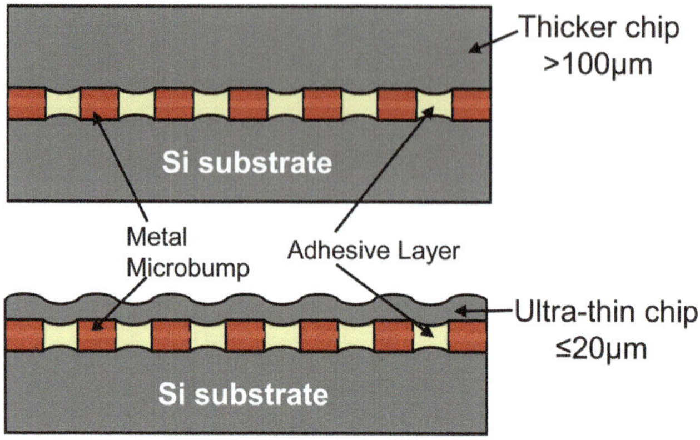

Fig. 11.17 Mechanical stress introduced into ultra-thin Si chip

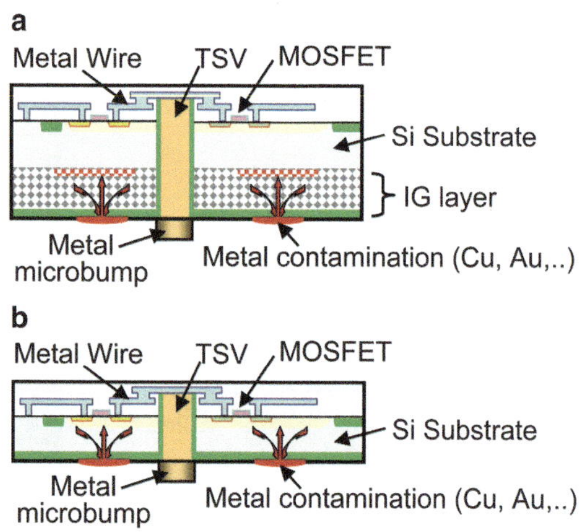

Fig. 11.18 Metal contamination in thinned Si chips with intrinsic gettering (IG) (**a**) and without IG (**b**)

thinned Si substrates on device characteristics are another concern in 3D-ICs with TSVs. Influences of mechanical stress and strain on device characteristics become more serious by thinning a chip, as shown in Fig. 11.17 [27]. We confirmed by a micro-Raman spectroscopy that TSVs and metal micro bumps introduce mechanical stress and strain in thinned Si substrates. In addition, the IC chip with thinned Si substrate is more easily affected by metal impurity contamination and crystal defects, as shown in Fig. 11.18. Usually intrinsic gettering (IG) layer and extrinsic gettering (EG) layer are formed in Si substrates of IC chips to minimize the influences of metal impurity contamination and crystal defects. These gettering layers might be removed by thinning the Si substrate. We confirmed by a transient capacitance measurement that a minority carrier lifetime in a thinned Si substrate without intrinsic gettering layer is seriously degraded by Cu contamination [27].

References

1. Koyanagi M (1989) Symposium on Future Electron Devices, Tokyo, Japan, pp 50–60
2. Matsumoto T, Koyanagi M et al. (1995) International conference on solid state devices and materials (SSDM), Osaka, Japan, pp 1073–1074
3. Koyanagi M et al. (1998) IEEE Micro 18(4):17–22
4. Igarashi Y, Koyanagi M et al. (2001) International conference on solid state devices and materials (SSDM), Tokyo, Japan, pp 34–35
5. Fan A et al. (2001) Electrochemical society: ULSI process integration symposium, Washington DC, USA, pp 124–128
6. Burns J et al. (2001) IEEE ISSCC, San Francisco, USA, pp 268–269
7. Lu J-Q et al. (2002) IEEE IITC, San Francisco, USA, pp 78–80
8. Klumpp A et al. (2003) IEEE ECTC, New Orleans, USA, pp 1080–1083
9. Knickerbocker JU et al. (2006) IEEE J Solid-State Circuits 41:1718–1725
10. Patti R (2006) Proc IEEE 94(6):1214–1222
11. Enquist P (2006) Proceedings of the international conference on 3D architecture for semiconductor integration and packaging San Francisco, USA, pp 5–6
12. Morrow P et al. (2006) IEEE Electron Device Lett 27(5):335–337
13. Swinnen B et al. (2006) IEEE IEDM, San Francisco, USA, pp 371–374
14. Temple D et al. (2006) IEEE IEDM, San Francisco, USA, pp 145–146
15. Koyanagi M et al. (2006) IEEE Trans Electron Devices 53(11):2799–2808
16. Koyanagi M et al. (2009) Proc IEEE 97(1):49–59
17. Fukushima T, Koyanagi M et al. (2005) IEEE IEDM, Washington DC, USA, pp 359–362
18. Fukushima T, Koyanagi M et al. (2007) IEEE IEDM, Washington DC, USA, pp 985–988
19. Fukushima T, Koyanagi M et al. (2008) IEEE IEDM, San Francisco, USA, pp 499–502
20. Fukushima T, Koyanagi M et al. (2009) IEEE IEDM, Baltimore, USA, pp 349–352
21. Lee K-W, Koyanagi M et al. (2009) IEEE IEDM, Baltimore, USA, pp 531–534
22. Kurino H, Koyanagi M et al. (1999) IEEE IEDM, Washington DC, USA, pp 879–882
23. Lee KW, Koyanagi M (2000) IEEE IEDM, San Francisco, USA, pp 165–168
24. Koyanagi M et al. (2001) IEEE ISSCC, San Francisco, USA, pp 270–271
25. Tanaka T, Koyanagi M (2007) IEEE IEDM, Washington DC, USA, pp 1015–1018
26. Ono T, Koyanagi M et al. (2002) IEEE COOL Chips, Tokyo, Japan, pp 186–193
27. Murugesan M, Koyanagi M et al. (2009) IEEE IEDM, Baltimore, USA, pp 361–364

Chapter 12
Substrate Handling Techniques for Thin Wafer Processing

Christof Landesberger, Sabine Scherbaum, and Karlheinz Bock

Abstract This chapter describes the necessity that carrier techniques be developed for thin wafer handling and processing. After an explanation on the main requirements for handle substrate techniques, we give an overview on the following temporary bonding techniques and mechanisms: thermoplastic adhesives, release layers, soluble glues, reversible tapes, mechanical debonding and electrostatic bonding. Finally, we conclude that the development of appropriate carrier techniques for ultra-thin wafers represents a key element of future thin wafer technology.

12.1 Need for Support: Thinner Devices in Semiconductor Industry

Most of today's semiconductor products undergo a back thinning process in the final step of the manufacturing process. A general reason for this is that thinner chips allow for thinner chip packages. However, thinner wafers are more fragile, which raises the question of minimum tolerable wafer thickness.

Over the last decade a simple rule of thumb regarding this question was found: The risk for wafer breakage is low as long as wafer thickness measured in micrometers is in the range of wafer diameter measured in millimetres. So, wafers of 200-mm diameter can be securely handled for a wafer thickness down to 200 μm. Thinner wafers generally require modifications of the wafer handling tools (grippers, robot end-effectors, cassettes and so on). However, a variety of chip devices exist that actually need to become distinctly thinner.

In order to understand the specific requirements for thin wafer handling it is useful to distinguish two different classes of thin semiconductor devices: The first group is given by chip products that do not need any further processing steps after

C. Landesberger (✉)
Fraunhofer Research Institution for Modular Solid State Technologies (EMFT),
Hansastrasse 27d, 80686 Munich, Germany
e-mail: christof.landesberger@emft.fraunhofer.de

J.N. Burghartz (ed.), *Ultra-thin Chip Technology and Applications*,
DOI 10.1007/978-1-4419-7276-7_12, © Springer Science+Business Media, LLC 2011

wafer thinning and die separation. Well-known examples are digital memory chips for mobile data storage and radiofrequency identification (RFID) devices for chip card applications. They are offered at chip thicknesses in the range of 50–150 μm.

The second class of semiconductor devices comprises such product wafers as those which initially need to be thinned and then require further processing steps at the rear side of the thin wafer. The most prominent examples in this product class are power devices (like the insulated gate bipolar transistor, IGBT), opto-electronic devices (LEDs, laser bars), discrete devices (single transistors, diodes) and solar cells. These products have a characteristic electronic property: An electrical current runs perpendicular through the chip from the front to the rear side. As semiconducting materials show comparatively high electrical resistivity, the current path through the chip produces heat and power losses. Wafer-thinning allows for reducing the length of the conductive path inside of the device and thereby improves efficiency and performance of the devices.

High frequency devices prepared on gallium arsenide (GaAs) substrates represent further important examples for semiconductor products, which require backside processing. In this case improved heat dissipation is the reason for wafer thinning.

Furthermore, most concepts for 3D-system integration require processes for material deposition and patterning at the rear side of ultra-thin device wafers. For instance, backside processes may be necessary to prepare metal contact pads for through-silicon via (TSV) or redistribution layers at the backside of thinned substrates or interposers.

It can be concluded that the functionality, performance or integration density of many microelectronic products could be improved by further reducing chip thickness and adding specific process steps at the rear side of device wafers. Therefore, handling schemes are required that enable secure processing of ultra-thin wafers in a semiconductor wafer fabrication performed at high throughput.

Within the first class of semiconductor products mentioned above, carrier substrates may be used to enable secure wafer handling during backside thinning. For the second class of devices, the technical aspects for carrier systems show far more complex requirements so that thin wafer breakage may be avoided in the various process environments. Typical wafer backside processing steps include ion implantation, thermal annealing, metal deposition, sintering of metal layers, resist coating and lithography, plasma etching and deposition of insulating thin film layers (SiO_2, Si_3N_4).

The technical solution which will be focused on in this chapter is the introduction of handle substrates (= carrier wafers), which have the outer dimensions of a standard wafer and which are temporarily bonded to the thin device wafer.

12.2 General Requirements

Carrier substrates enable both secure handling of thin device wafers independent of any restriction to large wafer diameters and the possibility for manufacturing equipment to be used that is already available in the fabrication lines.

Main requirements for thin wafer processing by means of temporary bonding are briefly summarised in the following list:

- Controllable reversibility of the bonding process
- Uniform thickness of the bond layer
- Capability to embed surface topography of device wafers
- Secure protection of the wafer edge
- Secure obviation of voids between the two wafers
- Mechanical stability against pressure during wafer grinding
- Chemical stability against solvents or etchants (liquid or gaseous)
- Thermal stability in the case where annealing steps or plasma etching are part of the backside treatment
- Necessity to leave clean and noncontaminated surfaces after debonding of the wafer stack
- Technique for thin wafer handling after debonding
- Low cost for adhesives, processes and equipment

Principal mechanisms and technical solutions for reversible bonding will be described in detail in the next chapters. The other requirements listed above will be explained in this section in order to allow the reader to understand their critical influence on thin wafer processing.

Uniform thickness of the bond layer is highly important during the wafer thinning processes; any tilt or any total thickness variation (TTV) of an adhesive film will be transferred into the final thickness profile of the device wafer that is to be thinned. This is a consequence of industrial grinding processes that are practically always used as the first step in wafer thinning and which typically remove more than 95% of wafer thickness. Material removal by grinding is not done parallel to the rear side of the device wafer but parallel to the surface of the vacuum chuck that holds the wafer stack during back grinding. The situation is different for typical stress-relief processes like etching or polishing. Here, the material is actually removed in a more or less parallel manner with respect to the rear side of the device wafer.

Surface topography of a semiconductor device wafer may be in the range of 2–5 μm for a standard complementary metal oxide semiconductor (CMOS) wafer. Critical locations in this context are the dicing lanes between the active chip areas. Much higher topography is present in the case of under-bump metallisation (2–20 μm) or even by solder balls (up to 200 μm). Of course, an adhesive bond layer must be thicker than the surface topography of the substrates and still must be as plane-parallel as possible. Voids at the bond interface often result in wafer breakage during grinding or in the generation of bubbles when the wafer stack is introduced into a vacuum chamber during subsequent processing. In order to prevent such problems, substrate bonding is generally done in a vacuum chamber.

Specific attention must be directed to the wafer edge. Due to the rounded shape of a wafer edge, a stacked wafer pair shows a gap at the edge. After back thinning one wafer the wafer edge becomes a sharp and narrow blade. If the overhanging part of the wafer edge is not supported then some parts of the bevel will brake away and cause unwanted particle contamination inside various processing equipment. In some cases the use of carrier substrates that are slightly larger in diameter than the device wafer may facilitate secure wafer handling. However, state of the art wafer production tools often require narrow specs for the wafer diameter allowed. Furthermore, bonding of two substrates of different diameter requires wafer aligning or careful wafer centring before they make contact. Other issues may arise with the orientation of the wafers with respect to each other. A wafer stack may show two notches or twisted flats. Optical tools for wafer alignment generally do not accept such ambiguity. All these examples should show that various equipment parameters need to be considered as well before set up of a manufacture flow for ultra-thin wafers.

The present discussion must extend to the consideration of the stage at which all the processes of thin device wafer fabrication are accomplished and the carrier substrate is finally removed. If adhesive bonding materials were used, cleaning steps could be necessary after debonding, and here the question of an appropriate handling scheme comes up again!

Last but not least we need to have a look into the type of material of the handle substrate. In principle, carrier substrates can be made of a variety of materials, e.g., silicon, glass, ceramic or polymer. However, there are some important requirements that need to be taken into account when a specific carrier material is chosen. First of all the geometric dimensions of the carrier substrate and the bonding layer must not have a deleterious effect on thinning. The most critical parameter in this context is the value of total thickness variation. Any thickness variation in the carrier system (= substrate plus bonding layer) will be transferred onto the device wafer which is to be thinned. Generally, polymeric plates or sintered ceramic substrates show TTV values far above 10 μm and therefore cannot be recommended for carrier substrates. Glass carriers offer the interesting option of making it possible for one to inspect the bond interface as well as allow for optical alignment of wafer patterns in the case that lithographic processes should be carried out at the backside of an ultra-thin wafer. However, glass plates of large diameters and with TTV values below 5 μm are quite expensive. Further questions concerning glass carriers are related to their compatibility with standard CMOS technology and their thermal properties. Boron silicate glass wafers would offer a solution as they are cost effective and show a coefficient of thermal expansion (CTE) that is close to the one of silicon. However, the presence of alkali ions in wafer fabs often is prohibited. Alternatively, alkali-free glasses may be used, taking into account that their thermal properties do not fit so perfectly to silicon wafers. However, larger differences in CTE will result in strongly bowed wafer stacks at elevated process temperatures.

So, there is no general recommendation for the type of carrier substrate to be used. It will depend on the specific production process. At least it can be stated that silicon wafer substrates offer optimum properties in terms of geometric uniformity, availability for all wafer diameters, compatibility with CMOS fab environment and cost.

12.3 Overview on Temporary Bonding Techniques

Temporary bonding of a device wafer and a carrier substrate has been used since many years in manufacture technology for compound semiconductors. High frequency devices based on GaAs wafers and opto-semiconductors prepared on indium phosphide (InP) wafers require handle substrates because of their mechanical fragility and the necessity for reduced material thickness. Still widely used bonding materials are natural or refined waxes that can be remelted at low temperatures in order to allow debonding after processing of the stacked wafer pair.

Meanwhile, various techniques for reversible bonding have been proposed and are being used for different applications. They can be categorised by the specific principle of their release mechanism. An overview of these techniques is given in Table 12.1.

The following section explains the most common temporary bonding techniques in more detail. Afterwards, the different bonding concepts are compared with respect to their technological applicability.

12.4 Thermoplastic Materials

The most commonly used bonding techniques are based on materials that melt at a specific temperature. In the case of wax, the melting temperature is in the range of 50–150°C. Industrial polymers show rework temperatures up to approximately 250°C; examples of thermoplastic materials can be found in [1–4]. The application of bond material can be done by dispensing the hot fluid and then pressing both substrates against each other or by spin-coating the dissolved polymer and subsequent drying of the film followed by bonding of the substrates.

Table 12.1 Basic mechanisms for reversible bonding techniques

Basic mechanism of release process	Release process type	Bonding principle
Thermal treatment	Melting of bond layer	Wax, thermoplastic adhesives, solder
	Reduction of adhesive force	Thermal release tapes
	Thermal decomposition of the bond layer	Adhesives that sublime or burn, e.g., transfer into CO_2 gas
Chemical treatment	Dissolve bonding material in solvents	Polymeric adhesives
	Wet- or dry-chemical etching of a sacrificial bond layer	Inorganic bond materials, e.g., SiO_2
Radiation-induced release	UV light or laser irradiation decomposes the bond layer	Adhesive tapes or glues
Mechanical release	Applied mechanical force separates wafer pair	Release layer at bond interface Low adhesion bonding
Controllable physical forces	Switch off electrical field	Electrostatic forces
	Switch off magnetic field	Electromagnetic forces
	Release in vacuum chamber	Difference pressure

The most important advantage of thermoplastic materials is their capability to level the surface topography of a CMOS device wafer as the bond material is in a liquid state during the thermally assisted bonding. Furthermore, there is a second very beneficial property resulting from the liquid state of the bond layer: the capability to fill the gap at the edge of a stacked wafer pair. Polymeric bonding materials that allow filling of the edge gap result in high mechanical robustness of the wafer edge as well as stability against penetrating liquids in wet-chemical bathes.

Beside these positive aspects of thermoplastic adhesives there are some disadvantages as well. The main drawbacks are related to the process steps for removing the carrier substrate and stripping of polymeric residues. If the device wafer shows a final thickness of just some tens of micrometres, then separating both substrates by sliding off or tilting is a critical issue. In principle, the thinned wafer can be transferred to another handle system in order to support the full wafer area. However, residuals of the molten material will adhere to the wafer surface and need to be removed completely. Such a cleaning step must be carried out very carefully, or the very thin device wafer will break.

12.5 Debonding After Laser Treatment

This briefly summarises a concept developed and industrialised by the 3 M Company [5]. It is based on a combination of thermal debonding after laser irradiation and mechanical removal of the bond layer due to its low adhesion to the device wafer frontside. First, the handle wafer is coated by a thin "LTHC release layer" (LTHC stands for "light to heat conversion"). At the surface of the device wafer a UV curable adhesion layer is applied by spin-coating. This ensures levelling of the surface topography of the product wafer and secure filling of the wafer bevel. After vacuum bonding the wafer stack is ready for further processing at temperatures up to some 250°C. Debonding is initiated by exposing the LTHC to a UV laser beam, which is scanned over the full wafer area. Of course, such a debond concept requires transparent glass carriers. The laser treatment reduces the bond adhesion and the glass substrate can be removed by tilting of the carrier and device wafer. Then, a tape is laminated onto the thick acrylic adhesive layer and subsequently peeled off again. Thereby, the adhesive layer is removed from the surface of the device wafer. During removal of carrier and bond layer the thinned device wafer may be attached to a film-frame-holder in order to supply a handle for further manufacture processes.

12.6 Soluble Bonding Materials

Many adhesives can be dissolved chemically in organic solvents. A first approach for dissolving the adhesive layer might be to place the wafer stack into a solvent bath and wait until the adhesive layer is removed. However, dissolving the few-micrometre thin bond layer between two wafers of 200-mm or even 300-mm diameter will take many hours or days until the bond layer is completely removed. So, this simple debond approach will be limited to small-sized substrates.

A more advanced and also widely industrialised concept is the introduction of perforated handle substrates. Available on the market are, for instance, sapphire substrates with hundreds of tiny holes, which are prepared by ultrasonic drilling. Perforated handle substrates allow for the penetration of solvents through the holes and thereby enable a much faster removal of the bond layer. Temporary bonding by means of perforated substrates represents a proven solution for manufacture processes of thin wafers which are still robust enough to allow thin wafer handling without rigid support substrate.

Finally, the concept dicing-by-thinning (see Chap. 4) offers a further option for chemically soluble adhesives. In this case the solvent can penetrate through all dicing trenches of the device wafer and then strip the thin bond layer beneath each chip. Depending on chip size such a process can be accomplished in less than 1 h. However, the chips will swim away in the solvent bath unless an appropriate second carrier system is introduced.

12.7 Reversible Adhesive Tapes

Polymeric tapes are well known in the semiconductor industry. They have been used for many years for mounting wafers on frame holders for subsequent dicing and also to protect the wafer frontside during backside grinding. Automated equipment for tape mounting, tape lamination, cutting and delamination is available from various suppliers. Tapes typically consist of a base film (e.g., PET), which is coated on one or both sides with specific adhesive layers. Thickness of base film is some 50 μm, thickness of coatings may be in the range of 15 μm up to 200 μm. Thick and soft adhesive layers enable levelling of highly topographic surfaces, for instance, bumped wafers. In order to protect the functional coatings, so-called liners are laminated onto the surface of the tapes. Application of adhesive tapes shows several advantages: the material is supplied from rolls, lamination is done in a dry process at room temperature and trapping of air between tape and wafer easily can be prevented by roller coating.

Reversible adhesive tapes are offered in two main variations: release by UV irradiation or by thermal annealing. UV tapes reduce their initial adhesion force by some 95%. This is sufficient to peel off an adherent tape. However, the remaining bonding force generally is still too high to debond two rigid substrates by a perpendicular acting force. UV tapes are widely used for wafer dicing; their application for temporary wafer bonding is yet not established.

Thermal release tapes completely lose bonding force within a minute when a certain temperature has been reached. Various suppliers offer such tapes for different debonding temperatures, typically in the range between 60°C and 180°C.

Reversible adhesive tapes offer a further advantage related to their debonding behaviour: Delaminating the tape causes the adhesive layer to be removed as well. Application of wet-chemical strippers generally is not required.

Besides the simplicity, the technological limitations of adhesive tapes need to be considered as well. First, tapes cannot fill the gap at the edge of a stacked wafer pair. This results in both a loss of mechanical support at the outer rim of a thin wafer and the lack of a secure sealing at the wafer edge against the penetration of wet-chemicals or water. Second, tapes consist of quite thick and more or less soft polymers. Consequently, a locally nonuniform mechanical pressure during back grinding may result in a deformation and finally cracking of a very thin wafer. Such issues may occur at a wafer thickness below some 50 μm. They can be circumvented by choosing a thinning sequence that combines grinding with nonmechanical thinning techniques, for instance, etching or polishing.

Another topic relating to wafer surface cleanliness after tape delamination concerns the type of surface material and its topography. Critical issues with polymeric tape residual concern the aluminium pads of the chip devices. One possible way to overcome these problems is the combination of two tapes within the bond layer. This means you first laminate a tape for front side protection and then a second tape, which delivers the required debond functionality (e.g., thermal release) within the wafer stack. According to the authors' experience, such an approach generally enables the successful preparation of 20–30-μm thin device wafers.

The application of adhesive tapes as a temporary bonding technique for thin wafer processing after back thinning also needs to be regarded. Introduction of tape-bonded wafer stacks in vacuum or plasma chambers may be complicated by outgassing of the polymeric composites. The generation of gaseous molecules at the bond interface generally results in a bulged wafer surface if its thickness is below 50 μm. Such problems may become really critical if the wafer temperature is raised inside a vacuum chamber because then the pressure tends to enlarge any bubbles.

12.8 Mechanical Debonding

Reversibility of a bonding process can also be achieved by use of a bond layer of low adhesion force. Actually, bond forces don't need to be very strong because the mass of a 200-mm silicon wafer of 50-μm thickness is just a few grams. Low bond strength in the adhesion layer enables debonding by permitting a moderate force to be applied to separate the substrates mechanically.

An interesting concept now discussed is temporary bonding technique, which allows for debonding of the carrier substrate at low tensile forces. A technical solution was developed by the company Thin Materials [6]. A proprietary release layer of a thickness of some 100 nm is coated onto the surface of a device wafer by spin-coating and a subsequent PECVD process. The handle wafer (silicon or glass) is spin-coated with a solvent-free silicone layer, which thickness may be chosen in a

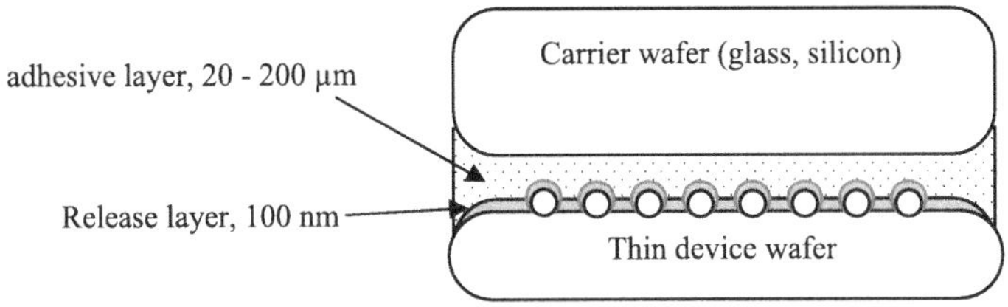

Fig. 12.1 Sketch of a temporary bonding technique that uses a release layer for debonding [6]. The spheres indicate a solder ball topography that is on top of the device wafer

wide range up to some 200 μm (Fig. 12.1). During vacuum bonding and curing of the bond layer the surface topography is well-levelled. The wafer stack can be separated by tilting the handle and device wafer whereas both substrates are fixed to specific vacuum holders. Thereby, the tensile force needs to overcome the bond adhesion only along a narrow line. This debonding line then propagates over the wafer surface from edge to edge. As the separation step also removes the release layer the device wafer is again free of polymeric material. The bonding polymers used can withstand temperatures up to 300°C. This opens a broad spectrum of possible process steps, which can be accessed by using such a handle technique.

12.9 Electrostatic Bonding

Electrostatic forces offer the unique opportunity for a manufacturer to realise a reversible bonding technique without adhesive polymers.

Oppositely charged materials attract each other by electrostatic forces. This physical principle has been used in the semiconductor industry for many years. So-called electrostatic wafer chucks are widely introduced in plasma etching equipment to reversibly affix a wafer substrate onto a pedestal. Interesting benefits of electrostatic attraction techniques are their applicability to high temperature processes, capacity for handling wafers under vacuum conditions and, of course, the lack of any adhesive bonding material. Electrostatic mechanisms allow for switch-on-and-off bonding forces just by charging and discharging electrostatic electrodes. Therefore, bonding and debonding of thin wafers onto electrostatic carriers can be achieved within very short time, in a repeatable manner and without any constraints regarding surface contaminants from bonding agents.

The aim of a mobile electrostatic carrier system for thin wafer handling and processing can reached through if one prepares a carrier plate that shows the dimensions of a standard semiconductor wafer and keeps its electrostatic status over a long period of time after disconnection of an external power supply [7, 8].

The basic concept for handling thin wafers by means of electrostatic carrier plates (e-carrier) is shown in Fig. 12.2. At the frontside of the e-carrier identical pairs of large electrode areas are formed. Then a multilayer of electrically

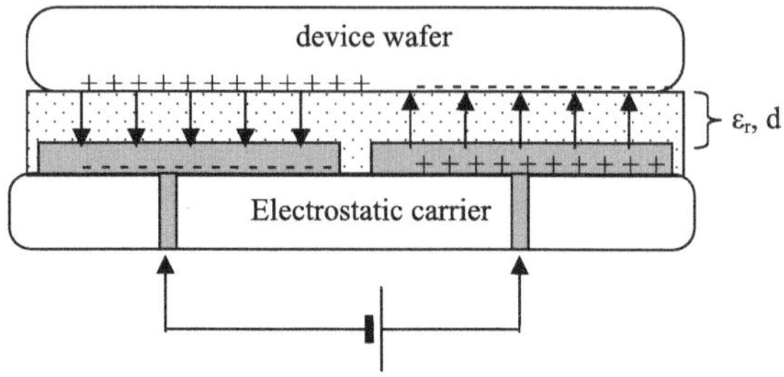

Fig. 12.2 Working principle of temporary wafer bonding by electrostatic forces

insulating material with dielectric constant ε_r and thickness d is deposited on top of the electrodes. Voltage is represented by U.

A thin semiconductor wafer can be placed on top of the e-carrier substrate and then the electrodes charged by an external power supply. The resulting electrostatic fields provoke a separation of charge carriers (electrons and holes) at the backside of the semiconductor wafer (see Fig. 12.2) and thereby cause an attractive force between carrier and thin wafer.

After initial charging the external power supply is disconnected. The electrostatic forces remain active for a longer period of time.

The configuration of the stacked wafer pair is similar to a plate capacitor. The attractive force between wafer and carrier can be calculated as:

$$F = \varepsilon_0 \varepsilon_r{}^2 A\, U^2/8d^2$$

The holding force F depends on the dielectric constant ε_r of the dielectric layers, the electrode area A (approximately half the wafer surface), the distance d between electrodes and wafer and the applied voltage U.

In principle, the base carrier plate can be made of different materials and also by different manufacturing technologies, for instance, thin film technology on silicon or glass wafers or thick film technology on ceramic plates. Choosing silicon as the base material for the carrier plate offers several advantages: high thermal conductivity, same coefficient of thermal expansion when thin silicon wafers are to be processed, full compatibility with common fabrication technology and availability of a large variety of high quality thin film layers. In particular, the argument of dielectric layer composition is of strong relevance with respect to the functional performance of e-carriers for two reasons: First, long duration times of electrostatic attraction require perfect electrical insulation between the electrodes and the thin wafer and also between the electrodes and the carrier substrate. Second, high attractive forces can be achieved when the insulating layers are very thin. Thin film technology on silicon wafer substrates with thermally grown oxide layers and plasma or CVD deposited dielectric layers fulfills these two requirements (Fig. 12.3). The picture in Fig. 12.3 shows frontside and backside of a mobile electrostatic carrier substrate prepared on a silicon wafer substrate and comprising through substrate vias and backside contact pads.

Applicability of mobile electrostatic carriers has already been demonstrated for a variety of process steps; for instance, bumping of very thin wafers [9] and backside

Fig. 12.3 Example of a mobile electrostatic carrier substrate with backside contacts prepared on silicon wafers

metallisation for thin wafers [10]. First evaluations of introducing mobile e-carriers into the manufacture process of thin GaAs wafers are reported [11].

12.10 Thin Die Handling

Handling single dies is of course a necessary step in chip assembly. One example would be manufacture of flexible electronic systems on foil substrates. In this case a flexible IC of a thickness of 10–30 μm needs to be attached onto a polymeric foil substrate or laminated between two foils. Technical solutions for pick-and-place processes for thin dies are available from various equipment suppliers.

Furthermore, there are also new manufacture steps that require reversible bonding of separated thin dies onto a carrier substrate. One application example is 3D stacking of a whole set of thin dies onto another device wafer. A principal sketch of this bonding situation is shown in Fig. 12.4.

3D production processes require the preselection of functional chips ("known good dies" or KGDs), as otherwise the yield of electrically working 3D chip-stacks would drastically decrease. In order to enable the parallel transfer of a larger number of chips each die must be placed with respect to specific alignment marks on the carrier. During the transfer process all chips need to be released from the carrier substrate and simultaneously be bonded to the base wafer. One approach for such temporary chip bonding is the use of thermoplastic adhesives. However, in this case, placement of dies on the carrier requires a specific thermal cycling step at each chip position. This will take a significant amount of time and also creates the risk for misalignment of neighbored chip devices.

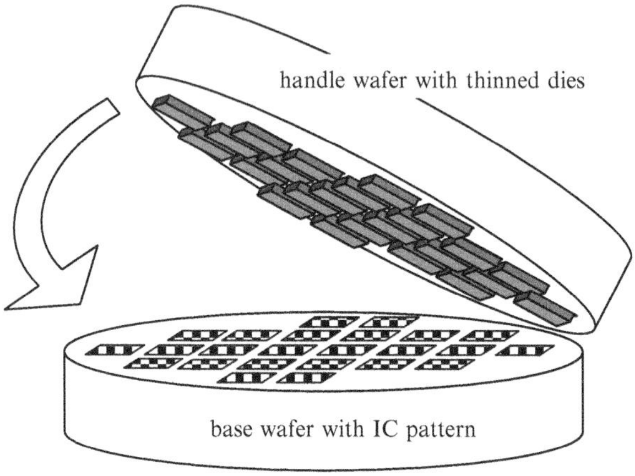

Fig. 12.4 Transfer of temporarily bonded thin chip devices from a handle substrate onto a base device wafer

A new approach would be offered by electrostatic bonding of the single dies on the carrier substrate. Aligned placement would be much faster as the whole sequence can be done at room temperature. Furthermore, after transferring the chips onto the base wafer the final bond process can be performed at high temperatures (up to 400°C) because no polymeric adhesives are used for temporary die attach. Electrostatic handling of single dies requires fine patterning of pairs of electrode areas, which need to be located beneath each single die. The feasibility of electrostatic die handling has already been proven [12].

12.11 Comparison of Different Carrier Techniques

The description of existing handle substrate techniques shows that each reversible bonding technique has specific advantages and disadvantages. For instance, polymeric bond layers allow for secure protection of the wafer edge but generally cannot afford high temperature processes above 300°C. On the other hand, adhesive free bonding, like electrostatic attraction, would need additional polymeric sealing materials if an e-carrier wafer stack were immersed into wet-chemical baths.

The characteristic properties of state-of-the-art carrier techniques are summarised in Table 12.2.

From the comparison of handle techniques given in Table 12.2 it can be concluded that up to the present there isn't just one carrier technique for all purposes in thin wafer technology. Therefore, it is of particular importance to look for useful combinations of different carrier techniques in order to enable sophisticated backside processing of ultra-thin wafers.

Table 12.2 Comparison of different carrier techniques for thin wafer processing

Temporary bonding technique	Advantages	Difficulties
Adhesive tapes	• Easy to apply and remove • Bonding at room temperature, fast bonding cycles • Good support during wafer thinning	• Low temperature stability • No sealing at wafer edge against wet chemicals • Low heat transfer • CTE not matched to semiconductor
Spin-coated adhesive	• Uniform thickness of bond layer • Capability to embed topography and to support wafer edge • Good edge sealing during wet-chemical processes	• Stripping of adhesive after debonding required • Bonding often requires heat, pressure and vacuum, longer bonding cycles • Limited temperature stability
Release layer and spin-on adhesive	• Cleaning of thin wafer is not required or facilitated • Enables embedding of surface topography	• Mechanical force for debonding must be controlled carefully
Electrostatic bonding	• No adhesive required, no need for cleaning • Bonding and debonding very simple • Temperature stability > 400°C, thermal properties equal device wafer • Good compatibility with other carrier techniques, allows for simple thin wafer transfer • Capability for processing of single dies	• Edge sealing required to enable wet-chemical processes • Strongly bowed device wafers are difficult to attract

Until today thin wafer support concepts are often evaluated as expensive techniques. However, it should be mentioned that the breaking of a completely processed device wafer in a fabrication line creates, probably, the worst impact on production cost. So it is concluded that introduction of handle substrates and thereby reduction or elimination of wafer breakage provides a positive outlook for future manufacturing concepts applied to ultra-thin semiconductors.

References

1. Combe S, Cullen J and O'Keefe M (2006) Reversible wafer bonding: challenges in ramping up 150 mm GaAs wafer production to meet growing demand. CS Mantech Technical Digest, Vancouver, 24–27 April, 2006, pp 193–196
2. Mould D and Moore J (April 2002) A new alternative for temporary wafer mounting. GaAs Mantech, Technical Digest, pp 109–112

3. Puligadda R, Pillalamarri S, Hong W, Brubaker C, Wimplinger M, Pargfrieder S (2007) High-performance temporary adhesives for wafer bonding applications. Mater Res Soc Symp Proc 970
4. Pargfrieder S, Burggraf J, Burgstaller D, Privett M, Jouve A, Henry D, Sillon N (March 2009) 3D integration with TSV: temporary bonding and debonding. Solid State Technol 52(3)
5. Webb R (February 2010) Temporary bonding enables new processes requiring ultra-thin wafers. Solid State Technol 53(2). www.3M.com/wss
6. Richter F (2009) Carrier technology for wafer thinning developed by thin materials AG. 10th international workshop 'be-flexible', organized by Fraunhofer IZM, Munich, Germany, 25 November 2009. www.thin-materials.com
7. US patent #7,027,283 B2; European patent EP-1 305 821 B1
8. Landesberger C, Wieland R, Klumpp A, Ramm P, Drost A, Schaber U, Bonfert D, Bock K (June 2009) Electrostatic wafer handling for thin wafer processing. European microelectronics and packaging conference & exhibition, Rimini
9. Landesberger C, Scherbaum S, Bock K (2007) Carrier techniques for thin wafer processing. Conference on compound semiconductors manufacturing technology, Austin, 14–17 May 2007
10. Raschke R (2009) New transferable electrostatic carriers for applications in vacuum up to 400°C. 10th International workshop 'be-flexible, organised by Fraunhofer IZM, Munich, 25 November 2009
11. Stieglauer H, Nösser J, Miller A, Jonson G, Behammer D, Landesberger C, Spöhrle H-P, Bock K (2010) Mobile electrostatic carrier (MEC) evaluation for a GaAs wafer backside manufacturing process. Conference on compound semiconductors manufacturing technology, Oregon, 17–20 May 2010
12. Wieland R, Hacker E, Landesberger C, Ramm P, Bock K (2008) Thin substrate handling by electrostatic force. Conference smart systems integration, Barcelona

Part IV
Assembly and Embedding of Ultra-Thin Chips

Add-on processing and ultra-thin chip assembly and embedding are tasks that are closely associated. In the fabrication of 3D systems, in particular, some approaches (e.g. TSV first; Chap. 10) fall into the area of add-on processing, while others (e.g. TSV last; Chap. 10) are clearly related to assembly and packaging. Therefore, some issues of this Part IV are also addressed in Part III. Nevertheless, it makes sense to distinguish between add-on processing and assembly and embedding of thin chips because the process equipment relates to semiconductor manufacturing in the first case and to packaging in the second case.

Table IV.1 and Fig. IV.1 illustrate the technologies, their applications and the book chapters, in which they are discussed. References are made not only to the chapters of this part (Chaps. 13 to 16) but also to Chaps. 10 and 11 of Part III for mentioned reasons. Only some applications are addressed in those chapters here. Applications will be treated exhaustively in Part VI.

Table IV.1 Assembly and embedding of ultra-thin chips

Technology	Applications	Related chapters
3D package-to-package assembly	3D IC 3D Microsystem	Chap. 10 Chap. 11 Chap. 16
3D wafer level chip assembly	3D IC 3D Microsystem	Chap. 10 Chap. 11 Chap. 15
Chip embedding in PWB	RF systems	Chap. 14
Chip embedding in foil	RF ID tag Flexible display Biomedical Security	Chap. 13 Chap. 14

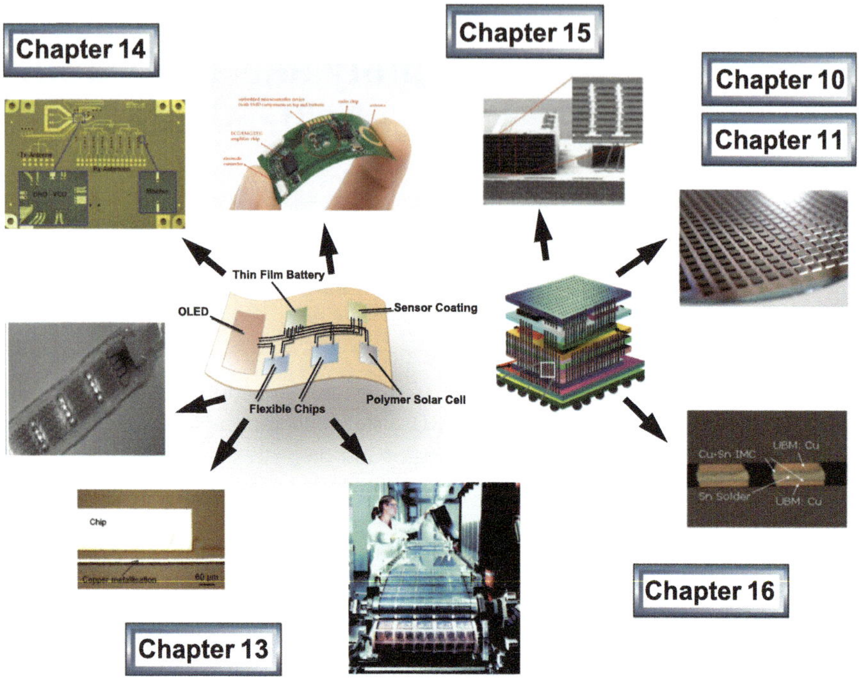

Fig. IV.1 Illustration of (**a**) the three generic concepts for the implementation of through-silicon-vias (TSVs) to build 3D ICs or 3D micro systems and of (**b**) wafer backside add-on processing by using handle carriers

Chapter 13
System-in-Foil Technology

Andreas Dietzel, Jeroen van den Brand, Jan Vanfleteren, Wim Christiaens, Erwin Bosman, and Johan De Baets

Abstract Systems-in-foil are an emerging new class of flexible electronic products in which complete systems are integrated in thin polymeric foils. First, a brief overview of the basic material choices and fabrication concepts will be given. Then, more detailed attention will be paid to the approaches of heterogeneous integration of thin chips. The different integration process flows and challenges and application fields will be described for three distinct areas: (1) ultra-thin chip packages with high density interconnects; (2) embedded circuitries for low cost flip-chip attachment to flexible substrates; and (3) embedded optical chips. It is important also to balance secondary material properties like thermal expansion, elastic moduli and adhesion strengths of different materials used to facilitate reliable operation of systems-in-foil under mechanical bending and at varied temperatures.

13.1 Introduction

Electronics on thin substrates such as foils will create a revolution in the electronics industry, enabling ultra-light and ultra-thin, flexible, easy-to-wear electronic products such as lighting and signage devices, reusable and disposable sensor devices, foldable solar panels and displays. An example of foil-based electronics used for many years are flexible printed circuitries. They are widely applied in printing heads, mobile phones and laptops. Recently, flexible electronic products have emerged in which a complete system is integrated in a thin polymeric foil to give a flexible end product [1–3].

Flexible systems-in-foil often have a larger surface area, which is required for interaction with the outside world (display, keyboard, large area sensor arrays), and the consumer experiences more freedom in usage when the device is bendable. Such devices are also suitable for conformal applications where they have to follow

---

A. Dietzel (✉)
Holst Centre, Eindhoven, KN 5605, The Netherlands
e-mail: A.H.Dietzel@tue.nl

J.N. Burghartz (ed.), *Ultra-thin Chip Technology and Applications*,
DOI 10.1007/978-1-4419-7276-7_13,

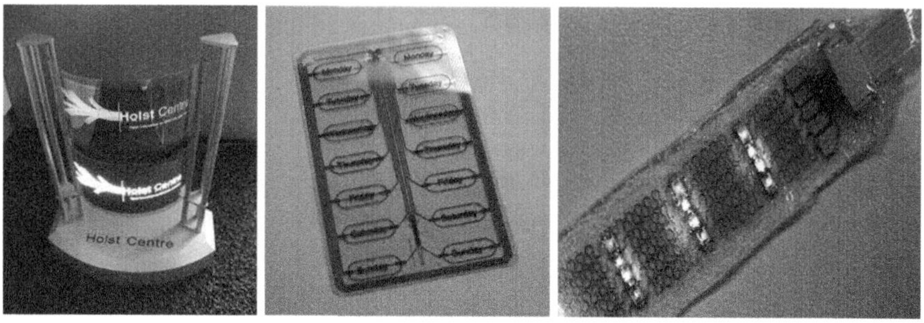

Fig. 13.1 *Left*: A flexible OLED device. *Middle*: A flexible smart medicine blister. *Right*: LEDs incorporated in a stretchable circuit (made from PDMS)

the shape of a given rigid housing or body. Flexible devices can either be used as one-time-flexible, so bended once to fit in their final application. Or, alternatively, they find applications in which they experience multiple bendings during their lifetime. Especially in the latter case, a reliable mechanical integrity of the device is challenging. An example of a dynamically flexed end product is the rollable display made by Polymer Vision, which is currently being applied in the Readius device [1]. Even though fully functional polymer ICs have been demonstrated, the mobility of charge carriers in polymers is not yet comparable to silicon. Many applications still require the integration of flexible ultra-thin silicon IC circuitry or opto-electronic devices made from III-V semiconductor materials in flexible foils without any alternative. Examples of systems-in-foil are shown in Fig. 13.1. Not only flexible but even stretchable electronic products would allow for ultimate design freedom. Recent developments in this field have, for example, been performed within the Stella European framework project [2].

13.2 Concept, Materials and Processes

A system-in-foil will be flexible if a flexible carrier is used and, possibly, flexible components as well. Of course, many components are rigid, but by using them in the naked die form and by thinning down to 25 μm or less, they become slightly flexible (typically 1-cm bending radius for 25-μm thick chips). Thin chips also offer the opportunity for a manufacturer to create very flat and lightweight systems. If the silicon chip is embedded inside a polymeric film, the overall thickness of the system can be less than 100 μm.

To make the flexible system-in-foil products as cost effective and reliable as possible, it is necessary one look in detail at the materials and processes that are used. The cost of the base material, i.e., the flexible foil substrate, represents a major contribution to the cost of the final product. Therefore, the products should be made, preferably, using low cost substrate materials. Polyimide foils provide beneficial properties, like good thermal and dimensional stability. However, polyimide is also a very expensive material. Alternatives for the substrate material are

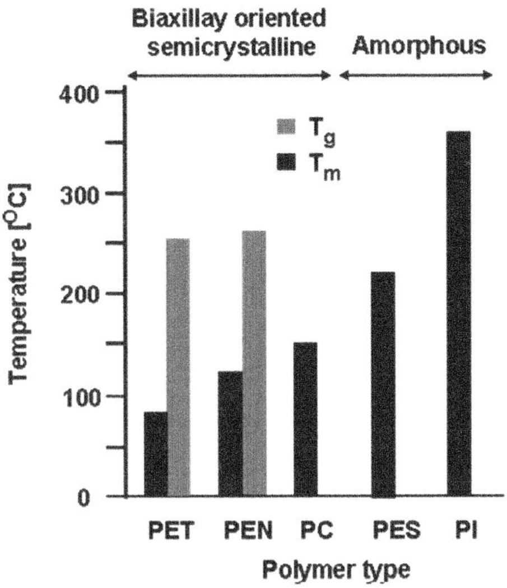

Material	Young's mod. E [GPa]	Therm. Exp. Coeff. CTE [ppm/K]
Hardcoat	6	60
Base polymer	3	65
Gas Barrier	150	10
ITO	120	8

Fig. 13.2 *Left*: Comparison of organic substrates materials by glass transition temperature. *Right*: Mechanical properties of typical layer components in flexible electronic systems

compared in Fig. 13.2. PEN and PET substrates allow a reduction of base material costs by a factor of 5 (PEN) or 10 (PET) [4]. A serious disadvantage though is their lower thermal stability; PEN has a Tg of 120°C and PET of 80°C while poly(imide) has a Tg of 350°C. This limitation in thermal stability excludes some well-established processes for fabricating electronic products and/or renders the use of existing processes much more challenging.

Despite the higher cost some applications will still need polyimide foils because of its better mechanical and thermal properties, resulting in a finer interconnection pitch and a higher reliability.

Furthermore, gas diffusion barriers are also essential to protect functional organic or inorganic materials, which are susceptible to moisture [5]. Their mechanical properties (see Fig. 13.2) may differ from the base material, which can induce stress. Conducting materials used include transparent ITO, metals and conducting lines made from composite materials.

High-volume production of the foils and a further reduction of the costs of the final product can be achieved through the manufacturing process itself. Reel-to-reel (R2R) manufacturing is considered the ultimate route for cost effective production of large quantities of flexible system-in-foil products. The combination of R2R manufacturing with low cost base materials opens up a completely new and interesting field of manufacturing challenges. Table 13.1 gives an overview of the most relevant technologies for R2R manufacturing of flexible system-in-foil products.

At the Holst Centre in Eindhoven large area coating and large area lamination is currently developed in the reel-to-reel laboratory (see Fig. 13.3). In the years to come laser processing and heterogeneous assembly will have to be transferred from current batch processing to R2R processing.

Table 13.1 Overview of technologies needed to manufacture cost effective electronic products using R2R technology

Technology	Purpose
Large area coating	Patterned large area coating technologies for depositing organic active layers, protective layers, adhesives
Laser processing	Selective ablation of different materials and substructures, micro-via drilling for interconnects
Lamination	Lamination of different functional foils with high overlay accuracy
Interconnection technology	Foil and chip electrical interconnect technologies using conductive adhesives
Fine structure printing and patterning technologies	Manufacturing of (sub)micron-sized structures using printing technologies/lithography on foil for making conductive circuitry, shunting lines for OLEDs
Heterogeneous assembly	Assembly of ultra-thin chips or other components on/in foils

Fig. 13.3 R2R lamination line (at Holst Centre, Eindhoven)

13.3 Embedding of Chips

13.3.1 Ultra-thin Chip Package (UTCP)

This section presents a polyimide-based embedding technology for 3D integration of very thin chips: the ultra-thin chip package (UTCP). Silicon chips, thinned to 15–25 μm, are packaged in between two spin-on polyimide layers, resulting in a total thickness of just 50–60 μm. Chip, PI and metal layers are so thin that the whole package is even bendable [6].

These very thin, very flexible packages can be used not only as a package for second level assembly on a printed circuit board or flex print, but the UTCP is also suitable for 3D embedding inside rigid or flex boards, as an alternative for bare die embedding. A fan-out pattern on the thin chip package relaxes required board interconnect pitch and the alignment constraints for the embedded package, and it allows testing before embedding.

UTCP thinness makes it also a proper technology for very compact package-on-package (PoP) configurations, by laminating several UTCP packages one on top of another, offering all the advantages of PoP technologies but limiting the total thickness of the package stack comparable to – or even smaller than – typical stacked die configurations.

An overview of the process flow of the UTCP technology is given in Fig. 13.4. The base material for the package is polyimide spin-coated on a rigid carrier. On this base polyimide layer, the chips are placed face up, using benzocyclobutene (BCB) as adhesive material. After curing the BCB, a top polyimide layer is spin-coated on top of the fixed dies. Finally vias are opened by laser drilling and metallised by sputtering.

The base material for the UTCP is a spin-on polyimide (PI2611-HD Microsystems). Polyimide is a suitable base material because of its excellent mechanical and thermal properties; the selected polyimide also has a thermal expansion coefficient (CTE) similar to silicon.

UTCP technology uses glass as a temporary carrier during processing. An easy release of the cured spin-on polyimide film from the glass carrier can be obtained by selective application of adhesion promoter on the carrier substrates. Well-chosen sticky areas keep the polyimide layer in place during the whole process and are cut out for the release once the package is finished. The use of a stable rigid carrier makes fine-pitch thin film techniques possible and allows easy testing of the package.

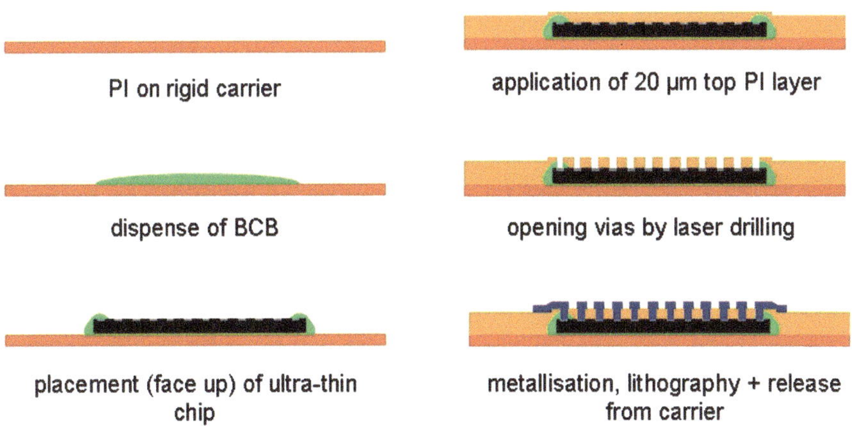

Fig. 13.4 Overview UTCP technology process flow

Typical layer thickness of the base polyimide layer is 20 μm. Very thin silicon devices, with thicknesses down to 15 μm, are placed face up on the base polyimide layer. The adhesive material for the fixation of the dies on the substrate has to be resistant to the high curing temperature (350°C) of the subsequent polyimide layer. Therefore, polyimides or BCB are possible adhesive layers. Because curing of polyimides frees a considerable amount of water, resulting in bubble formation at the interfaces, BCB has been selected as bonding material. BCBs of different viscosities were tested for optimisation of the placement. Best results were achieved with Cyclotene 3022–46 with a viscosity of 52 cSt at 25°C [7].

A void-free bond interface after placement of the chips is critical. Voids can be caused by evaporated water and solvents during curing, but also by trapped air at the bonding surface. For small chips these bubbles can be prevented by the dispensing of a well-controlled amount of the adhesive, followed by a spread of the adhesive under the die during placement. Larger dies need a vacuum-assisted placement to prevent any bubble formation.

Curing of the BCB adhesive layer occurs in vacuum and needs a temperature of 250°C, but as the top polyimide layer will be cured at 350°C, the BCB is cured at 350°C, too, to prevent further outgassing of BCB during the following curing steps.

In the next step a second PI2611 spin-on polyimide layer is spin-coated on top of the chip. This top polyimide has to adhere well on the cured base polyimide layer, on the cured BCB and on the surface of the die. Without treatment, this spin-on polyimide has a poor adhesion on all three of these surfaces. Adhesion is improved by a combination of an RIE (reactive ion etching) treatment and the application of a spin-coated adhesion promoter. After curing at 350°C a typical top polyimide thickness of 20 μm is obtained.

Once the chip is embedded in between the two polyimide layers, vias to the contacts of the chip are opened by CO_2 laser drilling. CO_2 laser ablation offers a highly selective process, as the IR laser energy is stopped at a metal interface, provided a moderate power level is used. Via ablation down to the contact pad can hereby be accomplished without damaging the contact pad material. Due to the laser wavelength (10.6 μm) and the diffraction at the laser optics, the minimum feasible via size is limited to about 40 μm. When using the UTCP technology for finer pitch contact pads, the intrinsic spot size of the CO_2 laser is no longer suitable, but needs to be combined with a sputtered metal via mask. Cu is a suitable masking material, because it reflects the IR light of the CO_2 laser very well. The fine openings in the metal mask are photolithographically defined and wet etched. Finally the polyimide is removed by the CO_2 laser through the small via openings. This technology allows further reduction of the via size to 15 μm.

Although CO_2 laser ablation offers a very selective process, the thermal ablation mechanism at that wavelength leaves a residual layer of (carbonised) polyimide on the contact pad in the via. Photochemical ablation at UV wavelengths would not have this problem, as it offers a clean process due to the breaking of chemical bonds and consequent evaporation of the material. But UV laser ablation would be not selective enough on the thin metal contact pads of the chip.

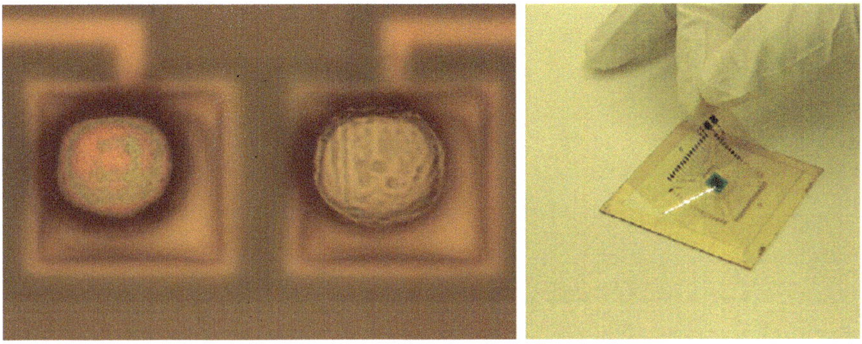

Fig. 13.5 *Left*: A via after CO_2 laser drilling only (*left* via) and a via after an extra excimer laser clean step (*right* via). *Right*: Release of a UTCP package after processing and testing

The residual film is then removed by a fast excimer gas laser clean step (wavelength 248 nm). The process window for the laser power level in this step is tuned to avoid too much damage to the Al contact pad and still have ablation of the debris layer. After laser drilling the metal mask is removed by wet etching. Stripping the metal mask also removes possible laser debris from the surface next to the via. Figure 13.5 gives a more detailed view of the vias: the left via still has the very thin residual polyimide film, the right via had an extra excimer clean step, exposing the pad's surface. Size of contact pads in this picture is 70×70 μm, via bottom diameter is 45 μm and contact pitch is 100 μm.

Once the dies are embedded in polyimide and the vias to the contacts of the chips are opened, the top metal layer is applied. A thin sputtered TiW/Cu (50 nm/1 μm) layer ensures a good contact to the chip pads and provides low ohmic interconnects. To ensure a good adhesion of the metal layer on the underlying polyimide, a plasma treatment (CHF_3/O_2) is applied, providing a very good mechanical and chemical bond. The peel strength of the sputtered metal layer, electroplated to a thickness of 25 μm, is higher than 1.6 N/mm after this plasma treatment. After processing, the chip package can easily be cut out from the carrier, either manually or by laser cutting.

Finally, the polyimide releases easily from the glass substrate (Fig. 13.5). A cross-section of a UTCP packaged silicon device with a total package thickness of only 60 μm is shown in Fig. 13.6, providing a very flexible chip package.

13.3.2 Embedded Circuitry for Ultra-thin Chip Assembly

Since the use of PEN and PET substrates allows a drastic reduction of base material costs, novel processes are developed that are compliant with the lower mechanical and chemical stabilities and the limited thermal stability as compared to polyimide foils. Although techniques like screen printing or inkjet printing are readily

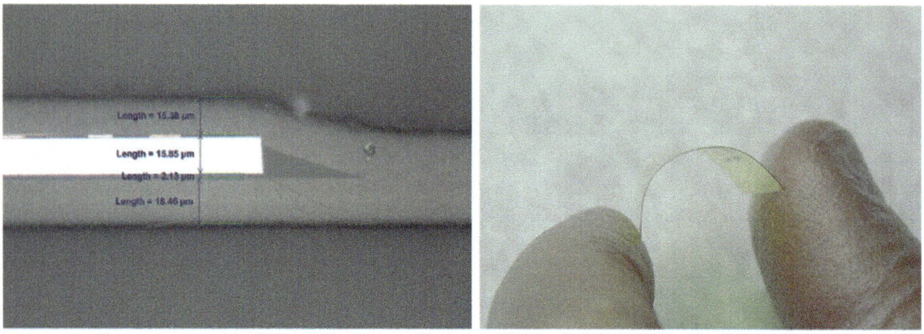

Fig. 13.6 *Left*: Cross section of UTCP with total package thickness of only 50 µm. *Right*: All layers of the UTCP are so thin that the whole package is bendable

available for cost effective electronic circuitry printing, they are limited in resolution and/or result in poor conductivity of the lines that can be made. An alternative is provided by embedding the circuitry in the depth of a polymer foil [9]. In a first step a laser is used to photo-ablate the circuitry pattern in the foil. This can either be done with a broadened beam through a mask or by direct writing with a focused laser beam. In a next step the resulting groove pattern is filled with an electrically conductive composite paste using a squeegee process as it is used in screen or stencil printing. As a final step, the circuitry is dried and cured and can be used for further processing. The advantage is that much finer features can be made than with traditional techniques like screen printing. The line widths and pitches that are possible are only limited by the resolution of the laser and by the size of the conductive particles in the conductive paste. Because the thickness of the foil can be used, the lines can be made much thicker than would normally be possible with screen printing or inkjet printing, thereby giving lower line resistivity. Figure 13.7 shows some photographs of a typical embedded circuitry in 125-µm thick PEN foil.

A fan-out test circuitry made of nanocomposite paste material is shown in Fig. 13.8, left side. Circuitry lines connect the four corners of the chip to a connector area. The circuitry lines have a width of 50 µm, a minimum pitch of 100 µm and a depth of 15 µm. Using these lines it is possible to measure the bump-to-filling resistance at the chip corners using four point electrical measurements. The chip that was used was a 50-µm thick chip with 8-µm bumps. Bonding was performed using a low temperature curing isotropic conductive adhesive (ICA).

The right hand side of Fig. 13.8 shows a microscopic cross-section image of a thin chip bonded on this circuitry. There is good and intimate contact between the chip bumps and the filling. It should be noted that measured contact resistances are negligible in comparison with the resistances of the conducting lines. To ensure a reliable, flexible circuit it is important to carefully choose the mechanical properties of the embedded material. Because the circuitry will experience deformation in its application, an important selection criterion for the embedded material will be its elastic modulus. Finite element modelling (FEM) is useful for the study of the effect of the elastic modulus of the embedding material on the stresses in the final circuitry. As input parameters for the modelling, the mechanical properties of

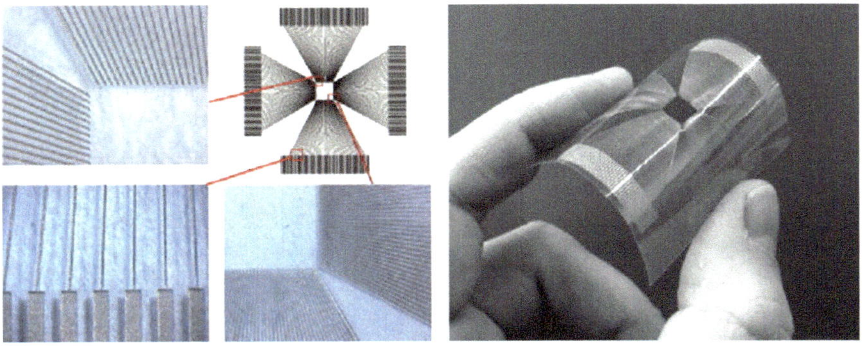

Fig. 13.7 Microscopic images of various parts of an embedded circuitry demonstrator. The total size of the demonstrator is 45 × 45 mm. The conductive lines have a width of 50 μm and a thickness of around 30–50 μm. The right-hand photograph shows an overview picture with a Si die bonded in the centre of the circuitry

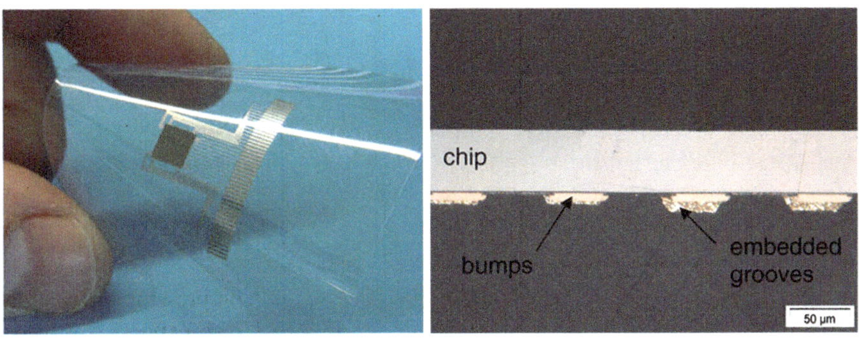

Fig. 13.8 *Left*: Fan-out circuitry used to test direct chip attachment. *Right*: Cross-section view microscopic image of the 50-μm thick chip bonded on fan-out circuitry [9]

PEN and the ICA were used. ICA was chosen as the filling material because detailed bulk mechanical properties could be readily retrieved from the supplier. The tensile and shear strengths of the PEN–ICA system were experimentally determined using a compact tension shear test. Failures were typically found to be adhesive in nature for both tensile and shear loading of the system. This indicates that the PEN–ICA interface is the weakest link in the system. A maximum interface shear strength of 28 MPa and a maximum interface tensile strength of 13 MPa were experimentally obtained.

To evaluate the stress levels as a function of bending radius, test structures can be virtually bended in FEM simulations the same as they would experience in a four point bending test. The lowest stress levels occur when the moduli of the substrate and the filling are similar. Here, the ICA composite material has an elastic modulus of 4.7 GPa while the PEN foil has an elastic modulus of 6 GPa. Figure 13.9 shows plots of the interface shear stress and interface tensile stress as a function of bending

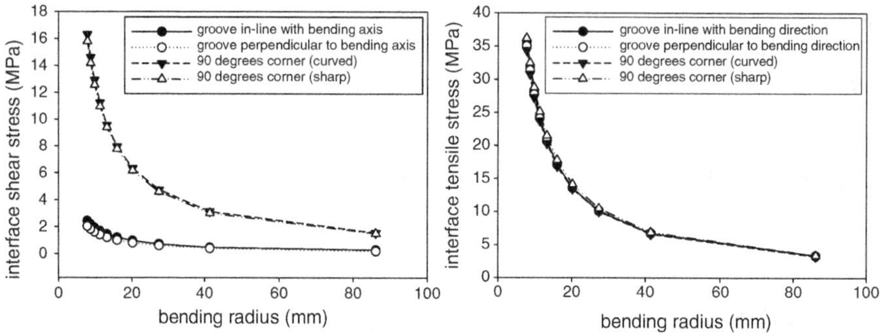

Fig. 13.9 Interface shear stresses (*left*) and interface tensile stresses (*right*) for different straight and corner line geometries as a function of bending radius for ICA-filled grooves in PEN [9]

radius of the foil for different geometries of the conductive lines. The maximum tensile interface stress was always found perpendicular to the bending axis and at maximum distance from the neutral bending line. The maximum shear interface stress was always found at a 45° angle to the bending axis. This explains the higher interface shear stresses observed for corner geometries. The stresses rapidly increase with decreasing bending radius. The interface tensile strength is the most critical, since at a bending radius of around 20 mm it surpasses the experimentally determined interface tensile strength of 13 MPa. The interface shear stress however does not exceed the experimentally determined shear strength of 28 MPa for the considered range of bending radii. The static bending radius limit of around 20 mm could also be verified in bending experiments.

13.3.3 Embedded Optical Chips

Optical data transmission has become the obvious choice for communication over longer distances, but new trends compel optical interconnections to bridge short distances also. The opto-electronic printed circuit board, also referred to as the OPCB, is an addition to the existing electrical PCB, containing optical sources (side- and surface-emitting), optical detectors (side- and surface-viewing) and optical waveguides as light transporting medium. This onboard optical add-on finds its application in high speed interconnections and optical sensing systems. This section presents a technology platform for the full integration of these components inside a 150-μm flexible foil [10]. Twenty-micrometre thin VCSELs (vertical cavity surface emitting laserdiodes) and photodiodes are embedded in a layer build-up of optical transparent material and polyimide mechanical support layers. Optical waveguides are structured in the optical layers and pluggable mirror components couple the light from the embedded opto-electronics in and out of the waveguides. The embedded opto-electronics are electrically connected with driver and amplifier circuitry by embedded copper tracks. Figure 13.10 shows a schematic cross-section of the technology.

The process for the embedding of opto-electronics, optical waveguides and coupling structures inside a 150-μm thin foil is summarised in the process flow in Fig. 13.11. The base material is a transparent optical material, preferably as flexible as possible. But many optical materials are brittle. Thin layers will be bendable, but easy to break. To increase the reliability of these layers polyimide mechanical support layers are provided at top and bottom, so the optical materials are sandwiched within the neutral axis for bending. All the processing is performed on a temporary rigid glass carrier using the same release technique as described in the section on UTCP processing. The bottom polyimide layer is spin-coated on the glass carrier. On top of this layer, 10-μm thick copper islands are fabricated as a heat sink for the opto-electronics, which will be positioned on the copper islands later on in the process. Next, a first layer of optical material (20 μm thick) is applied and cavities are laser-ablated using a excimer laser (248 nm wavelength), well-aligned with the heat sinks (Fig. 13.11a). The copper islands also act as a laser ablation stop for the formation of the cavities.

The thinned opto-electronics are then placed face up in the cavities using dedicated glue with a pick-and-place tool for accurate positioning (Fig. 13.11b).

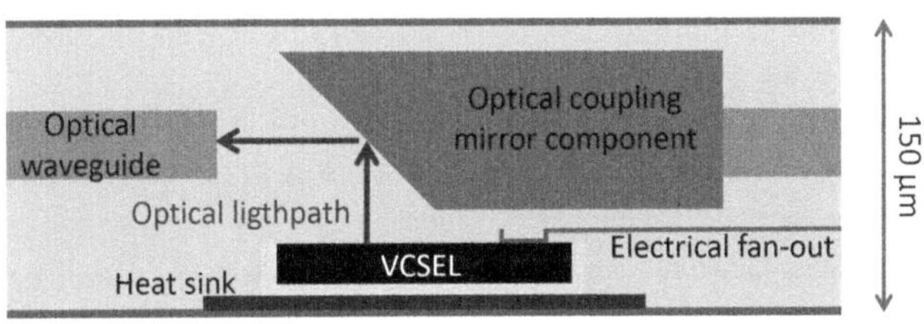

Fig. 13.10 Schematic cross-section of the optics embedding technology

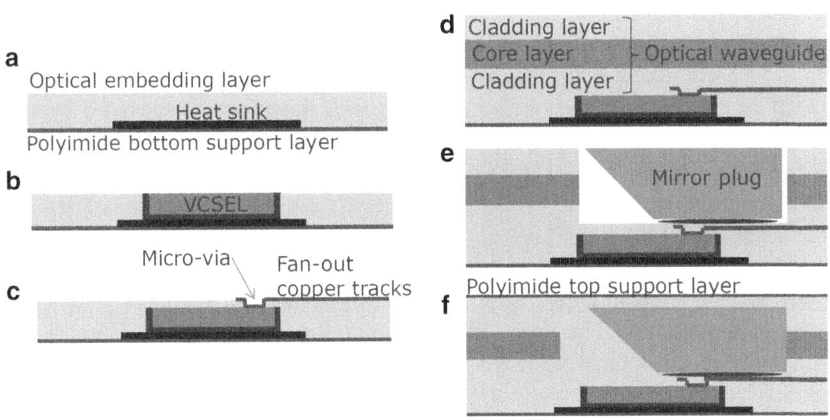

Fig. 13.11 Process flow for the embedding of optics in a flexible foil

Another layer of optical material (10 μm) covers the die on the top surface, so that it is completely embedded. Laser-drilled blind micro-vias are made towards the bond pads of the embedded dies with a excimer laser. Sputtered copper tracks (1 μm thick) fan-out the small pitch (125 μm) bond pads on the embedded dies towards larger pitch bond pads on the outer surface of the final foil (Fig. 13.11c).

On top of this 40-μm thin opto-electronic component package, the waveguide stack is fabricated. This stack consists of a core layer (50 μm) in between two cladding layers (50 μm). The core material and the cladding material have the same basic characteristics but differ in optical refractive index. The core layer is patterned using a photolithographic process resulting in 50 × 50 μm cross-sectional waveguides. Alignment of the waveguide can be achieved during photolithographic mask alignment using the active areas of the embedded VCSELs and photodiodes as alignment marks (Fig. 13.11d).

The use of vertical emitting and top-illuminated opto-electronics (like VCSELs and photodiodes) demands special attention to the optical out-of-plane coupling from the embedded opto-electronic components with the waveguides. This kind of coupling is achieved with a pluggable, separately fabricated degree mirror. The mirror plug is a flat insert with a 45° (±1°) mirror facet at the end. The mirror plugs are inserted inside the optical waveguide stack by the creation of cavities (Fig. 13.11e). Finally, an optical material layer is applied to embed the mirror plugs and to fill the gap underneath the 45° facets, followed by the top polyimide support layer (Fig. 13.11f).

State-of-the-art flexible electronics and opto-electronics have no common standard on testing equipment and procedures. Therefore the flexibility of the proposed waveguide foil and the embedded opto-electronics was tested with the set-up depicted in Fig. 13.12 (top). The flexible waveguides and embedded opto-electronics were bent around two identical cylinders and consecutively bent over the left and the right cylinder. This was done for several cylinder bending radii.

The minimum bending radius of the foil is less than 2 mm before the substrate becomes damaged. Figure 13.12 (bottom) shows an embedded VCSEL array bent by hand and a close-up of the VCSEL during bending. The bending endurance of the waveguides and embedded opto-electronic components has been tested by applying 1,000 consecutive bends at a bending radius of 2 mm. No damage or loss of functionality of the waveguides and embedded opto-electronics occurred and no additional optical propagation loss was induced in the waveguides.

The total optical link loss in a 2-cm long waveguide link between an embedded VCSEL and photodiode is 6.4 ± 1.5 dB. The crosstalk between two neighbouring links (250-μm pitch) is −17 dB.

The proof-of-principle of the embedded link for higher frequencies is demonstrated using a simple layout with side launch coaxial SMA connectors and SMD components, assembled on a rigid FR-4 substrate for handling reasons. A very clean eye at 1.2 Gb/s is obtained (Fig. 13.13), showing excellent operation of the optical link and the associated opto-electronic and electro-optic convertors, including the high frequency interconnects and circuits required for the characterisation.

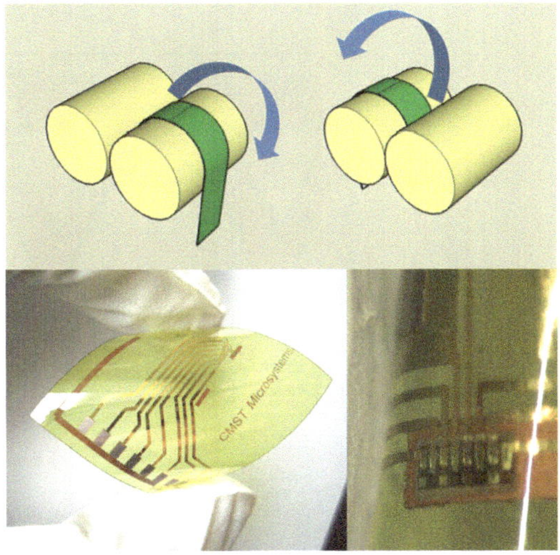

Fig. 13.12 Bending test set-up (*top*) and photographs of an embedded 1 × 4 VCSEL array, fabricated in Truemode Backplane™ polymer

The mechanical flexibility of the embedded optical links increases the reliability significantly. Temperature cycling tests are performed on both flexible and rigid embedded optical links. The rigid version is using the same optical materials, but is fabricated on top of a standard FR-4 PCB.

The samples were passed through 300 temperature cycles with a different temperature profile for each 100 cycles (cycle 0–100: 0–85°C; Cycle 101–200: −30–85°C; and cycle 201–300: −40–125°C). A clear deterioration of the rigid links is noticeable and a total failure occurs at temperature cycles down to −30°C. The optical layers are showing cracks and delamination. The flexible links show no change in optical transmission due to the cycling, even up to temperature ranges of −40–125°C.

Humidity resistance was tested in an environment with a constant 85% relative humidity at a constant temperature of 85°C. Figure 13.13 shows the evolution of the average total optical link loss during the humidity exposure. The flexible links resist the humidity much better, as can be seen in the graphs.

13.4 Applications

A possible application of UTCP packaged devices is to embed them into a conventional flexible multilayer circuit [8]. Ultra-thin active devices are first integrated in a flexible ultra-thin chip package, which is in turn embedded inside a standard double-layer flex print.

The UTCP packages serve as an interposer, bringing the fine pitch on the die to a contact pitch that can be handled by the standard flexprint technology. The UTCP

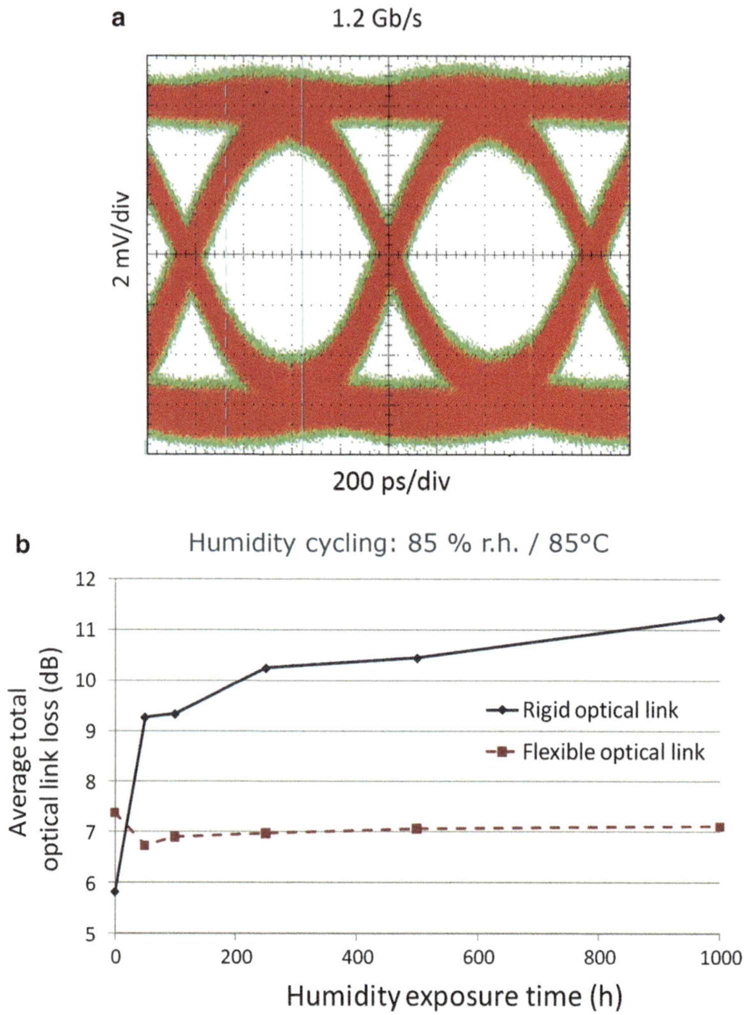

Fig. 13.13 *Left*: Eye diagram at 1.2 Gb/s of a fully embedded optical link. *Right*: Evolution of the total optical loss of embedded flexible and rigid optical links during temperature cycling

packages are first aligned and fixed (heat tack) on a patterned inner layer, followed by lamination and conventional through-hole interconnect.

This 3D integration technology has been successfully demonstrated for the embedding of a functional microcontroller device for a wireless ECG application. A Texas Instruments microcontroller MSP430F139 was successfully integrated inside a standard double-layer flex PCB, with even smaller surface mount components mounted above and below the embedded chip, realising a fully functional wireless biopotential system. The 3D integration of UTCP packages leads to high density integration, since SMD components can be mounted on the top and bottom of the integrated devices. In addition, the UTCP packaged thin silicon devices are mechanically flexible themselves, leading to an increased total flexibility of the resulting system. The UTCP packaged microcontroller and the wireless ECG node are depicted in Fig. 13.14.

A second application of the UTCP technology is the package-on-package approach: The extremely thin chip packages can be stacked, offering all advantages of PoP architectures, but with a very small total thickness. This technology is currently under further development and targets a 4-chip stack with a total thickness below 400 um.

Optical data communication on board has drawn a lot of attention and has proven to bring a solution to the emerging bottlenecks of electrical interconnects [11], but

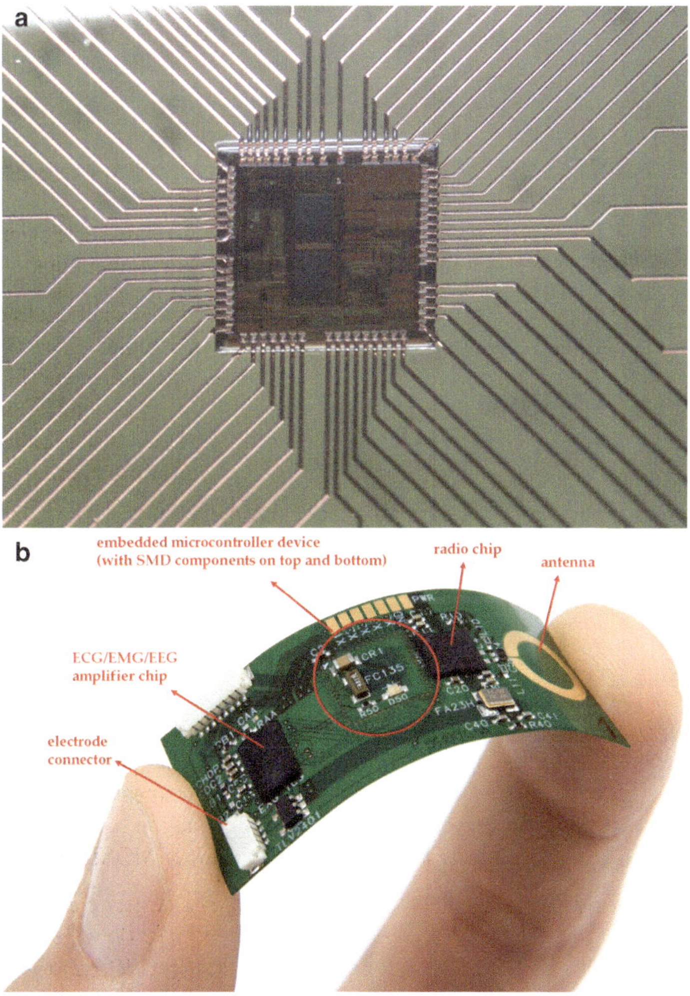

Fig. 13.14 *Left*: UTCP packaged microcontroller MSP430F149. *Right*: Wireless ECG module with 3D integrated microcontroller

adaption of this technology on a commercial scale has still not been conveyed. Though the demonstrated performance of optics on board improves every year, its adaption in the industry is still limited to niche applications. The technology discussed here addresses the introduction of on-board optical communication in the fastest growing market of flexible substrate electronics [10]. By making the optical interconnects and every accompanying active and passive feature very thin and flexible, we can integrate everything in a thin bendable foil. This opens a world of opportunities in portable and wearable applications, where flexible electronics are currently dominant [4]. In automotive, avionics, aerospace and medical applications, optical fibre communication and flexible electronics made their break-through long ago, mainly driven by the low weight and compactness of both and the immunity to electromagnetic interference and reliability in the harsh environments of optics. Flexible optical interconnections combine the electrical and optical network into one system.

Light is presented here as the carrier of bit data, but from an analogue point of view, light can be used for sensing a wide range of chemical, biological and physical parameters. Optical sensor systems have shown some major advantages over their electrical counterparts, with a large growth in this research area as a consequence. Sensors often need to be as compact and unobtrusive as possible; the high level of integration reached in this technology can significantly reduce the size, measurement point pitch and cost of existing optical sensors.

13.5 Conclusion and Summary

Flexible electronic devices form a new class of products in which a full electronic system can be integrated in a flexible sheet. Bare die integration into flexible materials offers some unique features such as minimum overall packaging thickness, bendability, ultra-low weight and fine-pitch interconnects. High-end products, lightweight portable products, odd-shaped products or 3D stackable packages will profit from these. These applications require high reliability and have to be built on dimensionally and thermally stable materials, for example, polyimide. For low cost applications it is necessary to use low cost base materials. The use of base materials like PEN or PET has several implications. First of all, these materials are less thermally stable than the conventional flex base materials like polyimide, which renders many established processes as not suitable. Therefore, a new embedded circuitry process has been developed on which ultra-thin bare dies can be flip-chipped. A further reduction of the costs of the final product can be achieved through the manufacturing process itself. Roll-to-roll manufacturing is probably the most efficient and effective method of manufacturing flexible electronic devices. The combination of low cost base materials with R2R manufacturing processes for flexible system-in-foil devices will open up a whole new arena of reliability challenges.

Also from the device design perspective, there will be many reliability relevant choices to be made. As discussed with several examples, the devices will likely be built from a range of different materials, both organic and inorganic. These materials have to support certain functions like environmental protection, transportation of current or integration of computing power as can well be realised by an embedded ultrathin Si chip. To ensure a reliable operation of the flexible device under various circumstances, it will be necessary to carefully match the secondary material properties like coefficient of thermal expansion, elastic moduli and internal and adhesion strengths, and to balance these with the material's primary functional properties.

References

1. Polymer Vision product announcement. www.readius.com
2. Stella project home page www.stella-project.de
3. van den Brand J, de Baets J, van Mol T et al (2008) Systems-in-foil: devices, fabrication processes and reliability issues. Microelectron Reliab 48(8–9):1123–1128
4. Fjelstad J (2007) Flexible circuit technology. BR Publishing, Seaside
5. Greener J, Ng KC, Vaeth KM et al (2007) Moisture permeability through multilayered barrier films as applied to flexible OLED display. J Appl Polym Sci 106(5):3534–3542
6. Govaerts J, Bosman E, Christiaens W, Vanfleteren J (2010) Fine-pitch capabilities of the flat ultra-thin chip packaging (UTCP) technology. IEEE Trans Adv Packag 33(1):72–78
7. Niklaus F, Enoksson P, Kälvesten E, Stemme G (2001) Low-temperature full wafer adhesive bonding. J Micromech Microeng 11:100–107
8. Christiaens W, Torfs T, Huwel W, Van Hoof C and Vanfleteren J (2009) 3D integration of ultra-thin functional devices inside standard multilayer flex laminates. 2009 European microelectronics and packaging conference (EMPC 2009), vol 1 and 2, Rimini, 16–18 June, pp 671–675
9. van den Brand J, Kusters R, Barink M, Dietzel A (2010) Flexible embedded circuitry: a novel process for high density, cost effective electronics. Microelectron Eng 87(10):1861–1867. doi:10.1016/j.mee.2009.11.004, available online
10. Bosman E, Van Steenberge G, Van Hoe B et al (2010) Highly reliable flexible active optical links. IEEE Photonics Technol Lett 22(5):287–289
11. Bona GL, Offrein BJ, Bapst U, Berger C, Beyeler R, Budd R, Dangel R, Dellmann L, Horst F (2004) Characterization of parallel optical-interconnect waveguides integrated on a printed circuit board. Proc SPIE 5453:134–141

Chapter 14
Chip Embedding in Laminates

Andreas Kugler, Metin Koyuncu, André Zimmermann, and Jan Kostelnik

Abstract This chapter provides an overview of the technology of embedding chips into laminates. The motivation, an overview of technological approaches and the challenges will be outlined. One of the technologies will be described in more detail through a case of an application. Finally a technological outlook is given in consideration of the innovations the embedded technology could make possible.

14.1 Introduction

The motivation for embedding chips into printed wiring board (PWB) stems from 3D integration efforts whose aim is an increased level of miniaturisation. Resulting higher packaging densities serve to reduce the gap between Moore's law and the packaging technology [1].

Various 3D-Integration concepts exist such as vertical system integration with through-silicon vias (TSVs), chip-on-chip and package stacking [1–4]. Main drivers for embedding chips in PWBs include attainment of a high level of miniaturisation and the shorter signal paths that result. Increased heat dissipation efficiency, better reliability compared to soldering on PWB, electromagnetic compatibility and the potential to reduce costs are other factors that fuel research in this field.

Embedded chip technologies can be classified depending on either their process flow or the utilised assembly and interconnection technologies. The latter classification reveals two main technological trends: interconnections formed by the galvanic process or interconnections based on conventional bonding techniques (e.g. flip chip). Naturally technologies based on conventional processes demand less development effort and offer the advantage of established technology platforms – not only for the

A. Kugler (✉)
Robert Bosch GmbH, Corporate Research and Development–Dept.
Plastics Engineering (CR/APP4), 71301 Waiblingen, Germany
e-mail: andreas.kugler@de.bosch.com

J.N. Burghartz (ed.), *Ultra-thin Chip Technology and Applications*,
DOI 10.1007/978-1-4419-7276-7_14, © Springer Science+Business Media, LLC 2011

PWB but also for establishing interconnections to the chip. However, galvanic processes, which are standard for PWBs but unconventional for contacting the chips, have the potential to achieve a higher level of miniaturisation.

In terms of technology, embedding chips into foils or PWBs transforms the wiring board from a carrier for packaged components into a system that includes first level interconnects. The business model in this configuration is yet to be defined and the diffuse borders of responsibility in the supply chain impose an important challenge. The system design and process flow have to ensure proper testing of ICs after embedding. Another important point is the limited capability to repair or exchange an embedded chip. The overall yield depends on that of the PWB process and the known good die (KGD) yield plays an increased role.

The following sections will describe the main technological trends for embedding chips into laminates. Emphasis will be given to the CHIP + technique with a functional demonstrator from the automotive industry.

14.2 Concepts for Embedding Chips in PWB

There are various ways to embed chips into laminates. In an approach developed by Texas Instruments [5] chips are placed into cavities inside a laminate layer and covered by additional layers, eventually laminating the whole assembly. In the Chip-in-Polymer (CiP) approach developed by Fraunhofer IZM [6], chips are thinned down to 50 μm and bonded onto the core substrate of the PWB using an adhesive with the active side up. RCC (Resin-Coated Copper) layers are used for the lamination. Connections to the bond-metallisation of chip and to the copper routing are obtained by laser drilled micro vias that are metallised through the conventional PWB process. Finally the top Cu layer is patterned.

A further development of CiP is the so called CHIP + approach developed in a joint industry project KRAFAS (funded by the German Ministry for Education and Research) for a high frequency application. The process flow is shown in Fig. 14.1. A main difference of CHIP + is that the chip is placed onto the polymer side of an RCC foil in the flip-chip configuration. Mechanical bonding of the chip to the B-stage polymer of the RCC, which provides a well-defined and uniform insulation layer (Fig. 14.2a), is achieved during lamination with the FR4 core. An optional process step before lamination is encapsulation of the chip(s) with an epoxy resin either by moulding or dispensing. This serves to even out the height difference in case of multiple chips and stabilise their positions. Similar to the CiP process, chip pads are contacted via metallised micro vias (Fig. 14.2b) followed by patterning of the top metallisation. Several such layers can be combined by lamination to form multilayer structures as shown in the last step in Fig. 14.1.

One possible approach to cost reduction is the use of a blank copper instead of RCC layers and mounting the chip onto the copper foil with an NCA (nonconductive adhesive). In this way the structure can be stabilised without an epoxy

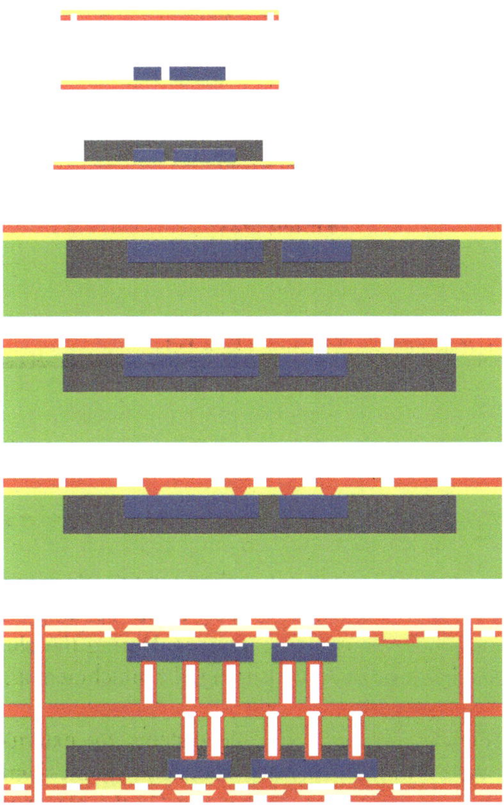

Fig. 14.1 Process flow of the CHIP + process

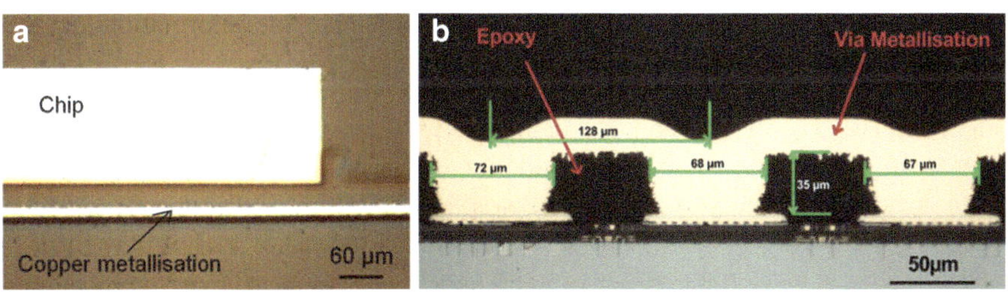

Fig. 14.2 Cross-sectional view showing (**a**) zoomed view of the embedded chip and (**b**) vias in the first embedding layer with approximately 70-µm diameter and approximately 130-µm pitch (Fraunhofer IZM)

encapsulation [7]. The NCA is removed during laser drilling of the micro vias, and a tight control of the bond line thickness becomes essential.

Another alternative is the IBOARD concept [8] from Schweizer Electronic AG. The copper layer of the RCC is patterned to form an interposer on which the flip chip is mounted using solder or conductive adhesive as the interconnect material.

In this case the micro vias connect the metallisation pads on the interposer to the top metallisation of the PWB. This approach offers a reduced processing risk because it separates the well-established processes of chip interconnection and PWB production.

CiD (Chip-in-Duromer) [9] is another approach that enables a high integration level of multiple chips without the need for an interposer. Chips are placed onto a common removable carrier substrate and moulded with an epoxy resin such that the active surfaces of the embedded chips lie on a common plane. This surface is then coated by a dielectric, e.g. BCB (benzocyclobutene), providing a rewiring layer. Chip-to-chip interconnections are provided through conductive traces deposited on the dielectric using photolithographic techniques. The epoxy casing around the chips makes the subsequent lamination and registration processes easier by compensating for the height differences and stabilising their position. Furthermore, very thin chips can be singulated and handled much easier in other processes such as embedding in PWBs.

The different approaches in embedding chips into laminates have the common goal of reaching a higher level of integration. PWBs populated with embedded chips bring into focus the importance of micro vias and patterning in the sub-100-μm region. Associated challenges are increased accuracy in component placement and via alignment with increased assembly speed and higher yield to realise the potential cost reduction.

14.3 Realisation of Embedded Technology in the Production Process: A Practical Approach

A high frequency driver assistance device was produced using the CHIP + concept [10]. First tests of implementation were done on small size PWBs in modular form. To render the process cost effective, a reel-to-reel production concept in the large panel format was also developed. The most important process-related challenge is the effect of production tolerances in the large format PCB production, where the component placement accuracy proved to be the key factor. The good news is that equipment manufacturers consider these requirements feasible and development is underway for a beneficial combination of processes for large area substrates. Using the latest technologies, registration of ICs in the single-digit micrometre range with acceptable repetition accuracy is possible. Consequently, electrical connections to pads as small as 100 μm in size are compatible with series production.

A material-related challenge is the relative displacement between the polymers and silicon that occurs during the production steps. This has to be taken into account throughout the whole process chain as it might have consequences on establishing interconnects and their reliability in the field.

Figure 14.3a shows a printed circuit board with the embedded 77-GHz SiGe chip, the sending antenna (Tx) and the receiving antennas (Rx). Insets zoom onto

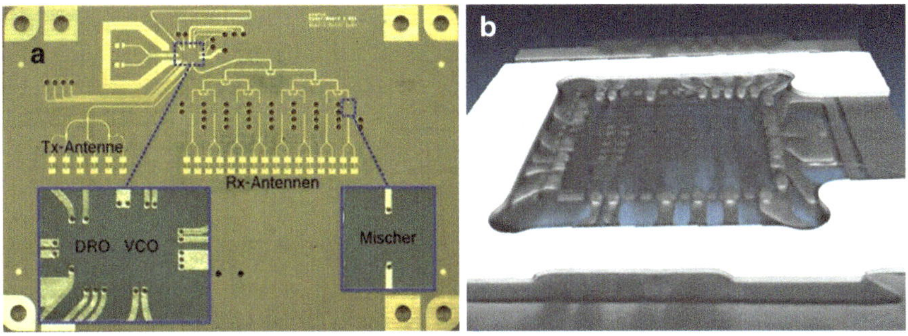

Fig. 14.3 (**a**) HF-side of the printed wiring board (*PWB*) showing the radar front ends with embedded SiGe chips (dielectric resonator-oscillator chip [DRO]), voltage regulated oscillator chip (*VCO*) and eight mixer chips (From UNI Bremen). (**b**) 3D X-ray scan of the same device (Bosch)

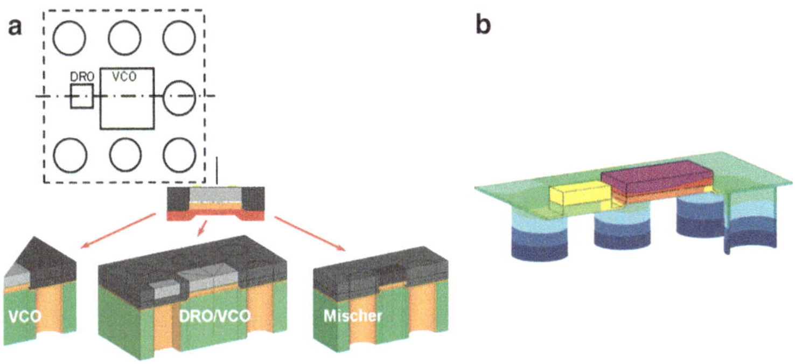

Fig. 14.4 (**a**) Design of thermal vias around VCO and DRO and the finite element model; (**b**) calculated heat distribution at operating conditions

the contact areas for the DRO, VCO and one of the mixer chips, where normally the packaged chips would have been mounted occupying a much larger PWB area. Figure 14.3b is a 3D X-ray scan of the same demonstrator showing the interior interconnection structure. The repeatability of this process flow was shown at IZM and Würth Elektronik under laboratory and real manufacturing conditions, respectively. Assembly of test vehicles for reliability tests helped to prove the process compatibility of foil pick-and-placement, transfer moulding and PCB embedding.

Heat dissipation plays a crucial role in the functionality of the high frequency driver assistance device. Substantial modelling effort was invested in the design of the device to ensure adequate heat dissipation and mechanical robustness. After an optimisation study using finite element analysis, the VCO chip generating 2.75 W average dissipated power was mounted on a copper plate with a highly conductive adhesive. This plate was connected to a heat sink by thermal vias. Seven vias around the dielectric resonator and voltage regulated oscillator chip and the VCO and one via per mixer chip provided sufficient heat removal (Fig. 14.4). The performance of the demonstrators shows good electrical functionality in high frequency applications, and the initial reliability results are promising for future applications. No electrical failure after 1,000 TC (–55/125°C) indicates reliable chip interconnections and μ-vias. Electrical functionality is also shown after 1,000-h storage at 85°C/85% RH. End-of-Life (EoL) tests are underway with a larger sample size to evaluate robustness.

Finite Element Analysis (FEA) was also employed in optimising design and process parameters for reliability. The production process of interconnects (electrical vias on the active side of the VCO) was examined step by step. Stresses at the end of each processing step with two subsequent temperature cycles after the last process step were analysed. The results showed that plastic deformation of copper takes place during lamination. Largest stresses show up as circumferential stresses within the copper layer after laser drilling and do not intensify in the subsequent processing steps.

14.4 Outlook: Flexible Systems with Embedded Actives

A reliable chip embedding process into laminates and advancing miniaturisation in the z-direction will not only increase packaging density but also offer physical flexibility. Carrying this over to other components such as power supply (thin film batteries, solar cells), sensors (foil-based capacitive sensors) and passives (integrated into foils) will pave the way to further innovations. The system will be flexible and have the potential of forming conformable electronic surfaces while physically disappearing into the background (ubiquitous computing). One should not disregard the challenges associated with interconnections of the involved components, which should withstand and not limit the flexibility of the structure. The interconnections have to be compatible with flexible components and flexible foils requiring innovative approaches in materials and processes. Manufacturing equipment should be able to handle thin and flexible components and substrates, process steps should be carefully designed taking interactions (e.g. between processing steps or assembly sequence of components) into account. Modelling will be instrumental in all aspects. Standardisation of components, interconnections and testing in joint development efforts will play an important role in defining the mainstream technologies and help their spread beyond niche markets. Entire value and supply chains have to be involved in such development efforts. Projects funded by the EC and BMBF (German Ministry for Education and Research) such as Interflex [11] and PRONTO are examples of such collaborations in this direction. Inclusion of organic electronics in these systems will aid cost reduction and contribute to the integration of this technology into everyday objects.

References

1. Tumala R, Swaminathan M (2008) Introduction to system-on-package. McGraw-Hill, New York
2. Reichl H et al. (2009) Heterogeneous integration: building the foundation for innovative products. In: Zhang QG, Roosmalen A (eds.) More than Moore: creating high value micro/nanoelectronics systems, 1st edn. Springer, Berlin
3. Barton J (2008) Embedded microelectronic subsystems. In: Delaney K (ed.) Ambient intelligence with microsystems: augmented materials and smart objects, 1st edn. Springer, New York
4. Chanchani R (2008) 3D integration technologies – an overview. In: Lu D, Wong CP (eds.) Materials for advanced packaging, 1st edn. Springer, New York
5. US Patent #6,400,573 B1, 2002
6. Ostmann A et al. (2002) Realization of a stackable package using chip in polymer technology. Polytronic Conference, Zalaegerszeg, Hungary, June 2002, pp 160–164
7. German Patent No. DE102008009220A1, 2008
8. German Patent No. DE102005032489B3, 2005
9. German Patent No. DE102007024189A1, 2007
10. German Patent No. DE102008009220A1, 2008
11. www.project-interflex.eu

Chapter 15
Handling of Thin Dies with Emphasis on Chip-to-Wafer Bonding

Fabian Schnegg, Hannes Kostner, Gernot Bock, and Stefan Engensteiner

Abstract Thin dies with all their advantages bear various new challenges for die handling during die attach or flip-chip processes. The changed properties, like increased brittleness as well as higher flexibility of the dies, have to be addressed accordingly with new concepts for effective detachment from the wafer tape, low stress handling during transfer and adhesion processes that need to be chosen more carefully for each application.

In this chapter various ejection principles and handling tool examples are described together with a few product applications that have gained in importance for mass production because of the development towards ultra-thin dies. Three-dimensional integration using through silicon vias and dies embedded in polymer are two examples of applications that became possible only by the introduction of thin dies.

15.1 Introduction to the Die Bonding Process

As part of the first level packaging the die bonding process has the task of picking up a die and placing it on its target position in the package. There are two main die bonding approaches. Either the die is placed directly onto a substrate with the active side up (direct die placement), or flipped by 180° and placed with its active side down (flip chip). In both cases there are similar but different die handling steps with different implications, advantages and challenges. In this section the basics of both processes are described, while Sect. 15.2 focuses on specific thin die challenges and dedicated solutions. The dies to be placed come either as diced wafers mounted on a wafer tape, which itself is held by a wafer frame, or in waffle packs, pocketed tapes or other input material. In the following sections only processes using wafer tapes are described as they are most challenging for thin die handling.

F. Schnegg (✉)
Datacon Technology GmbH, Innstr. 16, 6241 Radfeld, Austria
e-mail: fabian.schnegg@besi.com

J.N. Burghartz (ed.), *Ultra-thin Chip Technology and Applications*,
DOI 10.1007/978-1-4419-7276-7_15, © Springer Science+Business Media, LLC 2011

The technique of packaging thin and ultra-thin dies is especially relevant for electronics with increased performance and small form factors. Figure 15.1 shows an example of a stack of 20 ultra-thin memory dies with wire bonds.

Upcoming generations of 3D-integrated packages will benefit even more from the technique. In this respect, chip-to-wafer placement has a high potential as a competitive production technology. Correspondingly, the authors detect a fast-growing demand and interest in the chip-to-wafer die bonding processes.

15.1.1 Direct Die Placement

Figure 15.2 shows schematically all relevant process steps for direct die placement. The first step is the adhesive/solder paste application onto the target position of the die on the substrate (Fig. 15.2a). There are different techniques available like dispensing, stamping or screen printing. Figure 15.2b depicts the step of presenting and aligning the diced wafer in the pick and place machine in a way that the next die can be picked up accurately. Visual methods are used for exact alignment. The diced wafer is mounted on a sticky carrier tape where all dies are face up. Figure 15.2c shows the subsequent step of picking up the die using a vacuum gripper, often called a "pick&place" tool. The detachment of the die, a process also named "ejection," is assisted from below with an ejection tool that has needles to push up the die while the tape is held down by vacuum. This reduces the contact area between die and carrier tape, thus making proper detachment possible. In Fig. 15.2d the manipulator is shown moving the pick&place tool with the die on it over a vision system to detect misalignment, which may have occurred during the pickup step. This optional check is to improve the placement accuracy, but has to be utilized with caution because – with direct die placement – the visual alignment is carried out by the machine's visual system looking at the back of the

Fig. 15.1 Multi chip package with 20-30-μm-dies (Courtesy of Elpida Memory, Inc. [JP])

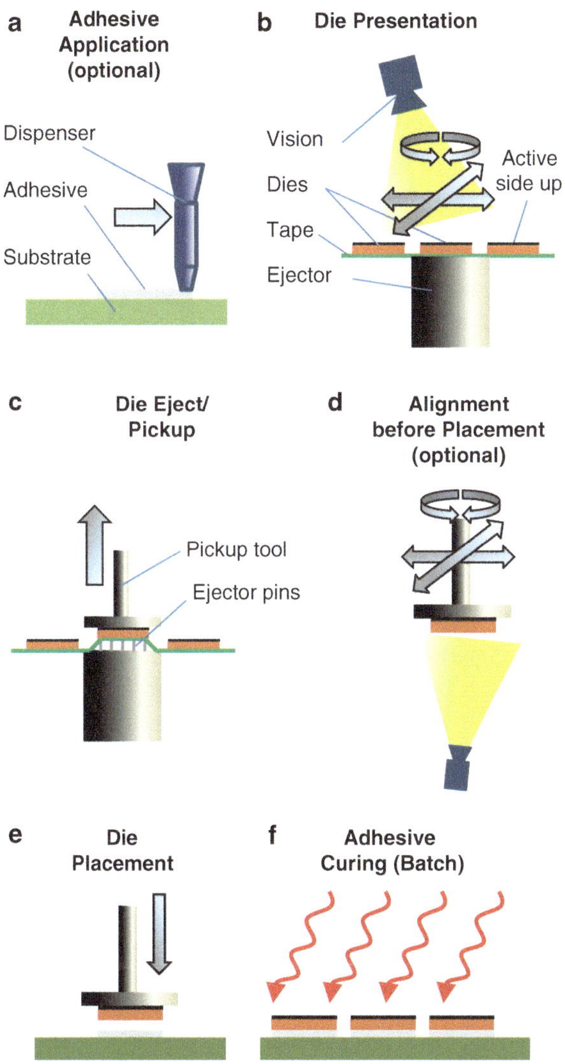

Fig. 15.2 Standard process steps for direct die placement. (**a**) Adhesive application (optional), (**b**) die presentation, (**c**) die eject pickup (**d**) alignment before placement (optional), (**e**) die placement, (**f**) adhesive curing (batch)

die. Potential offsets between front and back edge of the die as a consequence of the dicing process can introduce considerable alignment errors. Figure 15.2e, finally, depicts the die being placed onto the substrate and held in its position by the adhesive/solder paste. After placement, the adhesive/solder paste is cured/soldered in an external batch process, e.g., a reflow oven (Fig. 15.2f). The active side of the die is facing up to be electrically connected, e.g., with wire bonds.

15.1.2 Flip Chip Placement

Figure 15.3 shows a typical flip chip process scheme. In contrast to direct die placement the dies are usually bumped. Die presentation and alignment is shown in Fig. 15.3a. After pickup (Fig. 15.3b) the die is flipped by 180° with a flipping mechanism (Fig. 15.3c) and transferred to the placement tool (Fig. 15.3d). After transfer the bumps are dipped into a flux (Fig. 15.3e). In the following alignment step (Fig. 15.3f), image recognition can be carried out directly on the bumps or alignment marks on the active side. The achievable placement accuracy is therefore significantly higher than for direct die placement, where the alignment has to be done on the die edges and thus depends on the accuracy of the dicing process. After die placement (Fig. 15.3g), the die is temporarily fixed by the flux until the bumps are mechanically and electrically connected to the substrate pads by reflow soldering (Fig. 15.3h).

15.2 From Thick Die to Thin Die

While the thick die processes described in Sect. 15.1 are established in mass production, some steps become difficult when it comes to dies below 300 μm thickness. Thinner dies are more fragile and need to be handled more carefully. Below a certain thickness they are significantly flexible and tend to warp. Extremely thin dies are highly flexible and thus less fragile. Additionally, bumps on the dies or through-silicon vias (TSVs) act like local mechanical reinforcements, causing imbalanced stress when the dies are handled – making fractures more likely.

In principle, no standard dicing tape will work in combination with thin dies. Two alternative dicing tapes are available for thin die handling: ultraviolet (UV) curable tape ("tape 1") and thermo-release tape ("tape 2"). For both tape types the die adhesion can be reduced using UV light exposure for tape 1 or heat input for tape 2. Curing parameters like wavelength, UV light intensity, exposure time or temperature required are tape-dependent and precisely described in tape data sheets. Due to a low remaining tackiness of tape 1, UV curing can be applied to the whole wafer at once upstream of the pick&place process. Thermo release tape on the other hand loses all tackiness upon heat induction and therefore has to be processed locally during ejection. Due to this difference and the better price performance, nearly all thin die production uses UV curable dicing tapes.

Of course the chosen method and quality of preceding process steps like wafer dicing and wafer stress release have an effect on handling properties. Due to the focus on pick&place issues, these influences are not covered in this section. Therefore Table 15.1 gives an overview of known handling challenges with thin dies in comparison to thick dies.

The problems and challenges mentioned above are described in more detail in the following sections for each process step.

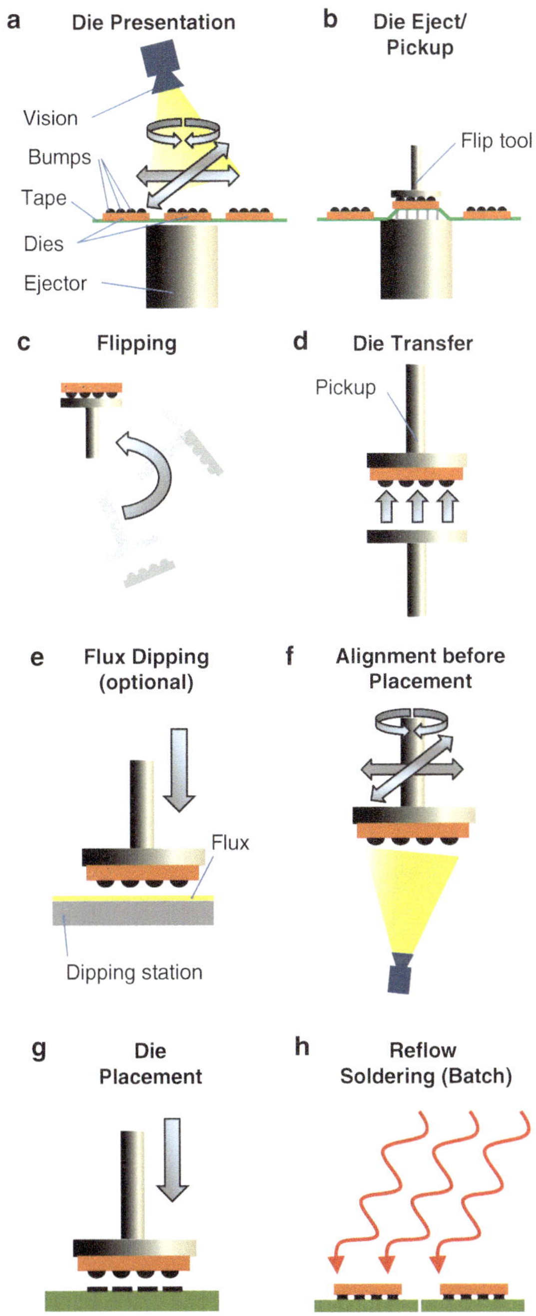

Fig. 15.3 Standard process steps for flip chip placement. (**a**) Die presentation, (**b**) die eject/pickup, (**c**) flipping, (**d**) die transfer, (**e**) flux dipping (optional), (**f**) alignment before placement, (**g**) die placement, (**h**) reflow soldering (batch)

Table 15.1 Overview of challenges and problems for the different process steps

Die handling step	Substep	Thick die process	Thin die process	Challenges/problems
Pickup	Wafer to ejector angular Alignment	Not critical	Very critical	Die cracking No pickup Impact on neighboring die
	Visual die alignment	Die is flat, illumination quality is good	Die can be warped, illumination is critical	Dies peel off prematurely Recognition quality Eject tool topography design
	Die ejection	Ejection done with needle(s)	New ejection concepts are necessary	Die cracking No pickup Slow ejection
Transfer	Handover to pickup or flip tool	Tool selection is not critical	Tool selection is critical, whole die area to be supported	Die bending Die cracking
	Visual die alignment	Die is flat, illumination quality is good	Die can be warped, illumination is critical	Warped die on tool Recognition quality Accuracy impact
Placement	Die release on substrate	Dispensed conductive/non-conductive pastes can be used	Other application techniques must be used, e.g., printing or pre-applied adhesive film (DAF)	Die cracking Die warpage Die surface contamination Tool contamination

15.2.1 Die Pickup from Wafer Tape

15.2.1.1 Wafer to Ejector Angular Alignment

In this process step the die is positioned over the ejection system. In thick die applications, conventional needles are used for ejection. As thick dies have a very high stability, few needles located far from the die edges are sufficient for a safe ejection. Thin dies need to be supported throughout the whole die area, especially at the die edges from where the peeling process naturally starts. For this reason, the die-to-ejector alignment becomes more crucial, since even small misalignments of die-to-ejector lead to a situation where the supporting area of the ejection system deviates from the optimal position. In this case, parts of the die to be ejected are not sufficiently supported and parts of the ejection system can even have an impact on neighboring dies, leading to die cracks in both.

To prevent this effect, different measures can be taken such as more accurate wafer mounting or a wafer angle correction mechanism in the handling equipment. Also, the dimensions of the ejector have to be in a well-defined proportion to the die size.

15.2.1.2 Visual Die Alignment

The optical inspection of thin dies prior to pickup has its own problems. Mechanical stress introduced into the die by upstream processes causes internal forces, which may curl up the die edges. When it is mounted on a tape, the adhesion is mostly strong enough to hold the dies down, but sometimes the die edges warp up just enough to make a visual alignment using bright field illumination impossible. Special lighting solutions like LED domes become necessary since they increase the dihedral angle of illumination and therefore yield a homogeneous picture brightness.

In conventional ejection processes the wafer tape is sucked down by vacuum grooves or holes. This tool topography affects the flatness of the die surface, which again has an influence on picture quality or can lead to die cracks. New approaches like sintered tool caps, new tool cap designs or different sequences for visual alignment help this issue to be overcome.

15.2.1.3 Die Ejection

The ejection concept of a few needles (or only one) pushing towards the die is no longer suitable for thin die handling. The brittleness and flexibility of thin dies demand a larger supporting area during the ejection process. At the same time the ejection concept has to enable the peeling of the tape from the underside of the die. The goal is to reduce the contact area of die and tape until the remaining adhesion is lower than the vacuum force of the pickup tool. Only then is a successful pickup possible.

To prevent curling during the release from the tape, the pickup tool needs to hold the die from the beginning of the ejection process. Another reason for the support of the die by the pickup tool is that a flexible die would assume the shape of the tape so that either the peeling cannot be initialized or the die is bent until it cracks. The peeling always starts from the die edges. It can start from all edges at the same time travelling towards the center or from one edge across the die to the opposite edge.

Table 15.2 shows a selection of different ejection tool concepts suitable for thin die handling. In general, a process window needs to be found in which the stress on the die is minimized while the pickup speed remains as high as possible. The multi-pin tool takes the conventional needle system a step further and offers a large configurable pin matrix. With this tool the use of sharp needles is not recommended because the punctual stress to the die easily causes die damage. To achieve a better force distribution larger needle tip diameters are used. The balance between die support in order to prevent die cracks and enough space between the needles for peeling has to be found for each individual die and tape.

The multistage tool works on the principle of reducing the contact area between die and tape in several discrete steps. This tool starts the peeling process by raising a stamp having nearly the size of the die. This stamp consists of several concentric rectangular frames that are then lifted in a telescopic manner. In the final extension the center piece of the stamp sticks out the highest, leaving a minimal contact area between die and tape. To further minimize the contact area within the stamp, other concepts can be used instead of a flat surface, e.g., a matrix of small pyramid shapes. The die remains only on the pyramid peaks, while the tape can additionally peel from the die into the pyramid valleys.

If the discrete steps between the stages still introduce too much stress into the die, the use of a membrane tool may become suitable, where the peeling takes place in a continous manner. A membrane is filled with pressurised air while the pickup tool slowly travels upwards with the die. As the membrane tends to form a hemispherical shape the tape starts to peel from the edges of the die according to the space which the pickup tool offers until, at the full extent of the membrane, only a punctual contact remains between die and tape. In this approach the peeling is controlled by the movement of the pickup tool.

Another good option is the peeling principle offered by a slider tool: Here a slider of the same width as the die is pulled away under the die, opening a vacuum chamber into which the tape is sucked. This way the tape is peeled off from one edge to the other. One big advantage of this concept is that the peeling speed and thus the amount of stress on the die can be controlled by the speed of the slider.

A completely different approach is the use of a heated tool with a structured surface (e.g., pyramid peaks). The adhesion of most wafer tapes increases upon heating, but at the same time the tape becomes softer and more elastic, which has the effect that it can assume the shape of the tool surface more easily. As long as this positive effect dominates over the negative effect of higher adhesion, dies can be picked with little stress and high speed as the heating can be started during the placement of the preceding die.

Table 15.2 Tool concepts with different ejection principles

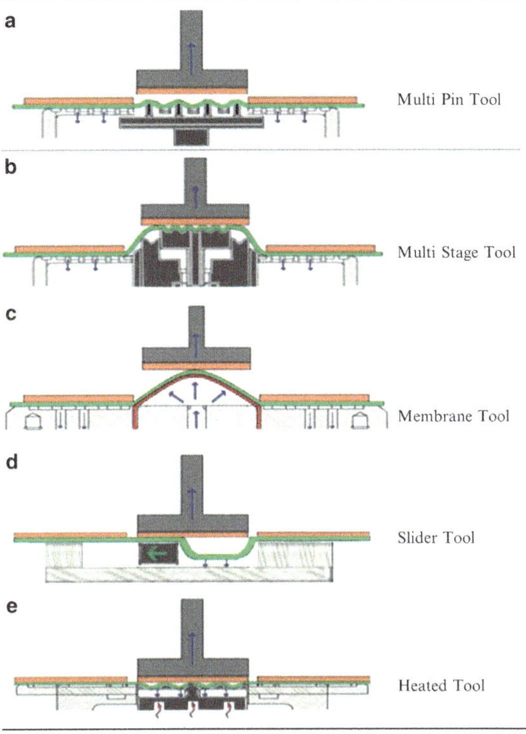

(a) Multi pin tool
(b) Multi stage tool
(c) Membrane tool
(d) Slider tool
(e) Heated tool

15.2.2 *Die Transfer*

15.2.2.1 Pickup Tools

All the ejection processes described above need to have the die supported with a pickup tool from above. This tool has to support the whole die area, offering as much vacuum as possible to hold the die flat while the tape is peeled off. The use of a die collet would offer very good vacuum but only support the die edges, resulting in die warpage. On the other hand, a flat tool having only one vacuum hole in the center would provide very good support to the die surface but would not hold the outer regions of the die flat, which is necessary due to the low stiffness of thin dies. Soft tools are an option, but depending on the chosen peeling process they can also allow too much bending of the die, which may lead to die cracks. Some examples of suitable pickup tools are shown in Fig. 15.4.

As mentioned before, die warpage can result in die cracking during the touch-down of the tool onto the die. For this reason it is very important to hold the die flat on all handling tools (flip and pick&place tool). The tool surface has to provide enough support for the die, e.g., to prevent damages during placement, and the vacuum needs to hold the die flat against the tool.

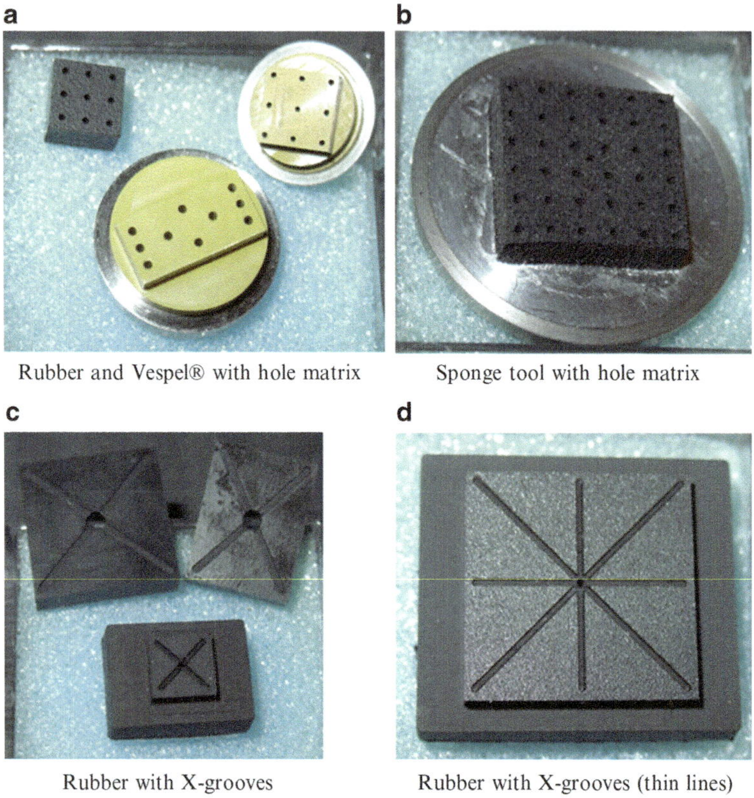

Fig. 15.4 Pickup tools for thin die handling. (**a**) Rubber and Vespel® with hole matrix, (**b**) Sponge tool with hole matrix, (**c**) Rubber with X-grooves, (**d**) Rubber with X-grooves (thin lines)

For flip chip processes, an additional challenge needs to be addressed: The flip tool handles the die from the active side, which is usually bumped. To achieve enough vacuum force, very soft tools like a sponge tool may be used where the bumps can be pressed into the tool surface so that the vacuum leakage is minimized.

15.2.2.2 Visual Die Alignment

Another visual inspection of the die on the tool may be necessary in a subsequent step. In principle, the same illumination issues caused by die warpage as described in Sect. 15.2.1 (die-on-wafer alignment) have to be addressed. As in this case the die is on the tool, it can easily be moved close to a vision system where both the distances to the camera optics and the lighting are much smaller than during the alignment on the wafer. Therefore the illumination angle is much wider and more light is captured by the optics, so that the die warpage has less impact on the recognition quality.

15.2.3 Die Placement

Besides die pickup and transfer, the placement of thin dies also involves difficulties. For direct die placement, mainly liquid adhesives are used to fix the die to the substrate. Process requirements are the bond line thickness (BLT, the final thickness of the adhesive layer below the die), the fillet height (vertical wetting height of the adhesive) and prevention of entrapped air between die and adhesive. The use of liquid adhesives becomes more critical when there is a changeover to thin die applications. The most common problems are:

- Wetting and contamination of the bond tool by the adhesive due to the lower thickness of the dies.
- The flexibility of thin dies may result in warpage and voids (trapped air) after placement of the dies.
- Liquid adhesives without instant curing possibility may result in accuracy problems due to "swimming" dies.
- Using an adhesive with UV snap cure possibility helps to achieve a smaller BLT and less warping, but increases the problem of cleaning a wetted bond tool.

Some of these problems are shown in Fig. 15.5.

An alternative to liquid adhesives is die attach film (DAF). DAF is a dual-layer dicing tape. The top layer is diced together with the wafer and sticks to the dies when being picked up. When the die with DAF is placed on a heated substrate, the DAF either cures right away or at least provides sufficient tackiness to hold the die in place until curing in a batch oven takes place. Using DAF addresses most of the problems described above: Die warpage, displacement on top of the liquid layer and tool contamination are not an issue any more. Only the pickup tool has to be selected carefully to control the formation of voids. Small voids that are formed during placement can still be removed during a curing step in a batch oven.

Figures 15.6 and 15.7 show a comparison of warpage measurement results created during the "Hiding Dies" project, supported by the European Commission under project number IST 507759 within the 6th Framework Programme. Two dies were placed on liquid adhesive (Fig. 15.6) and two with DAF (Fig. 15.7).

Fig. 15.5 Thin die problems with liquid adhesives (from *left* to *right*): Warped die, lack of fillet height control, surface contamination

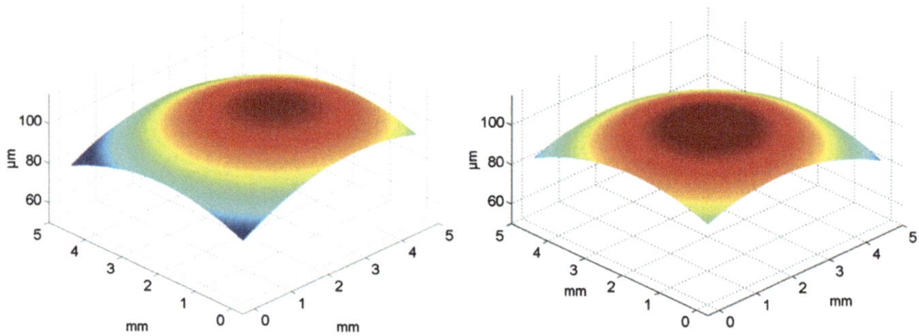

Fig. 15.6 Warpage measurement results: pre-applied liquid adhesive

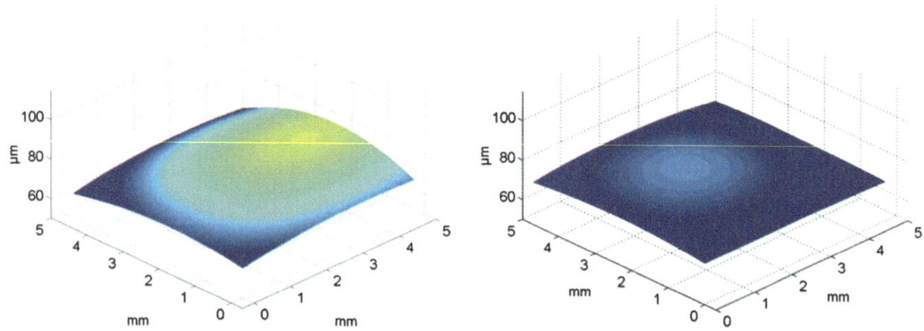

Fig. 15.7 Warpage measurement results: die attach film (*DAF*)

The major problems with liquid adhesives on the one hand and the good performance of DAF on the other have led to the conclusion that, for direct attach of thin dies, DAF has proved to be the best option as long as no thermal or electrical conductivity is needed on the die backside.

15.3 Application Examples

The modern market of mobile and communication applications demands more functionality within ever smaller presentations. This means that microelectronic packages have to offer more complex components with higher integration at chip level, higher frequencies, more functionality and increased performance in even smaller and cheaper packages.

The following two application examples show different ways of dealing with these challenges by exploiting the advantages of thin dies in addition to 3D integration:

1. Die embedding helps to optimize the form factor by laminating the bare die directly into the printed circuit board (PCB) substrate. Figure 15.8 shows a

cross-section of such an embedded die. Individual packaging of each die is not necessary, which lowers cost and gives the opportunity to shrink the PCB by using the third dimension for component placement.

2. 3D integration using chip-to-wafer stacking combines the advantages of low form factor and the possibility of higher frequencies as a consequence of very short signal paths. The 3D assembly shown in Fig. 15.9 was made in a chip-to-wafer process using TSV technology for interdie connection.

15.3.1 Embedded Thin Dies

Chip scale packages or flip-chips can help to reduce the lateral space requirement of active components on a circuit board to a minimum. For more space optimisation, 3D integration of components becomes an option. The availability of ultra-thin dies makes it possible to embed these bare dies directly into a resin-coated copper (RCC) layer of the PCB.

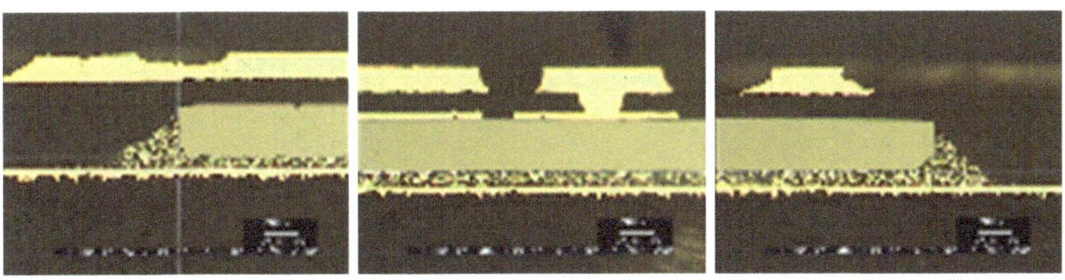

Fig. 15.8 Embedded die in a resin-coated copper (*RCC*) layer with thermally conductive adhesive

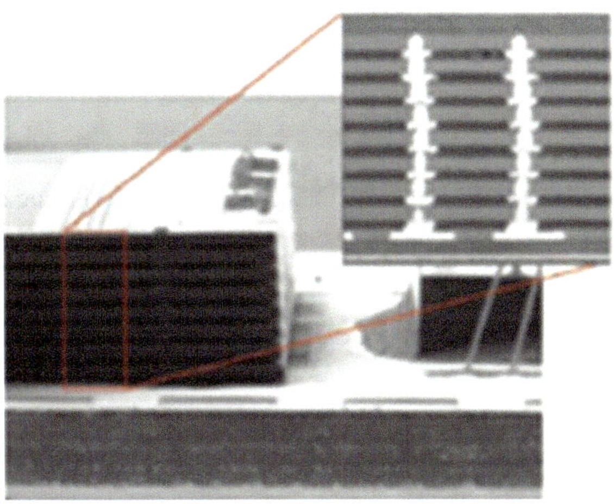

Fig. 15.9 Multi-layer stack with TSVs (Courtesy of Samsung)

The development described below has been achieved in the "Hiding Dies" project (FP6, IST 507759). Instead of flip-chip or wire bonds another basic interconnect structure has been established, which is shown in Fig. 15.10.

Figure 15.11 shows the main process steps of die embedding. Dies of 50-μm thickness are (a) bonded to the core of a PCB to be (b) embedded by vacuum lamination of RCC afterwards. This is followed by (c) laser drilling micro vias down to the bond pads and (d) subsequent Cu metallization of the vias to connect the die I/Os with the PCB.

In the next section only the die-attach steps for embedding are described. To gather information on the whole embedding process, the authors refer to [1].

As substrates, double-layered PCB cores are recommended. This particular project covered reliability tests for die sizes from $1 \times 1\ \text{mm}^2$ up to $10 \times 10\ \text{mm}^2$. The dies should be presented on UV dicing tape (e.g., Lintec D-175). For ejection,

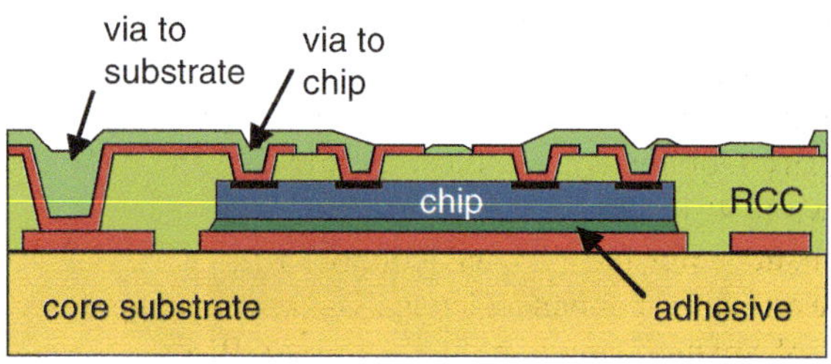

Fig. 15.10 Interconnect principle of an embedded die in a PCB build-up layer [1]

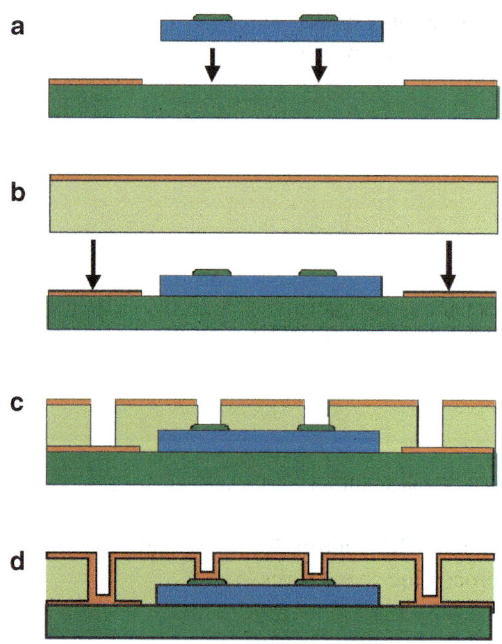

Fig. 15.11 Main process steps of die embedding [1]: (**a**) Die placement, (**b**) vacuum lamination, (**c**) laser drilling of micro vias, (**d**) via metallization

one of the thin die tools described above can be used. As the die needs to be embedded into the dielectric layer, the overall assembly height, i.e., die thickness plus bond line thickness, has to be less than the resin thickness. Three possible adhesion processes were developed to meet this requirement: Screen printing, the use of b-stage paste or the use of DAF.

The first of these processes was screen printing. Where dispensing reaches its limits of volume control, screen printing achieves not only a precise control of volume but also of location of the adhesive. When bonding large dies (>5 mm) on screen printed adhesive, the formation of voids from entrapped air became an issue, which was caused by the remaining screen topography on the paste's surface. Using stencil printing instead of screen printing produced a smooth paste surface and led to a good void control.

The second was the use of b-stage paste, which brought some additional advantages: After printing and drying the paste, storage or transportation of the boards was not an issue. Also the bonding results showed fewer voids compared to standard adhesives.

While paste (thermally conductive adhesive) is mostly used when heat dissipation is an issue, the third process – baseline embedding for applications with low thermal leakage power – was set up with DAF. Figure 15.12 shows a cross section of an embedded die which was bonded with thermally conductive adhesive.

15.3.2 Chip-to-Wafer

As mentioned above, the optimum of form factor together with short signal paths can be reached with 3D chip-to-wafer assemblies. To get an idea of the form factor possibilities opened up by thin die technology, Fig. 15.13 depicts a base wafer (280 μm thick) with 50-μm thin dies bonded to it and two thick dies (~725 μm) placed on the wafer for reference.

The recommended chip-to-wafer process consists of two steps, the first of which is the placement and temporary fixation of the dies onto the base wafer; the other is the final bonding in a batch oven with or without pressure applied to the dies [3].

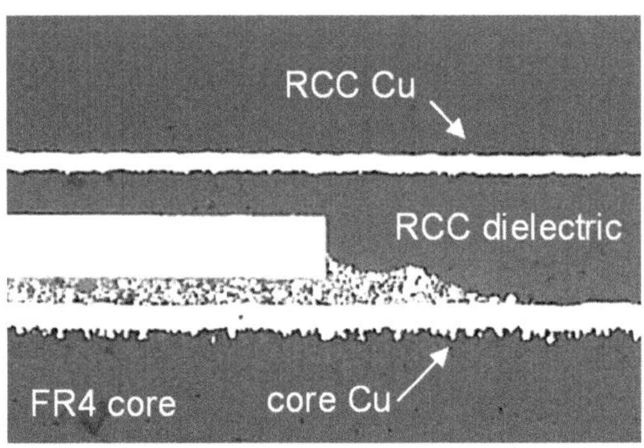

Fig. 15.12 Embedded chip in a RCC layer with thermally conductive adhesive [1]

Fig. 15.13 Chip-to-wafer assembly with thick dies for Reference [2]

Table 15.3 Four different packaging possibilities for chip-to-wafer

-	Single Stack	Multi Stack
Base wafer w/o TSVs		
Base wafer with TSVs		

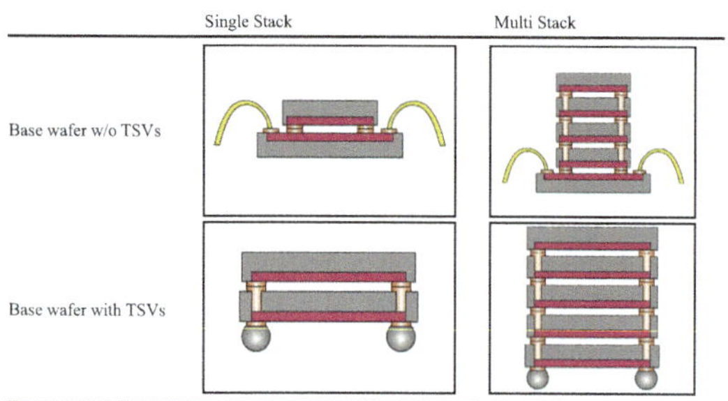

The main advantage of the separation consists of a gain in process speed as the soldering time does not add to the placement time of every single chip. In addition and in contrast to wafer-to-wafer applications, a higher yield is achieved by placing only known good dies in known good positions. There are different approaches and technologies used in chip-to-wafer applications. The most common ones are shown in Table 15.3.

For single stacks, conventional flip chip processes can be employed, but thin dies show their big advantages when it comes to multi stacks. TSVs become practicable where electrical connections are made from the front to the back of a die through its silicon body. This way, the use of wire bonds between the individual dies is no longer necessary. Stacks of very small outer dimensions – especially height – can be manufactured.

If the base wafer is also thinned, TSV technology becomes an option even here, making the wire bonds completely unnecessary. Thin wafers need to be mounted on carrier wafers for handling purposes during chip-to-wafer placement. Temporary wafer mounting and demounting technology is available using thermoplastic adhesive films on a thick silicon carrier wafer.

The electromechanical interface between two dies can be achieved by soldering with solder bumps and sticky flux as temporary adhesive or a solid–liquid interdiffusion (SOLID) process with either a temporary adhesive, which is removed during final bonding, or no-flow underfill (NFU) or ultrasonic tacking.

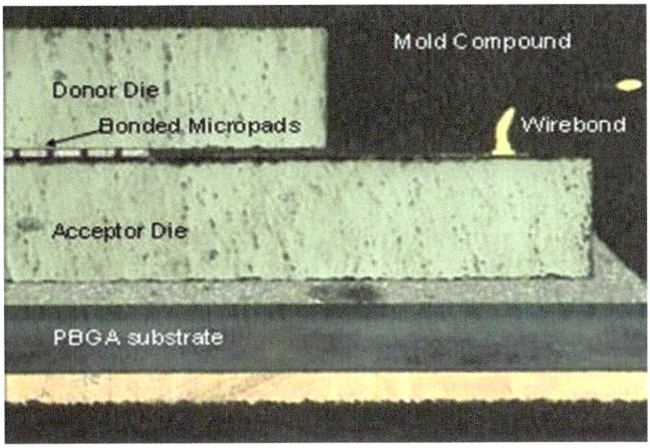

Fig. 15.14 Cross-section of a packaged single stack assembly [2]

As NFU is a very useful agent for temporary adhesion and soldering, it would be the material of choice for many of these applications. One disadvantage though for thin die handling is that most NFUs are dispensable materials, a fact that includes all the problems described above with thin dies in combination with liquid adhesives. A work-around for single stacks is to effect the placement with thick dies, avoiding the problems with liquid adhesives, and to do the thinning after final bonding. Successful trials with thick dies bonded to the base wafer using the SOLID process, which after final bonding were thinned to 50 μm, showed this approach to be feasible. This process was demonstrated in [2].

Figure 15.14 finally shows a cross-section single stack, which has been placed on a plastic ball grid array substrate, connected with wire bonds and molded.

References

1. Ostmann A, Kriechbaum A, De Baets J, Kostner H (2005) Chip embedding in printed circuit board production. *IMAPS Nordic Conference 2005*, Tønsberg, Norway
2. Pozder S, Jain A, Chatterjee R, Huang Z, Jones RE, Acosta E, Marlin B, Hillmann G, Sobczak M, Kreindl G, Kanagavel S, Kostner H, Pargfrieder, S (2008) 3D die-to-wafer Cu/Sn microconnects formed simultaneously with an adhesive dielectric bond using thermal compression bonding. *IITC 2008*, Colombo, Srilanka
3. Scheiring C, Kostner H, Lindner P, Pargfrieder S (2004) Flip chip-to-wafer stacking: Enabling technology for volume production of 3D system integration on wafer level. *IMAPS Conference 2004*, Long Beach, California, US

Chapter 16
Micro Bump Assembly

Paresh Limaye and Wenqi Zhang

Abstract One of the key technologies for enabling 3D integration of semiconductor devices is the micro bump based interconnect joining that allows silicon dies with through-silicon vias (TSVs) to be stacked on each other. In order to achieve a high level of integration, the pitch and dimensions of these micro bumps also need to be reduced. Significant assembly challenges need to be overcome in order to accomplish successful yielding micro bump assemblies. In this chapter, the assembly aspects of micro bump fabrication, joining and underfill integration are discussed. The metallurgical fundamentals of micro bump formation are discussed in order to give insight into the choice of interconnect bonding method, the under bump metallisation and the micro bump as such and the impact of these on the stacking approach. Finally, various stacking schemes for 3D integration are considered and their relative advantages and drawbacks are discussed.

16.1 Introduction

Scaling of micro bumps to enable higher interconnect density requires a reduction in micro bump pitch and standoff gap. In order to understand the challenges connected with this dimensional downscaling of the bumps we must understand the structure of micro bump connections. Figure 16.1 shows the cross-section of a flip chip bump connecting a die to die/substrate. The structure of the bump consists of an under bump metallisation (UBM) on the die, the solder bump and the UBM on the landing substrate.

This UBM comprises metal layers such as Cu, Ni, Au etc. and the metal stack is organised in such a way as to allow on the one hand a good adhesion between the final metal finish on the die I/O pad and the UBM and on the other hand a good solderable surface on the other end of the UBM (Fig. 16.1). The soldering reaction

P. Limaye (✉)
Interuniversity Microelectronics Centre (IMEC), Kapeldreef 75, 3001 Leuven, Belgium
e-mail: limaye@imec.be

J.N. Burghartz (ed.), *Ultra-thin Chip Technology and Applications*,
DOI 10.1007/978-1-4419-7276-7_16, © Springer Science+Business Media, LLC 2011

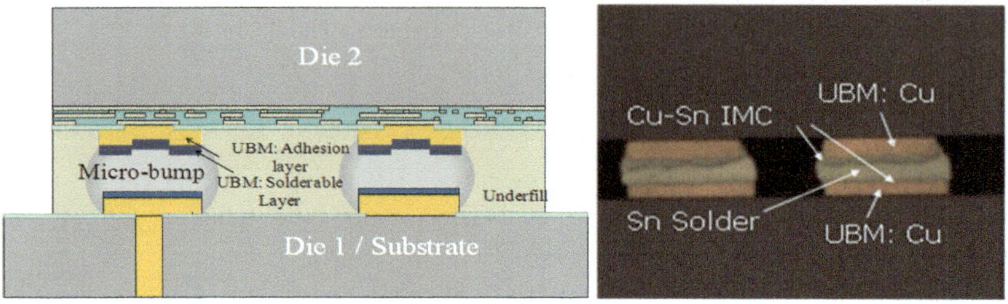

Fig. 16.1 Typical cross-section of a micro bump assembly

is a result of intermetallic formation between the solderable metal of the UBM and the solder material itself; this reaction must take place if a solder connection is to be formed and during this reaction, part of the UBM is 'consumed', i.e., converted to an intermetallic state. The amount of UBM that is consumed is determined by the reaction kinetics and is dependent on the combination of the solder-UBM temperature to which the micro bump is exposed. The growth of the intermetallic (and the corresponding UBM consumption) is a dynamic process and continues to occur over the lifetime of the micro bump. When all of the UBM material is consumed, the micro bump will no longer provide a physical connection between the die and the substrate and will result in an electrically open connection. Thus, the key to forming any solder-based interconnection lies in finding a balance between sufficient intermetallic formation to provide a good joint and having appropriate UBM stack to prevent UBM consumption.

When the dimensions and the pitches of the micro bumps are being scaled down, it is important to remember that this scaling down will predominantly occur in the x, y-dimensions of the bumps and bump pitches. Corresponding to the reduction of the pitch, the amount of solder that can be available needs to be reduced in order to prevent shorts between adjacent bumps during the assembly process. However, the height of the micro bump has an absolute minimum, which is determined by the solid state diffusion kinetics of the solder/UBM combination and the expected thermal and electrical load the micro bump is expected to see in its lifetime.

16.1.1 Micro Bump Formation

When one considers the formation of micro bumps (below 100 μm pitch), the more traditional methods of flip chip-type solder bump formation such as screen printing may not allow a sufficient dimensional accuracy for the bump formation. Other approaches such as evaporation and lift off are relatively slow and may, potentially, have yield issues. Hence, for these reasons electroplating with a photo-patternable resist is the most preferred approach for micro bump formation. The electroplated micro bump formation starts with the deposition of a plating seed layer. Typically, a plasma vapour deposition (PVD) process is used to deposit an adhesion and seed layer. Typical examples of adhesion layer materials are Ti or TiW, while the common seed layer material is Cu. Next a photo-patternable resist is deposited on the wafer with either spin-on polymer resists or laminated dry film resists. The next step involves photo-patterning of this resist and resist development to expose the locations on the wafer where micro bumps are to be formed. Following this, the micro bump is deposited by means of electrochemical deposition. Typical combinations of material to form the under bump metallisation include Cu, Ni and Au. Following this, the solder, typically Sn or SnAg, is electroplated on the UBM. Next, the resist is 'stripped' or removed, typically with wet solvents. The final step involves a wet or dry etch of the seed layer and the adhesion layer to remove all the excess metal (Fig. 16.2a). Depending on the solder volume and the type of micro bump connection, the solder may or may not be reflowed. The counterpart die/substrate may receive a similar bumping process if a 'symmetric bump' is to be formed, i.e., bumps with solders on both sides being joined. In the case where indium is used, a resist lift off process may be employed. In this process, the resist is photo-patterned, followed by evaporation of indium. Following this, the resist is removed along with the indium layer that is not deposited inside the pattern (Fig. 16.2b)

16.1.2 *Metallurgical Fundamentals for Micro Bump Assembly*

The common (UBM) materials are Cu, Ni, Au and Ag, while In, Sn and SnAgCu (SAC) often serve as the Pb-free solder materials. A small amount of Ag and Cu additives in Sn can lower its melting point and surface tension. Both Sn and In can react with the aforementioned UBM materials to form binary intermetallic compounds (IMCs). For example, the Ni-Sn binary phase diagram shows three intermetallic phases of Ni_3Sn, Ni_3Sn_2 and Ni_3Sn_4 at room temperature. However, more

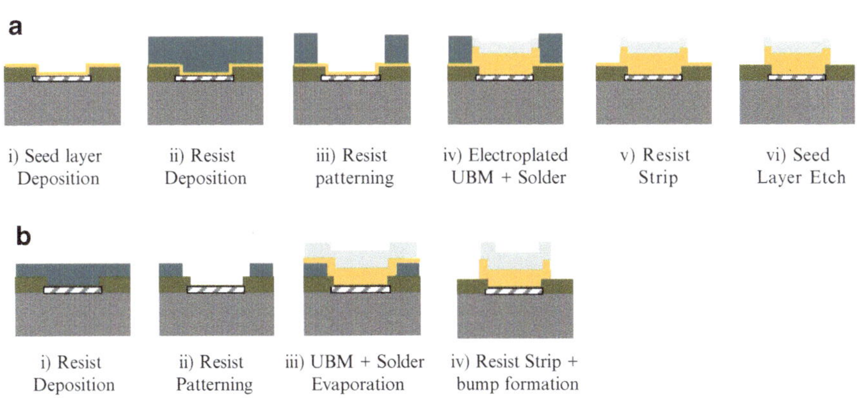

Fig. 16.2 (**a**) Electroplated micro bump process and (**b**) evaporation and lift off micro bump formation process

complex ternary IMCs such as $(CuNi)_6Sn_5$ and $(CuNi)_3Sn$ can be formed when SAC solder reacts with Ni.

The formation of a continuous IMC layer indeed indicates a good metallurgical bonding. In fact, control of the growth of IMCs is critical to good bonding. Such control requires a good understanding on solder/UBM materials' reactions at different conditions. The interfacial reactions between Sn-based solder and common UBM materials have been extensively investigated. The main IMCs formed in the Cu/Sn, Ni/Sn Au/Sn and Ag/Sn systems are summarised in Table 16.1.

Several published sources exist that give the growth kinetics of Cu-Sn intermetallic systems. Diffusivity D is given by the relation:

$$D = D_0 e^{-\frac{E}{kT}}$$

where D_0 is the diffusion coefficient, E is activation energy, k is the Boltzmann constant and T is the thermodynamic temperature in degrees Kelvin. For D_0 and E, there are large variations in literature as listed in Table 16.2. Unlike the various Sn/UBM systems, the reaction between In and the common UBM materials generates complex IMCs. In term of the kinetics of In/UBM reaction, the activation energy is generally smaller than that of the Sn/UBM system. As expected, there is also a large variation in the literature of the description of diffusivity D.

The growth kinetics of solder/UBM materials' reactions helps one optimise the solder/UBM stack in micro bumps in order to meet some critical bonding requirements. The diffusivity at different temperatures can be calculated from the aforementioned equation, and one can approximately estimate the IMCs thickness for a certain bonding or reflow profile. Based on this information, one can design the solder and UBM stack. Indeed for fine pitch micro bump, the solder volume is much smaller than the conventional solder ball. Moreover, a certain amount of solder will also be transformed into IMCs during room temperature storage (like Au/In and Cu/Sn systems) and temperature ramp-up of the bonding process. Therefore, the multilayer stack in micro bump needs to be optimised in order to keep sufficient solder in the stack before reaching the bonding or reflow peak temperature (Fig. 16.3).

Table 16.1 Main IMCs identified in Cu/Sn, Ni/Sn Au/Sn and Ag/Sn systems

UBM	Cu/Sn	Ni/Sn	Au/Sn	Ag/Sn
IMCs	Cu_6Sn_5, Cu_3Sn	$NiSn_3$, Ni_3Sn_4	$AuSn_4$, AuSn	Ag_3Sn

Table 16.2 Diffusion parameters for the Cu/Sn system

Reaction couples				
Solder	UBM	E (eV)	D_0 (cm^2/s)	Remark
Sn	Electroplated Cu	0.086	3.4	Solder reflow process [5]
Sn 3.5%Ag	Electroplated Cu	0.142	37.3	Solder reflow process [5]
Sn	Cu	0.65	5.92×10^{-5}	For Cu_3Sn [6]
Sn	Cu	0.73	1.1×10^{-3}	For Cu_6Sn_5 [6]

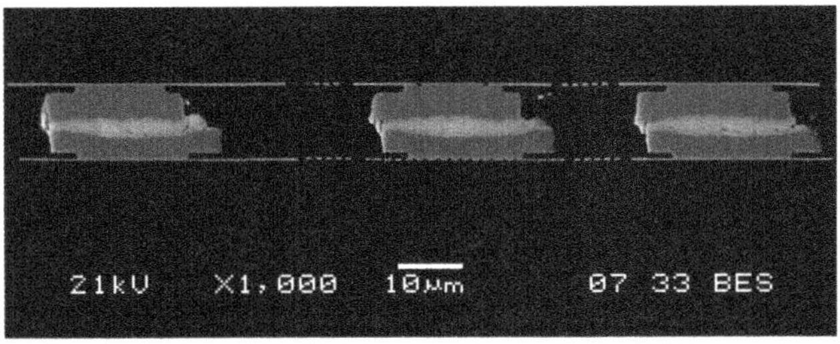

Fig. 16.3 40 μm pitch Cu/Sn microbump assembly

16.2 Micro Bump Joining Methods

16.2.1 Reflow Solder Joining

Reflow soldering is a classical method used to attach solder flip chip bumps and may be employed for joining micro bumps as well, provided sufficient solder volume is available. The die with micro bumps is first dipped into a thin layer of flux and then aligned and placed on the landing substrate or die (depending on the stack being constructed). Next this assembly is sent through a reflow oven to melt the solder, which results the intermetallic reaction with the landing pad. At the end of the reflow, the solder solidifies to provide a solid electrical and mechanical connection. For reflow soldered assemblies, capillary underfill is the main approach for underfilling the die to provide extra mechanical rigidity and to improve the robustness of the assembly. This approach is the same as used in classical flip chip assemblies. However with reduced bump dimensions, the gap between the die and substrate/landing die also shrinks, giving rise to two main issues. The first issue involves getting good control on the amount of flux deposited. As the bump dimension (hence height) decreases, the thickness of the flux also needs to reduce. This process is hard to control and can result in excessive flux entrapment in the gap. This may result in extra underfill voiding during the underfill step. Furthermore, the reduction in gap height also requires the filler particle size in the underfill to dramatically reduce. This can result in underfills that have adequate filler particle content for CTE matching but sluggish flow or underfills that fill fast but have lower filler content and hence potentially lower reliability.

16.2.2 Thermo-compression Joining

An alternative approach to assembling micro bump dies is the use of thermo-compression bonding. In this approach, flux may still be used for to provide in situ cleaning. As with the reflow process, the die is dipped into a thin layer of flux, aligned and placed on the landing substrate. Unlike the reflow soldering though, the next step involves an in situ application of pressure and reflow of the solder resulting in a micro bump formation.

Two types UBM metal to solder bonding processes can be used for 3D integration. One is transient liquid phase (TLP) bonding, in which all solder is transformed into IMCs that have higher melting points than that of solder itself, e.g., 232°C, 217°C and 156°C for Sn, SAC and In, respectively [1–4]. During this process, the solder is initially molten and starts forming intermetallics. By controlling the stoichiometry available for soldering, i.e., relative volumes of solder and UBM, the process can be designed so that at the end of the process, the solder is fully converted to high melting point intermetallics. This enables repeated stacking of additional layers without remelting of the joints at lower levels of the stack.

Another approach is solid state diffusion bonding processes, where the bonding temperature can be well below the melting point of solder [4]. During this bonding the intermetallic compounds are formed by the solid state interdiffusion instead of the liquid–solid reaction in the TLP bonding. Also, unlike TLP bonding, solid state bonding typically requires a high bonding pressure in order to make intimate contact between the solder and UBM. Since the solid state bonding temperature is below the melting point of solder, it does not remelt the solder joints at lower levels of the stack either. This also facilitates multilevel 3D interconnects.

Yet another approach for solid state diffusion bonding with thermo-compression process can be achieved without use of any solder. In this approach, Cu bumps are formed on both the dies and substrates being stacked, and a thermo-compression bonding at relatively high temperature (300–350°C) facilitates a Cu–Cu bond formation. An extension of this approach, in the case where Cu TSVs are used, is for one to directly bond the TSV to a corresponding landing pad [7].

16.2.3 Underfill Integration

Integration of the underfill is an important aspect of the micro bump assembly process. As in the case of classical flip chip assemblies, a combination of flux for soldering and post solder capillary underfill dispense can be used. Alternatively, a class of underfill materials referred to as nonconductive paste (NCP) or no-flow underfill (NUF) may be used. These materials typically contain some flux and provide the function of cleaning and underfilling in the same step. The material is first dispensed on the landing substrate and then the thermo-compression bonding is carried out. The main difference between NCP and NUF is that the NCP materials typically contain filler particles and are more viscous. In the current set of materials available, materials classified as NCP or NUF can be found both with and without filler particles. Another class of materials that is also of interest is wafer-applied underfills or wafer level underfills (WUF=WLUF). These materials have similar characteristics to NCP/NUF materials except they are deposited at a wafer level by liquid material spin-on or with dry film lamination. The b-staged materials thus deposited have a much tighter volume distribution and hence an overall improved manufacturability.

16.3 Micro Bump Stacking Approaches

The main categories of 3D stacking approaches with micro bumps can be categorised as

- Die-to-package stacking
- Die-to-wafer stacking
- Wafer-to-wafer stacking

A given set of stacked dies may utilise more than one approach to deliver the final stacked dies in the package. In the die-to-packaging stacking approach, the bottommost die is first bonded onto the package substrate. Following this, the next layer of die is stacked and all the subsequent dies are stacked sequentially in this manner. One approach involves use of solder reflow bumps. In this the different dies to be stacked are placed on top of each other with flux in between to secure the alignment, then they are reflowed. Following this, the entire stack is underfilled with capillary underfill. The main advantages of this scheme is that a high throughput can be achieved, but the drawbacks include loss of alignment in the multitier pick and place as well as underfill controllability when underfilling multiple gaps is done at the same time. Furthermore, in the case of thinned dies with relatively thick back-end-of line-metallisation (BEOL), warpage can be a significant issue. Another approach involves solder reflow of the bottommost die, followed by underfill. The next tiers of dies are then bonded using thermo-compression bonding and NCP underfill. The main advantage of this scheme is that alignment and underfill control can be maintained as the flatness of the dies is during bonding, but this is at the cost of throughput. In general, die-to-package stacking faces challenges in the control of the warpage of the dies after placement. In the case of multitier stacking each tier of dies bonded increase the coupling between the stack of dies and the package substrate and cause severe planarity issues in the bonding of subsequent dies. Lower throughput of the process is another concern. The key advantage of die-to-package stacking is that the existing packaging infrastructure can be utilised and the yield loss managed because one is able to stack known good dies (KGDs).

In the die-to-wafer stacking approach, the KGD-enabled stacking advantage of the die-to-package is maintained. In this approach, known good dies are placed on the landing wafer with flux/NCP/NUF/WLUF materials predispensed on either of the dies. The complete wafer is populated with a rapid pick, align and place operation. Following this, the whole wafer is collectively bonded in a modified wafer bonding tool. This approach combines the high throughput of the collective bonding with the advantages of the thermo-compression bonding. Figure 16.4 shows schematics and images of the die-to-wafer stacking approach.

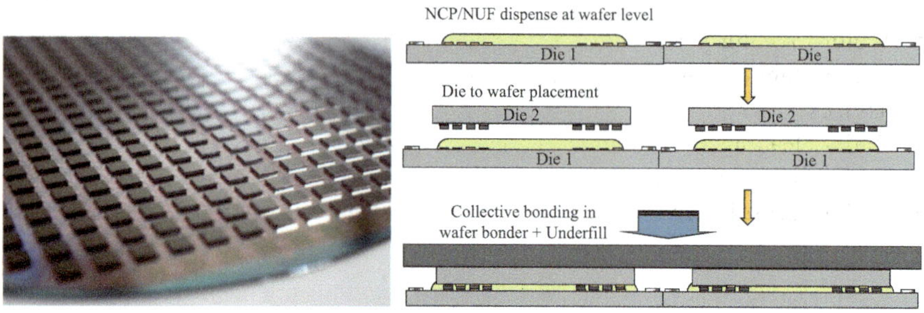

Fig. 16.4 Die-to-wafer stacking

In the wafer-to-wafer stacking approach, the wafers being bonded are aligned and bonded primarily by the thermo-compression bonding process. Wafer-to-wafer alignment bonder tools have a much higher alignment accuracy (sub-1 μm) compared to the die to die/die-to-wafer pick-and-place tools (1–3 μm). This stacking approach has a very high throughput. However, this approach can be used only in the cases where the dies being stacked are of the same size. Furthermore, this approach does not allow KGD-enabled stacking, which may have a dramatic impact on the overall stack yield. Wafer-to-wafer stacking may be beneficial in the cases where the die size is very small (hence expensive for die-to-wafer pick-and-place) and where the wafer yield is very high.

References

1. Roman JW, Eagar TW (1992) In: Proceedings of the international society for hybrid microelectronics (ISHM), Reston, p 1. "Low Stress Die Attach by Low Temperature Transient Liquid Phase Bonding," 1992 International Symposium on Microelectronics, 52, (sponsored by the International Society for Hybrid Microelectronics, ISHM), San Francisco, CA, 81, 1992.
2. Munding A, Kaiser A, Benkart P, Heittmann A, Hübner, Ramacher U, Kohn E, (2006) In: Proceedings of the IEEE 2006, Montreux, p 262. Scaling aspects for microjoints for 3D chip interconnects. ESSDERC 2006, European Solid-State Device Research Conference, Montreux, Switzerland, September 18-22, 2006.
3. Zhang W, Ruythooren W (2008) Study of the Au/In reaction for transient liquid-phase bonding and 3D chip stacking. J Electron Mater 37(8):1095–1101
4. Agarwal R, Zhang W, Limaye P, Ruythooren W (2009) High density Cu-Sn TLP bonding for 3D integration. In: 59th electronic components and technology conference (ECTC), San Diego, pp 345–349
5. Blair HD, Pan T, Nicholson JM, Cooper RP, Oh S, Farah AR (1996) Manufacturing concerns of the electronic industry regarding intermetallic compound formation during the soldering stage. In: IEEE/CPMT international manufacturing technology symposium 1996, Austin, pp 282–292
6. Oh M (1994) Ph.D. dissertation, Growth Kinetics of intermetallic phases in the Cu-Sn binary and the Cu-Ni-Sn ternary system at low temperatures Lehigh University
7. Jourdain A et al (2009) Electrically yielding collective hybrid bonding for 3D stacking of ICs. In: Proceedings of the ECTC 2009, San Diego, pp 11–13

Part V
Characterization and Modelling

Compared to thick conventional chips, ultra-thin silicon dies exhibit a physical behaviour that is either different or more pronounced. This part of the book addresses these differences, which are summarised in Table V.1.

Chapter 17 deals with the mechanical characteristics of ultra-thin, flexible chips, which depend considerably on the quality of the chip edge and plane surfaces. Also, the difficulty in quantitatively characterising the mechanical properties of thin chips is explained. Chapter 18 adds to this with a discussion of the detrimental effect that structural features at the chip edges and surfaces have on the chip's mechanical properties. Chaps. 19–22 focus on the electrical aspects, starting with a description of the well-known piezoresistive effect in MOS transistors (Chap. 19) and how this effect can be altered in a chip-on-foil assembly (Chap. 20). Chapter 20 also highlights the difficulty involved in characterising electronic effects – namely, the piezoresistive effect in devices on ultra-thin chips. Chap. 21 presents compact MOS transistor models and provides insight on how to incorporate the piezoresistive effect in such models. Chap. 22 alerts the reader to the piezojunction effect in bipolar transistors. Thermal effects in thin dies and their modelling are explained in Chap. 23. Finally, optical effects in thin silicon material are discussed in Chap. 24, with the focus on solar cell devices.

Table V.1 Physical characterisation of, and device modeling on, ultra-thin chips

Characterisation and modelling	Related chapters
Mechanical characterisation	Chap. 17
	Chap. 18
Electrical effects and device modelling	Chap. 19
	Chap. 20
	Chap. 21
	Chap. 22
Thermal effects and modelling	Chap. 23
Optical characterisation	Chap. 24

Chapter 17
Mechanical Characterisation and Modelling of Thin Chips

Stephan Schoenfelder, Joerg Bagdahn, and Mattias Petzold

Abstract In order to ensure reliable products, electronic, as well as mechanical, properties of thin chips must also be characterised. In particular, the strength of silicon devices is an important key for improvement of the yield and avoidance of the fracture of silicon chips and devices in manufacturing and application, respectively. This chapter discusses strength parameter in detail and how strength of thin silicon chips can be analysed. Besides theoretical relations, examples of strength behaviour of thin silicon chips are presented regarding different manufacturing steps.

17.1 Fracture and Strength of Brittle Materials

Strength of materials is defined as resistance to cleavage. From an atomistic point of view, strength defines the limit of energy before two single atoms are separated. In this case, the strength is interpreted as the theoretical strength of a material. For uniaxial tensile stress conditions, the Orowan equations [4] can be used to estimate the theoretical strength

$$\sigma_{\text{th}} = \sqrt{\frac{\gamma E}{a}}. \tag{17.1}$$

Herein, the theoretical strength σ_{th} is defined by the Young's modulus E, the surface energy γ and the lattice constant a of the material. Silicon has two cleavage planes, the {110} and {111} planes with <110> as propagation direction. For both planes the theoretical strength is summarised in Table 17.1 using (17.1) and the material parameters of silicon [12, 17, 37]. In more recent works the theoretical strength was calculated using ab initio calculations [7, 49]. The values are also shown in Table 17.1. Although there is a large variation in values, it can be

S. Schoenfelder (✉)
Fraunhofer Institute for Mechanics of Materials, Halle, Germany
e-mail: stephan.schoenfelder@csp.fraunhofer.de

J.N. Burghartz (ed.), *Ultra-thin Chip Technology and Applications*,
DOI 10.1007/978-1-4419-7276-7_17, © Springer Science+Business Media, LLC 2011

Table 17.1 Theoretical strength of silicon following Eq. (17.1) for the different cleavage planes {110} and {111} in comparison to ab initio simulations

Plane	–	Young's Modulus (GPa)	Lattice Distance (Â)	Surface Energy (J-m^{-2})	Theoretical Strength σ_{th} (GPa)
{111}	Eq. (17.1)	187.67	3.13	1.44	29.38
–	ab initio [7]	–	–	–	21.00
{110}	Eq. (17.1)	168.96	3.84	1.73	27.59
–	ab initio [7]	–	–	–	17.00
–	ab initio [49]	–	–	–	15.80

concluded that both planes show very similar theoretical strength, whereas the weakest plane in silicon is the {110} plane with a theoretical value of 15.8 GPa. In practice, both planes can be observed as cleavage planes in fractured silicon chips. In chips from a {100} wafer, often the {111} plane can be identified macroscopically as the cleavage plane due to easier crack propagation in the <112> direction compared to <100> [37, 44].

Although theoretical strength is a real material property, these values are only reached for very small samples [9, 19, 28]. For silicon chips in microelectronics strength values of multiple orders of magnitude lower than the theoretical values can be observed, e.g. in [26, 43, 47, 50]. This behaviour can be explained by defects of different types and sizes within the material. Due to the brittle nature of silicon at room temperature, it is very sensitive to single cracks or stress-concentrating defects. Brittle materials cannot relief high stresses, which are concentrated at the crack tip, by plastic deformation as ductile materials can do. Consequently, cracks will propagate through the material and the silicon chip will fracture, if a critical stress values is reached at a crack. Hence, the strength of silicon chips is a parameter that depends on the defect state caused by the manufacturing process. In the following, deterministic and probabilistic analyses of the strength parameter are introduced.

The field of fracture mechanics can be considered as a deterministic description of the stress state in the presence of cracks. The beginnings of the theory of fracture mechanics can be found in the work of Griffith, who, in 1920, based his description of a crack in a solid on the energy concepts of mechanics and thermodynamics [23]. Following this, the energy needed to create two new surfaces during crack propagation is equal to the released mechanical energy in the body.

Besides the energy concept, the K-concept is widely used in fracture mechanics. The parameter K represents the stress intensity factor, which is calculated at the crack tip as a quantification of the stress concentration depending on the loading case. In general, the stress intensity factor is defined as

$$K = Y\sigma\sqrt{\pi a} \tag{17.2}$$

with crack length a, applied stress σ and the geometry function Y for a certain load case. The stress intensity factor is divided in different modes of crack opening, which is (1) the pure tensile mode, (2) the shear mode and (3) the tearing mode [23]. With these concepts, the stresses in the vicinity of crack can be analysed. If the stress reaches a critical value, defined by a critical material-dependent energy release rate or stress intensity factor, the crack propagates. Thus, single crack geometries and fracture of the device can be evaluated. Unfortunately, crack sizes and geometries have to be known in order for one to determine the fracture stress. Hence, fracture mechanical analysis is used for special problems or qualitative analysis.

Besides the fracture mechanics, failure of a material can also be described with respect to a continuum stress state. By neglecting local defects and stress concentrations, one defines a stress threshold value at which the material fractures. For brittle materials, the maximum stress criterion is often used. Here, the material fails if the maximum or minimum principal stress (σ_{I}, σ_{II}, σ_{III}) reaches the uniaxial tensile strength value σ_{t} or compressive strength value σ_{c}, respectively.

$$\begin{aligned} \max(\sigma_{\mathrm{I}}, \sigma_{\mathrm{II}}, \sigma_{\mathrm{III}}) &= \sigma_{\mathrm{t}} \text{ for } \sigma_{\mathrm{I}}, \sigma_{\mathrm{II}}, \sigma_{\mathrm{III}} > 0, \\ |\min(\sigma_{\mathrm{I}}, \sigma_{\mathrm{II}}, \sigma_{\mathrm{III}})| &= -\sigma_{\mathrm{c}} \text{ for } \sigma_{\mathrm{I}}, \sigma_{\mathrm{II}}, \sigma_{\mathrm{III}} < 0 \end{aligned} \tag{17.3}$$

For most materials the critical compressive strength is much larger than the tensile strength. Thus, this criterion is commonly used for tensile stresses. Further strength criteria can be found in literature [2, 11], whereas most of them are only valid for isotropic materials, for example, the maximum stress criterion. Though, these criteria can describe a fracture in a deterministic way, they can be insufficient for describing brittle materials. Brittle materials show large scattering in strength values and thus it is difficult to determine a single strength value, for instance, σ_{t}. The mean strength values from strength testing cannot be used as design parameters. Due to the large variation, there are a lot of samples with much lower strength values. Hence, a probabilistic model for the strength behaviour is needed.

The weakest link model itself was first discussed by Peirce [36]. A probabilistic model to describe the strength of materials, which is based on the weakest link model, was introduced from Weibull in 1939 [53]. It assumes that a chain can only survive if all links survive. This represents the failure mode in brittle materials, where one critical defect can lead to fracture of the whole sample. Weibull introduced a statistical distribution, which assumes that the volume of the loaded body consists of small volume elements ΔV_i, which fail independently from each other. Hence, the volume survives if all volume elements survive. It can be shown that the probability of failure P for this case can be written as

$$P = 1 - \exp\left[-\int_V N(\sigma) dV\right] \tag{17.4}$$

with a function $N(\sigma)$, which describes the defect distribution and strength behaviour of the volume element ΔV_i. As the simplest approach for the function $N(\sigma)$ Weibull introduced [53, 54]

$$N(\sigma) = \left(\frac{\sigma - \sigma_u}{\sigma_0}\right)^m, \tag{17.5}$$

where σ_u is the minimum threshold strength, σ_0 the scale parameter and m the Weibull modulus. The scale parameter represents the material strength at a failure probability of 62.3%. The Weibull modulus is a parameter of variation, whereas a large modulus means small scattering and vice versa. The variable σ defines the actual stress in the volume, whereas the stress value can be defined by a failure criterion. Because the minimal threshold strength is hard to determine, it is often set to $\sigma_u = 0$. Thus, the Weibull distribution is used as a two-parameter distribution for $\sigma \geq 0$

$$P = 1 - \exp\left[- \int_V \left(\frac{\sigma}{\sigma_0} \right)^m dV \right]. \tag{17.6}$$

In Fig. 17.1 the density and distribution function are shown for different values of the Weibull modulus.

The strength of a sample does not only depend on the defects in the volume, but also on the defects at the surface. In cases where the influence of the volume can be neglected and the surface dominates the failure, the integration in (17.6) is performed for the surface area A. In general cases, which are influenced by both volume and area, the overall probability of failure can be expressed by

$$P = P_{\mathrm{V}} + P_{\mathrm{A}} - P_{\mathrm{V}} P_{\mathrm{A}}, \tag{17.7}$$

where P_{V} is the probability for the volume and P_{A} the probability for the surface of the sample. With the Weibull distribution the fracture probability can be calculated for a certain stress level in the sample.

As can be seen in Eq. (17.6), failure probability also depends on sample volume. This dependency is known as the 'size effect of strength'. Hence, samples with

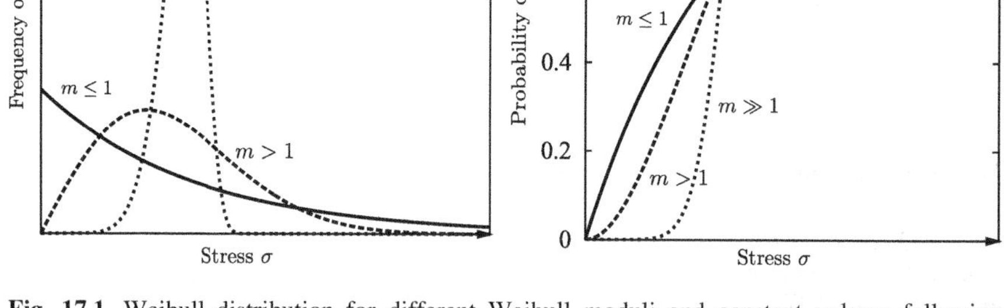

Fig. 17.1 Weibull distribution for different Weibull moduli and constant volume following Eq. (17.6): (**a**) density function, (**b**) distribution function

large volumes show smaller strength than the same kinds of samples with a smaller volume. Assuming a homogenous defect distribution, the probability of finding a critical crack is higher in a larger volume. The size effect must be considered, when strength properties need to be applied on practical loading cases such as the bending of thin chips in applications. Using (17.6), the scalar stress value σ can be written as a comparison stress for an uniaxial or multiaxial stress state defined as $\sigma_{\max}$ and a function $g(x, y, z)$, which represents the change of $\sigma_{\max}$ in the sample

$$\sigma = \sigma_{\max} g(x, y, z). \tag{17.8}$$

Then (17.6) can be written as

$$P = 1 - \exp\left[-\int_V \left(\frac{\sigma_{\max} g(x, y, z)}{\sigma_0}\right)^m dV\right] = 1 - \exp\left[-V_e\left(\frac{\sigma_{\max}}{\sigma_0}\right)^m\right]. \tag{17.9}$$

The parameter V_e is called the effective volume:

$$V_e = \int_V g(x, y, z) dV. \tag{17.10}$$

It represents an equivalent volume of a tensile sample with the same probability of failure as the analysed load case. Furthermore, the effective volume and the scale parameter can be replaced by one parameter σ_θ, the characteristic fracture stress, as long as the effective volume is constant for all samples or devices in a strength analysis

$$\sigma_\theta = \sigma_0 {V_e}^{1/m}. \tag{17.11}$$

Then follows from (17.6) that

$$P = 1 - \exp\left[-\left(\frac{\sigma}{\sigma_\theta}\right)^m\right], \tag{17.12}$$

which is the familiar form of the Weibull distribution function. It is noted that (17.12) can only be used if the same effective volume or surface can be assumed for all samples. The characteristic fracture stress, which is commonly determined by strength testing, is not a size-independent parameter and cannot be used to analyse arbitrary load cases of devices.

The probabilistic approach to determine the strength of brittle materials like thin silicon chips is easy to perform and thus widely used in industry. The stress evaluation is based on continuum mechanics and is much simpler than a fracture mechanical analysis, which requires good information about defect sizes and geometries. Nevertheless, the probabilistic analysis needs to consider the size effect of strength, which can be a more expensive analysis [3].

17.2 Strength Issues in Silicon Chip Manufacturing

As shown in the previous section, strength of silicon chips is correlated to defect distributions and defect sizes in the material. Hence, various process steps influence the strength of the chips in manufacturing. On the frontside, semiconductor patterning or etching steps in MEMS (micro-electro-mechanical systems) structure the surface with sharp notches or corners, which can act as stress concentrators. The patterning with different materials induces residual stress in the silicon material, which decreases the resulting strength. On the backside, grinding and polishing techniques are responsible for different strength behaviour [16, 26, 57]. The abrasive grinding process induces cracks, plastically deformed silicon and residual stress [56]. These defects can be removed by polishing steps to reach a defect-free surface [58]. Besides the front- and backside, the strength of the edges of a chip is important for the device's reliability. Different dicing technologies as sawing, dicing-by-thinning or laser technologies have been developed and investigated to ensure high strength edges [13, 15, 22, 24, 42, 43]. It is important to note that for many dicing technologies the strength is different for the edges on the frontside and the edges on the backside. In Fig. 17.2 the different regions of strength properties on a silicon chip are shown.

It can be concluded from these examples of process steps that the strength of chips is driven by different strength-limiting defects on the frontside, backside and the edges. Thus, a comprehensive understanding of the strength behaviour is important to design and to ensure reliable products. Furthermore, the strength can be used to characterise different process steps and maintain high quality chips.

Although, the strength of thick chips has been an important parameter in the past, it becomes even more important as new technologies emerge, like RFIDs or stacked dice with very thin chips. Due to new applications and more extreme load cases,

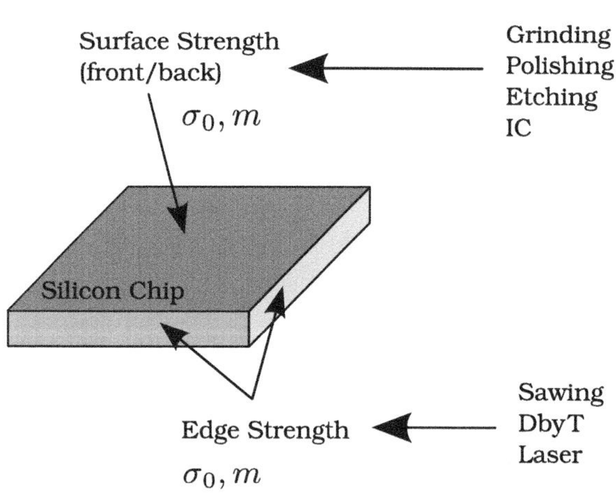

Fig. 17.2 Regions of different strength properties on a silicon chip, caused by different process steps

the reliability of semiconductor products will remain an important question, which is directly connected to the strength of the devices.

17.3 Strength Testing Methods for Thin Samples

In Sect. 17.1 it was mentioned that the probabilistic strength analysis can be performed easily for determination of strength parameters. This section will discuss what kind of test methods can be used to derive fracture stress, especially for very thin chips. The procedure to characterise strength can be summarised in three steps:

1. *Strength testing experiment.* For a given sample type an appropriate testing method has to be found and the experiment has to be performed for a statistically sufficient number of samples. After testing, fractographic analysis can be performed to identify the fracture origin. This information helps one understand the fracture mechanism and choose an appropriate failure criterion for stress calculation. As a result, fracture force and fracture displacement for every sample are derived.
2. *Calculating fracture stress.* Based on the experimental data, the fracture stress for every sample has to be calculated by using an appropriate failure criterion. Often, nonlinearities have to be considered for thin structures. As a result, the fracture stress for every sample is known.
3. *Statistical evaluation.* Once the fracture stresses are known for the batch, the results can be evaluated statistically by the Weibull distribution. The Weibull parameters have to be estimated based on the experimental values. In consideration of the effective size (volume or area), the size-independent Weibull parameters can be calculated. As a result, the statistical parameters of the batch are known and can be compared with other types of samples or can be used for a reliability analysis.

Since steps 2 and 3 depend on step 1, the choice of the test method is the essential point in characterising strength. The experimental method to test the samples must be chosen with consideration for the sample geometry and, first of all, the aim of the analysis. What kind of strength needs to be analysed? Is either the surface strength or the edge strength of interest? Additionally, the sample geometry also influences possible test methods. Thin chips are shaped as thin plate structures. Thus, pure tensile tests are difficult to perform. For chips, mostly bending experiments are used. They can be divided in uniaxial and biaxial bending tests. For glass sheets these test methods are standardised [31]. For thin silicon chips such standards are still missing.

It should be noted that the influence of residual stress can also be determined indirectly by strength characterisation. In order to separate the effect of residual stress from material defects, additional residual stress analysis can be performed using micro-Raman spectroscopy, electron backscattering diffraction (EBSD) or photoelasticity. Thus, the role of residual stress in the samples can be quantified and fracture mechanisms can be analysed in detail.

17.3.1 Uniaxial Bending Tests

Uniaxial bending tests cover all test methods that bend the sample about one axis. The 3-point or 4-point bending setups represent this type of bending, as shown in Fig. 17.3. In these setups the sample is placed on parallel support rollers and bended by another set of rollers. During bending, the top surface (loading rollers) is in a compressive stress state and the bottom surface (support rollers) is in a tensile stress state. Different standards can be found for ceramics [29, 34], glass [33] and silicon chips [35] using uniaxial bending methods. In general, long and slim structures are recommended for this kind of bending, following the assumptions of the linear beam theory. Then, the stress is constant along the width b of the sample. Thus, the surface, as well as the edges, is loaded with equal stress. If it can be assumed that the edges are damaged more than the surface these methods can be used to determine edge strength.

The stress along the sample axis differs compared to both 3-point and 4-point bending. The 3-point bending shows a maximum of bending stress in the centre of the sample, whereas the sample in the 4-point bending is equally stressed within the loading rollers. Thus, in 4-point bending a larger surface area and volume is tested with a constant stress. Furthermore, this region is free of shear forces. The bending stress can be calculated for small deflections from [33, 35],

$$\sigma_{3\mathrm{PB}} = \frac{3Fl}{2bh^2} \text{ and } \sigma_{4\mathrm{PB}} = \frac{3F}{2bh^2}(l_2 - l_1), \qquad (17.13)$$

with load span l or l_2, the width b, the thickness h and the applied force F. For the 4-point bending also the inner load span l_1 is needed. By using these test methods, different aspects have to be kept in mind [27], like the influence of changing contact position on the sample, Hertzian pressure below the rollers or additional bending moments due to friction.

If the uniaxial bending methods are used for very thin chips, different problems must be tackled. At first the large deflection will be discussed. Large deflection

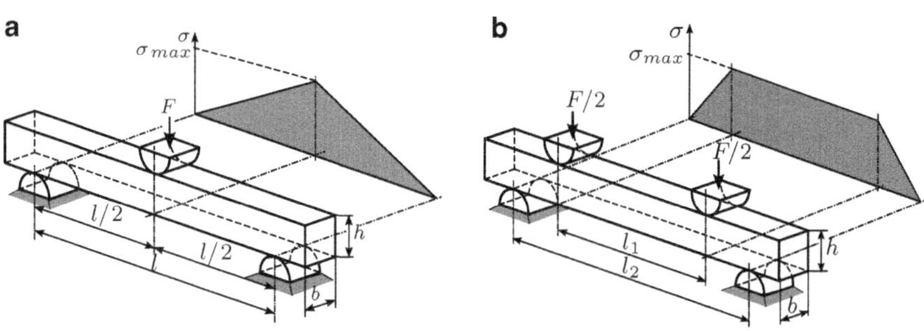

Fig. 17.3 Principle of (**a**) 3-point bending and (**b**) 4-point bending with projected bending stress along the sample axis

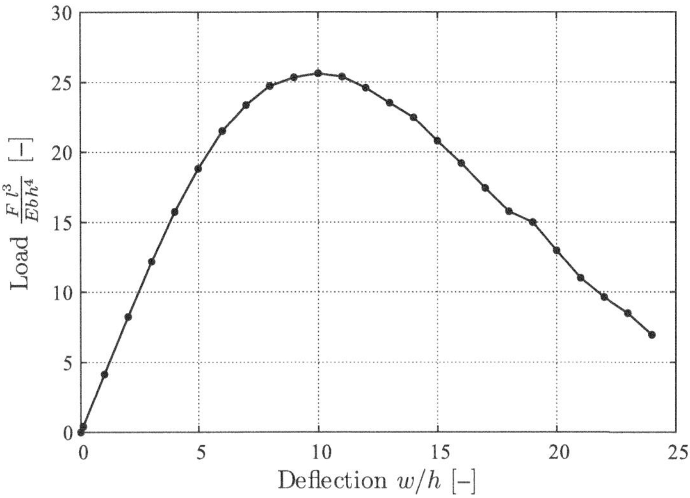

Fig. 17.4 Force-deflection chart for 3-point bending for thin samples for isotropic material ($\nu = 0.25$) calculated with a finite element (*FE*) model considering large deflections and contact between the sample and the rollers [41]

occurs in uniaxial bending due to the high flexibility of the samples. The deflection can be several times larger than the thickness until the samples fracture. The force-deflection chart for samples with large deflection is strongly nonlinear as shown in Fig. 17.4 for the 3-point bending. The loading conditions are changed and the sample is sliding on the support due to the large deflection. Thus, the vertical force reaches a maximum, while the deflection increases further. The mechanism of this behaviour is explained in [43].

This strongly non-linear behaviour causes also a non-linear relation between force and stress or deflection and stress. The assumptions for (17.13) are violated, and only numerical methods can be used to determine the stress state in the sample. In Fig. 17.5 the maximum principal stress, which is dependent on the force and the deflection, is shown. The maximum of the force can be also found here. A linear calculation using (17.13) would underestimate the real stress values.

Furthermore, it can be seen that the stress reaches a maximum at nearly $w/h = 20$. Beyond that point the stress decreases in the sample. This can be explained by a relaxation of the sample in case of very large deflection. Thus, there is a maximum stress in uniaxial bending. If the sample can withstand even more stress it will not fracture in this test setup.

As a second problem of thin chips the plate structure is discussed. As introduced, uniaxial bending methods are usually applied for testing slim structures, for which the width is rather small compared to the length. For chip samples the width is often the same size as the length of the sample. Hence, a bending moment perpendicular to the main bending moment is induced on the order of magnitude of Poisson's ratio. This second bending moment causes stress perpendicular to the main bending

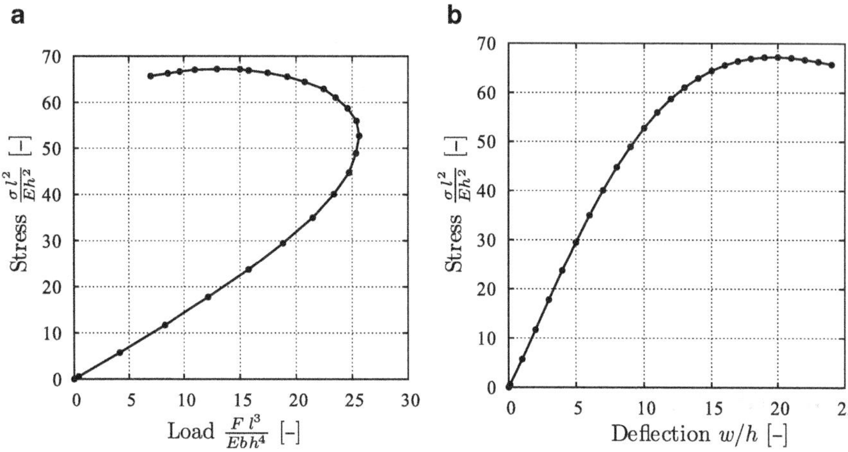

Fig. 17.5 Relation between stress and (**a**) force and (**b**) deflection for a thin sample in 3-point bending for isotropic material ($n = 0.25$) calculated with an FE model considering large deflections and contact between the sample and the rollers [41]

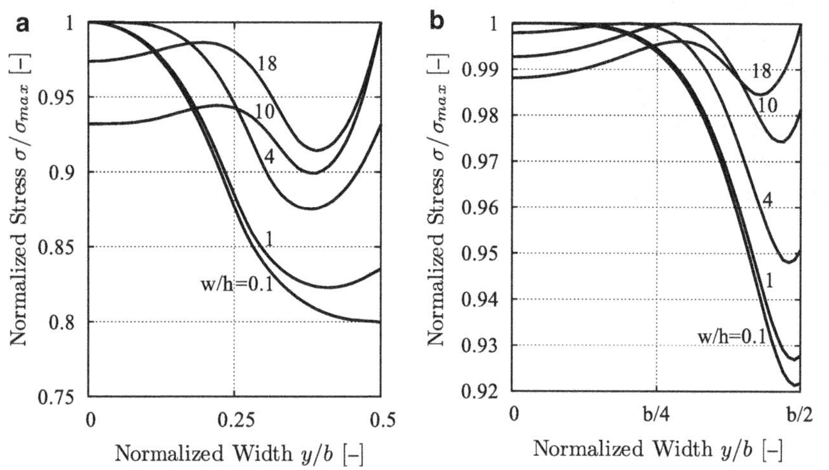

Fig. 17.6 Normalized maximum principal stress in a 3-point bending sample in the centre at the bottom side (tensile stress) of the sample for different deflections and crystal orientations of silicon: (**a**) edge of sample along [100] direction; (**b**) edge of sample along [110] direction. All charts calculated with FE model considering large deflections and contact between sample and rollers. Geometry: $b/l = 0.5$ [41]

stress, resulting in a biaxial stress state in the centre of the sample and a uniaxial stress state at the edge. Thus, the stress is no longer constant along the width of the sample as shown in Fig. 17.6. In linear cases for small deflection the surface is loaded up to 20% more than the edge for this geometry and crystal orientation (Fig. 17.6a). If the deflection increases, the stress distribution is changing and

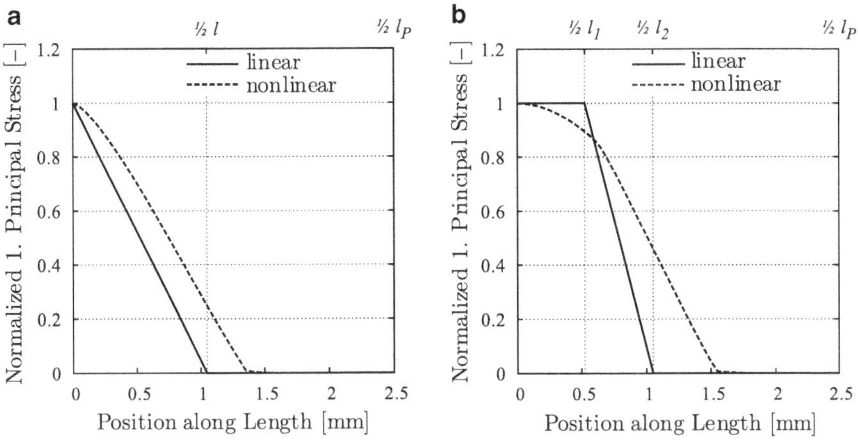

Fig. 17.7 Theoretical normalized maximum principal stress in (**a**) 3-point bending and (**b**) 4-point bending for linear and nonlinear case. All charts calculated with FE models considering large deflections and contact between sample and rollers for $l = 2$ mm, $l/h = 40$, $b/h = 20$ and $w/h = 18$ [41]

finally the edge is more loaded than the surface. The influence of this effect depends on Poisson's ratio and the width of the chip. Consequently, the results can be influenced by the surface defects to a different extent for samples with small and large deflection.

As it is shown in Fig. 17.6, the large deflection does influence the stress field in the sample. But the stress field is not only changing across the sample width, but also along the sample. In Fig. 17.7 the maximum principal stress is shown along the sample for the 3-point and 4-point bending for small deflections (linear case) and large deflection (nonlinear case). For the 4-point bending, the constant stress state within the inner load span changes into a nonlinear shape with a maximum in the sample centre. Thus, the advantage of a constant stress region cannot be found in samples with large deflection. The stress distribution also changes in 3-point bending, but the qualitative distribution is very similar in the linear and the nonlinear case. Finally, the 3-point bending is a more robust testing mode than the 4-point bending if nonlinear behaviour is expected.

17.3.2 Biaxial Bending Tests

In biaxial bending tests the stress state is also biaxial or axisymmetric. In comparison to uniaxial stress, the biaxial stress often occurs in application, e.g. due to temperature loads and thermal strain. Furthermore, the biaxial bending setups load mainly the surface and defects at the edges are not considered in testing. There are various setups of biaxial bending tests, whereas the ring-on-ring test [40, 51] and the

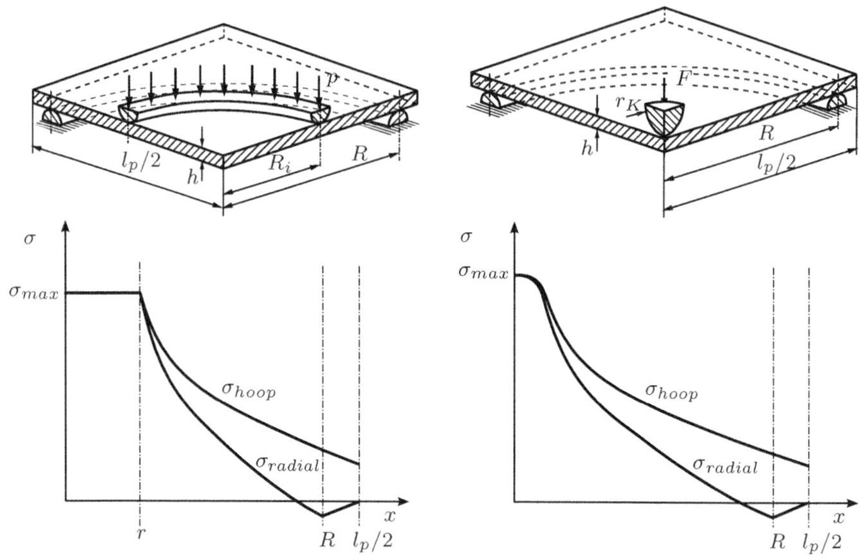

Fig. 17.8 Principle of (**a**) ball-on-ring test and (**b**) ring-on-ring test with radial and hoop stress distribution on the bottom side (tensile stress)

ball-on-ring test [6, 25] are commonly used. In the literature other biaxial test methods can also be found, as the piston-on-3-balls test [30, 52], the ball-on-3-balls test [5], the 3-balls-on-3-balls test [10] and the ball-breaker test [14, 18, 48].

Both ball-on-ring test and ring-on-ring test are shown in Fig. 17.8. The sample is placed on a support ring and loaded with a concentric ring for the ring-on-ring test and a concentric ball for the ball-on-ring test. Hence, the sample is bent and an axisymmetric stress state is induced with tensile stresses on the bottom side (support ring) and compressive stresses on the top side (load ring/ball). The edges of the sample are loaded with much smaller stress than the surface, depending on the geometry [39, 40, 51]. The ring-on-ring test is standardised for glass sheets in [32]. The ring-on-ring test results in a constant biaxial stress field within the inner loading ring and loads a larger area and volume than the ball-on-ring test. There, only the centre of the sample is highly stressed and thus only a small volume or area is loaded. For small deflections the maximum stress can be calculated for the ring-on-ring test and ball-on-ring test, respectively from [40, 45]

$$\sigma_{\max} = \frac{3(1+\upsilon)F}{4\pi h^2}\left(2\ln\frac{R}{R_i} + \frac{1-\upsilon}{1+\upsilon}\frac{R^2-R_i^2}{r_m^2}\right)\text{and} \tag{17.14}$$

$$\sigma_{\max} = \frac{3F}{4\pi h^2}(1+\upsilon)\left(1 + 2\ln\frac{R}{c'} + \frac{1-\upsilon}{1+\upsilon}\left(1-\frac{c'^2}{2R^2}\right)\frac{R^2}{r_m^2}\right). \tag{17.15}$$

Herein, the loading force F, the thickness of the sample h, Poisson's ratio ν, the support radius R, the loading radius R_i, the contact radius c' and the radius of the sample r_m are used as parameters. Since these equations are based on plate theory for axisymmetric problems, the sample shape is assumed to be circular. In practice, quadratic samples are used for convenience. Hence, an equivalent sample radius has to be determined from the quadratic shape. It was shown in [40, 51] that it can be used with the sample length l_p:

$$r_m = \frac{l_p(1+\sqrt{2})}{4}. \tag{17.16}$$

The stress calculation in the centre for the ball-on-ring test following the plate theory is difficult because a point load will cause infinite stress. Thus, it is assumed that the load is distributed on a small contact area with the radius c'. In [45] the contact radius was found to be $c' \sim h/3$ for sufficient results. While this is an approximation, one can use the equation for the maximum stress from Woinowsky–Krieger [55], which is based on the theory of elasticity for axisymmetric problems:

$$\sigma_{\max} = \frac{F}{h^2}\left[(1+\upsilon)\left(0.485\ln\frac{R}{h}+0.52\right)+0.48\right]. \tag{17.17}$$

If very thin samples are used in biaxial bending tests, one has to deal with large deflections. Unfortunately, biaxial bending is much more sensitive for large deflection than uniaxial bending. If the deflection exceeds the half of the thickness of the sample ($d_{\max} > h/2$) the linear assumptions are violated and nonlinear equations must be used. Equations 17.14–17.17 cannot be used to determine the maximum fracture stress. The ring-on-ring test does not even show a constant biaxial stress field for large deflection anymore [20]. Thus, this test setup cannot be used for thin samples undergoing large deflections. In Fig. 17.9 the relationship of force and deflection is shown for the ball-on-ring test. A strongly nonlinear behaviour of the sample can be seen for a deflection only a few times larger than the thickness. Thus, the nonlinear behaviour of thin samples in biaxial bending tests has to be considered for fracture stress evaluation.

Usually, the nonlinearities in biaxial bending tests due to large deflection can be modelled with state of the art simulation software using shell models. Though, for shell models the deformation and stress is not well calculated in the region of the loading ball in ball-on-ring tests. The assumptions of the plate models, which are based on plate theory, are violated in this region due to the very local concentrated loading. Thus, the fracture stress cannot be determined using shell models, even though they can consider large deflections and nonlinear behaviour. The alternative, using a 3D FE model is too expensive for calculation. In literature a submodelling approach is used [8, 41]. In this approach the nonlinear deformation of the thin sample is calculated using shell model in finite element simulations. In a second step the solution of deformation in the centre region is applied on a 3D submodel of this region, which can handle all local mechanical effects. The stress distribution in the submodel can be solved analytically [8] or numerically [41]. In a numerical

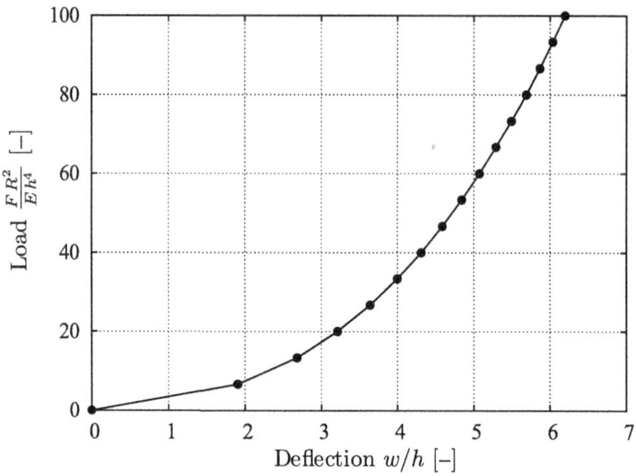

Fig. 17.9 Theoretical analysis of a force-deflection chart for ball-on-ring tests using finite element (*FE*) models [41]

solution also anisotropic material behaviour and changes of the contact between ball and sample surface can be considered. In Fig. 17.10 the difference in deflection and stress is shown for the numerical shell model and corresponding submodel. As it can be seen, the shell model overestimates the values of deflection and stress in the centre region.

In conclusion, biaxial bending tests are used if the influence of the edges needs to be neglected. The ball-on-ring test seems to be appropriate for very thin samples, although it concentrates the load in the centre of the sample. The ring-on-ring test loses the advantage of a constant large area biaxial stress field in the case of large deflection and cannot be used for thin samples. Other biaxial bending tests can be analysed similar to the ball-on-ring test, since they are dealing with similar problems in fracture stress calculation.

17.3.3 Statistical Evaluation

Once a representative number of samples per batch are fractured in the experiment and the fracture stresses are calculated, considering all nonlinearities occurring for thin samples, the values can be statistically evaluated. As introduced in Sect. 17.1 the Weibull distribution is used to represent the statistical distribution. For small deflections in the experiment, linear relationships and a constant size of the effective volume or area of the sample can be assumed. Then, (17.12) can be used to estimate the characteristic fracture stress and the Weibull modulus. It is recommended one use the maximum likelihood estimation to determine the parameters [1]. For the Weibull distribution the following equations have to be solved:

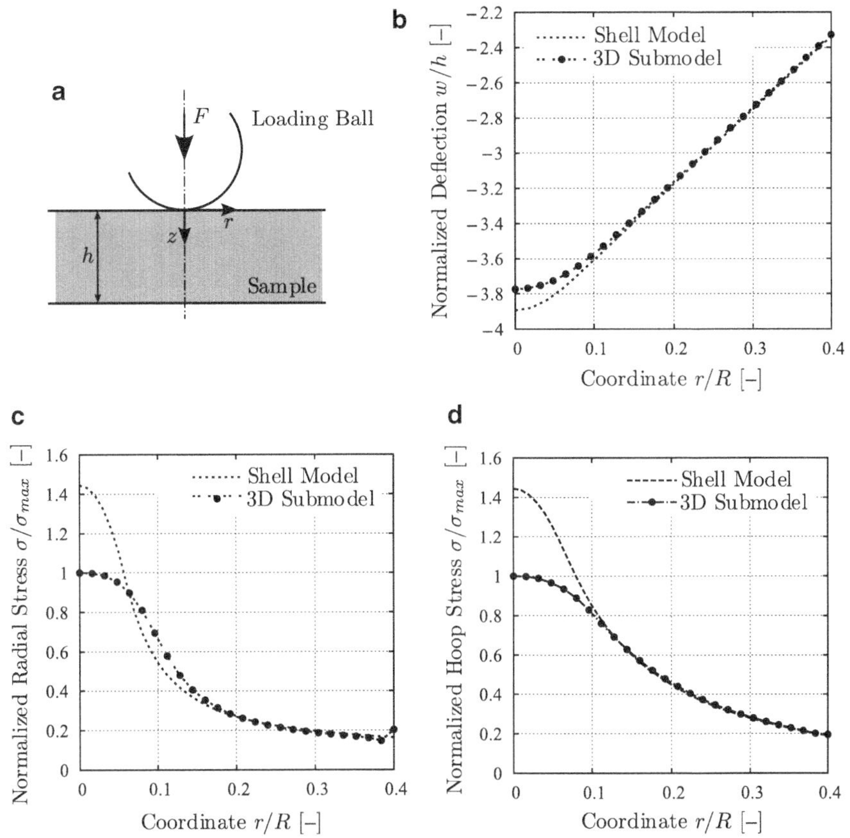

Fig. 17.10 Comparison of different models, shell model for the whole sample and submodel for the region under the loading ball at $z = h$: (**a**) sketch of geometry, (**b**) deflection, (**c**) radial stress distribution, (**d**) hoop stress distribution. Calculations of FE models with $R/h = 20$ are from [41]

$$0 = \frac{n}{m} + \sum_{i=1}^{n} \ln \sigma_i - n \frac{\sum_{i=1}^{n} \sigma_i^m \ln \sigma_i}{\sum_{i=1}^{n} \sigma_i^m} \tag{17.18}$$

$$\sigma_\theta^m = \frac{1}{n} \sum_{i=1}^{n} \sigma_i^m. \tag{17.19}$$

Equation 17.18 determines the Weibull modulus m and needs to be solved numerically, whereas n is the number of samples and σ_i the single fracture stress value. The result can be used to solve (17.19). Finally, if the effective size of the sample in the experiment is known, the scale parameter σ_0 can be calculated with (17.11).

If thin chips show large deflections during testing in uniaxial or biaxial bending tests, the nonlinearities can cause changes in the stress distribution. Hence, the effective size of volume or area can be changed. Because the samples show large scattering in strength, the stress distribution and effective size also vary. As a result, every sample has been fractured with a different fracture stress and effective size. It was shown elsewhere [41] that this effect can be found for uniaxial as well as biaxial bending tests. The latter shows a strong size effect in experiment for thin samples. In order to consider the size effect in the experiment, the Weibull distribution has to be used in the form of (17.6), as it is introduced in [41]. Every sample has to be evaluated regarding the fracture stress and, additionally regarding the corresponding effective size. Both values – fracture stress and effective size – are used to estimate the Weibull parameters. For this the equation for the maximum Likelihood estimation has to be modified [41]:

$$0 = \frac{n}{m} + \sum_{i=1}^{n} \ln \sigma_i - n \frac{\sum_{i=1}^{n} \sigma_i^m A_{\mathrm{e},i} \ln \sigma_i}{\sum_{i=1}^{n} A_{\mathrm{e},i} \sigma_i^m} - n \frac{\sum_{i=1}^{n} \sigma_i^m \frac{\partial A_{\mathrm{e},i}}{\partial m}}{\sum_{i=1}^{m} A_{\mathrm{e},i} \sigma_i^m} + \sum_{i=1}^{n} \frac{1}{A_{\mathrm{e},i}} \frac{\partial A_{\mathrm{e},i}}{\partial m}, \quad (17.20)$$

$$\sigma_0^m = \frac{1}{n} \sum_{i=1}^{n} A_{e,i} \sigma_i^m. \quad (17.21)$$

In these equations additional terms with effective size $A_{e,i}$ for every sample appear. After numerical parameter estimation the scale parameter σ_0 and the Weibull modulus m are determined. In order to derive reliable strength results, it is recommended one use these modified equations for parameter estimation if thin samples with large deflections are tested.

17.3.4 Importance of Considering Nonlinearities

As described in the previous sections, nonlinearities in testing thin silicon chips must be considered in the calculation of the fracture stresses and the statistical evaluation. Especially, the fracture stresses can be misinterpreted if linear approximations are used. In the following example the difference in linear and nonlinear stress calculation will be demonstrated. Data are taken from a ball-on-ring test with 48-μm-thick silicon chips ((3 × 3) mm^2) with a polished surface [41]. The experiment was performed for 40 chips. The fracture stress was calculated with a numerical FE model on one side and the linear stress formula (17.15) on the other side. In Fig. 17.11 the fractures stresses are compared to linear and nonlinear calculations. It can clearly be seen that the linear calculation shows much higher stresses, which are even larger than the theoretical strength of silicon (cf. Table 17.1). Thus, the linear equations overestimate the fracture stress in this case. Furthermore, the scattering is much larger for the linear evaluation than it is

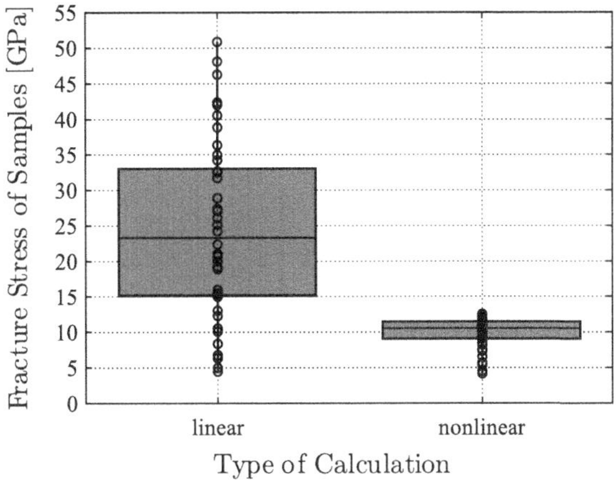

Fig. 17.11 Box plots for fracture stresses calculated with linear and nonlinear methods. Data were taken from [41] for 48-μm silicon chips in ball-on-ring testing

for the nonlinear evaluation. Consequently, the fracture stress and the variation of stress will be misinterpreted if linear equations are used for thin silicon chips with large deflections.

17.4 Examples of Strength of Thin Chips

Because mainly theoretical aspects of strength testing of thin silicon chips have been described, in the following section some examples of strength characterisations will be given. On one hand, dicing technologies such as sawing, dicing-by-thinning and laser cutting are analysed. On the other hand, the strength of the chip surface influenced by grinding is shown. Besides the strength distributions, microstructural results are also shown and correlation between strength and microstructure can be found.

17.4.1 Strength of Chips Edges: Dicing Technologies

Dicing technologies are one of the key factors regarding the strength behaviour of the chip. The processes to separate the wafer into single chips can damage the chip edge and lower the critical load for the devices. As a conventional technology sawing is used for different thicknesses of wafers. For thinner wafers improved technologies are used, e.g. dicing-by-thinning. Furthermore, laser technologies become more interesting for cutting wafers if the thickness is reduced.

In [41, 43], the capabilities of dicing-by-thinning processes were investigated in detail. Monocrystalline silicon chips with different thicknesses were tested on the

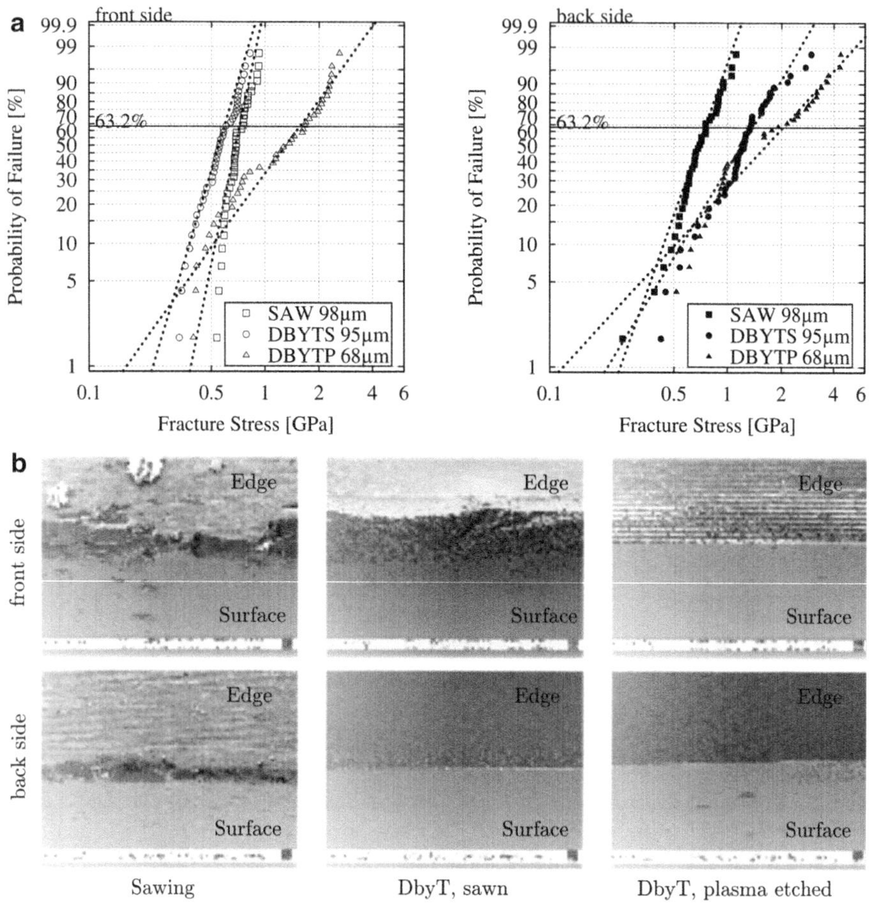

Fig. 17.12 (**a**) Weibull distributions of strength characterisations of 3-point bending tests regarding sawing and dicing-by-thinning processes. The samples were monocrystalline silicon sample without semiconductor patterning. (**b**) SEM images of the diced edges for different separation technologies corresponding to the strength distribution are shown in (**a**). SAW, sawing; DBYTS, dicing-by-thinning with sawed grooves; DBYTP, dicing-by-thinning with plasma etched grooves (Data were taken from [41])

front-and backside in 3-point bending tests to determine the influence of the edge damage. In Fig. 17.12a the strength distributions of the different batches are shown. The microstructure was investigated using SEM as shown in Fig. 17.12b. Clearly, the sawed samples had the lowest strength on both front- and backside due to the highly damaged edges of the chips. If the dicing-by-thinning process is used, the strength of the backside can be improved significantly. The microstructure shows a very smooth edge. The frontside remains similar to the sawed sample since this damage has not been modified by the dicing-by-thinning process. For thin silicon chips of 48 μm the dicing-by-thinning process was modified using a plasma etching process to induce the grooves. Here, very high strength values can be reached for the frontside as well as the backside. In the microstructural image no larger defects could be found for both edges, which explains the very high strength. It should be noted that the scattering of the batches increases when the characteristic strength reaches high values. Due to the very good edge quality the samples become more sensitive for handling and contact damage. Thus, a few big defects can drastically reduce the strength and increase the variation in the batch.

Laser technologies can be an interesting alternative for wafer dicing; such technologies are widely used in the photovoltaic or microeletronic industry today. On one hand new innovative technologies are developed using the laser, such as the stealth dicing process from Hamamatsu [21], or similar technologies, such as thermal laser separation (TLS) from Jenoptik and the LIS technology from Laser-zentrum Hannover [13]. In these technologies the material is not evaporated. The stealth dicing only damages the material for crack initiation. In the TLS and LIS technology a crack is propagated by a thermally induced stress field. On the other hand, the majority of applications use conventional laser technologies evaporating and melting the silicon material. In recent studies the influence of the laser technology on the strength of silicon chips was investigated [13, 24, 42, 46]. Examples of such analyses from [42] are given in Fig. 17.13 for the comparison of a continuous 1,064-nm dry laser with a Laser MicroJet process [38] using the same laser. Since in this investigation the same sample sizes were used as in Fig. 17.12, the strengths can be compared. The strength after cutting with laser

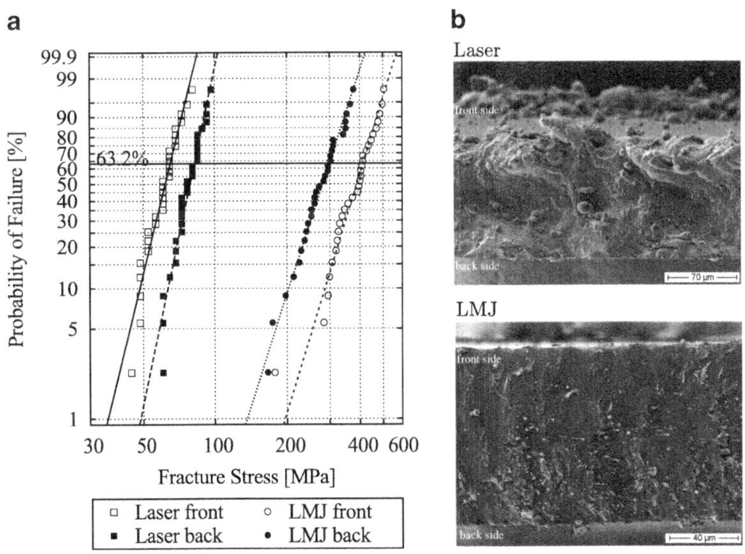

Fig. 17.13 (**a**) Weibull distributions of strength characterisations using 3-point bending tests for monocrystalline silicon chips (100-μm thickness) for standard (dry) laser process (laser) and laser microjet process (*LMJ*). (**b**) SEM images of sample edges for different laser processes with molten and recrystallised material, which can induce cracks and residual stresses [42]

processes is significantly lower than for conventional sawing, whereas a conventional continuous laser shows the lowest strength. For these samples, the microstructural investigation revealed a high amount of molten material (Fig. 17.13b) and induced cracks of 100 μm in the length [42]. The samples being cut with the LMJ process show much higher strength and less molten material at the edges (Fig. 17.13b). It should be noted that the strength of laser-diced chips can be improved with optimal laser parameters. But as a thermal process, strength is mostly reduced due to crack nucleation and residual stress caused by laser dicing.

17.4.2 Strength of Chip Surface: Grinding Technologies

The surface of the chips is treated with different processes in chip manufacturing. On the frontside different layers of material are patterned for the electronic function of the device. These layers also influence the strength of the chip, but they cannot be easily modified regarding an optimised strength property. On the other side, the backside of the wafer is only grinded to reduce thickness. Thus, the grinding process and the subsequent stress relief and polishing steps give the chip its capabilities for strength optimisation. In Fig. 17.14 the strength of thin silicon chips is shown for different surface properties tested in ball-on-ring tests. While the frontside has been kept polished, the backside was ground and chemically polished using a spin etching process. The amount of removed material in spin

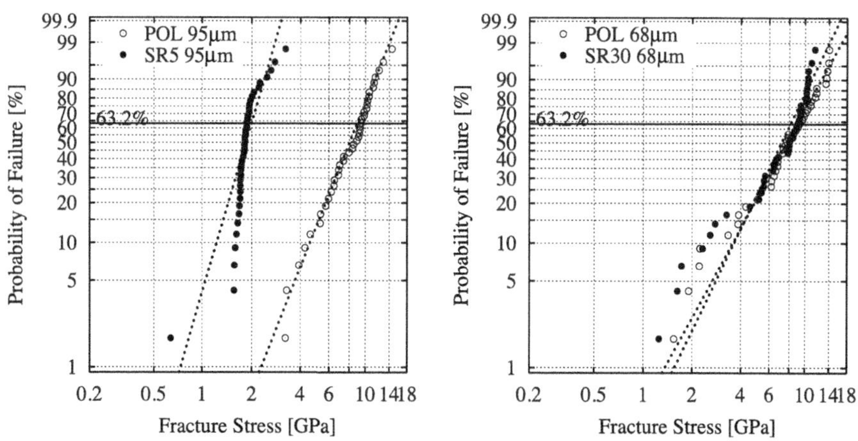

Fig. 17.14 Weibull distributions of the strength of monocrystalline silicon chip surfaces tested in ball-on-ring tests for different surface qualities. Sample size were 3 × 3 mm^2 on a support ring of 2 mm. *POL* polished surface, *SR5* stress relief of 5 μm after grinding, *SR30* stress relief of 30 μm after grinding (Data were taken from [41])

etching differed, with 5 and 30 μm. As it can be seen, the polished frontsides reach characteristic stress values of 10 GPa. If the surface is grinded and only 5 μm are removed by chemical etching, the strength is significantly reduced by a factor of 5 (Fig. 17.14, left). If 30 μm are removed after grinding the strength can reach the strength of the polished surface (Fig. 17.14, right). Hence, strength properties of thin chips can be optimised by adjusting process parameters. The final strength of the chip is significantly influenced by the quality of the thinning process.

17.5 Conclusions

In this chapter the strength of brittle materials such as thin silicon chips and wafers was introduced. Different methods for strength testing were discussed and important issues regarding testing very thin chips highlighted. It could be shown in some examples that the strength of thin chips can be measured and that the manufacturing steps strongly influence the strength of the chip or the device. In combination with microstructural investigations the fracture mechanisms and causes can be identified.

In order to ensure reliable strength values, different issues have to be considered for thin silicon chips. At first, the strength test has to fit the question to be answered. Once the test setup is clear, the experiments have to be analysed regarding non-linearities, like nonlinear force-deflection charts or sliding of the sample on the supports. Considering the nonlinearities, which mostly occur during testing of thin samples, the fracture stresses have to be evaluated. Then, the statistical parameters of strength can be estimated. If also the effective sizes of the samples are determined the size-independent strength parameters can be calculated and used for a reliability analysis.

Though current chip thicknesses can be characterised in strength by conventional strength testing methods like 3-point bending test or ball-on-ring test, these tests are challenging for even thinner chips. In biaxial bending tests the samples tend to buckle if the thickness is too small regarding the setup geometry. Due to buckling the problem of stress calculation is strongly nonlinear and the stress field is changed. If very thin chips are tested in uniaxial bending tests, the stress generated in the setup can be insufficient for fracture. Then the chip is too flexible to break and the fracture force cannot be determined. Thus, for future chip thicknesses (below 20 μm) strength testing can be challenging and new test methods have to be developed to ensure a reliable strength testing for thin silicon chips.

Acknowledgements The results being presented here has been enabled by different public funded projects during the last few years in the field of microelectronics and photovoltaics. The authors gratefully acknowledge the funding from the Kultusministerium Sachsen-Anhalt (contract no. 0037 KL/0903B) and the Federal Ministry of Education and Research within the Innoprofile initiative 'SiThinSolar' (contract no. 03IP607).

References

1. Abernethy RB (2000) The new Weibull handbook, 4th edn. Abernethy, North Palm Beach
2. Altenbach H (1993) Werkstoffmechanik. Deutscher Verlag für Grundstoffindustrie, Stuttgart
3. Bagdahn J, Sharpe W, Jadaan O (2003) Strength of polysilicon at stress concentrations. J Microelectromech Syst 12(3):302–312
4. Clarke DR (1992) Fracture of silicon and other semiconductors. Semiconductor Semimetals 37:79–140
5. Danzer R, Supancic P, Harrer W (2006) Biaxial tensile strength test for brittle rectangular plates. J Ceram Soc Jpn 114(1335):1054–1060
6. de With G, Wagemans HHM (1989) Ball-on-ring test revisited. J Am Ceram Soc 72 (8):1538–1541
7. Dubois SMM, Rignanese GM, Pardoen T, Charlier JC (2006) Ideal strength of silicon: an ab initio study. Phys Rev B Condens Matter Mater Phys 74(23):235203
8. Duderstadt F (2003) Anwendung der von Karman'schen Plattentheorie und der Hertz'schen Pressung für die Spannungsanalyse zur Biegung von GaAs-Wafern im modifizierten Doppelringtest. Ph.D. thesis, Technische Universität Berlin
9. Ericson F, Schweitz JA (1990) Micromechanical fracture strength of silicon. J Appl Phys 68 (11):5840–5844
10. Fett T, Rizzi G, Ernst E, Müller R, Oberacker R (2007) A 3-balls-on-3-balls strength test for ceramic disks. J Eur Ceram Soc 27(1):1–12
11. Gross D, Seelig T (2007) Bruchmechanik mit einer Einführung in die Mikromechanik. Springer-Verlag, Berlin
12. Hall JJ (1967) Electronic effects in the elastic constants of n-type silicon. Phys Rev 161 (3):756–761
13. Haupt O, Siegel F, Schoonderbeek A, Richter L, Kling R, Ostendorf A (2008) Laser dicing of silicon: comparison of ablation mechanisms with a novel technology of thermally induced stress. J Laser Micro/Nanoeng 3:135–140
14. Hawkins G, Berg H, Mahalingam M, Lewis G, Lofgran L (1987) Measurement of silicon strength as affected by wafer back processing. In: 25th annual proceedings – reliability physics 1987, IEEE, San Diego, CA
15. Heinze P, Amberger M, Chabert T (2008) So macht man den perfekten Chip. Mirkoprodution 1:45–50
16. Hu S (1982) Critical stress in silicon brittle fracture and effect of ion implantation and other surface treatments. J Appl Phys 53(5):3576–3580
17. Hull R (1999) Properties of crystalline silicon. Institution of Engineering and Technology, London
18. Jiun H, Ahmad I, Jalar A, Omar G (2006) Effect of wafer thinning methods towards fracture strength and topography of silicon die. Microelectron Reliab 46(5–6):836–845
19. Johansson S, Schweitz JA, Tenerz L, Tiren J (1988) Fracture testing of silicon microelements in situ in a scanning electron microscope. J Appl Phys 63(10):4799–4803
20. Kao R, Perrone N, Capps W (1971) Large-deflection solution of the coaxial-ring-circular-glass-plate flexure problem. J Am Ceram Soc 54(11):566–571
21. Kumagai M, Uchiyama N, Ohmura E, Sugiura R, Atsumi K, Fukumitsu K (2007) Advanced dicing technology for semiconductor wafer–stealth dicing. IEEE Trans Semicond Manuf 20 (3):259–265
22. Landesberger C, Klink G, Schwinn G, Aschenbrenner R (2001) New dicing and thinning concept improves mechanical reliability of ultra thin silicon. In: International symposium on advanced packaging materials, Braselton
23. Lawn B (1993) Fracture of brittle solids, 2nd edn. Cambridge University Press, Cambridge
24. Li J, Hwang H, Ahn EC, Chen Q, Kim P, Lee T, Chung M, Chung T (2007) Laser dicing and subsequent die strength enhancement technologies for ultra-thin wafer. In: 57th electronic components and technology conference, 2007, Reno

25. McKinney KR, Herbert CM (1970) Effect of surface finish on structural ceramic failure. J Am Ceram Soc 53(9):513–516
26. McLellan N, Fan N, Liu S, Lau K, Wu J (2004) Effects of wafer thinning condition on the roughness, morphology and fracture strength of silicon die. J Electron Packag 126(1):110–114
27. Munz D, Fett T (2001) Ceramics: mechanical properties, failure behaviour, materials selection, 1st edn. Springer-Verlag, Berlin-Heidelberg/New York
28. Namazu T, Isono Y, Tanaka T (2000) Evaluation of size effect on mechanical properties of silicon by nanoscale bending test using AFM. J Microelectromech Syst 9(4):450–459
29. NORM ASTM C 1161–02c (2002) Standard test method for flexural strength of advanced ceramics at ambient temperature
30. NORM ASTM F 394-78 (1996) Standard test method for biaxial flexure strength (modulus of rupture) of ceramic substrates
31. NORM DIN EN 1288-1 (2000) Bestimmung der Biegefestigkeit von Glas – Teil 1: Grundlagen, 2000
32. NORM DIN EN 1288-2 (2000) Bestimmung der Biegefestigkeit von Glas – Teil 2: Doppelring-Biegeversuch an plattenförmigen Proben mit großen Prüfflächen
33. NORM DIN EN 1288-3 (2000) Bestimmung der Biegefestigkeit von Glas – Teil 3: Prüfung von Proben bei zweiseitiger Auflagerung
34. NORM DIN EN 843-1 (2006) Hochleistungskeramik – Monolithische Keramik – Mechanische Eigenschaften bei Raumtemperatur – Teil 1: Bestimmung der Biegefestigkeit
35. NORM SEMI G86–0303 (2003) Test method for measurement of chip (Die) strength by mean of 3-point bending
36. Peirce FY (1926) Tensile tests for cotton yarns v. – 'The weakest link' theorems on the strength of long and of composite specimen. J Text Inst 17:T355–T368
37. Pérez R, Gumbsch P (2000) Directional anisotropy in the cleavage fracture of silicon. Phys Rev Lett 84(23):5347–5350
38. Perrottet D, Housh R, Richerzhagen B (2006) Fast cutting and scribing of silicon PV cells using the water-jet-guided laser technology. In: 21st European photovoltaic solar energy conference and exhibition, Dresden
39. Ritter J Jr, Jakus K, Batakis A, Bandyopadhyay N (1980) Appraisal of biaxial strength testing. J Non-Cryst Solids 38/39(Pt 1):419–424
40. Schmitt RW, Blank K, Schoenbrunn G (1983) Experimental stress analysis for the coaxial ring bending test method. Sprechsaal 116(5):397–405
41. Schoenfelder S (2010) Experimentelle und theoretische Untersuchungen zur Festigkeit dünner Siliziumsubstrate. PhD thesis, Martin-Luther-Universität Halle, Wittenberg
42. Schoenfelder S, Bagdahn J, Baumann S, Kray D, Mayer K, Willeke G, Becker M, Christiansen S (2006) Strength characterization of laser diced silicon for application in solar industry. In: 21st European photovoltaic solar energy conference and exhibition, 2006,Dresden
43. Schoenfelder S, Ebert M, Landesberger C, Bock K, Bagdahn J (2007) Investigations of the influence of dicing techniques on the strength properties of thin silicon. Microelectron Reliab 47(2–3):168–178
44. Sherman D (2003) Hackle or textured mirror? Analysis of surface perturbation in single crystal silicon. J Mater Sci 38(4):783–788
45. Shetty DK, Rosenfield AR, McGuire P, Bansal GK, Duckworth WH (1980) Biaxial flexure tests for ceramics. Am Ceram Soc Bull 59(12):1193–1197
46. Theuss H, Koller A, Kröninger W, Schoenfelder S, Petzold M (2008) Assessment of a laser singulation process for Si-wafers with metallized back side and small die size. In: 58th electronic components and technology conference, 2008, Orlando
47. Toftness R, Boyle A, Gillen D (2005) Laser technology for wafer dicing and micro via drilling for next generation wafers. In: Proceedings of SPIE–the international society for optical engineering, vol 5713, SPIE, Xsil Ltd., Loveland/Dublin
48. Tsai M, Chen C (2008) Evaluation of test methods for silicon die strength. Microelectron Reliab 48(6):933–941

49. Umeno Y, Kushima A, Kitamura T, Gumbsch P, Li J (2005) Ab initio study of the surface properties and ideal strength of (100) silicon thin films. Phys Rev B 72(16):165431
50. Vedde J, Gravesen P (1996) The fracture strength of nitrogen doped silicon wafers. Mater Sci Eng, B 36(1–3):246–250
51. Vitman FF, Pukh VP (1963) A method for determining the strength of sheet glass. Zavodskaya Laboratoriya 29(7):863–867
52. Wachtman J, Capps W, Mandel J (1972) Biaxial flexure tests of ceramic substrates. J Mater 7(2):188–194
53. Weibull W (1939) A statistical theory of the strength of materials. Ingeniörsvetenskapsakademiens Handlingar Nr. 151. Generalstabens Litografiska Anstalts Förlag, Stockholm
54. Weibull W (1951) A statistical distribution function of wide applicability. J Appl Mech 18:293–297
55. Woinowsky-Krieger S (1933) Der Spannungszustand in dicken elastischen Platten. Ing Arch 4(4):305–331
56. Yang Y, Munck KD, Teixeira RC, Swinnen B, Verlinden B, Wolf ID (2008) Process induced sub-surface damage in mechanically ground silicon wafers. Semicond Sci Technol 23(7):075038
57. Yang Y, Teixeira RC, Roussel P, Swinnen B, Verlinden B, Wolf ID (2009) Statistical analysis of the influence of thinning processes on the strength of silicon. In: Materials research society symposium proceedings, Warrendale 1112:E03–E09
58. Zarudi I, Zhang L (1996) Subsurface damage in single-crystal silicon due to grinding and polishing. J Mater Sci Lett 15(7):586–587

Chapter 18
Structure Impaired Mechanical Stability of Ultra-thin Chips

Tu Hoang, Stefan Endler, and Christine Harendt

Abstract Flexibility of ultra-thin chips is an important condition for applications in flexible electronics. Especially, mechanical reliability of ultra-thin chips is a key factor generally determining the final performance and reliability of a flexible electronics product. Structures and designs implemented on the chips are the elements that weaken the mechanical stability of the final product.

In this chapter the excellent mechanical strength of 20-μm thin silicon chips produced by the Chipfilm™ technology is discussed. The influences of different parameters, such as the number of anchors per chip edge, the backside porous silicon and the topography formed by CMOS integration, on their mechanical reliability are investigated by means of uniaxial bending tests. The experimental results are statistically evaluated by the Weibull theory, in which the resulting failure stress data are used to obtain the fracture strength and estimate failure origins of the tested ultra-thin chips.

18.1 Concepts

For ultra-thin chips with integrated circuits the mechanical stability of the base silicon material is only one factor influencing the device performance. The mechanical properties of surface layers, their structure and topography as well as the backside of the chips become more important with reduced thickness devices and might even dominate the mechanical stability of the chips. Thus, the investigation of these effects is essential for device optimisation.

In general, when we analyse the mechanical stability of ultra-thin chips it is necessary to examine the specified mechanical limits of the applied environment, meaning that environment in which the chips can react to any employed operation without mechanical failure. The ultra-thin chips used in this context have some

T. Hoang (✉)
Institute für Mikroelektronik Stuttgart (IMS CHIPS), Stuttgart, Germany
e-mail: hoang@ims-chips.de

J.N. Burghartz (ed.), *Ultra-thin Chip Technology and Applications*,
DOI 10.1007/978-1-4419-7276-7_18, © Springer Science+Business Media, LLC 2011

specific features like anchor designs on chip edges, porous silicon (PS) layers at the chip backside and CMOS circuitry on the chip surface. These features are believed to impair overall mechanical stability of ultra-thin chips.

18.2 Experimental Methodologies to Examine Structure Impaired Mechanical Stability of Ultra-thin Chips

The ultra-thin chips used in our investigations were produced using Chipfilm^{TM} technology, which was recently introduced [1, 2]. The thickness of chips fabricated in this technology is controlled prior to the CMOS integration process. This is a big advantage over the conventional backside thinning technique [3, 4].

The mechanical strength and defect formation of the produced chips are investigated by means of uniaxial bending tests. With these tests it is feasible to examine both the origin of the mechanical defects and their propagations. Flexural tests, compared with a pure tensile test, insert the maximum stress on the chip surface, resulting in a higher sensitivity against surface defects. Applying a 3-point bending test (3PB), uniaxial stress will be introduced, its maximum along the symmetry line parallel to the load's central axis. Surface and edge defects that act as rated breakpoints are distributed randomly in the stressed region of the chips. Thus, it is necessary to investigate a large number of samples and use a statistical method for analysing fracture probability. In principle, a 4-point bending test could be used alternatively. This would introduce the maximum tensile stress not only along a line but over a defined part of the chip surface so that it is more probable that at least one weak point exist in the maximum stress region. However, due to the small dimensions of the test chips and the resulting requirements on the test setup, this approach was not applied.

Guidelines to perform a uniaxial test setup for thin silicon dies are described in the SEMI standard G86-0303 [5] or ASTM standard C1161-02c [6]. However, these standards are not appropriate for ultra-thin silicon chips. Based on linear bending theory their application is limited to homogeneous materials and small deflections. The ultra-thin silicon dies produced by the Chipfilm^{TM} technology are composed of at least two different layers. Due to their high flexibility it is difficult to avoid large deflections during the flexural test. The relation between bending force and deflection is expected to be nonlinear. Therefore, it is not possible to calculate the mechanical stress from the bending force by using linear bending theory.

Experimental results are statistically evaluated by the two-parameter Weibull theory, which is commonly used to investigate the reliability of monolithic and brittle materials like silicon [7]. In fact, the measured data provide the failure force set of the tested chips, which is dependent on the geometric dimension of the specimens and the test apparatus. Thus, to be able to compare the reliability of different chip sets it is necessary to calculate the mechanical stress at failure instead of using the failure force.

In addition, also due to the mentioned nonlinear behaviour, analytical and numerical calculations are performed using the finite element method (FEM) to determine fracture stresses from fracture forces [8].

18.2.1 Chipfilm™ Process and Sample Preparations

The Chipfilm™ technology features an additive approach in defining the thickness of a thin chip rather than reducing the thickness of a conventional chip by back-thinning techniques, which are subtractive by nature.

In the Chipfilm™ process [2] buried narrow cavities within the defined chip areas are created in the silicon substrate. The cavities are fabricated by exposing a conventional bulk silicon wafer to a two-step anodic etching process in hydrofluoric acid (HF) after defining the chip areas through an n^+ mask.

The wafer is then treated with a high temperature sintering step that transfers the two-layer porous silicon region into a nano cavity-rich single crystalline silicon layer with underlying continuous cavities of ~200 nm height. Due to the elevated temperature in sintering a closed and uniform silicon surface with minimum topography is formed, enabling the simultaneous deposition of a high quality epitaxial silicon layer. By this the device quality and the final chip thickness are determined.

After the CMOS integration process, the chips are separated by trench etching along their sides, leaving a number of lateral anchors at a desired place to keep the chips in a weak attachment to the substrate. Finally, the chips are taken off by mechanical breaking of these supporting anchors. This is done by attaching a conventional vacuum tool to the chip surface in order to break off, support and transport the thin chips into the packaging cells by a post-process named Pick, Crack & Place™.

Anchor structures suffer from shear stresses at picking. They are more likely to produce cracks, which may induce failure. The anchor design is a trade-off: on the one hand they have to be strong enough to keep the chips attached to the substrate during trenching and handling processes and on the other hand sufficiently weak to be able to break off without causing defects. An evaluation process has been carried out to identify the most suitable anchor structure which increases the overall strength of Chipfilm™ chips [3], [9]. This optimised anchor structure was implemented in all samples under the test presented in this section.

There were two sets of samples used for our mechanical reliability investigation. The first sample set consisted of blank Chipfilm™ chips without a pattern. This set was used to evaluate influences of the porous silicon layer at the chip backside and of the number of anchors per chip on the mechanical strength of the chips. The second set of samples, including Chipfilm™ IC chips with completed CMOS circuits on them, was used to investigate the effect of the chip surface topography and the internal stress formed due to the CMOS integration processes on chip stability.

Details of characterised data and discussions on these aspects are illustrated in Sect. 18.3.

All tested Chipfilm™ chips were 4.63 × 4.63 mm^2 and 20 μm thick, including 18 μm epitaxy and 2 μm porous silicon beneath the thicker layer. The IC chips had additional top layers due to the CMOS circuitry.

18.2.2 3-Point Bending Test

In our investigation the mechanical strength of ultra-thin chips was carried out by means of a uniaxial bending test using the 3-point bending (3PB) setup. A general mechanism of the 3PB setup is presented in Fig. 18.1a, and the mechanical parts, including the loading roller (a knife) and horizontal slit (supporters) of the 3PB assembly used, are shown in Fig. 18.1b.

The sample is located in the horizontal slit and bent upwardly towards a small vertical slit built up by two interchangeable metal plates. The rounded edges of the narrow slit form the supports. When pushed by the loading roller the chip is able to bend into this horizontal slit of 2 mm height and 1 mm width.

The bending tests were performed using a Dage 4000 Multifunction Bond Tester. The used load cartridge is able to apply and measure a maximum load of 10 N within a total accuracy of ±0.25% (±24.25 mN). Moreover, the testing machine captures the displacement of the loading roller continuously accurate to a micrometre per millimetre of traversed path.

The loading roller moves in the vertical slit with a slackness of 50 μm, ensuring a frictionless and straight-line movement. The knife is 0.8 mm thick and tapers towards the tip. The tip fillet radius is about 50 μm. A bearing roller helps the setup meet the parallelism requirements between load and sample surface as it permits self-adjustment of the knife in the horizontal plane.

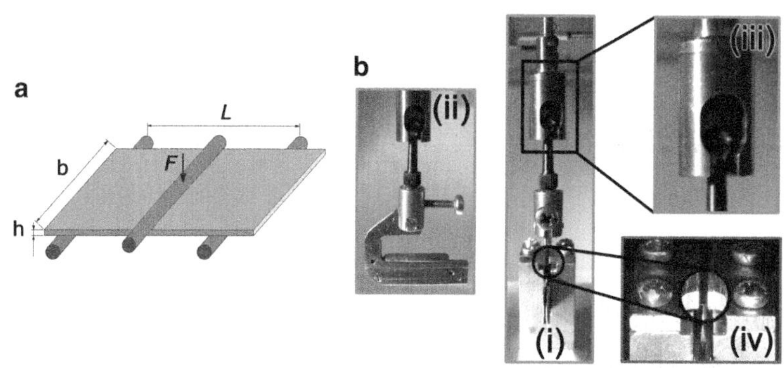

Fig. 18.1 (**a**) General 3-point bending test (*3PB*). (**b**) The 3PB test assembly: mechanical parts in 3PB (i) are associated with the bond tester; the loading roller (ii) is directly connected to the load cartridge via roller bearing (iii); and magnification of the horizontal slit with a specimen (iv)

18.2.3 Weibull Statistics and FEM Analysis

Due to the fact that the defects and microscopic irregularities are distributed randomly over the specimen volume, there is a high scattering in the measured force at mechanical failure. Thus, a suitable statistical method to evaluate the experimental results is necessary. According to the ASTM standard C1238-07 Weibull statistics are such a suitable approach to derive reliable material parameters for advanced ceramics [7].

The two-parameter Weibull theory describes the probability of failure P_{f} that depends on uniaxial fracture stress σ.

$$P_f = 1 - \exp\left[-\left(\frac{\sigma}{\sigma_0}\right)^m\right] \text{ with } \sigma, \sigma_0 \text{ and } m > 0. \tag{18.1}$$

The Weibull parameter σ_0 is the characteristic stress where mechanical failure occurs with probability of 63.2% (CDS = characteristic die strength). Beyond that the minimal die strength (MDS) indicates the failure stress with probability of 1%. The Weibull modulus m is a value that is related to the scattering level of measured data. The higher the Weibull modulus the smaller is the scattering of the data.

During the bending test, the applied force and displacement relationship is measured. The fracture force depends on geometrical aspects like sample parameters and on the span width of test setups. Therefore, in order to compare the mechanical strength of different geometries and loading conditions one has to determine fracture stress from the measured fracture force.

In the classical beam theory for 3PB the bending stresses on the sample surfaces decrease linearly towards the outer edges with a maximum value at break that can be obtained by:

$$\sigma_1 = \frac{3FL}{2bh^2}, \tag{18.2}$$

where σ_1 is the first principal stress at the chip surface at the moment of failure, F is the applied load at failure, h is the die thickness (20 μm), b is the die width (4.63 mm) and L is the span width (1 mm).

Ultra-thin silicon chips are extremely flexible. Large displacements during bending cause geometrical nonlinearity effects. As a consequence, a nonlinear relationship between force and stress occurs. The nonlinearity increases with decreasing die thicknesses, increasing displacements, or increasing loads. Moreover, the nonlinear effects vary with specimen and test geometry [9].

Numerical investigations using finite element analysis have been performed to determine the correct relationship between fracture stresses and fracture forces when geometrical and structural nonlinearity effects are considered. A 3D finite element model (ANSYS) can be seen in Fig. 18.2. Due to geometrical symmetry

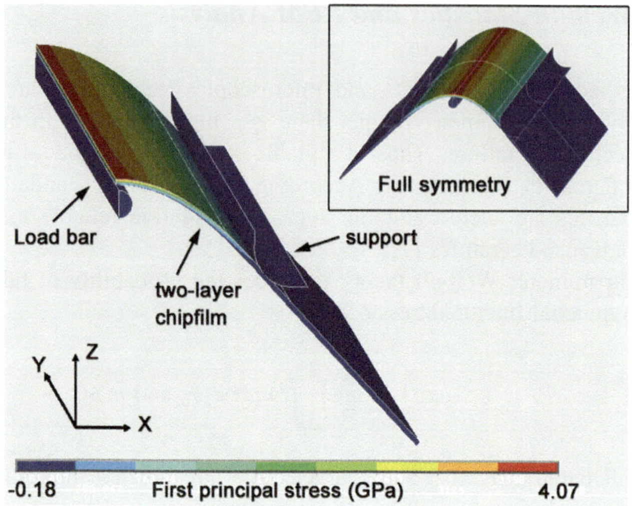

Fig. 18.2 Quarter symmetry of 3D ANSYS FE model: cylinder load bar of 50 μm diameter; thin chip sectioning of 18 μm epitaxy Si and 2 μm porous Si; and supporter having a filet radius of 120 μm. Applying a force of 2.56 N on the load bar results in a 364 μm displacement and a maximum tensile stress of 4.07 GPa on Si surface

Table 18.1 Input parameters used for the ANSYS FEM model: *E* stands for Young's modulus measured in GPa and υ stands for Poisson's ratio of materials

Assembly parts	Simulation parameters	
Loading roller and supporters	$E_{steel} = 210$ GPa; $\upsilon_{steel} = 0.3$ Element type: high order solid186 element	
Chipfilm^{TM} chips	Element type: high order solid-shell (solidsh190) element Sectioning of 18-μm epitaxial Si and 2-μm sintered porous Si	
	Epitaxial Si Anisotropic stiffness matrix of crystalline Si as <110> direction is along *x*-axis (GPa): $\begin{pmatrix} \mathbf{194.4} & \mathbf{35.2} & \mathbf{63.9} & \mathbf{0} & \mathbf{0} & \mathbf{0} \\ \mathbf{35.2} & \mathbf{194.4} & \mathbf{63.9} & \mathbf{0} & \mathbf{0} & \mathbf{0} \\ \mathbf{63.9} & \mathbf{63.9} & \mathbf{165.8} & \mathbf{0} & \mathbf{0} & \mathbf{0} \\ \mathbf{0} & \mathbf{0} & \mathbf{0} & \mathbf{79.6} & \mathbf{0} & \mathbf{0} \\ \mathbf{0} & \mathbf{0} & \mathbf{0} & \mathbf{0} & \mathbf{79.6} & \mathbf{0} \\ \mathbf{0} & \mathbf{0} & \mathbf{0} & \mathbf{0} & \mathbf{0} & \mathbf{50.9} \end{pmatrix}$	**Sintered porous Si** Isotropic material with $E_{PS} = 70$ GPa and $\upsilon_{PS} = 0.22$

only a quarter of the system is modelled to save the computational time. The loading knife was modelled by a simple cylinder (radius 50 μm).

High order 3D solid-shell elements with sectioning were used for the silicon specimen. Table 18.1 lists all input parameters used for the FEM model. There are

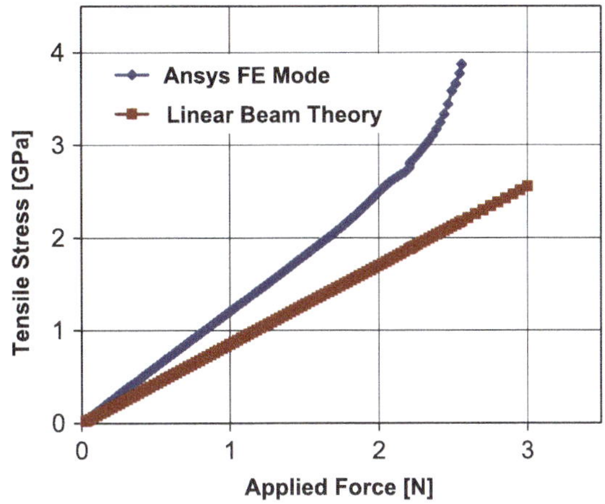

Fig. 18.3 The first principle stress (tensile stress) on silicon versus the applied force obtained by ANSYS finite element (*FE*) model compared to that calculated by the linear beam theory

two contact pairs in this model: One is flexible-to-flexible contact between the load and chip surface and another is the rigid-to-flexible contact between the support and other side of the chip allowing for the consideration of dynamic friction effects (structural nonlinearity). Due to sliding and large displacement effects the Lagrange and penalty method is a proper solution for this analysis [10].

The static analysis for considering stress stiffening and large deflection effects was used in our FEM model because in ultra-thin structures the bending stiffness is very small compared to axial stiffness, leading to large strain or large deflection effects, which in turn cause nonlinearities in stiffness.

In order to determine reliable fracture stresses one has to find the most suitable model (for each experiment set) that fits the measured force-displacement relationship best. The model fit can be achieved by modifying several parameters, such as the friction between supporters and chip surface. Figure 18.3 presents the obtained simulation data, which show the best fit to experiment, for the case of porous silicon under compression (see Sect. 18.2.1) compared to the data calculated by the linear beam theory.

18.3 Experimental Results and Discussions

The ultra-thin chips produced by the Chipfilm™ technology have several interior characteristics, which have particular influences on their overall mechanical strength, such as a sintered porous silicon layer of 2 μm at the chip backside, a certain number of remained parts of anchor structures on the chip-edges after the Pick, Crack & Place™ post-process and the quality of the epitaxy process. Since the Chipfilm™ layer is always grown by a high quality and-well controlled epitaxy process, minor effects of this epitaxy process were ignored in this experiment. In the following subsections the influences of anchors, porous silicon layer and IC chip surface topography are presented.

18.3.1 Evaluation of Anchor Effects on Chip's Mechanical Strength

As described previously, the anchor designs must fulfil two contradictive requirements: on the one hand the anchor connection of the chip to the substrate has to be strong enough to keep the chip on its position during the trenching and handling processes but on the other hand the anchors must be sufficiently weak to enable the chip picking process while minimising the resulting mechanical defects. It is assumed that the character of these defects after the Pick, Crack & Place^{TM} post-process will affect the chip mechanical strength. To investigate this effect it is necessary to separate it from the influence of the porous silicon layer. Therefore, the test for the difference of fracture stress between five anchors and six anchors per edge was initially carried out by applying compressive stress on the PS-layer.

In Fig. 18.4 the Weibull distribution of failure probabilities of 20-μm blank chips for two cases of five and six anchors per chip edge is presented. By comparing the maximum failure stresses of chips with six anchors per edge (diamond symbols) with those of five anchors per edge (square symbols) one can see that the characteristic strength of chips with six anchors per edge (3,305 MPa) seems to be much higher than that of chips with five anchors per edge (918 MPa). This can be explained as follows.

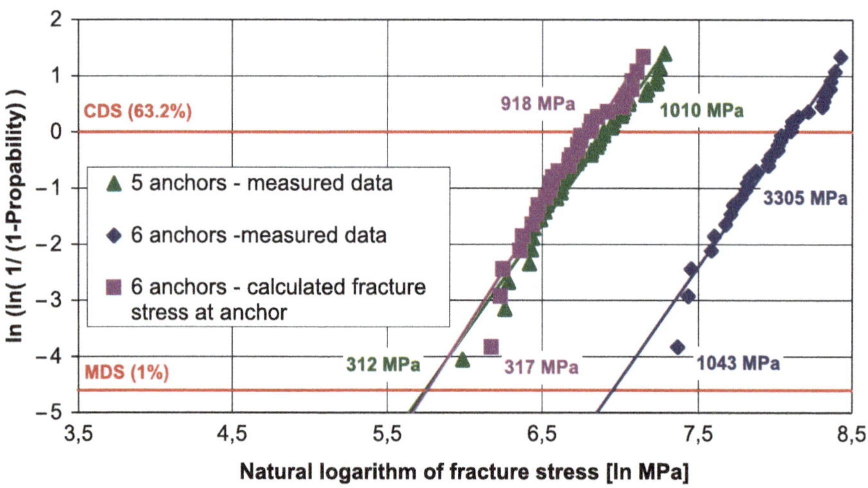

Fig. 18.4 Evaluation of anchor effects with porous silicon under compression: If the chips have five anchors per edge (*triangle*) one anchor stays within the maximum stress region, so the chips always break at the middle (or at anchor); however, if the chips have six anchors per edge with no anchor sitting in the maximum stress region the chips can break in the middle or at the anchor. The fracture stress at the anchor position can be calculated (*squares*) from the measured data gathered from the samples with six anchors (*diamonds*). *CDS*, characteristic die strength of the measured specimen set with a failure stress at 63.2% of the total failures; *MDS*, minimum die strength, which corresponds to a probability of 1%

From the 3-point bending theory it becomes evident that the applied force can be correlated to the stress at the load position (18.2). Due to the symmetrical test setup (Fig. 18.1) for chips with six anchors per edge, two anchors are inside the stress region but located at a distance to the loading force that depends on the geometry. With the assumption that the anchor acts like a rated breakpoint, it has to be stated that for six anchors per edge there are two possibilities of locations where failures can occur: (1) at the chip centre, where highest tensile stress exists; or (2) at the anchor position (the weak point). On the other hand, when five anchors per edge are present, one anchor is located at the load position (the middle anchor), where most breakage occurs. Furthermore, if the anchor is likely a rated breakpoint, the bending moment to break the five-anchor chips (at the middle) should be the same as that needed to break six-anchor chips at the anchor.

When force is applied to the load bar with the specimen set of six anchors the maximum stress distributed to the chip centre is much higher than that distributed to the anchor position. This relation can be obtained by a linear model, as presented in Fig. 18.5. The dependence between the maximum bending moment (M_{max}) and the bending moment at the anchor (M_{eff}) can be derived from the linear bending theory using the geometrical dimension of the chip with six anchors per edge. With the geometrical parameters of the measured chips and the used 3PB setup, it is easy to see that M_{eff} is just about 17% of the maximum value M_{max}.

However, due to the nonlinear effects the real value of M_{eff} is higher than the calculated value. This is approved by ANSYS FEM analysis. The extracted data from the simulation show that M_{eff} is about 32% of M_{max} which means it is almost two times the value obtained by linear theory. Taking these aspects into consideration, one can acquire data of the fracture stress at the anchor position for six-anchor-per-edge chips from the measured data (see the square symbol data in Fig. 18.4). It shows that the measured characteristic die strength of five-anchor-per-edge chips is just a little bit higher than the calculated strength of six-anchor-per-edge chips, assuming the break at the anchor (1,010 MPa compared to 918 MPa). This can be explained from the fact that the six-anchor-per-edge chips bring four anchors into the stressed area compared to just two anchors for

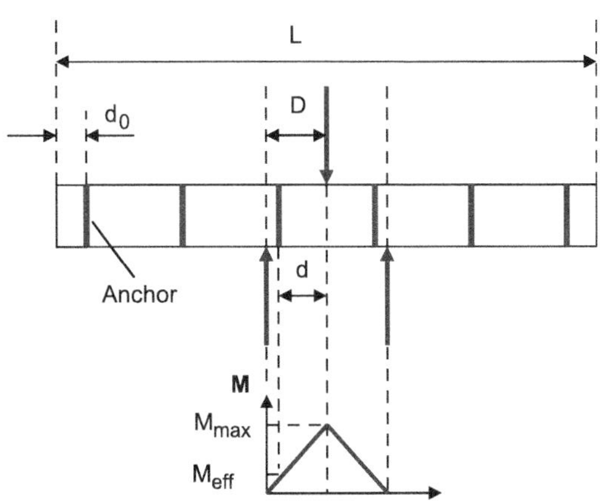

Fig. 18.5 The linear model used for calculating bending moment at anchor position (M_{eff}) from the applied maximum moment (M_{max}) for the case of six anchors per chip edge

the five-anchor-per-edge chips, leading to a lower failure probability for the later. This supports the assumption that the failures on six-anchor-per-edge samples started at the anchor, which acts as a rated breakpoint.

The influence of the anchor on the chip's mechanical strength can be further investigated if we use the same specimen set but set the PS layer under tensile stress. However, in this case the effect of the PS layer must also be considered, as it is discussed in detail in the next subsection.

The Weibull distribution of measured data for the chips with six and five anchors per edge is shown in Fig. 18.6. The difference of the characteristic die strength for five- and six-anchor samples is now just 114 MPa (399 –285 MPa) compared with 2,295 MPa (3,305 –1,010 MPa) in the previous experiment. This indicates that the anchor effect is less important when the PS is in tension. Moreover, the Weibull distribution of five-anchor samples shows a multi-slope curve, which indicates different failure mechanisms occurred this experiment. It also means that when the PS layer is under tension we cannot distinguish influences of anchors from those of the PS layer on the mechanical strength of the ultra-thin chips. Therefore, the previous calculation of the fracture stress at the anchor from the measured data is invalid here.

In the next subsection the effects of the porous silicon layer on the mechanical strength of Chipfilm^{TM} chips are investigated.

18.3.2 Evaluation of Influences of the Porous Silicon Layer

As mentioned, the ultra-thin chips produced by the Chipfilm^{TM} technology consist of at least two layers, a porous silicon layer and the epitaxial layer. Due to this

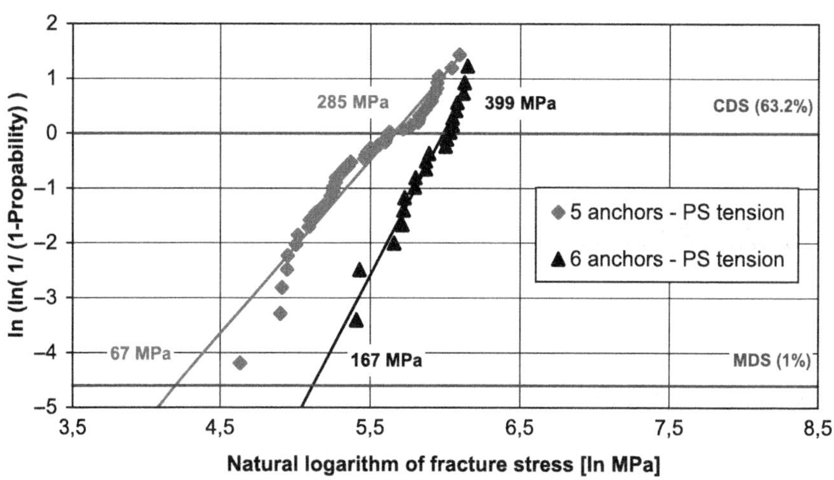

Fig. 18.6 The Weibull distribution of five- and six-anchor specimens with the PS layer under tension. It cannot be distinguished between the influences of anchors and those of the PS layer on the mechanical strength of Chipfilm^{TM} chips when PS is under tension

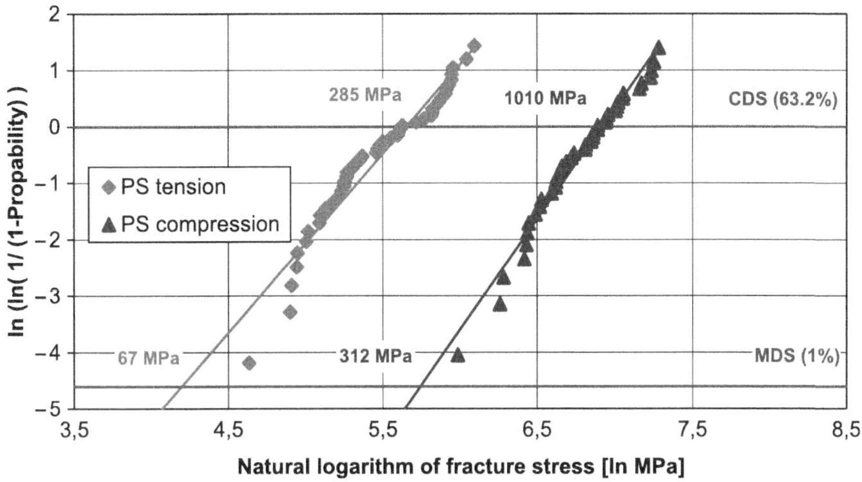

Fig. 18.7 Weibull distribution of failure probabilities of blank Chipfilm™ chips with five anchors per edge when the porous Si is under compressive stress (*triangle*) and under tensile stress (*diamond*). It shows that the tensile-stressed PS layer further reduces the characteristic die strength by about 72% (From 1,010 MPa into 285 MPa)

material inhomogeneity, an anisotropic behaviour regarding the mechanical strength is expected. To investigate this effect the characteristic die strength of the sample set with the PS layer under tension was compared with that one with PS under compressive stress. To ensure that the anchor effects are always included in this experiment the specimen set of five-anchor-per-edge chips was used.

The Weibull distributions of failure probabilities for both circumstances when the PS layer is compressive and tensile are displaced in Fig. 18.7

Where the anchor effects are included the extracted CDS for the PS under tension (285 MPa) is rather small compared to that for the PS under compression (1,010 MPa), indicating that the PS layer further reduces the mechanical strength of the Chipfilm™ chips.

Summarising, one can conclude that the anchors located at the edges of the chips have a big influence on the mechanical strength of the chips if the epitaxial layer is tensile stressed. Apart from that the number of anchors per chip edge does not affect the mechanical stability significantly. If the epitaxial layer is under compressive stress (which means the PS layer is under tensile stress) the dominating effect on the chip stability is probably caused by the PS layer itself, which reduces the mechanical strength of the whole chip.

Therefore, optimising the anchor design and improving the PS layer quality (the homogeneity and pore size) are important to increasing mechanical reliability of the ultra-thin Chipfilm™ chips. Beyond that, for some applications it is possible to avoid external mechanical stress in both directions. Armed with the knowledge about dependence of mechanical stability of Chipfilm™ chips on stress direction, we can improve the strength of the complete system.

18.3.3 Evaluation Effects of Ultra-thin IC Chips

Ultra-thin Chipfilm™ chips after CMOS integration processes (ultra-thin IC chips) were also evaluated for mechanical strength using the mentioned 3PB setup. In this subsection the evaluation data are presented.

Since the ultra-thin IC chips have many additional layers and include different patterns on their surface, after several high temperature processes the final topography of the chip surface and internal stresses should reduce their overall mechanical stability. To investigate this, the 3PB test was performed with a specimen set of IC chips, with the PS layer under compression. This was done only because PS layer effects may be ignored if the porous silicon is in compressive stress. Moreover, to avoid the anchor effect in this analysis only four anchors distributed equally per chip edge were used and, thus, no anchor was located within the stressed volume of the specimens.

The Weibull distribution of the measured data is shown in Fig. 18.8. The Weibull modulus of 2.3 indicates a high scattering data. This can also be seen in the multi-slope distribution. The CDS value of 690 MPa is much lower than that of blank Chipfilm™ chips presented previously. This all can be explained by the complicated surface features and internal stress of the tested IC chips. Different origins of failures are the reason for different slopes of the Weibull distribution.

Based on this experiment it becomes evident that the CMOS circuitry on the IC chips considerably reduces the mechanical stability of the ultra-thin Chipfilm™ chips, and the effects of the porous silicon and anchors on chip mechanical strength become less important compared to effects of the CMOS circuitry.

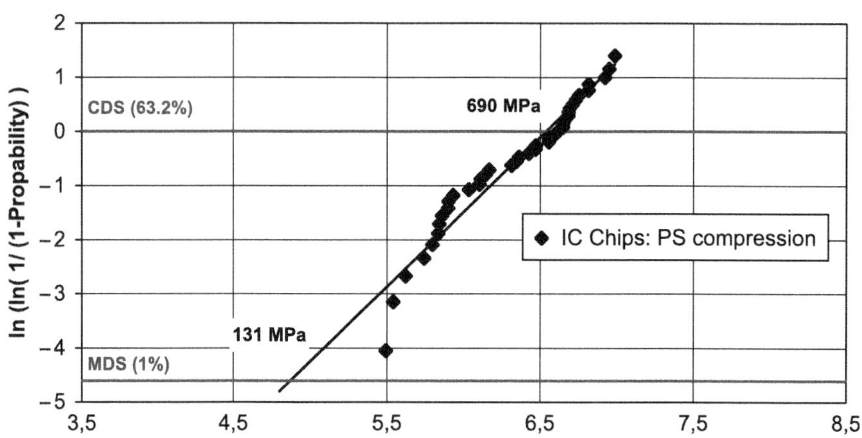

Fig. 18.8 Probability of failure for ultra-thin IC chips when the porous silicon is under compressive or the chip surface is under tensile stress. The multi-slope distribution indicates that many failure mechanisms are involved in the entire mechanical stability of the IC chips

18.4 Conclusions

The structure impaired mechanical stability of ultra-thin Chipfilm™ chips was investigated thoroughly with respect to the different effects from the anchor structures on the chip edges, the porous silicon layer at the chip backside and the CMOS circuitry on the chip surface. This was done using the uniaxial bending test through the 3-point bending setup. Several conclusions can be drawn from this:

- Ultra-thin Chipfilm™ chips show excellent mechanical strength when the PS layer is under compression.
- The nonlinear force-stress behaviour of ultra-thin chips derives from many sources, such as the two-layer system, the anisotropic property of crystalline silicon and geometrical and structural properties. Therefore, an improvement of the uniaxial bending test assembly intended to reduce the uncertainties stemming from these nonlinearities is essential.
- The effects of both the PS layer and the anchors significantly impair the overall mechanical stability of the ultra-thin chips. Therefore, optimising the anchor designs and improving PS layer quality will enhance the mechanical reliability of ultra-thin Chipfilm™ chips.
- CMOS circuitry and the internal stress on the IC chips are elements that also dramatically reduce the mechanical stability of blank Chipfilm™ chips. Improved materials and special designs for stress relief have to be developed to achieve better mechanical behaviour of ultra-thin IC chips.

Acknowledgments The authors would like to thank Astrid Kiss for her contributions to this work and also Lutz Diesing and Peter Ostrowski for their assistance in carrying out the mechanical experiments. The authors also thank the Landesstiftung Baden-Württemberg for the financial support (project Chipfilm™).

References

1. Zimmermann M, Burghartz JN, Appel W, Remmers N, Burwick C, Würz R, Tobail O, Schubert M, Palfinger G, Werner J (2006) A seamless ultra-thin chip fabrication and assembly process. IEDM Tech Dig 2006:1010–1012
2. Burghartz JN, Appel W, Rempp HD, Zimmermann M (2009) A new fabrication and assembly process for ultrathin chips. IEEE Trans Electron Devices 56(2):321–327
3. Burghartz JN, Harendt C, Hoang T, Kiss A, Zimmermann M (2008) Ultra-thin chip fabrication for next-generation silicon processes. In: Proceedings of the IEEE BCTM, Capri, Italy, pp 131–137
4. Landesberger C, Klink G, Schwinn G, Aschenbrenner R (2001) New dicing and thinning concept improves mechanical reliability of ultra-thin silicon. In: Proceedings of the International Symposium on Advanced Packaging Materials, Braselton, GA, pp 92–97
5. Global Assembly and Packaging Committee (2003) Test method for measurement of chip (die) strength by mean of 3-point-bending. SEMI Standard G86-0303, Japan
6. ASTM International (2008) Standard test method for flexural strength of advanced ceramics at ambient temperature. ASTM Standard C1161–02c, West Conshohocken, PA

7. ASTM International (2007) Standard practice for reporting uniaxial strength data and estimating Weibull distribution parameters for advanced ceramics. ASTM Standard C1239-07, West Conshohocken, PA
8. Schoenfelder S, Ebert M, Landesberger C, Bock K, Bagdahn J (2007) Investigations of the influence of dicing techniques on the strength properties of thin silicon. Microelectron Reliab 47:168–178
9. Kiss AV (2009) Mechanische Eigenschaften von Duennschichtsilizium. Diploma thesis, University of Stuttgart, Germany
10. Release 12.0 Documentation for ANSYS Contact technology guide. ANSYS Inc. www.ansys.com Last accessed on May 2010

Chapter 19
Piezoresistive Effect in MOSFETS

Nicoleta Wacker and Harald Richter

Abstract Mechanical deformation changes the silicon material's electrical resistivity (Smith, Phys Rev 94(1):42–49, 1954). This is an important property of the material. The effect known as 'piezoresistive' has found many applications over time. In micro-electro-mechanical systems (MEMS), for example, it has enabled the integration of pressure sensors with complementary metal-oxide semiconductor (CMOS) circuits that translate the stress into voltage signals.

The piezoresistive effect has been also observed and studied in the metal-oxide-silicon field effect transistor (MOSFET), the principal component of a CMOS integrated circuit (IC). In transistors the strain affects the mobility of the carriers and as a result the drain current and the speed of the transistor.

Although scaling tends to achieve its limitations, the need to further increase the MOSFET performance by having more power and speed remains. Based on mobility enhancement due to the stress/strain applied to the MOSFET channel, the 'strained Si' technique has been chosen as a new degree of freedom to further improve MOSFET performance and to defeat scaling limitations.

In this chapter we discuss aspects related to the piezoresistive effect in bulk Si and its modifications in MOSFET Si inversion layer. We analyse the strain effects on Si electronic band structure and its transport properties. Piezoresistive coefficient values from literature for bulk Si and MOSFETs under uniaxial strain are presented and compared.

19.1 Strain Effects: From Si to MOSFETs

In this section we introduce important concepts of mechanics such as stress and strain. Then, we present an overview of the stress effects on the physical and transport mechanisms of the Si crystal. Further, we discuss the influence

N. Wacker (✉)
Institut für Mikroelektronik Stuttgart, Allmandring 30a, 70569 Stuttgart, Germany
e-mail: wacker@ims-chips.de

J.N. Burghartz (ed.), *Ultra-thin Chip Technology and Applications*,
DOI 10.1007/978-1-4419-7276-7_19,

of stress on the conductivity of Si-based metal-oxide-silicon field effect transistor (MOSFETs).

19.1.1 Stress and Strain

Any deformation [1] of a material induced by stretching, compressing or twisting can be described by the concepts of stress and strain. Stress (σ) is expressed as force per unit area of material. Its unit in SI is Pascal (Pa). Strain (ε) quantifies the change of the relative positions of the material atoms or of the lattice constant subjected to stress. Strain can be generated either by applied stress or by stress arising from lattice-mismatch film growth and is a unitless parameter.

The Si crystal has cubic symmetry and lattice constant $a = 0.54$ nm [2], corresponding to minimum interatomic energy (IAE). An applied axial force can stretch/compress the lattice, increasing the attractive/repulsive IAE (Fig. 19.1).

Different types of stress applied to the Si crystal (uniaxial, biaxial, hydrostatic) are converted into mechanical strain. Strain can change both bond length and angle between bonds. For example, [110][1] applied tensile strain along bonds pulls atoms apart, elongating the interatomic bonds along the same direction (Fig. 19.2.). This leads to compression along [$\bar{1}$10], shortening and rotating of the corresponding bonds [3].

Strain causes not only the variation of the relative positions of the atoms but also changes the Si electronic band structure, as discussed in the next subsection.

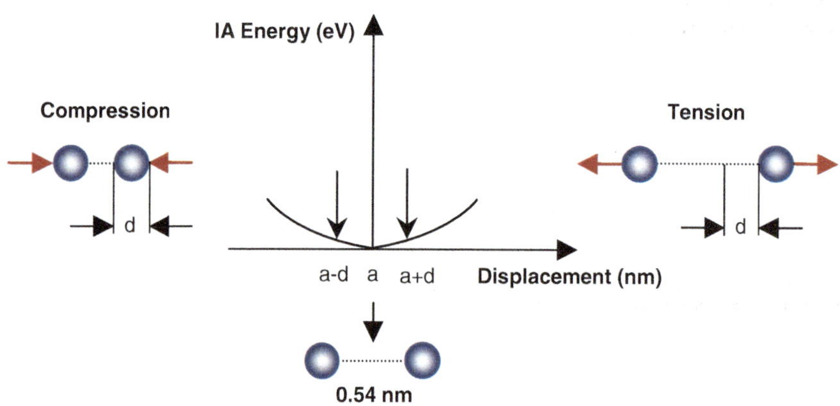

Fig. 19.1 Stretching and compression of Si interatomic bond

[1] [110] denotes a crystallographic direction, unlike <110> that denotes a family of directions related by symmetry operations.

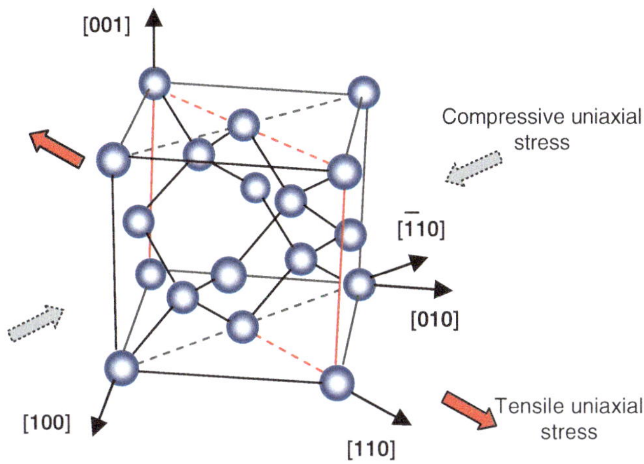

Fig. 19.2 Example of tensile uniaxial stress along the [110] Si crystallographic direction, inducing compressive uniaxial stress along the [$\bar{1}$10] direction

19.1.2 Strain-Induced Enhancement of Si Electron Mobility

Strain influences differently electrons and holes. In this section we describe briefly the strain effect on charge transport at room temperature.

In a crystalline solid like Si the relationship between energy and momentum $E(k)$ is given by the band structure of the crystal. The conduction band structure of crystalline Si along the [100], [010] and [001] directions comprises six degenerate valleys with minima at Λ points [2]. The degeneracy is not accidental but reflects the cubic symmetry of Si crystal. The ellipsoidal shape of the valleys has different curvatures and thus two effective masses [4]: a transverse $m_t = 0.19\,m_0$, and a longitudinal one, $m_l = 0.97\,m_0$, where $m_0 = 9.11 \times 10^{-31}$ kg is the free electron rest mass. The total effective mass for conductivity ($0.26\,m_0$) is a function of both transversal and longitudinal effective masses [5]:

$$m^* = \left[\frac{1}{6}\left(\frac{2}{m_l} + \frac{4}{m_t}\right)\right]^{-1} \tag{19.1}$$

Strain lifts the degeneracy and determines the repopulation of the valleys with electrons. This repopulation is accompanied by m^* variation.

For a better understanding, we present the case of $<110>$ applied uniaxial strain. The conduction band position (at 300 K), calculated from the valence band (E_v) and the band gap (E_g) positions [6] is

$$E_c(\Delta_6) = E_v + E_g \Rightarrow E_c = (-7.03 + 1.17)\text{eV} = -5.85\text{eV} \tag{19.2}$$

To account for the energy shifts of the ellipsoids induced by strain, linear deformation potential theory [7] can be applied, based on the Δ band Si deformation potentials and on strain components. Uniaxial strain $\varepsilon_{<110>}$ splits the conduction band into two sub-bands Δ_2 and Δ_4 (Fig. 19.3b) with energies

$$\begin{aligned} E_c(\Delta_2) &= -2.83 \cdot \varepsilon_{<110>}(\text{eV}) \\ E_c(\Delta_4) &= +5.27 \cdot \varepsilon_{<110>}(\text{eV}). \end{aligned} \tag{19.3}$$

1% Strain leads to ~80 meV or ~3 kT (at room temperature) energy difference between the Δ_2 and Δ_4 sub-bands. As the tensile strain increases, more electrons move into the lowest energy sub-band Δ_2 (Fig. 19.3c). Electron repopulation has two effects: total effective mass m^* reduction and momentum relaxation time τ variation ($1/\tau$ is the intervalley scattering rate on phonons and impurities). Both contribute to mobility variation.

The mobility of a particle is the ratio between the reached constant average velocity $|\vec{v}|$ and the applied electric field $|\vec{E}|$, expressed as (Drude model)

$$\mu = \frac{|\vec{v}|}{|\vec{E}|} = \frac{q \cdot \tau}{m^* \cdot m_0} \tag{19.4}$$

where q is the electron charge. However, the main factor that leads to mobility variation is the conductivity effective mass m^* reduction [8].

Both <110> uniaxial and biaxial strain have similar effects on Si conduction band for the (001) Si surface plane [3], except <100> uniaxial strain that splits the conduction band into three sub-bands.

19.1.3 Strain-Induced Enhancement of Si Hole Mobility

The valence band of intrinsic, unstrained p-type Si in **k** space is very complex. It consists of three bands with energy maxima at the Γ point. The first two,

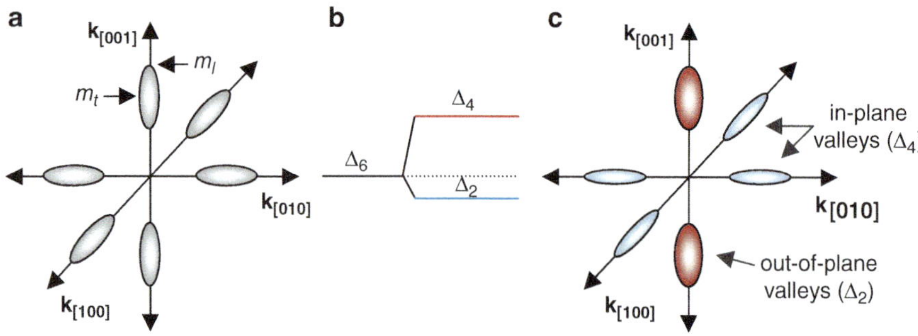

Fig. 19.3 Schematic of the Si conduction band structure in reciprocal **k**-space, for (**a**) unstrained Si and in the presence of <110> uniaxial tensile strain: (**b**) conduction band splitting and (**c**) variation of the valleys effective masses

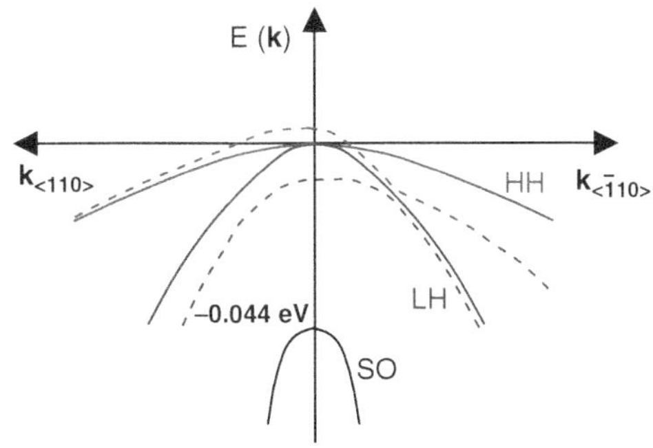

Fig. 19.4 Schematic of the Si valence band structure: without strain (*solid lines*) and with <110> uniaxial tensile strain (example: *dashed lines*) (After [4])

degenerated at $\mathbf{k} = 0$, are called the heavy holes (HH) and the light holes (LH) bands. The spin-orbit interaction shifts the split-off (SO) band at –0.044 eV ($k = 0$) (Fig. 19.4). This band is less important for hole transport [5].

The properties of each band depend on the crystallographic direction and on energy. For instance, the HH band effective mass is the largest of all three bands and therefore is preferentially occupied by holes. The effective conductivity hole mass ($0.36m_0$) is a combination of HH ($m^*_{\mathrm{HH}} = 0.49m_0$) and LH ($m^*_{\mathrm{LH}} = 0.16m_0$) effective masses [9]:

$$m^* \cong \frac{3}{\frac{1}{m^*_{\mathrm{HH}}} + \frac{1}{m^*_{\mathrm{LH}}}} \tag{19.5}$$

The application of strain lifts not only the degeneracy of HH and LH bands but also warps the valence band [10]. This results in the variation of the total effective hole mass. Band warping burdens the energy calculation.

As an example, at low stresses (<1 GPa), a valence band splitting of about 20–30 meV was reported [3]. Hole mobility enhancement is predominantly determined by the conductivity mass change. The investigations have shown that the highest holes' mobility enhancement was acquired for (110)/<110 > compressive strain [10].

19.1.4 Transport Properties in Strained Si

Strain changes the Si electronic band structure, thus having an effect on electronic transport properties, which are further addressed.

The Si crystal, with its high cubic symmetry, has isotropic conductivity. Hence, the current flow direction is parallel to the electric field. The ratio of the current density to the electric field is known as conductivity (σ). Resistivity (ρ) is the inverse of conductivity. Both are related to Si band structure through the concentration of electrons (n) and holes (p) and their mobilities μ_n and μ_p:

$$\sigma = \frac{1}{\rho} = q(n\mu_{\rm n} + p\mu_{\rm p}) \tag{19.6}$$

$qn\mu_n$ contains the contribution of all six ellipsoids of the conduction band (Fig. 19.3a) and $qn\mu_p$ the contribution of the HH and LH components [11].

Strain breaks Si cubic symmetry. As a result its conductivity is no longer isotropic. As a piezoresistive material, the resistivity of Si varies with stress (σ) [1] (both ρ and σ are second order tensors):

$$\frac{d\rho}{\rho} \cong -\frac{d\mu}{\mu} = -\Pi \cdot \sigma. \tag{19.7}$$

Π is the piezoresistivity matrix, defined by a set of three *fundamental* piezoresistive coefficients (Π_{11}, Π_{12}, Π_{44}). Each of them give us straightforward information about how much current enhancement can be achieved under particular stress.

These coefficients were determined for the first time by Smith [1] from resistance measurements under stress. Π_{11} represents the *longitudinal* coefficient (0° between the directions of the stress and that of current flowing along {100}), Π_{12} is the transversal coefficient (90° between the directions of the stress and of the current flowing along {100}),[2] and Π_{44} is the shear coefficient. They depend on material, its doping type and concentration. Their measured values are given in Table 19.1 together with the piezoresistive coefficients (Π_S, Π_L, Π_T), which represent linear combinations of the fundamental coeffitients.

(Π_L, Π_T) are often measured for MOSFETs inversion layer, as presented below.

19.2 Strain-Induced Effects in MOSFETs

In this section we discuss the physical aspects of carrier mobility variation in the inversion layer of strained-Si MOSFETs, underlying the differences to strained bulk Si. Relevant for ultra-thin and flexible chips are the effects induced by external applied stress, discussed here. Moreover, we summarise the piezoresistive

Table 19.1 Fundamental piezoresistance coefficients (Π_{11} (stress || current)$_{<100>}$, Π_{12} (stress ⊥ current)$_{<100>}$, Π_{44} (shear coefficient) of lightly doped Si (× 10^{-12} Pa^{-1}) at 298 K [1] and (Π_S, Π_L, Π_T) linear combinations of the fundamental coeffitients

Material	Resistivity (Ω-cm)	Π_{11}	Π_{12}	Π_{44}	$\Pi_S = \Pi_{11} + \Pi_{12}$	$\Pi_L = (\Pi_S + \Pi_{44})/2$	$\Pi_T = (\Pi_S - \Pi_{44})/2$
n-Si	11.7	1,022	–534	136	488	312	176
p-Si	7.8	–66	11	–1,381	–55	–718	663

[2] { } denotes a family of planes.

coefficients of bulk MOSFETs based on results from literature and compare them with those of bulk Si.

Unlike bulk Si, two new features must be considered for the carrier transport in MOSFETs: (1) the electric confinement and (2) Si/SiO_2 interface scattering.

1. The 2D potential well-created by the MOSFET gate bias [11] confines the charge carriers to the surface channel. This quantises the conduction band energies and changes the effective mass in the out-of-plane direction by removing the degeneracy between the in-plane and out-of-plane valleys (similar aspects are shown in Fig. 19.3c [12]. Therefore, in the inversion layer, even in the absence of strain, the energetic levels are nondegenerate (Fig. 19.5.a) due to the presence of the transverse electric field. The energy difference between the Δ_2 and Δ_4 sub-bands (ΔE_0) varies with the intensity of the transverse effective field.

   As example, for unstrained n-MOSFET, Sun [3] obtained $\Delta E_0 = \sim 121$ meV, split between Δ_2 and Δ_4, at $T = 300$ K, for an effective gate field of magnitude 1 MV/cm and $1 \times 10^{13}/\text{cm}^2$ inversion electron density. Moreover, 80% of the electrons occupy the Δ_2 valleys. Hence, the electron repopulation induced by any strain would not be significant because most electrons already occupy the low energy states.

   In the case of a p-channel MOSFET, the hole mobility depends on Si surface and channel orientation. Thus, the lowest hole mobility was obtained for (001)/<110> p-MOSFETs and the highest one for (110)/<110> [10]. The confinement field shifts the degeneracy of HH and the LH sub-bands (Fig. 19.4). A higher split leads to lower interband scattering. For example, for (001)/<110> at $T = 300$ K, 1 MV/cm surface field, induces a splitting between HH and LH bands of ~25 meV [13], the HH sub-band being the ground band. For (110)/<110> the same field determines a higher splitting between the HH and LH sub-bands, lower interband scattering and hence, higher hole mobility.
2. Besides scattering on phonons and impurities, the carriers in the inversion layer experience interface scattering not present in bulk Si. Its sources are Coulomb scattering on charges trapped at the oxide/Si interface and interface roughness

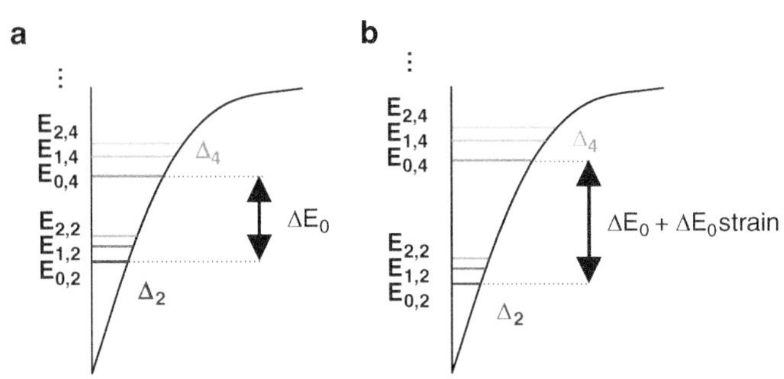

Fig. 19.5 Discrete energy levels formed within the 2D potential of a metal-oxide-silicon field effect transistor (*MOSFET*) inversion layer (**a**) unstrained and (**b**) <110> uniaxially strained

scattering [3]. The interface scattering increases with the transverse electric field, creating a limiting factor of the carrier's mobility at room temperature.

The presence of the strain in the MOSFET channel induces an additional shift of the energy levels $\Delta E_0^{\text{shift}}$. Depending on the type of applied stress (uniaxial, biaxial, shear) this shift is additive or subtractive to the existing shift due to confinement (Fig. 19.5b) [3]. This leads to the variation of the out-of plane effective mass and of density of states (DOS) in both conduction and valence bands accompanied by the scattering rate variation [8]. The electric confinement and the Si/SiO_2 interface scattering determine mobility enhancement factors with stress different from bulk Si.

The variation of the electron's mobility causes the variation of current I_{DS}. For long channel transistors their relation can be expressed as [14]

$$\frac{d\mu}{\mu} = \frac{\Delta I_{DS}}{I_{DS}} = \Pi \cdot \sigma, \tag{19.8}$$

where Π is the piezoresistive coefficient, introduced in Sect. 19.1.4.

In the case of short channel devices, source-drain parasitic resistances play a greater role in the total resistance. Hence, to obtain the piezoresistance coefficients, I_{DS} must be corrected for the parasitic resistance contribution [15].

The factors that influence the piezoresistive response of MOSFETs are:

1. Wafer surface orientation: (001), (110) or (111). The largest electron mobility variation was achieved for n- and p-MOSFETs built on (001) wafer surface [8]. The reason behind is that the carriers repopulation (Sect. 19.1.2) results in the highest effective mass variation only in the (001) case [16].
2. Stress application direction and drain current I_{DS} flow direction.

Most transistors are built such that I_{DS} flows parallel (in some cases orthogonal) to the wafer flat (Fig. 19.6). Although any type of stress can be applied to (001) Si plane, we further refer to the uniaxial one. Most investigated cases are:

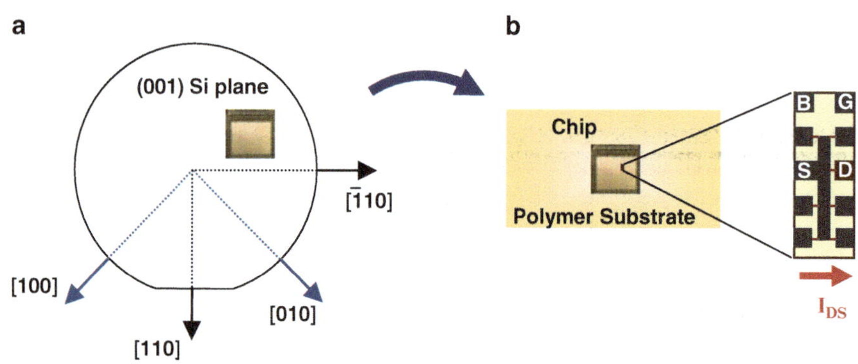

Fig. 19.6 (**a**) Schematic representation of (001) Si wafer crystallographic directions; (**b**) ultra-thin chip on polymer substrate (e.g. Kapton® foil) and a metal-oxide-silicon field effect transistor (*MOSFET*) test structure

Table 19.2 Piezoresistance coefficients of n-MOSFETs ($\times 10^{-12}$ Pa^{-1})

Reference	W × L (μm^2)	Π_{11}	Π_{12}	Π_{44}	Π_L	Π_T
Dorda [18]	400 × 40				300	200
Canali [19]	400 × 2,000	840	–340	170–120	335	190
Bradley [14]	IBM ($L \in \{15.1, 0.3\}$)			100	450	350
Bradley [14]	TI (L ∈ {15.1, 0.3})			70	320	250
Bradley [14]	LT (L ∈ {15.1, 0.3})			150	500	350
Gallon [15]	10 × 10, 10 × 0.13				485	212
Wacker [17]	16 × 16, 16 × 0.8				450	210

TI Texas instruments, *LT* Lucent technologies

Table 19.3 Piezoresistance coefficients of p-MOSFETs ($\times 10^{-12}$ Pa^{-1})

Reference	W × L (μm^2)	Π_{11}	Π_{12}	Π_{44}	Π_L	Π_T
Colman [20]	(0.254 × 2.29) mm^2	10	–238	–1,278	–753	525
Canali [19]	400 × 2,000	125	–280	(–1,050) – (–1,150)	–600	500
Bradley [14]	IBM (L ∈ {15.1, 0.3})			–950	–500	450
Bradley [14]	TI (L ∈ {15.1, 0.3})			–800	–415	385
Bradley [14]	LT (L ∈ {15.1, 0.3})			–1,000	–600	400
Gallon [15]	10 × 10, 10 × 0.13				–561	469
Wacker [17]	16 × 16, 16 × 0.8				–440	430

TI Texas instruments, *LT* Lucent technologies

- *Transversal* ($\sigma \perp I_{DS}$, $I_{DS} \parallel$ <110>)
- *Longitudinal* ($\sigma \parallel I_{DS} \parallel$ <110>)

For n-MOSFETs, tensile uniaxial stress leads to additive shift of the energy levels [3] (see Fig. 19.5b). This determines the enhancement of the electron mobility that results mainly from interband scattering suppression due to the large confinement-induced valley splitting at normal operation fields. Thus, ~100 MPa uniaxial tensile applied stress determines ~5% increase of electron mobility in the *longitudinal* case and ~2% in the *transversal* one [15, 17].

The longitudinal Π_L and transversal Π_T piezoresistive coefficients are determined from the slopes of I_{DS} variation versus stress or μ variation versus stress. Some published values for bulk MOSFETs are given in Table 19.2.

For p-MOSFETs, strain lifts the degeneracy of the HH and LH bands at Γ point and alters the effective mass. Thus, HH and LH sub-bands are replaced by the top (ground state) and second (first excited state) sub-bands [10]. For (001)/<110> pMOSFETs under uniaxial compressive stress, the top band is LH-like. The total effective mass decreases (since holes in the top sub-band are LH-like) as the stress increases, resulting in higher mobility (see Sect. 19.1.3). Additionally, strain alters the DOS and changes the phonon scattering rate.

Chu [16] showed that ~100 MPa uniaxial compressive stress results in ~7% mobility enhancement. On the other hand, ~100 MPa tensile uniaxial stress determined an increase of ~4.5% of hole's mobility in *transversal* case and a decrease of ~6% in *longitudinal* one [15].

Examples of piezoresistive coefficients of bulk p-MOSFETs from literature are given in Table 19.3.

We observe that the absolute values of Π_{11} and Π_{12} for MOSFETs are lower than those corresponding to bulk Si (Table 19.1.) but the values of Π_{44} are similar. The difference is due to the presence of the confinement field in MOSFETs.

While the effects of strain on bulk and thin-film Si [21] as well as on bulk MOSFETs have been investigated, the effects of strain on ultra-thin chip devices are the subject of current investigations.

19.3 Conclusion

In this chapter we introduced important aspects related to the piezoresistive effect in Si and in MOSFETs. The ongoing research is concerned not only with the effects caused by uniaxial strain in ultra-thin chips but also those due to biaxial and shear applied stress. The analysis of piezoresistive effect, especially in ultra-thin chips, is important for aspects related to yield and reliability.

The strain can be used to enhance the devices speed and performance without the need of scaling, as applied currently in industry. On the other hand, its effects resulting from chip-attachment and packaging processes, detrimental to MOSFETs operation, must be identified and minimised.

References

1. Smith CS (1954) Piezoresistive effect in germanium and silicon. Phys Rev 94(1):42–49
2. Kittel C (2005) Introduction to solid state physics. Wiley, New York
3. Sun Y, Thompson SE, Nishida T (2007) Physics of strain effects in semiconductors and metal-oxide-semiconductor field-effect transistors. J Appl Phys 101:104503, 2007
4. Bir GL, Pikus GE (1974) Symmetry and strain-induced effects in semiconductors. Wiley, New York
5. Mohta N, Thompson SE (2005) Mobility Enhancement; The Next Vector to Extend Moore's Law. EEE Circ Devices Mag 21(5):18–3, 18–23 Sept/Oct 2005
6. van de Walle CG (1989) Band lineups and deformation potentials in the odel-solid theory. Phys Rev B 39(3):1871–1883
7. Bardeen J, Shockley W (1 October, 1950) Deformation Potentials and Mobilities in Non-Polar Crystals. Phys Rev 80(1):72–80
8. Irie H, Kita K, Kyuno K, Toriumi A (2004) In-plane mobility anisotropy and universality under uni-axial strains in n- and p-MOS inversion layers on (100), (110), and (111) Si. In: IEEE IEDM Proceeding, San Francisco, pp 04-225–04-228, 2004
9. van Zeghbroeck B (2007) Principles of Semiconductor Devices. Electronic resource: http://ecee.colorado.edu/~bart/book/. Last access on Oct 2010
10. Sun G, Sun Y, Nishida T, Thompson SE (2007) Hole mobility in silicon inversion layers: stress and surface orientation. J Appl Phys 102:084501-1–084501-7

11. Ivanov T, Gotszalk T, Sulzbach T, Chakarov I, Rangelow IW (2003) AFM cantilever with ultra-thin transistor-channel piezoresistor: quantum confinement. Microelectron Eng 67–68:534–541
12. Dorda G, Eisele I (1973) Piezoresistance in n-type silicon inversion layers at low temperatures. Phys Stat Sol (a) 20:263–273
13. Fischetti MV, Ren Z, Solomon PM, Yang M, Rim K (15 July, 2003) Six-band kp calculation of the hole mobility in silicon inversion layers: dependence on surface orientation, strain and silicon thickness. J Appl Phys 94(2):1079–1095
14. Bradley AT, Jaeger RC, Suhling JC, O'Connor KJ (2001) Piezoresistive characteristics of short-channel MOSFETs on (100) silicon. IEEE Trans Electron Devices 48(9):009–2015
15. Gallon C, Reimbold G, Ghibaudo G, Bianchi RA, Gwoziecki R, Orain S, Robilliart E, Raynaud C, Dansas H (2004) Electrical analysis of mechanical stress induced by STI in short MOSFETs using externally applied stress. IEEE Trans Electron Devices 51(8):1254–1261
16. Chu M, Nishida T, Lv X, Mohta N, Thompson SE (2008) Comparison between high-field piezoresistance coefficients of Si metal-oxide-semiconductor field-effect transistors and bulk Si under uniaxial and biaxial stress. J Appl Phys 103:113704-1–113704-7
17. Wacker N, Hassan M-U, Richter H, Rempp H, Burghartz JN (2009) "Compact modeling of CMOS transistors under uniaxial stress." In: Proceedings of SAFE 2009, Rome, Italy, pp 179–181, 2009
18. Dorda G (14 August, 1970) "Effective mass change of electrons in silicon inversion layers observed by piezoresistance." Appl Phys Lett 17:406–408
19. Canali C, Ferla G, Morten B, Taroni A (1979) "Piezoresistivity effects in MOS-FET useful for pressure transducers." J Phys D Appl Phys 12:1973–1983
20. Colman D, Bate RT, Mize JP (1968) "Mobility anisotropy and piezoresistance in silicon p-type inversion layers." J Appl Phys 39(4):1923–1931
21. Maegawa T, Yamauchi T, Hara T, Tsuchiya H, Ogawa M (April 2009) "Strain effects on electronic bandstructures in nanoscaled silicon: from bulk to nanowire." IEEE Trans Electron Devices 56(4):553–559

Chapter 20
Electrical Device Characterisation on Ultra-thin Chips

Mahadi-Ul Hassan, Horst Rempp, Harald Richter, and Nicoleta Wacker

Abstract For ultra-thin chips (<20 μm) intended for use in flexible systems, external mechanical stress is inevitably relevant. Because mechanical stress can directly affect the electrical behaviour of semiconductor materials, it is vital to examine the effect of bending on different electrical devices and circuits to anticipate their behavior in flexible systems. Moreover, this effect can be well-used for sensor applications. Such applications also demand in-depth investigation of the mechanical deformation effect on different devices and circuits fabricated on ultra-thin chips. In this chapter various usable chip bending and electrical measurement techniques for devices are discussed. We also elaborate on limitations and error factors associated with applied characterisation methods, along with possible solutions apprehended from our experiences with ultra thin chip characterisations at Institut für Mikroelektronik Stuttgart (IMS Chips).

20.1 Introduction

Mechanical stress is directly responsible for changes in the electrical behaviour of semiconductor material, which eventually affects the electrical characteristics of devices and circuits. Stress can be introduced in the devices during processing steps, during various packaging steps and by mechanical deformation of chips. Depending on the type and source of stress, it can either have a detrimental or an enhencement effect on the device performance.

Both experimental and theoretical studies have shown that introduction of internal stress in the active layers of the devices during process steps can significantly improve overall device performance [1–3]. This method has already been implemented in remarkably small devices (<100-nm gate length) to enhance their performance by increasing electron and hole mobility [4].

M.-U. Hassan (✉)
Institute für Mikroelektronik Stuttgart (IMS CHIPS), Stuttgart, Germany
e-mail: mhassan@ims-chips.de

J.N. Burghartz (ed.), *Ultra-thin Chip Technology and Applications*,
DOI 10.1007/978-1-4419-7276-7_20, © Springer Science+Business Media, LLC 2011

In spite of the usefulness of internal stress for enhancement of device performance, external stresses arising from packaging or mechanical bending can turn out to be unfavourable for normal functionality of devices. During the chip packaging steps stress can be introduced through wire-bonding, die attachment and encapsulation steps [5].

For ultra-thin chips (<20 μm) intended for use in flexible systems, the external mechanical stress is inevitably relevant. The stress effect associated with mechanical bending or twisting can have two significant consequences; (1) the electrical properties of the devices as well as circuits can change in an unwanted manner or (2) the change can be utilised in terms of sensor applications. Both cases require a detailed knowledge and investigation of the influence of deformation on the electrical behaviour of devices.

In this section measurement techniques are described to investigate the effect of bending on different types of devices fabricated on ultra-thin chips. These measurements are also the basis of the determination of simulation parameters needed to extend existing device models to incorporate the stress effects so that they can be used to model applications where the deformation is included.

For this purpose, suitable bending techniques, different measurement procedures and feasible methods for evaluation of measurement data will be discussed. The section concludes with a short discussion of the difficulties and deficiencies that arise from methods and techniques discussed here.

20.2 Bending Techniques

A well-known method used in the case of nonflexible chips is 4-point bending technique [6–8], depicted in Fig. 20.1a. The key advantage of this method is its Ability to co-relate the applied mechanical stress to the resultant bending effects based on a simple mathematical model [8]. However, this technique is not very effective for chips of thickness <150 μm [9]. Other popular bending methods are the bending jig method, shown in Fig. 20.1b [10, 11] and the cantilever method [12], which have been shown to be functional for chips of thickness >35 μm. These methods are difficult to apply as sample sizes get thinner and smaller. So, for ultra-thin chips (thickness <20 μm) these methods cannot be realised any longer, as these chips are so small and extremely flexible.

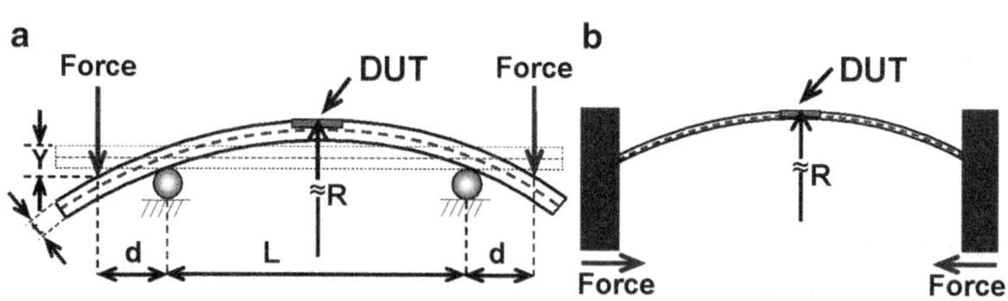

Fig. 20.1 Chip bending techniques: (**a**) 4-point bending and (**b**) bending jig method

Due to their small size and extremely small thickness, ultra-thin chips must be measured with great caution. A major challenge associated with handling ultra thin chips is the implementation of external electrical connections from the chip. A practical solution for these problems is to either glue the chips on flexible substrates or to embed the chips inside a flexible substrate. The electrical connections on the chip can be realised by means of bond wires or some other post-process technology [10]. This will allow bending of the chips to very small radii (even less than 7 mm) without breaking or disrupting external connections as well as the chips themselves.

To characterise devices, many factors have to be taken into account. The most important aspects can be placed into two categories. Primarily, these characterisations depend on the crystal orientation of silicon, the device type and device orientation with respect to bending direction. Secondly, they depend on the quantity, orientation and types of stress (for example, uniaxial, biaxial, torsional, etc.) caused by bending. The characterisation of devices can include the combination of all these factors.

A practical method for characterising devices on ultra-thin chips could be the gluzing of chips on a flexible foil. The flexible foil along with the chip can be strapped over cylindrical barrels of specific radii (Fig. 20.2a). In Fig. 20.2c a practical realisation of a possible measurement set-up is shown. The construction of the apparatus should guarantee a high reproducibility of bending conditions. In this case the desired bending radius is realised by means of different rolls. The electrical contacts do not affect the bending conditions and are easily accessible from outside.

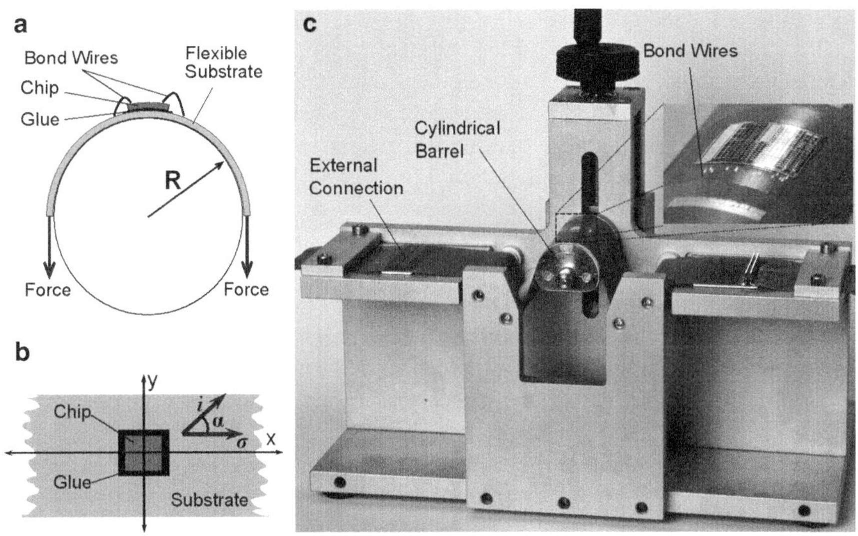

Fig. 20.2 Ultra-thin chip bending techniques: (**a**) strapping of chip along with the flexible foil around a cylinder; (**b**) orientation of chip glued on the foil. Here α represents the angle between stress (σ) and current (i) direction; (**c**) an in-house bending apparatus, along with chip bent over a bending roll (Courtesy of Institut für Mikroelektronik Stuttgart)

The principal disadvantage of this method is that a complete system made of different materials – rather than pure bulk silicon – is being characterised. In this case, in contrast to bulk material a system consisting of substrate, glue and thin silicon has to be taken into consideration. As the underlying glue layer can act as the stress-transferring media, possible visco-elastic or, in the worst case, plastic behaviour of the glue can introduce a long-term stress relaxation effect [13]. These effects are directly reflected in the electrical measurements causing nonlinear behaviour and relaxation effect over long time period. So, unlike the 4-point bending technique, calculation of stress acting on the devices (active area) is much more complicated in this case. These effects can hamper the measurement results enormously and make device characterisation and interpretation extremely difficult.

20.3 Measurement Techniques

The two most important factors associated with characterisation of devices on ultra-thin chips are the mechanical system used for bending and the system temperature. For characterising devices, uniaxial bending is the most desirable method; it is quite difficult to achieve this because of the chips' small dimension and extremely flexible nature. Moreover, a complex system of chip-glue-substrate is introduced rather than pure semiconductor materials of well-known mechanical properties. Therefore, the measurement becomes dependent on different factors – both known as well as not clearly defined.

Pronounced temperature sensitivity of the devices demands accurate temperature compensation of measurement data along with bending effect measurements so that accurate apparent bending responses may be achieved. For this purpose identical transistors with different orientations could be measured to allow for inherent temperature compensation. For example, in the case of current mirror circuits two orthogonally oriented transistors provide the desired inherent temperature compensation of measurement results. But, for single transistor measurements, an in situ compensation has to be implemented, as described later in this section. A proven technique to perform characterisation of devices with the least amount of error factors is the measuring of different transistors (different sizes, orientations) on the same chip simultaneously to ensure the same boundary conditions, such as temperature and applied stress, and thus to achieve reliable comparison of measured data.

A block diagram of a possible measurement setup is shown in Fig. 20.3. Several devices on the same chip for a particular bending condition are measured simultaneously. A Pt-100 temperature sensor is placed in the vicinity of the chip to record the temperature. The temperature is measured just before each characterisation measurement (e.g. input characteristic, output characteristic) and later these temperature data can be used to correct the characterisation data by eliminating the effect of temperature variation during the measurement (see Fig. 20.5). Different

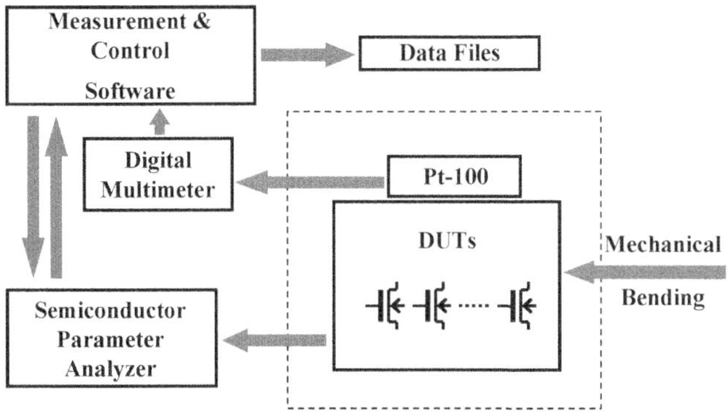

Fig. 20.3 Simplified block diagram of measurement setup

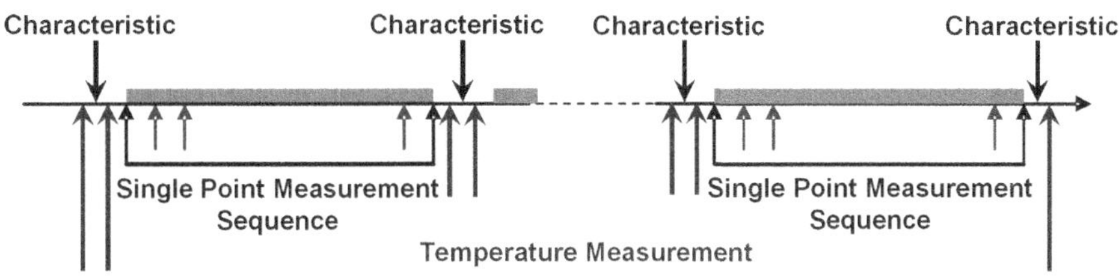

Fig. 20.4 Simplified measurement procedure

measurements of the devices on the chip can be performed with a semiconductor parameter analyser (SPA). The entire measurement procedure is executed by software that controls the communication between the equipment, carries out data collection and simultaneously performs several on-line calculations and monitors particular results.

In Fig. 20.4 the principle of the measurement procedure of Metal Oxide Semiconductor (MOS) transistors is shown, which allows an extensive evaluation of the relationship between bending and the electrical response of devices. The measurement sequence below is a defined combination of relatively long lasting characterisation measurements, short spanned spot measurements (one or several distinct operating points) and temperature measurements. Spot measurements data as well as the temperature data can be used for monitoring the entire measurement online, while the temperature data can be used later to correct the entire measurement data. A complete long-term characterisation might need several minutes upto many hours to accomplish depending on specific circumstances. Characterisation as well as single-point measurements can be performed for several transistors simultaneously to ensure identical measurement conditions.

The described procedure allows one to examine the measurement during runtime in terms of time-dependent variations of the system (e.g., glue) and mechanical problems caused by the system (e.g., sudden changes caused by the assembly).

These effects are entirely exceptional for a chip-glue-substrate system and are not present in the case of common bulk measurements.

One effective technique is that of measuring subsequent bending states with flat state measurements in between. This technique has proven to be very effective in case of glue-attached chips, where the glue layer plays a greater role in stress transformation to the chip. It can visibly show the level of reversibility of changes caused by bending and can suggest if there are other effects involved. Figure 20.5 shows an example of a single-point measurement sequence (drain current I_{DS} in the linear operating range at 100 mV of V_{drain} and 5 V of V_{gate}) lasting about 50 hrs. The red points are actual measured values, which include both bending and temperature effect. After modifying the current measurement data (red dots) using the temperature data (green dots), a compensated current data (blue dots) are achieved which now includes only the bending effects. Different radii are applied and the flat states between different bending radii represent a relatively stable reversible flat state of the device being tested. The time-dependence of the system can be clearly seen. The term "flat" in the figure refers to the way that the chip is strapped over a flat part of the roll placed in the bending equipment shown in Fig. 20.2. R_i indicate different applied roll radii.

20.3.1 Temperature Effect

As it was mentioned earlier, temperature dependence of MOS devices cannot be neglected in the discussion of bending characterisation. Either an extremely good controlled environment for measurement or an in situ compensation can be applied.

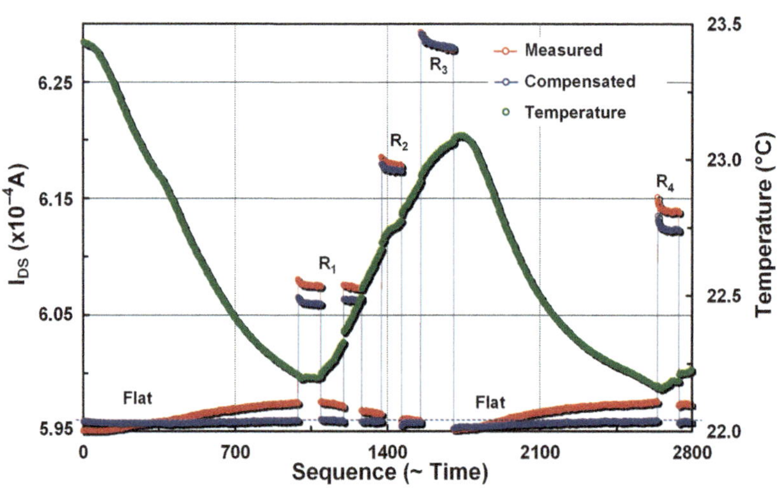

Fig. 20.5 Example of measurement results of an NMOS transistor on ultra-thin chip ($16 \times 0.8\ \mu m^2$)

The in situ compensation method is used in the measurement shown in Fig. 20.5. The temperature coefficients (TC) for each device that is measured are extracted using the long-term flat states of the experiment. This method has the advantage that the absolute value of the TC for a particular device can be used for compensation. The other method of compensating for temperature, i.e., dividing measured values, allows one to measure current differences instead of absolute values. For some purposes this might be sufficient.

20.4 Evaluation of Measurement Data

As described in the previous section, the measurement data set consists of:

- Temperature data
- Single-point measurement data at one or several operating points of interest
- Complete characterisation data (e.g., input, output characteristics)

The monitored temperature data in the data set are used for temperature compensation of the actual measurement data – both single-point and characterisation data. Later these measurement data can be used for evaluation purposes.

Single-point measurements can be used to monitor the experiment, to extract TCs, to record time-dependent variations of the device (e.g., MOS transistor) responses and to visualise sudden mechanical changes in the substrate-glue-silicon system. Also inherent degradation effects, such as hot carrier effect, can be observed by means of single-point measurements. Characterisation measurements can be used to extract transistor model parameters as a function of bending (e.g., mobility μ_0).

Apart from time-dependent effects described earlier, a nearly perfect staircase function of the recorded current should be observed in the single-point measurement sequence, which would depict the corresponding bending effect. In Fig. 20.6 measurement results for a MOSFET device on bulk silicon (thickness of ~600 μm) are shown, where the drain current I_{DS} can be used directly to extract the piezoresistive coefficient (π). Bending was performed with a 4-point bending setup, where the stress acting on the devices located on the bulk silicon was calculated with the help of mechanical parameters given in Fig. 20.1a. The relationship to determine π (α) is given by

$$\frac{\Delta I_{DS}}{\Delta \sigma} = \pi(\alpha) \quad \text{and} \quad \sigma = \sigma(Y, L, d, t, E_{Si}) \tag{20.1}$$

where Y, L, d, t are solely mechanical parameters and E_{Si} is the Young's modulus of silicon. ΔI_{DS} is the deviation of current from flat state corresponding to the related bending state.

The change in the temperature compensated I_{DS} (Fig. 20.6a) ends up in the plot in Fig. 20.6b, where the deviation of drain current is represented with respect to curvature values as well as calculated stress values based on (20.1). The slope of the

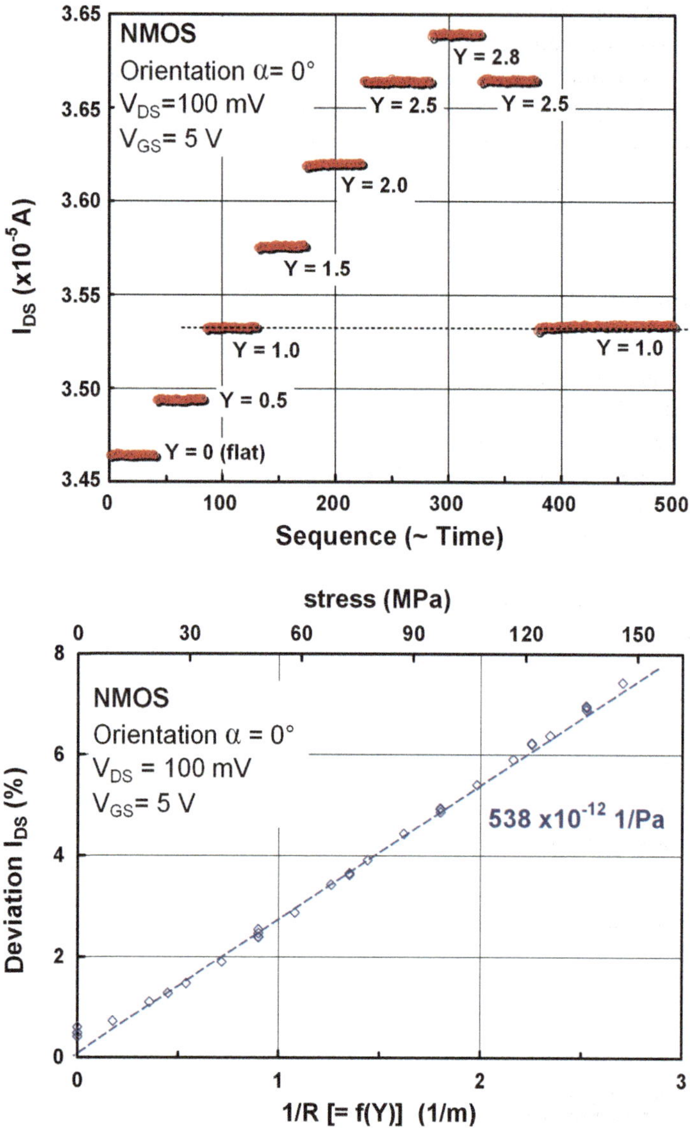

Fig. 20.6 Time-independent bulk silicon measurement. (**a**) Measurement result I_{DS} temperature compensated (Y = deflection; see Fig. 20.1). (**b**) Evaluation of (**a**) based on (20.1)

plot in Fig. 20.6b provides the piezoresistive coefficient value for the particular device. The method is straightforward in the case of simple systems, absolute stable conditions and uniaxial stress [14]. In the case of a system that includes ultra-thin chips and components with different mechanical properties, the measurement result can be quite different. The relationship between bending conditions and the applied stress in the active region of a device to be investigated is not clearly defined and cannot be modelled by a simple mathematical formula like the earlier one, (20.1). The mounting of chips on the substrate introduces some amount of stress on the chip. As a result, the alleged uniaxial bending of the chip-glue-substrate will not experience only the expected uniaxial stress rather a different quantity of stress due to the pre-stress situation denoted as 'pretreatment' in equation 20.2 will be experienced by the chip. This difference in the real acting stress can produce a significant difference in the measurements. Hence, the slope of deviation versus curvature values is not necessarily a straight line any more. It includes the complexity of the substrate-glue-silicon system outlined as system properties and can be described in a general way:

$$\frac{\Delta I_{DS}}{\Delta \sigma} = \pi(\alpha) \quad \text{and} \quad \sigma = \sigma(R, \text{system properties, pretreatment}) \tag{20.2}$$

This complex situation can result in nonlinear dependence between the pure bending parameter R (see Fig. 20.2a) and current deviation. Figure 20.7 shows corresponding results compared to the bulk material (Fig. 20.6) for an ultra-thin chip mounted on foil, as described above. It shows an example with extremely strong nonlinearities in the case of the $16 \times 0.8\ \mu m^2$ NMOS transistor ($\alpha = 90°$).

In Fig. 20.7a, the time-dependent behaviour can be seen with a time constant for smaller radii in the minute range, which is caused by non-elastic property of the glue. Figure 20.7b shows a distinct nonlinear behaviour for different orientations of short NMOS transistors and the result for a large $16 \times 16\ \mu m^2$ transistor. In contrast to what is possible for bulk material, we cannot extract piezoresistive coefficients without further investigation.

Another contrast to bulk material, ultra-thin chips show – depending on processing and layout – a certain geometric shape (see Fig. 20.8), which is the result of the equilibrium the chip experiences in all inherent stress conditions. When the chip is mounted on a substrate the shape will be changed and so internal induced stress offsets are introduced.

Nonhomogeneous layouts can show a much stronger self-bending effect, hence, a larger pre-stress acting in the active layer of the chip after the mounting process. This stress distribution is actually the so-called "flat state" of the system. This inherent existing stress will be superimposed with the stress caused by the experimental setup. The uniaxial applied stress to the substrate will be transferred by the glue layer to the silicon. As a consequence a biaxial stress distribution at the active layer of the device is expected. This distribution depends strongly on the bending radii R_i and can cause nonlinear behaviour response in the electrical measurements. Assuming a certain pre-stress (flat state) the I_{DS} response can be calculated to fit to the measured I_{DS} curve [15].

Using proper temperature compensation, one can extract transistor model parameters (e.g., mobility μ_0, threshold voltage V_{TH}) as functions of bending using the relevant characteristics measurements, bearing in mind the above mentioned obstacles. On the other hand, bulk material measurement data can be used to compare with ultra-thin chip results to draw conclusions from the substrate-glue-silicon system.

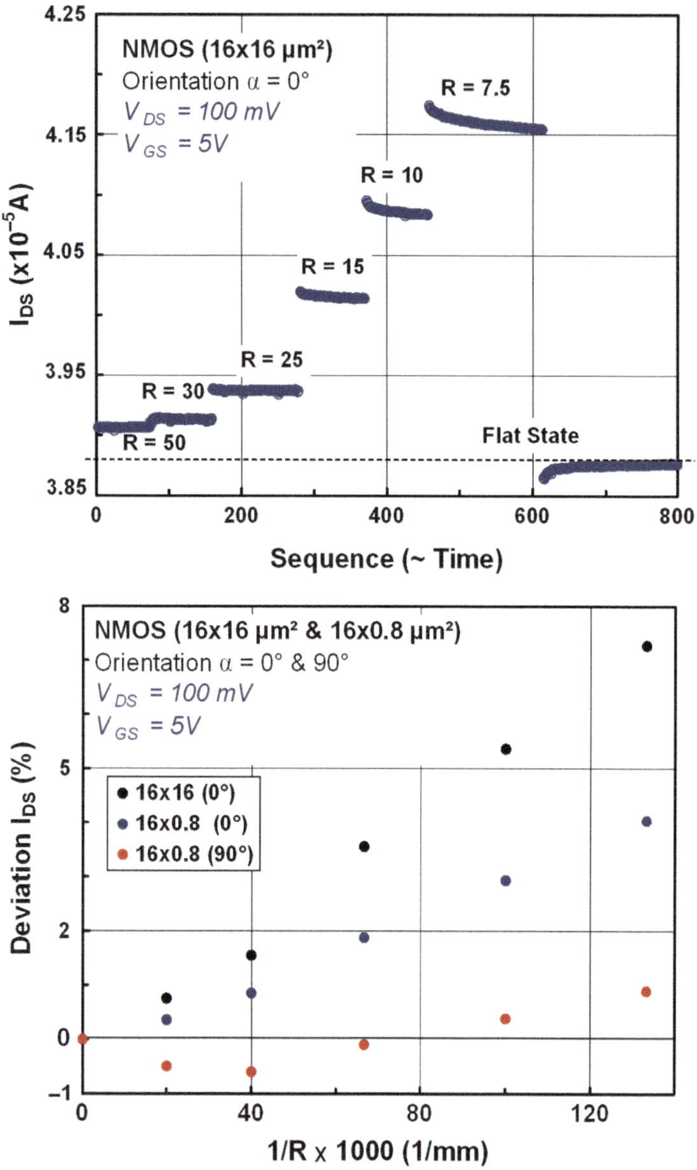

Fig. 20.7 Time-dependent measurement result of an ultra-thin chip (~20 μm)[$\bar{1}$; 10] orientation ($\alpha = 0°$). (**a**) R_i radius of applied rolls (in mm). (**b**) Deviations of I_{DS} from the flat state for different geometries and roll radii and orientations α

In the case of a certain distribution of the pre-stress and the position of the measured device on chip, linear behaviour may be experienced. Figure 20.9 shows an example of linear behaviour in the case of diffused resistors localised at a

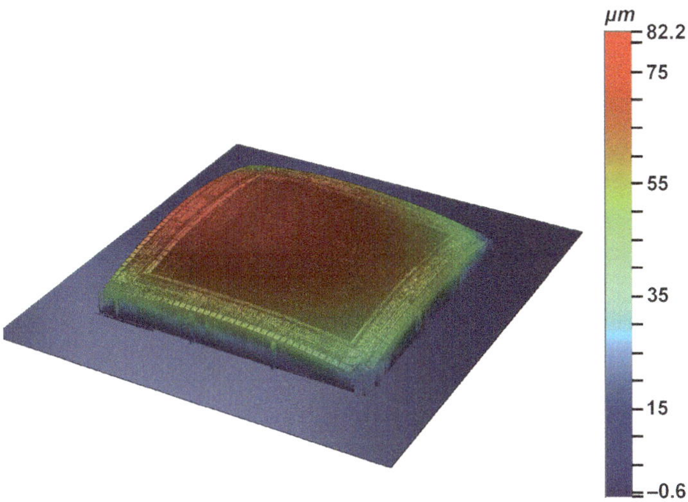

Fig. 20.8 Shape of an ultra-thin chip (~20 µm) with relatively homogenous layout

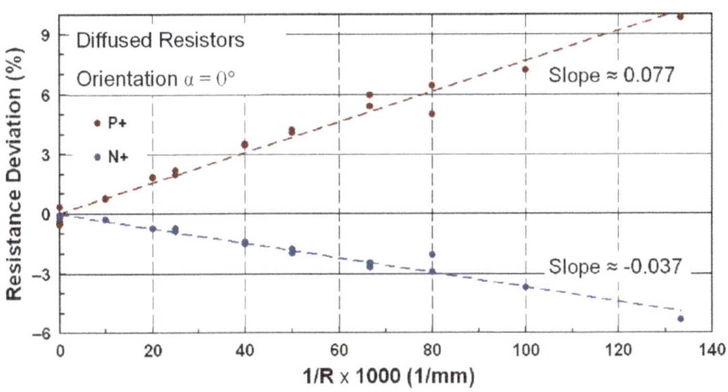

Fig. 20.9 Resistor measurement result exhibiting linear dependence over $1/R$ (curvature)

position with presumably low pre-stress conditions. In this case, a behaviour based on (20.2) can be assumed. Nevertheless, the σ in (20.2) cannot be calculated easily.

Based on the methods described, we can investigate elementary circuit blocks using few devices, e.g., current mirrors by means of bending [15]. In the case of current mirrors only small changes in currents due to temperature are expected based on inherent temperature compensation of the circuit function and layout (see Sect. 20.3). The response of current mirrors can be simulated and compared to measurement results. The simulation is based on change of transistor parameters and the operating conditions of the transistors in the particular circuits.

20.5 Conclusions

In contrast to bulk material, ultra-thin chips show – depending on processing and layout – a certain geometric shape (Fig. 20.7). Hence, mounting chips on a substrate will cause an inherent stress distribution, which can be rather complex. Both known and little known properties of different components in the system (chip-glue-foil), especially the effects caused by the glue layer (material properties, homogeneity, geometrical conditions) are effecting the measurement results. Uniaxial stress application to the substrate will be transferred to a biaxial stress distribution at the active layer of the chip surface caused by the system chip-glue-substrate. To avoid chip-to-chip variations caused by the assembly process, in particular the gluing process, one can measure several devices be in parallel on the same chip. This method allows an accurate investigation of time- and position-dependent effects of the substrate-glue-silicon system.

For detailed measurements, particular test structures are advantageous that can also be measured in parallel (e.g., different orientations in near neighbourhood), and these allow a direct comparison of results without uncertainties regarding system assembly. This allows also effective temperature compensation. A calibrated temperature measurement on the chip would be an optimum solution, but such is very difficult to realise.

Whenever possible, comparable measurements of bulk material and ultra-thin film material should be performed. This is especially true in the case of model parameter extraction purposes, because in the first order no effect of the thickness down to a few micrometres between both types of devices is expected. Wafer level measurements of ultra-thin chips (flat state before disassembly [16]) and comparable measurements on bulk wafer on a statistical basis have shown very similar results in terms of mobility and speed of devices [17].

In the case of ultra-thin chips one has to be aware of many accompanying effects that can influence the measured response compared to what is experienced in a pure silicon system.

References

1. Ismail K, Nelson SF, Chu JO, Meyerson BS (1993) Electron transport properties of Si/SiGe heterostructures: measurements and device implications. Appl Phys Lett 63:660–662
2. Vogelsang T, Hofmann KR (1993) Electron transport in strained Si layers on Si_{1-x}, Ge_x, substrates. Appl Phys Lett 63:186–188
3. Bufler FM (2004) Exploring the limit of strain-induced performance gain in p- and n-SSDOI-MOSFETs. In: IEDM Tech Dig., San Francisco, pp. 601–604
4. Thompson SE, Armstrong M, Auth C et al. (2004) A 90-nm logic technology featuring strained-silicon. IEEE Trans Electron Dev 51(4):191–193
5. Bradley AT, Jaeger RC, Suhling JC, O'Connor KJ (2001) Piezoresistive characteristics of short-channel MOSFETs on (100) silicon. IEEE Trans Electron Dev 48(9):2009–2015

6. Chen T, Huang Y (2002) Evaluation of MOS devices as mechanical stress sensors. IEEE Trans Compon Packag Technol 25:511–517
7. Shetty S, Reinikainen T (2003) Three- and four-point bend testing for electronic packages. J Electron Packag 125:556–561
8. Richter J, Arnoldus MB, Hansen O, Thomsen EV (2008) Four point bending setup for characterisation of semiconductor piezoresistance. Rev Sci Instrum vol.79, pp. 044703(1–10)
9. Richter J, Hansen O, Larsen AN, Hansen JL, Eriksen GF, Thomsen EV (2005) Piezoresistance of silicon and strained $Si_{0.9}Ge_{0.1}$. Sens Act A Phys vol. 123–124, pp. 388–396
10. Sakurai R, Hattori R, Asakawa M, Nakashima T, Tanuma I, Yokoo A, Nihei N, Masuda Y (2008) Ultra-thin and flexible LSI driver mounted electronic paper display using quick-response liquid-powder technology. Journal of the society for information Display vol. 16, issue 1, pp. 155–160
11. Ng DC, Isakari K, Uehara A, Kagawa K, Tokuda T, Ohta J, Nunoshita M (2003) A study of bending effect on pulse-frequency-modulation-based photosensor for retinal prosthesis. Jpn J Appl Phys 42:7621–7624
12. Watanabe N, Kojima T, Maeda Y, Nishisaka M, Asano T (2004) Breakdown voltage in uniaxially strained n-channel SOI MOSFET. Jpn J Appl Phys 43:2134–2139
13. Gunawan M, Davila LT, Wong EH, Mhaisalkar SG, Tsai TK, Osiyemi S (2004) Static and cyclic relaxation studies in nonconductive adhesives. Proc Int Conf Mat Adv Tech vol. 462–463, pp. 419–426
14. Timoshenko S (1976) Strength of materials, 3rd edn. Krieger Publishing Company, New York
15. Hassan M-U, Rempp H, Hoang T, Richter H, Wacker N, Burghartz, JN (2009) Anomalous stress effects in ultra-thin silicon chips on foil. In: IEEE electron devices meeting (IEDM), Baltimore, 2009, pp. 535–538
16. Burghartz JN, Appel W, Rempp H, Zimmermann M (2009) New fabrication and assembly process for ultra-thin chips. IEEE Trans Electron Dev 56(2):321–327
17. Rempp H, Burghartz JN, Harendt C, Pricopi N, Pritschow M, Reuter C, Richter H, Schindler I, Zimmermann M (2008) Ultra-thin chips on foil for flexible electronics. In: IEEE international solid-state circuits conference (ISSCC), San Francisco, p 334

Chapter 21
MOS Compact Modelling for Flexible Electronics

Slobodan Mijalković

Abstract Circuit design in flexible electronics should account for shifts in metal-oxide semiconductor (MOS) characteristics with the mechanical strain induced by substrate bending. Apart from the standard process and layout strain sources, externally applied strain in flexible electronics varies with bending direction and radius. The physical compact models for the strain effects on semiconductor band structure and carrier mobility due to substrate bending are derived in this chapter from basic deformation potential theory. It is demonstrated how existing MOS compact models could be extended for bending-induced strain effect by modification of the set of material model parameters controlling the semiconductor band structure and carrier mobility.

21.1 Introduction

The goal of compact modelling is to derive a computationally efficient representation of the device behaviour for all modes of operation, layout specifications and environmental conditions relevant to circuit design. Compact model parameters serve in process design libraries as an essential technology communication bridge between the circuit designers and process foundries. While MOS transistors continue to break records in gate length as well as maximum operation frequency and power, MOS compact models are continually improved in order to fully utilise the potential of present and future MOS technologies in circuit design.

With the emergence of ultra-thin single-crystal flexible electronics [1–3], circuit design consideration should also include the effects of mechanical strain due to the bending of the semiconductor substrates. It should be emphasised that strain effects are not a new topic in MOS technology and compact modelling. Process and layout-induced strain is systematically used today to improve the performance of

S. Mijalković (✉)
Silvaco Europe, Compass Point, St Ives, Cambridge, PE27 5JL, UK
e-mail: slobodan.mijalkovic@silvaco.com

J.N. Burghartz (ed.), *Ultra-thin Chip Technology and Applications*,
DOI 10.1007/978-1-4419-7276-7_21,

nanoscale MOS transistors on rigid substrate. Existing MOS compact models are typically applied without modification in the case of the fixed layout and process-induced strain. Some MOS compact models are extended to account for layout-dependent strain effects [4]. However, these empirical model extensions, usually based on the geometrical proximity of shallow-trench isolation areas, cannot account for general strain effects in flexible electronics.

This chapter presents a methodology that extends the existing MOS compact models to account for substrate bending-induced strain effects. Basically, mechanical strain alters MOS electrical characteristics by modifying the semiconductor band structure and carrier transport properties. Instead of using empirical relationships for the strain dependence of MOS electrical characteristics, the present approach focuses on the compensation of the material and mobility model parameters for the strain effects. Basic strain effects on band structure and carrier mobility are derived from the fundamental deformation potential theory in analytical form suitable for the straightforward incorporation into existing physical MOS compact models.

The rest of the chapter is organised as follows. State of the art in MOS compact modelling is presented in Sect. 21.2. Basic considerations specific to substrate bending-induced strain in flexible electronics are formulated in Sect. 21.3. The analytical models of semiconductor band structure and carrier mobility-dependence on the general externally applied strain are presented in Sects. 21.4 and 21.5, respectively. Finally, Sect. 21.6 demonstrates how material and mobility model parameters can be compensated for strain effects in the commonly used physical MOS compact model BSIM4 [4].

21.2 The State of the Art in MOS Compact Modelling

The MOS compact models most commonly used and implemented in commercial circuit simulators are presented Table 21.1 in groups defined by MOS technology and the underlying compact modelling concept.

The first generation of MOS compact models, represented by Spice Level 1–3, was mainly used for small digital circuits employing MOS technologies with a channel length greater than 2 μm. With the rapid evolution of the MOS and CMOS

Table 21.1 Metal-oxide semiconductor (*MOS*) compact models commonly implemented in commercial circuit simulators

Concept technology	Threshold voltage-based	Charge-based	Surface potential-based
Bulk MOS	Spice Level 1–3	EKV	PSP
	BSIM2		HISIM
	BSIM3v3		
	BSIM4		
SOI	BSIMSOI4		
Power MOS			HSIM-HV
Multiple gate			BSIM-CMG

technology in the 1980s, Spice Level 1–3 models became inappropriate for increasingly larger circuits with even smaller transistor sizes. A higher modelling accuracy was required and the second generation of MOS compact models was born with the introduction of the BSIM (Berkeley Short-channel IGFET Model) and later, its successor BSIM2. The second generation of MOS compact models provided a path into the submicron era but also an entry into the MOS analogue circuit design. The focus of the model development was on computational aids for faster and more robust circuit simulation, resulting in complex model implementations whose model parameters had unclear physical meaning.

Physical MOS compact modelling was reintroduced in the 1990s with the third model generation represented by BSIM3v3 and its successor BSIM4 [4]. The youngest member in this MOS model family is BSIMSOI4, extending the BSIM4 modelling concepts to silicon-on-insulator technology. The currents and charges in these models are defined as the explicit functions of the terminal voltages, but they are valid only in particular regions of operation. Empirical interpolating functions have been employed to provide the continuous and smooth behaviour of device characteristics across all the operation regions. In the first three MOS compact model generations, the threshold voltage (V_T) is the key parameter, and these models are also known as V_T-based models. However, the weaknesses of the piecewise regional V_T-based MOS compact models have became rapidly exposed in recent years with a steady downscaling of device sizes, reduced power consumption requirements as well as increasing demands for the accurate MOS RF and mixed-signal circuit design.

A remarkable effort in academia and companies has been made in the last 10 years to provide a new framework for compact model description of MOS transistor operation that will overcome the fundamental weaknesses of the three previous model generations and meet the requirements of future MOS technologies. In the fourth MOS model generation, the drift-diffusion channel current is expressed as a function of either inverse charge density (Q_i) or surface potential (φ_s) [5]. On the other hand, implicit equations for either Q_i or φ_s are formulated from the Poisson equation and gradual channel approximation. Depending on the selected control variable, MOS compact models are recognised as either Q_i-based (e.g., EKV [6]) or φ_s-based (e.g., HISIM [7]). Both Q_i-based and φ_s-based models provide accurate description of the intrinsic channel inversion. However, φ_s-based approach appears to be more general for application in the source-drain overlap regions and for channel accumulation where neither V_T nor Q_i are useful variables. Consequently, φ_s-based MOS models are predicted to be mainstream in production industry applications of the future.

Besides the radical change in the model core, the newly emerging MOS compact models have to also account for two-dimensional and quantum mechanical effects caused by aggressive downscaling below channel lengths of 0.1 μm. These models should provide an accurate physical link from material and fabrication process parameters to electrical device characteristics in order to account for the process variability in statistical modelling or for the ageing effects in the reliability modelling. There is also a continuous need for the development of the model variants specialised for power transistor applications (e.g., HSIM-HV [7]) or multi-gate and cylindrical gate nano-MOS structures (BSIM-CMG [8]). All these requirements have brought the MOS compact modelling closer to the field of technology of CAD (TCAD) [9], a source of the most fundamental device modelling concepts, providing insights that are difficult to obtain from measurements.

The need for the mechanical strain-aware compact models has appeared with the aggressive scaling of the standard MOS circuits on rigid substrates. Since the introduction of 90-nm node, strain silicon technology became the standard MOS process to further enhance semiconductor carrier transport properties. The amount of the applied strain is controlled by the processing conditions (e.g., the Ge composition in SiGe based technology) or circuits layout parameters (e.g., the STI-induced strain). The advanced MOS compact models (e.g., BSIM4 [4]) have been extended to account for layout-induced strain effects. However, these extensions are not general enough to handle the variable strain conditions imposed by the substrate bending in flexible electronics.

21.3 Strain Considerations in Flexible Electronics

Thin chip substrates in flexible electronics typically bend into a cylindrical shape, as shown in Fig. 21.1. The origin of the bending moment could be an externally applied force or misfit strain in multilayer substrate structures.

The bending-induced strain in the thin flexible substrate is the most conveniently defined in the coordinate system (x_i, $i = 1, 2, 3$), whose axis unit vectors (e_i, $i = 1, 2, 3$), are aligned with bending direction, through-thickness directions and substrate surface plane, respectively. The origin of the local coordinate system is positioned on the strain-neutral plane, which is denoted by the dashed line in Fig. 21.1. In a uniform substrate, the strain-neutral plane is in the middle of the substrate depth. In the multilayer structures its position depends on the depths and mechanical properties of the different material layers in the substrate [10].

The geometry in Fig. 21.1 dictates that the strain in the bending direction depend only on the distance x_2 from the strain-neutral plane as

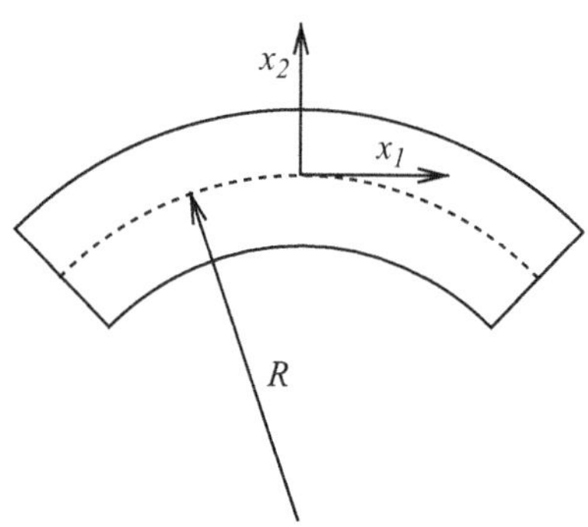

Fig. 21.1 The cylindrical shape of the substrate bending and the strain reference coordinate system

$$\varepsilon_{11} = \frac{x_2}{R}, \tag{21.1}$$

where R is the bending radius of the strain-neutral plane. Assuming that the substrate materials are elastic solids, the remaining strain tensor components are

$$\varepsilon_{22} = \varepsilon_{33} = -v(x_2)\varepsilon_{11},$$

$$\varepsilon_{12} = \varepsilon_{13} = \varepsilon_{23} = 0$$

where v is the Poisson ratio. Substrate-bending produces the equivalent strain effect as occurs in the application of in-plane uniaxial stress

$$\sigma = Y(x_2)\frac{x_2}{R} \tag{21.2}$$

in the bending direction along x_1 axis, where Y is the Young's modulus of elasticity. Notice that the elastic properties of the substrate can vary in x_2 direction.

The strain tensor can be decomposed into the hydrostatic (spherical)

$$\boldsymbol{\varepsilon}_S = \frac{1 - 2v(x_2)}{3}\varepsilon_{11}\mathbf{I} \tag{21.3}$$

and shear (deviatoric)

$$\boldsymbol{\varepsilon}_D = \boldsymbol{\varepsilon} - \boldsymbol{\varepsilon}_S \tag{21.4}$$

strain components, where $\mathbf{I}$ is the identity tensor. The hydrostatic strain component does not break the crystal symmetry and only shifts the bend energy levels. It affects bandgap, intrinsic carrier concentration, threshold voltage, leakage currents and any other device electrical parameter or characteristics depending on band edge energy levels [16, 17, 18, 19]. On the other hand, the deviatoric strain lowers the crystal symmetry, causing band degeneracy and warping that affects the carrier's effective mass and mobility [20, 21, 22, 23].

The impact of the bending-induced strain on semiconductor energy bands, and related MOS electrical characteristics, depends also on the relative position of the bending direction, MOS channel direction and crystal lattice directions. This is schematically shown for the case of silicon (001) substrate in Fig. 21.2.

The most convenient local coordinate system for considering strain effects on semiconductor band structure is the crystallographic (principal) coordinate system with the axes unit vectors $\mathbf{e}_1^c$, $\mathbf{e}_2^c$ and $\mathbf{e}_3^c$ in the lattice directions [100], [010] and [001], respectively. Given a strain tensor ε in the local strain coordinate system, the corresponding strain tensor in the crystallographic coordinates is obtained as

$$\boldsymbol{\varepsilon}^c = \mathbf{C} \cdot \boldsymbol{\varepsilon} \cdot \mathbf{C}^T, \tag{21.5}$$

where $\mathbf{C}$ the transformation matrix with elements $(\mathbf{C})_{ij} = \mathbf{e}_i^c \cdot \mathbf{e}_j$.

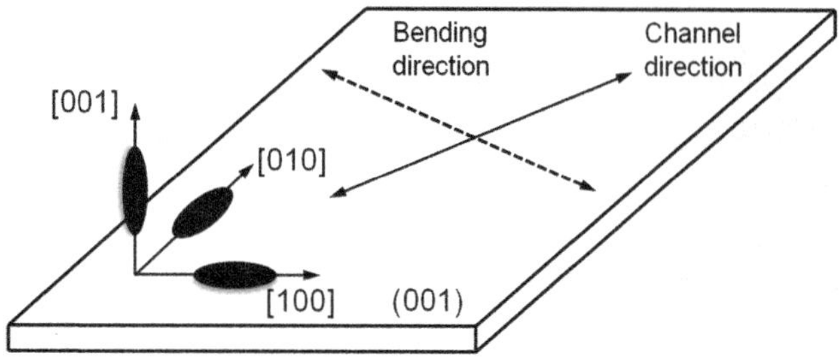

Fig. 21.2 The crystallographic coordinate system and relevant planar directions for stress and strain considerations in thin-chip substrates

21.4 Modelling of General Strain Effects on Band Edge Energies

Externally applied strain alters the silicon band structure by shifting the conduction and valence band edge energies. The model presented here is based on the deformation potential theory originally developed by Bardeen and Shockley [11].

The shift in the energy minimum of the conduction band valley pair corresponding to the ith lattice coordinate is obtained from the deformation potential theory as [12, 13]

$$\Delta E_c^{(i)} = \Xi_d\left(\varepsilon_{11}^c + \varepsilon_{22}^c + \varepsilon_{33}^c\right) + \Xi_u \varepsilon_{ii}^c, \tag{21.6}$$

where Ξ_u and Ξ_d are conditional band deformation potentials. On the other hand, the shift of the light and heavy hole bands (abbreviated here as l and h) is obtained as

$$\Delta E_v^{(l,h)} = -a\left(\varepsilon_{11}^c + \varepsilon_{22}^c + \varepsilon_{33}^c\right) \pm \sqrt{\xi + \eta} \tag{21.7}$$

with

$$\begin{aligned} \xi &= \frac{b^2}{2}\left[\left(\varepsilon_{11}^c - \varepsilon_{22}^c\right)^2 + \left(\varepsilon_{22}^c - \varepsilon_{33}^c\right)^2 + \left(\varepsilon_{33}^c - \varepsilon_{11}^c\right)^2\right] \\ \eta &= d^2\left[\left(\varepsilon_{12}^c\right)^2 + \left(\varepsilon_{13}^c\right)^2 + \left(\varepsilon_{23}^c\right)^2\right] \end{aligned} \tag{21.8}$$

where a, b and d are valence band deformation potentials. The positive sign in (21.7) applies to the light and the negative to the heavy hole band. The first terms in (21.6) and (21.7) are proportional to the hydrostatic strain component that shifts the energy levels of all the valleys equally. The splitting (degeneracy) of the conduction and valence bands arises only from the second terms in (21.6) and (21.7), which lowers the crystal symmetry. Table 21.2 shows the typical values of the deformation potentials used as default values in the Silvaco's device simulator Atlas [14].

Table 21.2 The default values of the deformation potentials used in the device simulator Atlas

Ξ_d	Ξ_u	a	b	d
1.1 eV	10.5 eV	2.1 eV	−2.33 eV	−4.75 eV

The shear deformation potentials (Ξ_u, b and d) are often obtained with a good accuracy from piezoresistance measurements [15]. On the other hand, determination of the hydrostatic deformation potentials (Ξ_d and a) is not so straightforward and their values vary widely in the literature.

In the standard TCAD and compact device models no distinction is made between different electron valleys or light and heavy hole bands. Instead, these models are based on common conduction and valence band edges. One must formulate a model for effective conduction and valence band edge energy shifts. Let $n^{(i)}$ be the partial contribution of the ith ellipsoid pair in the conduction band to the total electron concentration:

$$n = \sum_{i=1}^{3} n^{(i)}. \tag{21.9}$$

Assuming the Boltzmann statistics and equally distributed density of states (DOS) among ellipsoid pairs (strain-independent DOS), the effective shift of conduction band energy is obtained from (21.9) as

$$\Delta E_c = -kT \ln\left[\sum_{i=1}^{3} \frac{\exp\left(-\frac{\Delta E_c^{(i)}}{kT}\right)}{3}\right], \tag{21.10}$$

where $k$and T denote the Boltzmann constant and the absolute temperature, respectively. Similarly, from the splitting of the hole concentration among light and heavy bands as

$$p = p^{(l)} + p^{(h)}, \tag{21.11}$$

the effective shift of the valence band edge energy is obtained as

$$\Delta E_v = kT \ln\left[\frac{\eta}{1+\eta} \exp\left(\frac{\Delta E_v^{(l)}}{kT}\right) + \frac{1}{1+\eta} \exp\left(\frac{\Delta E_v^{(h)}}{kT}\right)\right], \tag{21.12}$$

where $\eta = (m_l/m_h)^{3/2}$ with m_l and m_h representing the DOS effective masses of the light and heavy hole bands, respectively. The typical value of the parameter η in unstrained silicon is 0.186.

Note that for pure spherical strain with

$$\Delta E_c = \Delta E_c^{(1)} = \Delta E_c^{(2)} = \Delta E_c^{(3)}$$

and

$$\Delta E_v = \Delta E_v^{(l)} = \Delta E_v^{(h)},$$

the relative contribution of the ellipsoid pairs to the total electron concentration, as well as relative contribution of the light and heavy hole bands to the total hole concentration, remain unchanged. The expressions (21.10) and (21.12) are derived assuming a rigid shift of the energy bands and neglecting the density of state effective mass strain dependences.

21.5 Modelling of General Strain Effects on Channel Mobility

The strain-induced shifts of the energy spectrum also affect the relative population of energy valleys and effectively modifies carrier mobility. Electron current density in silicon can be expressed as a superposition of longitudinal and transversal current density components summed over all energy ellipsoid pairs in the crystallographic coordinate system as

$$\mathbf{J}_n = n \sum_{i=1}^{3} \mu_n^{(i)} \frac{\partial \phi_n}{\partial x_i^c} \mathbf{e}_i^c, \tag{21.13}$$

where

$$\mu_n^{(i)} = \frac{n^{(i)}}{n} \mu_{\|} + \left(1 - \frac{n^{(i)}}{n}\right) \mu_{\perp} \tag{21.14}$$

is the effective electron mobility along the ith crystallographic axis, ϕ_n is quasi-Fermi potential while $\mu_{\|}$ and $\mu_{\perp}$ denote electron mobility in the longitudinal and transversal ellipsoid directions, respectively. Notice that in the absence of strain, $n^{(i)}/n = 1/3$ and the effective mobility becomes a scalar quantity

$$\mu_{n0} = \frac{\mu_{\|} + 2\mu_{\perp}}{3}. \tag{21.15}$$

The summation sign in (21.13) can be removed, giving the standard expression

$$\mathbf{J}_n(\boldsymbol{\varepsilon} = 0) = n\mu_{n0} \nabla \phi_n \tag{21.16}$$

for the electron current density.

For one-dimensional current flow along the channel direction, aligned with the reference coordinate axis x_l having unit vector $\mathbf{e}_l$, electron current density (21.13) can be expressed as

$$\mathbf{J}_n = n\frac{\partial \phi_n}{\partial x_l}\sum_{i=1}^{3}\mu_n^{(i)}\left(\mathbf{e}_l\cdot\mathbf{e}_i^c\right)\mathbf{e}_i^c \tag{21.17}$$

and effective electron mobility is

$$\mu_n = \frac{\mathbf{J}_n\cdot\mathbf{e}_l}{n\frac{\partial\phi_n}{\partial x_l}} = \sum_{i=1}^{3}\mu_n^{(i)}\left(\mathbf{e}_l\cdot\mathbf{e}_i^c\right)^2. \tag{21.18}$$

From (21.14) and (21.15), the relative change in the electron mobility due to externally applied strain can be expressed as

$$\frac{\mu_n}{\mu_{n0}} = 1 + M_n\sum_{i=1}^{3}\left[\exp\left(\frac{\Delta E_c - \Delta E_c^{(i)}}{kT}\right) - 1\right]\left(\mathbf{e}_l\cdot\mathbf{e}_i^c\right)^2, \tag{21.19}$$

where

$$M_n = \frac{1-\frac{\mu_\perp}{\mu_\parallel}}{1+2\frac{\mu_\perp}{\mu_\parallel}}$$

is the physical quantity that defines the sensitivity of the effective channel mobility to the redistribution of electron concentration among conduction band valleys. Notice that in the presence of the externally applied strain, the effective electron mobility is anisotropic and its strain dependence originates from the ellipsoid nature of the electron band structure in silicon.

On the other hand, the effective hole mobility can be directly expressed as

$$\mu_p = \frac{p^{(l)}\mu_l + p^{(h)}\mu_h}{p} = \mu_h + (\mu_l - \mu_h)\frac{p^{(l)}}{p}, \tag{21.20}$$

where μ_l and μ_h are the light and heavy hole mobility values. In the absence of strain the effective hole mobility becomes

$$\mu_{p0} = \mu_h + (\mu_l - \mu_h)\frac{\eta}{1+\eta} \tag{21.21}$$

and the relative change in the hole mobility can be expressed from (21.20) and (21.21) as

$$\frac{\mu_p}{\mu_{p0}} = 1 + M_p \left[\exp\left(\frac{\Delta E_v^{(l)} - \Delta E_v}{kT} \right) - 1 \right], \tag{21.22}$$

where

$$M_p = \frac{1 - \frac{\mu_h}{\mu_l}}{1 + \frac{\mu_h}{\eta \mu_l}}, \tag{21.23}$$

similar to M_n, defines the sensitivity of the hole mobility to the redistribution of the hole concentration among the light and heavy bands.

It should be emphasised that the sensitivity coefficients M_n and M_p could also depend on the bending stress direction. It is particularly emphasizised for the hole coefficient M_p that even changes sign with variation of the bending direction relative to the channel direction [20].

21.6 Compact Model Implementation

The change of the conduction and valence band edge energies with applied external strain can be considered as due to an effective change in the semiconductor material. Modern MOS compact models often provide a set of model parameters to support the impact of different material properties on electrical characteristics. Table 21.2 gives a list of BSIM4 material parameters and the corresponding correction terms that account for the externally applied strain.

In order to activate the impact of the BSIM4 material parameters on the electrical characteristics, the global model selector MTRLMOD should be set to 1. The first three strain correction terms in Table 21.3 are obtained in the straightforward way using (21.5) and (21.7) as

$$\begin{aligned} \Delta E_g &= \Delta E_c - \Delta E_v \\ \Delta n_i &= \text{NI0SUB} \cdot \left[\exp\left(-\frac{\Delta E_g}{2kT} \right) - 1 \right] \\ \Delta \chi_s &= -\Delta E_c \end{aligned} \tag{21.24}$$

Table 21.3 The BSIM4 model parameter affected by external strain

Parameter name	Description	Strain correction terms
BG0SUB	Bandgap of substrate at $T = 0$ K	ΔE_g
NI0SUB	Intrinsic carrier concentration at $T = 300.15$ K	Δn_i
EASUB	Electron affinity of substrate	$\Delta \chi_s$
PHIG	The gate work function	$\Delta \Phi_g$
VFB	Flat-band voltage	ΔV_{fb}

The change in the gate work function $\Delta\Phi_g$ depends on the gate material: It remains almost unchanged for a metal gate; it changes approximately to the same extent as the bulk work function $\Delta\Phi_s$ in the case of a poly-Si gate. An empirical relationship

$$\Delta\Phi_g = \alpha\Delta\Phi_s \tag{21.25}$$

with an additional parameter $0 \leqslant \alpha \leqslant 1$, provides the required model flexibility in the selection of the gate material. The change of the semiconductor work function is obtained as

$$\Delta\Phi_s = \Delta\chi \mp \Delta E_g, \tag{21.26}$$

with the minus and plus signs for N-type and P-type silicon bulk, respectively. Finally, the changes of the flat-band voltage in the channel is evaluated as

$$\Delta V_{fb} = \Delta\Phi_g - \Delta\Phi_s. \tag{21.27}$$

Note that in other to account for the corrected material model parameters in the threshold voltage, the BSIM4 model parameter VTH0 should not be given in the model card.

The strain-aware mobility model can be easily implemented in any MOS compact model by the relative correction of the low field mobility parameters (e.g., parameter U0 in BSIM4 mode) using expressions (21.19) and (21.22).

References

1. Kao HL et al (2005) Low noise RF MOSFETs on flexible plastic substrates. IEEE Electron Device Lett 26:489–491
2. Ahn J-H et al (2007) Bendable integrated circuits on plastic substrates by use of printed ribbons of single-crystalline silicon. Appl Phys Lett 90:213501
3. Li Y et al (2006) Bendability of single-crystal Si MOSFETs investigated on flexible substrate. IEEE Electron Device Lett 27:538–541
4. Hu C, Liu W (2010) BSIM4: theory and engineering of MOSFET modelling for IC simulation. World Scientific, New Jersey
5. Galup-Montoro G, Schneider MC (2007) MOSFET modelling for circuit analysis and design. World Scientific, New Jersey
6. Enz CC, Vittoz EA (2006) Charge-based MOS transistor modelling: the EKV model form low-power and RF IC design. Wiley, Chichester
7. Miura-Mattaucsh M, Mattausch HJ (2008) The physics and modelling of MOSFETSs: surface-potential model HiSIM. World Scientific, New Jersey
8. Dunga MV et al (2008) BSIM CMG: a compact model for multi-gate transistors. In: Colinge J-P (ed) FinFETS and other multi-gate transistors. Springer, New York
9. Amstrong GA, Maiti CK (2007) Technology computer aided design for Si, SiGe and GaAs integrated circuits. The Institution of Engineering and Technology, London

10. Ballas RG (2007) Piezoelectric multilayer beam bending actuators: static and dynamic behaviour and aspects of sensor integration. Springer, Berlin
11. Shockley W, Bardeen J (1950) Energy bands and mobilities in monatomic semiconductors. Phys Rev 77:407–408
12. Sun Y et al (2007) Physics of strain effects in semiconductors and metal-oxide-semiconductor field-effect transistors. J Appl Phys 101–104503:1–22
13. Egley JL, Chidambarrao D (1993) Strain effects on device characteristics: implementation in drift-diffusion simulators. Solid State Electron 36:1653–1664
14. Atlas Users Manual (2010), Silvaco, Santa Clara
15. Wang ZZ, Suski J, Collard E (1993) Piezoresistive simulations MOSFETs. Sensors Actuat A 37–38:357–364
16. Zhao W, Seabaugh A, Adams V, Jovanović D, Winstead B (2005) Opposing dependence of the electron and hole gate currents in SOI MOSFETs under uniaxial strain. IEEE Electron Device Lett 26:410–412
17. Nayfeh HM, Hoyt JL, Antoniadis DA (2004) A physically based analytical model for the threshold voltage of strain-Si n-MOSFETs. IEEE Trans Electron Devices 51:2069–2072
18. Ji-Song L, Thompson SE, Fossum JG (2004) Comparison of threshold-voltage shifts for uniaxial and biaxial tensile-stressed n-MOSFETs. IEEE Electron Device Lett 25:731–733
19. Venkataraman V, Nawal S, Kumar MJ (2006) A simple analytical threshold voltage model of nanoscale single-layer fully depleted strained-silicon-on-insulator MOSFETs. IEEE Trans Electron Devices 53:2500–2506
20. Matsuda K (2005) Strain-dependent hole masses and piezoresistive properties of silicon. J Comput Electron 3:273–276
21. Tan Y, Li X, Tian L, Yu Z (2008) Analytical electron-mobility model for arbitrary stressed silicon. IEEE Trans Electron Devices 55:1386–1390
22. Dhar S, Kosina H, Palankovski V, Ungersboeck SE, Selberherr S (2005) Electron mobility model for strain-Si devices. IEEE Trans Electron Devices 52:527–533
23. Ungersboeck E, Dhar S, Karlowatz G, Sverdlov V, Kosina H, Selberherr S (2007) The effect of general strain on the band structure and electron mobility of silicon. IEEE Trans Electron Devices 54:2183–2190

Chapter 22
Piezojunction Effect: Stress Influence on Bipolar Transistors

J. Fredrik Creemer and Patrick J. French

Abstract The saturation current of a bipolar transistor changes considerably under the influence of mechanical stress. This change is called the *piezojunction effect*. It is unwanted in precision circuits such as bandgap references, which usually become stressed during fabrication, packaging or use. However, it can be used to good advantage in mechanical sensors. The effect is strongly anisotropic, depends on the transistor type (*npn* or *pnp*) and is in many aspects similar to the piezoresistive effect for resistors. The piezojunction effect can be modelled similarly by a MacLaurin series expansion. For stresses below 200 MPa, an expansion up to the second order is appropriate. For lower stresses a first-order description can be used. This description can be displayed in polar plots that may be used to find optimum values. The coefficients of the expansion are determined by the piezojunction coefficients, which are material parameters. The piezojunction coefficients depends on stress-induced changes in both the mobility and the intrinsic carrier concentration. They can be determined from measurements but also from calculations using band deformation theory.

22.1 Introduction

Bipolar junction transistors are useful components in analog circuits where gain and precision are important. In addition, they are essential in subcircuits such as bandgap voltage references and bandgap temperature sensors. Those subcircuits can be realised in standard CMOS processes, where they often employ parasitic bipolar substrate transistors [1, 2].

Like other semiconductor components, bipolar transistors are very sensitive to mechanical stress. Stress influences the main characteristic of the transistor: the exponential relation between the collector current and the base-emitter voltage

J.F. Creemer (✉)
Delft Institute of Microsystems and Nanoelectronics (DIMES),
Delft University of Technology, P.O. Box 5053, 2600 GB Delft, The Netherlands
e-mail: j.f.creemer@tudelft.nl

J.N. Burghartz (ed.), *Ultra-thin Chip Technology and Applications*,
DOI 10.1007/978-1-4419-7276-7_22, © Springer Science+Business Media, LLC 2011

[3–5]. This effect is caused by the piezojunction effect. The effect is in many aspects comparable to the piezoresistive effect in semiconductors [6], but it is less well known. In magnitude, it causes about 5% current change per 100 MPa stress. It strongly depends on the transistor type (*npn* or *pnp*) and is highly anisotropic. In fact, the magnitude and sign depend on the orientation of both the stress and the current in the base with respect to the crystal lattice.

The piezojunction effect can be employed to good advantage for sensing stress [7–9]. Compared to piezoresistive sensors, bipolar transistors display only half of the noise and have a much better small-signal impedance. In addition, their sensitive volume (the intrinsic base) can be much smaller. On the other hand, the piezojunction effect should be avoided or compensated for in many circuits, especially in those relying on the apparent bandgap [1, 2]. Care should also be taken in subcircuits that need matching transistors such as differential stages, and translinear circuits such as multipliers.

Unintended stresses in circuits are usually introduced during chip fabrication, e.g., while dielectric layers are being deposited and trench isolations created [10, 11]. They may also result from packaging, when the circuit is overmoulded with a hot polymer. Finally, stresses may result from mechanical loading of the final chip, for instance by bending, gluing or thermal cycling. Stresses are higher when the chip becomes thinner, as it then offers less resistance to mechanical deformation.

This section first describes the piezojunction effect as it appears in current-voltage characteristics. Second, it derives a model that can quantitatively describe the effect for any orientation with a set of material parameters. Third, it shows how the model could be used for optimisation and design purposes. Finally, it sketches how the piezojunction effect can be calculated from first principles, i.e. the energy band diagram of the material.

The stresses under consideration are limited to ±200 MPa. These are roughly the maximum values for safe bending of a diced chip or a large MEMS structure. However, stresses up to 10 GPa have been demonstrated, especially by compression of a transistor with a stylus [3, 4]. Models for this range can often be simpler than the one presented here [12, 13], because changes in intrinsic carrier concentration overwhelm changes in mobility [14].

22.2 Current–Voltage Characteristics

For designers of analog circuits, a bipolar transistor is ideally a voltage-controlled current source with a purely exponential relation between the collector current I_c and the base-emitter voltage V_{be}. In forward bias it is:

$$I_c = I_S \left[\exp\left(\frac{qV_{be}}{k_B T} \right) - 1 \right] \tag{22.1}$$

where I_S is the saturation current, q the elementary charge, k_B the Boltzmann constant and T the absolute temperature (Fig. 22.1a). I_S contains all information on the transistor material and geometry. This relation holds for many practical bipolars to good accuracy over a very wide current range (e.g. seven orders of magnitude; see Fig. 22.1c). It is at the heart of virtually all compact transistor models.

Transistors are not only described by I_S, but by other parameters such as current amplification factor and junction capacitances. For circuit design, however, those are usually only of secondary importance. It is generally sufficient that the values of secondary parameters remain in a range where their influence on circuit behaviour can be neglected. For stress effects it is assumed here that this is the case.

Stress influences the I_c-V_{be} relation through I_S. For stresses between –200 and 200 MPa, this influence is approximately quadratic (Fig. 22.2) [15–18]. The shape of the quadratic curve depends strongly on the transistor type and current and stress orientation. However, for a particular situation it can be generalised into the following MacLaurin series:

$$\frac{I_S(X)}{I_S(0)} = 1 + aX + bX^2 \tag{22.2}$$

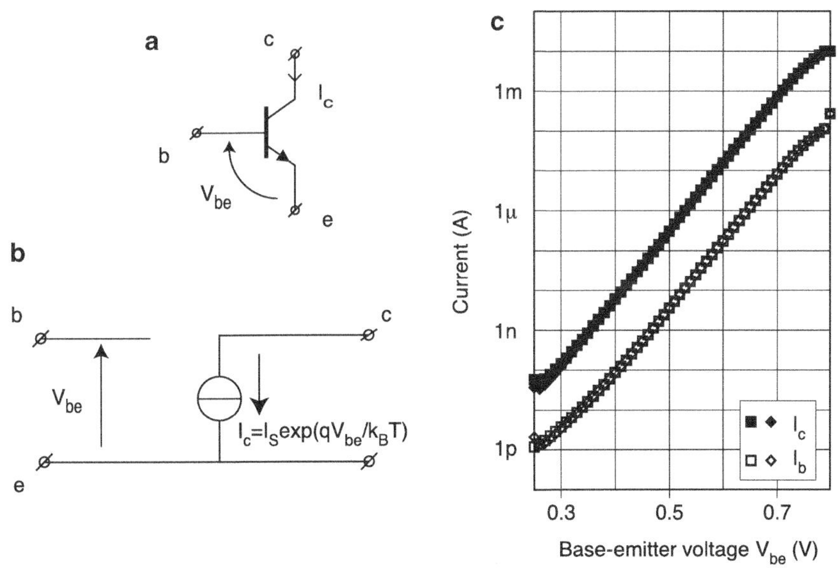

Fig. 22.1 Primary performance of a bipolar *npn* transistor in forward bias. (**a**) Network symbol, indicating the base, collector and emitter terminals *b*, *c* and *e*, respectively. (**b**) Basic equivalent circuit: a voltage-controlled current source. (**c**) Measured Gummel plot of a silicon bipolar transistor [16] showing the collector current I_c and the base current I_b as a function of the base-emitter voltage V_{be}. The transfer function $I_c(V_{be})$ is almost purely exponential over 7 decades of current

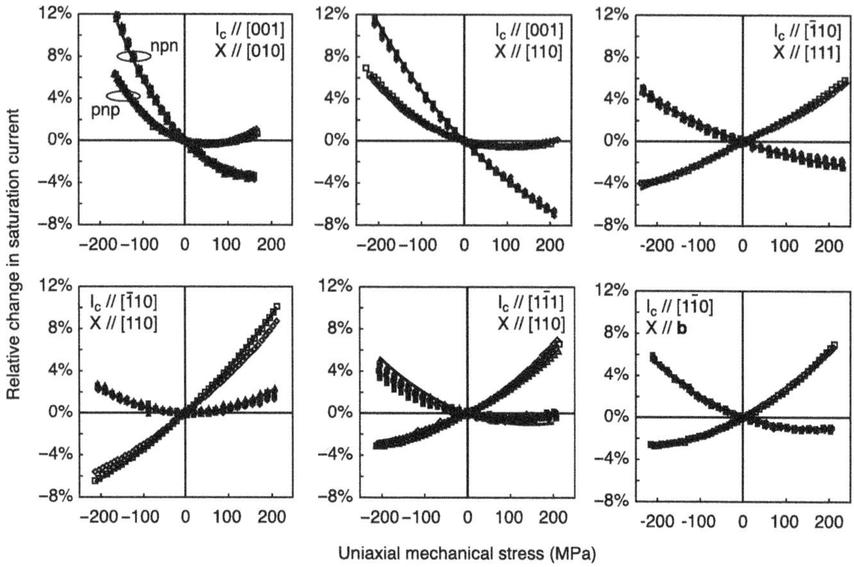

Fig. 22.2 (**a–f**) Changes in the saturation current I_S of silicon transistors as a function of uniaxial stress X, for different orientations of the stress and collector current I_c with respect to the crystal lattice. Closed symbols represent *npn* transistors; open symbols *pnps*. For each transistor type, data points 2–3 equivalent specimen are plotted in the same graph. The solid lines represent an overall fit made with the model of 22.9, yielding the piezocoefficients of Table 22.1. For the *npn* transistors, the changes are essentially constant for I_c between 0.14 nA and 0.8 mA (Adapted from [15, 16])

where X is the stress amplitude and a, b are polynomial constants. These constants can be calculated for each situation from a small number of material parameters, as will be demonstrated in the next section.

22.3 Charge Transport Parameters

The saturation current of most transistors can be related to charge transport parameters by using the Gummel approach [19]. This approach is based on a number of assumptions. In today's technology the most restrictive ones are that (a) the base material is nondegenerate, so not too heavily doped, (b) the charge carrier concentration in the base caused by the injection of current is inferior to the local doping concentration and (c) the base is long enough to neglect the effects of velocity saturation. In that case,

$$I_S \sim q\mu^{mi}n^2{}_{i0} = q\mu^{mi}np, \tag{22.3}$$

where q is the unit charge, μ^{mi} is the mobility of the minority charge carriers in the base, n_{i0} is the intrinsic carrier concentration in thermodynamic equilibrium and

n and p are the electron and hole concentrations, respectively, in thermodynamic equilibrium. It has been used here that ${n_{i0}}^2 = np$. The other factors determining I_S are base geometry and temperature.

Concentrations n and p can be expressed in terms of σ^{mi} and σ^{ma}, which are the contributions of minorities and majorities, respectively, to the total conductivity σ of the base material in thermodynamic equilibrium. For p-type material this yields:

$$\sigma = \sigma^{mi} + \sigma^{ma} = q(\mu^{mi}n + \mu^{ma}p), \tag{22.4}$$

where μ^{ma} is the local mobility of the majority charge carriers. Normally, the base doping type and temperature are such that the concentration of majority charge carriers (say, p) is entirely determined by the doping concentration N_A. Combining the above expressions yields:

$$I_S \sim \frac{q\mu^{mi}n_{i0}^2}{N_A} = \frac{\sigma^{mi}}{N_A}. \tag{22.5}$$

Mechanical stress particularly influences I_S through μ^{mi} and ${n_{i0}}^2$. It also has influence through N_A by a change in the geometry (strain). However, geometrical changes are usually a factor of 10^2 smaller than those in the charge transport parameters and can often be neglected [6, 16]. Combining this information with (22.2) yields the following expression for the stress-induced change in the saturation current:

$$\frac{I_S(X)}{I_S(0)} = \frac{\left(\mu^{mi}n_{i0}^2\right)(X)}{\left(\mu^{mi}n_{i0}^2\right)(0)} = \frac{\sigma^{mi}(X)}{\sigma^{mi}(0)} = 1 + aX + bX^2 \tag{22.6}$$

The interesting point to notice here is that change depends entirely on *material* parameters, namely the local mobility and intrinsic concentration, or, alternatively, the minority conductivity. At the same time the model equation is mathematically simple and easily invertible, facilitating its use with, e.g., compact transistor models.

22.4 Piezoresistance and Piezojunction Coefficients

In order to find the polynomial coefficients of (22.6), it should be noted that they depend on some quantities that are, in principle, tensors. Stress, e.g. is a tensor of rank 2 and can be described as X_{ij}, where i and j run independently over the x-, y- and z-dimensions, indicated by the numbers 1, 2 and 3, respectively [20]. Also mobility and conductivity are tensors of rank 2. Usually, this fact has no practical significance because in unstressed semiconductors with cubic lattice symmetry, such as Si, the tensors reduce to scalars. In the presence of stress, however, the symmetry is broken and the full tensor notation should be used. The minority conductivity tensor $\sigma_{ij}{}^{\mathrm{mi}}$ is related to the scalar values in (22.6) by:

$$\sigma^{\mathrm{mi}} = l_i l_j \sigma_{ij}{}^{\mathrm{mi}} \tag{22.7}$$

in which l_i is the unit vector of the direction of the current and the applied field, and where the Einstein convention has been used to indicate the summation of the elements with the dummy subscripts [20]. Thus, it is assumed here that the current in the considered transistor base has one dominant orientation with respect to the semiconductor crystal and that it is measured along the direction of the applied field. Also, the conductivity tensor is by definition the inverse of the minority *resistivity* $\rho_{ij}{}^{\mathrm{mi}}$, in the sense that

$$\sigma_{ij}{}^{\mathrm{mi}} \rho_{jk}{}^{\mathrm{mi}} = \delta_{ik}, \tag{22.8}$$

where δ_{ik} is the unity tensor of rank 2.

The relation between the total resistivity of a material and stress is well known and is called piezoresistance [21–24]. Usually, it is described as a MacLaurin series. Up to the second order it reads as

$$\rho_{ij} = \rho_{ij}{}^{0} + \pi_{ijkl} X_{kl} + \pi_{ijklmn} X_{kl} X_{mn}, \tag{22.9}$$

where $\rho_{ij}{}^{0}$ is the resistivity in the absence of stress and in thermodynamic equilibrium; π_{ijkl} and π_{ijklmn} and are the first- and second-order piezoresistive tensors, respectively, equally defined around zero stress and current. An analogous description can be used to describe the resistivity of *minority charge carriers*:

$$\rho_{ij}{}^{\mathrm{mi}} = \rho_{ij}{}^{0,\mathrm{mi}} + \zeta_{ijkl} X_{kl} + \zeta_{ijklmn} X_{kl} X_{mn}, \tag{22.10}$$

where ζ_{ijkl} and ζ_{ijklmn} are defined as the first- and second-order *piezojunction tensors*, respectively [15]. These tensors are also defined at vanishing stress and current.

The piezoresistive and piezojunction tensors are large, as the first- and second-order tensors contain 3^4 and 3^6 elements, respectively. However, this number can be greatly reduced if we employ the fact that they represent material properties and therefore obey the principle of Neumann [20]. This principle states that *the symmetry elements of any physical property of a crystal must include the symmetry elements of the point group of that crystal.* Crystals of Si and Ge, e.g. have cubic symmetry. This means that their resistivity tensor $\rho_{ij}{}^{0}$ has equal elements ρ^0 on the main diagonal and zeros elsewhere. It also means that their ζ_{ijkl} has only three independent nonzero values, called the *piezojunction coefficients.* They are usually written as ζ_{11}, ζ_{12} and ζ_{44} in the reduced-index notation, where pairs of subscripts from ζ_{ijkl} have been replaced as follows:

$$11 \rightarrow 1; 22 \rightarrow 2; 33 \rightarrow 3; 23, 32 \rightarrow 4; 13, 31 \rightarrow 5; 12, 21 \rightarrow 6. \tag{22.11}$$

In literature, however, other conversion rules can be found [25, 26]. The 729 elements of ζ_{ijklmn} reduce to only nine values: ζ_{111}, ζ_{112}, ζ_{122}, ζ_{123}, ζ_{144}, ζ_{166}, ζ_{414}, ζ_{616} and ζ_{456}. The piezoresistive coefficients reduce correspondingly. All these coefficients can be determined experimentally or calculated from first principles. In literature, however, second-order coefficients are difficult to find, and the set isn't known completely (Table 22.2). Much more common are values of the first-order piezoresistive coefficients (Table 22.1), although there is a non-negligible scattering in the reported values: up to 10s of percents for the most significant value.

The first-order coefficients can be expressed in terms of charge transport parameters by combining (22.8)–(22.10) with (22.3):

$$\pi_{\mathrm{OP}} = -\frac{1}{\sigma^0}\frac{\partial\sigma_{\mathrm{O}}}{\partial X_{\mathrm{P}}}\bigg|_0 = -\frac{1}{\mu^{0,\mathrm{ma}}}\frac{\partial\mu_{\mathrm{O}}^{\mathrm{ma}}}{\partial X_{\mathrm{P}}}\bigg|_0 \tag{22.12}$$

$$\zeta_{\mathrm{OP}} = -\frac{1}{\sigma^{0,\mathrm{mi}}}\frac{\partial\sigma_{\mathrm{O}}^{mi}}{\partial X_{\mathrm{P}}}\bigg|_0 = -\frac{1}{\mu^{0,\mathrm{mi}}}\frac{\partial\mu_{\mathrm{O}}^{\mathrm{mi}}}{\partial X_{\mathrm{P}}}\bigg|_0 - \frac{1}{n_{\mathrm{i0}}^2}\frac{\partial n_{\mathrm{i0}}^2}{\partial X_{\mathrm{P}}}\bigg|_0 \tag{22.13}$$

where σ^0, $\sigma^{0,\mathrm{mi}}$ and $\mu^{0,\mathrm{mi}}$ are the values on the main diagonal of the respective tensors in the unstressed case (strictly speaking, the trace). Seen in this way, the piezoresistive effect depends in the first order on the changes in the mobility, whereas the piezojunction effect depends on changes in the mobility *and* in the intrinsic carrier concentration. In the considered stress range, both contributions are

Table 22.1 First-order piezojunction coefficients PJ as extracted from the measurements of Fig 22.2 [15, 16], compared with two sets of piezoresistive coefficients PR1 [22] and PR2 [21]. All values are in 10^{-11} Pa^{-1}. The error intervals indicate the maximum bounds in which the values are expected to lie

	Electrons			Holes		
Coefficients	PJ	PR1	PR2	PJ	PR1	PR2
ζ_{11}	−28.4 ± 3.0	−77	−102.2	30.8 ± 2.6	0	6.6
ζ_{12}	43.4 ± 1.5	39	53.4	13.8 ± 1.3	2	−1.1
ζ_{44}	13.1 ± 4.3	−14	−13.6	119.8 ± 4.2	119	138.1

Table 22.2 Second-order piezojunction coefficients PJ as extracted from the measurements of Fig 22.2 [15, 16], compared with the corresponding piezoresistive coefficients PR [22]. All values are in 10^{-18} Pa^{-2}. The error intervals indicate the maximum bounds in which the values are expected to lie

	Electrons		Holes	
	PJ	PR	PJ	PR
$\zeta_{111} - 4\zeta_{616} + 2\zeta_{414} + 2\zeta_{456}$	−1.5 ± 1.0	0.97	−1.7 ± 1.3	0.22
$\zeta_{112} + \zeta_{166}/2 - \zeta_{414} - \zeta_{456}$	0.30 ± 0.43	−0.40	0.99 ± 0.53	0.81
ζ_{122}	−1.29 ± 0.21	−0.36	−1.21 ± 0.23	−0.08
$\zeta_{123} - \zeta_{166} + 2\zeta_{616} + \zeta_{414} + \zeta_{456}$	1.6 ± 1.0	0.68	0.3 ± 1.3	−1.63
$\zeta_{144} + 2\zeta_{166} - 4\zeta_{616} - 2\zeta_{414} - 2\zeta_{456}$	−2.1 ± 2.1	0.03	−0.7 ± 2.6	2.97

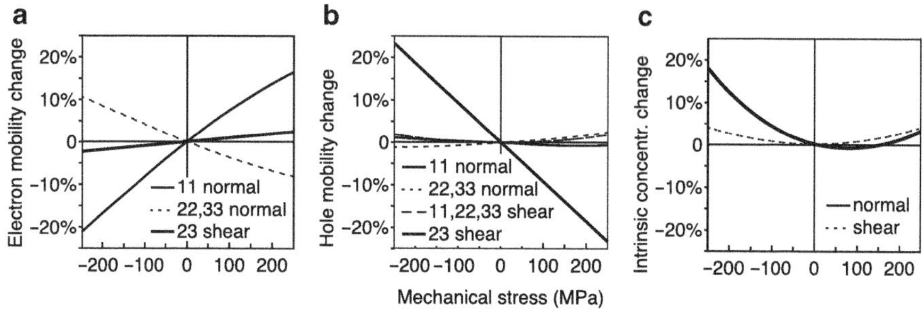

Fig. 22.3 Changes in charge carrier parameters of silicon as a function of stress, as calculated from energy band deformations [16, 27]. (**a**) Changes in the mobility tensor elements μ_{ij} of electrons, and (**b**) holes. (**c**) Changes in the intrinsic carrier concentration. Calculations have been done for both a normal uniaxial stress X_{11} and a pure shear stress X_{23}. Only nonzero changes are plotted. Changes by stress in other orientations can be derived from symmetry considerations

of comparable magnitude, as can be seen in Fig. 22.3. Only for pure shear stress, the first-order change in the intrinsic concentration vanishes, meaning that $\zeta_{44} = \pi_{44}$.

At very high stresses (1–10 GPa) the changes in mobility saturate [14], whereas the intrinsic carrier concentration continues to change exponentially. This can increase the saturation current by as much as a factor 10^3 [3, 4, 12, 13]. Above 10 GPa compressive stress, the material is damaged. For tensile stress, damage may occur much earlier (e.g. at 0.3 GPa), depending mainly on the smoothness of the surfaces.

22.5 Use of the Model for Design Purposes

With the aid of the previous sections the piezojunction effect can be predicted, provided that the stress tensor is known and the direction of the current in the intrinsic base of the transistor is also known. With this information the resistivity change can be calculated with the aid of (22.10). This can be converted into a conductivity change by (22.8), and finally into a change in the saturation current by (22.6) and (22.7).

It is also possible to use the equations for optimisation purposes, to choose orientations where the stress sensitivity is maximum or minimum [16, 28]. This poses the problem that the number of stress and current orientations is in principle infinite. However, it is possible to consider a few common cases that drastically reduce the degrees of freedom. Those common cases arise from the following observations:

- Transistors are usually made in wafers that are of the {100} or {111} crystal orientation; other orientations are much less common.
- Bipolar transistors are either vertical or lateral, meaning that the current in the intrinsic base is perpendicular to the wafer plane or in the wafer plane, respectively (see Fig. 22.4).

- Transistors in a sensor are often aligned with the cantilever beam in which they are integrated. This usually means that the stress is either longitudinal or transverse to the current direction in the base.
- Transistors in a circuit usually have a fixed orientation: They are aligned with respect to the wafer flat, so in a <110> direction (Fig. 22.4).
- Stresses usually arise from bending or from stretching. In this case they are in the plane of the wafer (Fig. 22.4).
- Stresses are often uniaxial, or can be considered as a superposition of uniaxial stresses.
- For many sensor applications stresses are quite small, and knowledge about the first-order sensitivity is already sufficient.

The stress sensitivities can now be displayed in a few polar plots, analogously to those established for the piezoresistive effect [29]. This yields some interesting results.

First, it appears that vertical transistors in {100} and {111} wafers are insensitive to the orientation of the in-plane stress. In a polar plot this leads to concentric circles. The magnitudes of the sensitivity are given in Table 22.3. It can be seen that in absolute values, the sensitivity is largest for vertical *npn*s in {100} wafers, and quite small for vertical *pnp*s in either wafer type. Especially the latter fact is welcome in circuits where substrate *pnp*s are used for, e.g. bandgap temperature sensors.

Second, it appears that lateral transistors in {111} wafers also have an isotropic sensitivity when the current is either longitudinal or transverse to the stress. This situation is sketched in Fig. 22.5a, whereas values are given in Table 22.3.

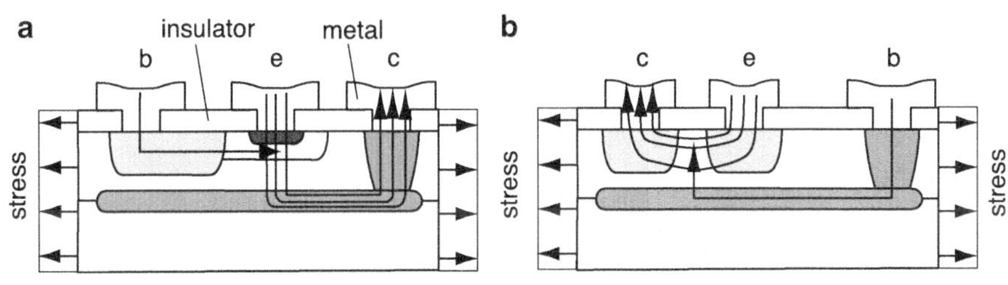

Fig. 22.4 Common orientations of stress and current flows in bipolar transistors under in-plane stress. (**a**) Schematic cross-section of a basic vertical transistor. The main current crosses the intrinsic base in a direction perpendicular to the wafer. (**b**) Schematic cross-section of a basic lateral transistor. The main current crosses the intrinsic base in a direction in the plane of the wafer

Table 22.3 First-order stress sensitivity of bipolar trainsistors in orientations with isotropy

Wafer plane	Current orientation	Stress orientation	First-order stress sensitivity	npn 10^{-11} Pa^{-1}	*pnp* 10^{-11} Pa^{-1}
{100}	Vertical	In-plane	$-\zeta_{12}$	−43.4	−13.8
{111}	Vertical	In-plane	$-(\zeta_{11} + 2\zeta_{12} - \zeta_{44})/3$	−15.1	20.5
{111}	Lateral	Longitudinal	$-(\zeta_{11} + \zeta_{12} + \zeta_{44})/2$	−7.6	−82.9
{111}	Lateral	Transverse	$-(\zeta_{11} + 5\zeta_{12} - \zeta_{44})/6$	−28.2	4.0

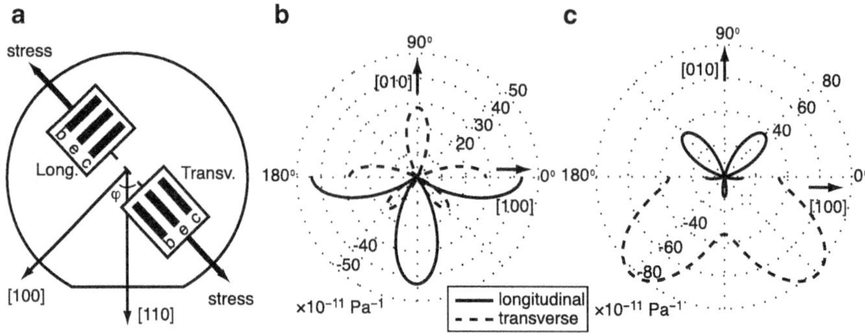

Fig. 22.5 (**a**) Lateral bipolar transistors in orientations longitudinal or transverse to an applied uniaxial stress of arbitrary orientation in the wafer plane. (**b**) Stress sensitivity of lateral npn transistors on a {100} wafer as a function of the stress angle φ that is defined with respect to the [100] direction. (**c**) Stress sensitivity of lateral *pnp* transistors on a {100} wafer as a function of the stress angle φ. The results are only plotted for φ from 0 to 180°, because of the symmetry of the stress tensor

Third, large anisotropic sensitivities are observed for lateral transistors in a {100} wafer. This is shown in Fig. 22.5 for longitudinal and transverse sensitivities. For lateral *npn*s the sensitivities are largest in the <100> directions, whereas for *pnp*s the sensitivities are largest in the <110> directions. They are about two times larger than for *npn*s.

Strong anisotropy is also shown in Fig. 22.6, where the stress angle is varied in the {100} plane but the current orientation is fixed in a standard <110> direction. For an increasing stress angle, the sensitivity for both *npn* and *pnp* transistors changes from negative to slightly positive to strongly negative again. So the sensitivity crosses zero twice. When the current is directed in the [110] direction, the stress sensitivity reaches a maximum (in absolute numbers) for stress in the $[\bar{1}, 1, 0]$ direction.

Finally, the anisotropy is still present but very different for lateral transistors of fixed <110> orientation in {111} wafers. Figure 22.7 shows that, most notably, *pnp* transistors are maximally sensitive to stress in the [110] direction, but almost negligibly in the directions $[1, \bar{1}, \bar{2}]$ and $[\bar{1}, 1, 2]$.

22.6 Conductivity Calculations from First Principles

The conductivity under stress and piezocoefficients of the previous sections can be determined from measurements. However, it is also possible to calculate them from first principles by using the semiclassical transport model for charge carriers and the deformation potential theory for energy bands. For this purpose various models have been developed, often of great sophistication and rather, mathematically complex [25, 30–34]. This section presents a sketch of such an approach. The

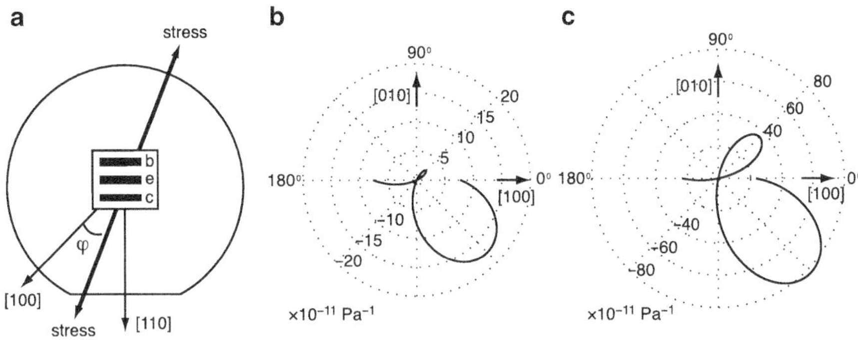

Fig. 22.6 (**a**) Lateral bipolar transistor in a {100} wafer in a standard <110 > orientation along the wafer flat, with an applied uniaxial stress of arbitrary orientation in the wafer plane. (**b**) Stress sensitivity of lateral *npn* transistors as a function of the stress angle φ that is defined with respect to the [100] direction. (**c**) Stress sensitivity of lateral *pnp* transistors as a function of the stress angle φ. The results are only plotted for φ from 0° to 180°, because of the symmetry of the stress tensor

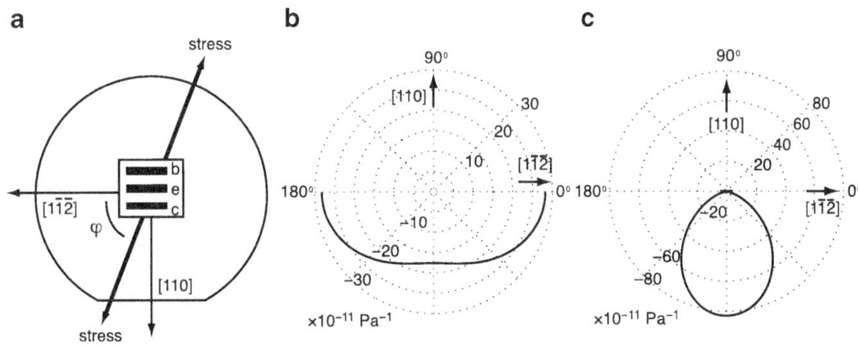

Fig. 22.7 (**a**) Lateral bipolar transistor in a {111} wafer in a standard <110 > orientation along the wafer flat, with an applied uniaxial stress of arbitrary orientation in the wafer plane. (**b**) Stress sensitivity of lateral *npn* transistors as a function of the stress angle φ that is defined with respect to the [100] direction. (**c**) Stress sensitivity of lateral *pnp* transistors as a function of the stress angle φ. The results are only plotted for φ from 0° to 180°, because of the symmetry of the stress tensor

main objective is to show the physical phenomena that underlie both the piezojunction and piezoresistive effects.

In the semiclassical transport model, in the relaxation-time approximation, the DC conductivity of a material close to equilibrium is given by [35–39]:

$$\sigma_{ij} = \sum_n \sigma_{ij}^{(n)} = \sum_n e^2 \int \frac{d\mathbf{k}}{4\pi^3} \tau^{(n)}(\mathbf{k}) v_i^{(n)}(\mathbf{k}) v_j^{(n)}(\mathbf{k}) \frac{\partial f}{\partial E}\bigg|_{E=E^{(n)}(\mathbf{k})} \tag{22.14}$$

where $\sigma_{ij}^{(n)}$ are the contributions to the conductivity of the involved energy bands, e is the unit charge, **k** equals the wave vector k_i, the integration runs over the first Brillouin zone, $\tau^{(n)}$ is the relaxation time, i.e. the average time between two scattering events of a charge carrier, $v_i^{(n)}$ the group velocity of a charge carrier, f the distribution function, E the energy and $E^{(n)}$ the energy in a specific band n. Equations for those quantities are often found as follows.

The group velocity of the charge carriers $v_i^{(n)}$ is defined by the energy band equations $E^{(n)}$ **k**), as $v_i^{(n)} = \partial E^{(n)}/\partial k_i$. The energy band equations can be found in literature in various forms. What is needed for this application, however, is a description near the band edges, because the bands of a semiconductor at room temperature are only filled up to a very shallow level. Such a description is provided by the **kp** method [25]. In the **kp** equations, stress-induced changes can be included by using deformation potential theory [34].

The distribution function for electrons or holes f is given by the Fermi–Dirac function [35, 36]. For nondegenerate material (with a doping concentration of, say, smaller than 10^{18} cm^{-3}), the Fermi–Dirac function can be replaced by the Maxwell–Boltzmann function, which simplifies mathematics [27].

The Fermi–Dirac function depends on the Fermi level E_F. In thermodynamic equilibrium, E_F is equal for electrons and holes. It can be found by solving the equation for the concentration of majority charge carriers (say, p), which at normal temperatures equals the doping concentration N_A:

$$p = N_A = \sum_n p^{(n)} = \sum_n \int \frac{d\mathbf{k}}{4\pi^3} f\left(E^{(n)}(\mathbf{k}), E_F\right) \tag{22.15}$$

In (22.14) the use of the relaxation time approximation assumes that (a) the energy distribution after a collision is independent of the distribution before, and (b) the collision does not influence the shape of the distribution at thermodynamic equilibrium. Quite often the equations are simplified further by assuming that (c) $\tau^{(n)}$ depends on **k** only through the kinetic energy, thus through $E^{(n)}$ **k**). Usually, this dependency is cast in a simple power law [37], which in the calculation of the piezocoefficients already gives quite reasonable results [27].

Schematically, the stress influences on the energy bands of silicon is represented in Fig. 22.6. In the unstressed case, the conduction band has six equivalent minima which, for a certain current direction, have different effective masses (m_l^* and m_t^*). These minima are separated by a bandgap E_G from two valence bands, the heavy hole band and the light hole band, which are degenerate in the origin.

Application of stress modifies the bandgap, the effective masses (i.e. the band curvatures) and the degeneracies. Due to breaking of the crystal symmetry, some conduction band extrema shift downwards and others upwards. As a result, the bands become populated differently and charge is transferred between bands. This leads to a change in the concentration of minority charge carriers changes as a whole. The concentration of majorities remains equal, as it is given by the doping concentration. However, there is a modification in the proportion between electrons

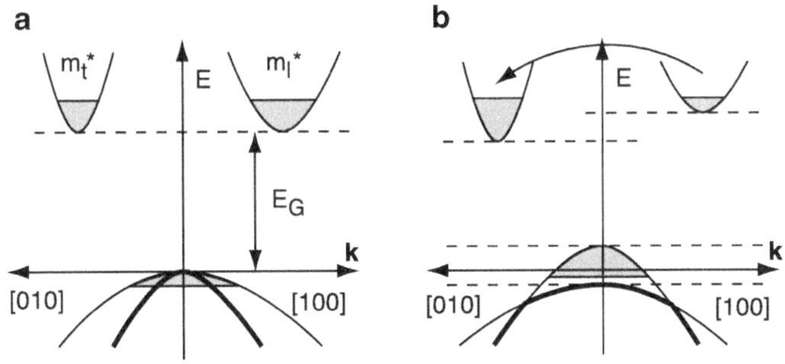

Fig. 22.8 Schematic of the effects of stress on the energy band edges of silicon. (**a**) Unstressed situation, sketching the bandgap E_G between the conduction and the valence bands, and the shallow filling with charge carriers. (**b**) Situation when a stress is applied that breaks the symmetry of the crystal lattice. Band edges are shifted and deformed, and charge is transferred from one band to the other

with heavy and light effective masses. The stress also lifts the degeneracy of the valence bands. As those bands are strongly coupled, this also changes their curvature. All those changes lead to modifications in the overall conductivity and include changes in the mobilities and the intrinsic carrier concentration.

22.7 Conclusions

Under the influence of mechanical stress, the saturation current of a bipolar transistor changes considerably. This effect is strongly anisotropic, depends on the transistor type (*npn* or *pnp*) and is in many aspects similar to the piezoresistive effect for resistors. For transistors, the effect is called the *piezojunction effect* and can be modelled by a MacLaurin series expansion. For stresses up to 200 MPa, a second-order expansion is usually sufficient. For lower stresses a first-order approach can be used, which can be displayed in polar plots and is of help in optimisation problems. The polynomial coefficients are determined by the piezojunction coefficients, which are material parameters. They depend on the mobility and the intrinsic carrier concentration under stress. The coefficients can be found from measurements, but also from calculations using band deformation theory.

The presented model is highly applicable in the development of mechanical sensors, where stress is often a parameter that needs to be detected. In addition, it can be used to good advantage in circuit design, to reduce the sensitivity of critical bipolar components to stress that is involuntarily present by fabrication, packaging or mounting.

Future work on this subject could entail further experimental determination of the piezojunction coefficients. Also, the theory could be extended to higher doping levels and to more accurate modelling from first principles. Study could be made of the stress sensitivity of other parameters of compact transistor models, such as the current amplification factor and the early voltage. Finally, the model could be added to existing compact-transistor models. In that case mechanical stress could be given as an input parameter, which would allow more accurate circuit simulations.

References

1. Sebastiano F, Breems LJ, Makinwa KAA, Drago S, Leenaerts DM, Nauta B (2010) A 1.2 V 10 μW NPN-based temperature sensor in 65 nm CMOS with an inaccuracy of ±0.2 °C (3σ) from –70 °C to 125 °C. In: Dig. IEEE international solid-state circuit conference (ISSCC 2010), San Francisco
2. Ge G, Zhang C, Hoogzaad G, Makinwa K (2010) A single-trim CMOS bandgap reference with a 3σ inaccuracy of ±0.15% from –40 °C to 125 °C. In: Dig. IEEE international solid-state circuit conference (ISSCC 2010), San Francisco
3. Rindner W (1962) Resistance of elastically deformed shallow p-n junctions. J Appl Phys 33:2479–2480
4. Bulthuis K (1966) Effect of local pressure on germanium p-n junctions. J Appl Phys 37:2066–2068
5. Monteith LK, Wortman JJ (1973) Characterization of p-n junctions under the influence of a time varying mechanical strain. Solid-State Electron 16:229–237
6. Creemer JF, Fruett F, Meijer GCM, French PJ (2001) The piezojunction effect in silicon sensors and circuits and its relation to piezoresistance. IEEE Sensors J 1:98–108
7. Schellin R, Mohr R (1993) A monolithically-integrated transistor microphone: modelling and theoretical behaviour. Sens Actuat A 37–38:666–673
8. Puers BL, Reynaert L, Snoeys W, Sansen WMC (1988) A new uniaxial accelerometer in silicon based on the piezojunction effect. IEEE Trans Electron Devices ED-35:764–770
9. Fruett F, Meijer GCM (2001) A new sensor structure using the piezojunction effect in PNP lateral transistors. Sens Actuat A 92:197–202
10. Smeys P, Griffin PB, Rek ZU, de Wolf E, Saraswat KC (1999) Influence of process-induced stress on device characteristics and its impact on scaled device performance. IEEE Trans Electron Dev 46:1245–1252
11. Egley L, Chidambarrao D (1993) Strain effects on device characteristics: implementation in drift-diffusion simulators. Solid-State Electron 36:1653–1664
12. Wortman JJ, Hauser JR, Burger RM (1964) Effect of mechanical stress on p-n junction device characteristics. J Appl Phys 35:2122–2131
13. Kanda Y (1973) Effect of compressive stress on silicon bipolar devices. J Appl Phys 44:389–393
14. Adams AR, Pickering C, Vinson PJ (1980) An apparatus for high uniaxial stress electrical investigations of semiconductors. J Phys E 13:1331–1335
15. Creemer JF, French PJ (2002) A new model of the effect of mechanical stress on the saturation current of bipolar transistors. Sens Actuat A 97–98:289–295
16. Creemer JF (2010) The effect of mechanical stress on bipolar transistor characteristics. Ph.D. thesis, Delft University of Technology (Eburon, Delft, 2002), http://repository.tudelft.nl/assets/uuid:ff17cb76-7599-486b-b287-3123c806ac0f/emc_creemer_20020114.PDF. Accessed April 20, 2010
17. Fruett F, Meijer GCM (2001) Experimental investigation of piezojunction effect in silicon and its temperature dependence. Electron Lett 37:1366–1367
18. Fruett F, Wang G, Meijer GCM (2000) Piezojunction effect in NPN and PNP vertical transistors and its influence on silicon temperature sensors. Sens Actuators A 85:70–74

19. Gummel HK (1970) A charge control relation for bipolar transistors. Bell Syst Tech J 49:115–120
20. Nye JF (1985) Physical properties of crystals, 2nd edn. Clarendon, Oxford
21. Smith CS (1954) Piezoresistance effect in germanium and silicon. Phys Rev 94:42–49
22. Matsuda K, Suzuki K, Yamamura K, Kanda Y (1993) Nonlinear piezoresistive coeficients in silicon. J Appl Phys 73:1838–1847
23. Durand S, Tellier CR (1996) Linear and non-linear piezoresistance coefficients in cubic semiconductors I. Theoretical formulations. J Phys III France 2:237–266
24. Mason WP, Thurston RN (1957) Use of piezoresistive materials of displacement, force and torque. J Acoust Soc Am 29:1096–1101
25. Bir GL, Pikus GE (1974) Symmetry and strain-induced effects in semiconductors. Wiley, New York
26. Bittle DA, Suhling JC, Beaty RE, Jaeger RC, Johnson RW (1991) Piezoresistive stress sensors for structural analysis of electronic packages. J Electron Packag 113:203–214
27. Creemer JF, French PJ (2004) The saturation current of silicon bipolar transistors at moderate stress levels and its relation to the energy-band structure. J Appl Phys 96:4530–4538
28. Creemer JF, French PJ (2002) Anisotropy of the piezojunction effect in silicon transistors. In: Proceedings of the IEEE micro electro mechanical systems conference (MEMS 2002), Las Vegas, pp 316–319
29. Kanda Y (1982) A graphical representation of the piezoresistance coefficients in silicon. IEEE Trans Electron Devices ED-29:64–70
30. Nakamura K, Isono Y, Toriyama T, Sugiyama S (2009) Simulation of piezoresistivity in n-type single-crystal silicon on the basis of the first-principles band structure. Phys Rev B 80, 045205:11
31. Kozlovskiy SI, Boiko II (2005) First-order piezoresistance coefficients in silicon crystals. Sens Actuat A 118:33–43
32. Fischetti MV, Laux SE (1996) Band structure, deformation potentials, and carrier mobility in strained Si, Ge, and SiGe alloys. J Appl Phys 80:2234–2252
33. Herring C, Vogt E (1956) Transport and deformation-potential theory for many-valley-semiconductors with anisotropic scattering. Phys Rev 101:944–961
34. Luttinger JM (1956) Quantum theory of cyclotron resonance in semiconductors: general theory. Phys Rev 102:1030–1041
35. Marshak AH, van Vliet CM (1984) Electrical current and carrier density in degenerate materials with nonuniform band structure. Proc IEEE 72:148–164
36. Beisswanger J, Jorke H (1996) Electron transport in bipolar transistors with biaxially strained base layers. IEEE Trans Electron Devices 43:62–69
37. Lundstrom M (1992) Fundamentals of carrier transport (Modular series on solid state devices), vol X. Addison-Wesley, Reading
38. Ashcroft NM, Mermin ND (1976) Solid state physics. Saunders College, Philadelphia
39. Gough PA (1994) Fundamental equations. In: Hart PAH (ed) Bipolar and bipolar-MOS integration. Elsevier, Amsterdam

Chapter 23
Thermal Effects in Thin Silicon Dies: Simulation and Modelling

Niccolò Rinaldi, Salvatore Russo, and Vincenzo d'Alessandro

Abstract Detailed three-dimensional (3D) numerical thermal simulations are employed to clarify the influence of the key technological parameters and material properties on the thermal behaviour of ultra-thin chip stacking modules, wherein the heat removal from the thinned active dies is considerably hampered by benzocyclobutene layers. In particular, the impact of heat source area and thickness of silicon dies and benzocyclobutene layers is investigated in structures with one, two, and three thin silicon dies. The adoption of compact thermal models is then suggested as a solution to reduce the computational resources required to numerically analyse stacked-die modules.

23.1 Introduction

Three-dimensional (3D) stacked-die architectures are devised to effectively increase the integration density of semiconductor systems, thereby leading to smaller, lighter and cheaper products. However, stacked modules are particularly subject to thermal effects since the increase in dissipated power is not accompanied by a corresponding improvement in cooling efficiency. Cooling strategies are therefore sought to increase the heat removal capability from the dissipating regions [1]. Thermal effects are even more exacerbated in ultra-dense packaging technologies resulting from the recent advances in thinning and attachment techniques: in ultra-thin chip stacking (UTCS) technology modules, this is mainly due to the low thermal conductivity of benzocyclobutene (BCB) layers adopted to electrically insulate silicon chips thinned up to 10 μm, which are vertically integrated in the structure on the same inactive host silicon substrate [2, 3].

N. Rinaldi (✉)
Department of Biomedical, Electronics and Telecommunications Engineering, University of Naples Federico II, Via Claudio 21, 80125 Naples, Italy
e-mail: nirinald@unina.it

J.N. Burghartz (ed.), *Ultra-thin Chip Technology and Applications*,
DOI 10.1007/978-1-4419-7276-7_23, © Springer Science+Business Media, LLC 2011

To improve the cooling capability of UTCS structures, the use of numerical thermal simulation tools has become essential [2]. Numerical thermal simulation makes it possible to evaluate the temperature distribution under both steady-state and transient conditions, identify critical heat flow paths, determine the effect of technology and layout parameters, and optimise the thermal architecture.

In this chapter we present a detailed 3D thermal simulation study based on the commercial finite-element method (FEM) Comsol software package [4]. The impact of various technological parameters and material properties is investigated, as, e.g. heat source area, thicknesses of silicon die and adhesive/planarisation BCB layers, as well as thermal conductivity of the package header. This analysis is presented in Sect. 23.2.

One critical issue in the thermal analysis of UTCS modules is related to the inherent complexity of the structure. To solve this problem, compact thermal models (CTM) have been developed [5–7]. The formulation of CTMs and their application to UTCS modules are addressed in Sect. 23.3.

23.2 Numerical Analysis of UTCS Modules

23.2.1 *Single-Level Module*

The 2D representation of the single-level UTCS module under analysis is depicted in Fig. 23.1. W_{die}, L_{die} and t_{die} denote, respectively, the width, length and thickness of the thinned silicon die containing the dissipating circuitry; W_{HS} and L_{HS} represent the width and length of the indefinitely thin heat source, located on the die top; W_{sub}, L_{sub} and t_{sub} designate, respectively, the width, length and thickness of the inactive silicon substrate; $t_{BCB\text{-}ad}$ is the thickness of the BCB layer adopted as adhesive between active die and substrate; W_{header}, L_{header} and t_{header} are the width, length and thickness of the AlN PGA header; $t_{Pb/Sn}$ is the thickness of the thermally conductive grease that solders the silicon chip to the package. A 'reference' single-level structure is considered, the geometrical parameters of which are reported in Table 23.1. In order to quantify the impact of the layers overhung by the host silicon substrate, a 'reduced' UTCS structure is also defined (Fig. 23.2), where these layers are removed and the substrate bottom is assumed at ambient temperature. All simulations were performed by disregarding nonlinear thermal effects, that is, the thermal conductivities of all materials were considered as temperature-insensitive and equal to their values at $T = 300$ K (listed in Table 23.2). This approximation was found not to appreciably affect the simulation accuracy: a maximum error amounting to less than 1% was determined at a dissipated power of 20 W. Metallic interconnections were also neglected since their small size does not sensibly favour heat removal [2]. The thin heat source was accounted for by enabling a uniform heat flux entering the domain. Adiabatic boundary conditions were assumed for both top and lateral faces of the structure.

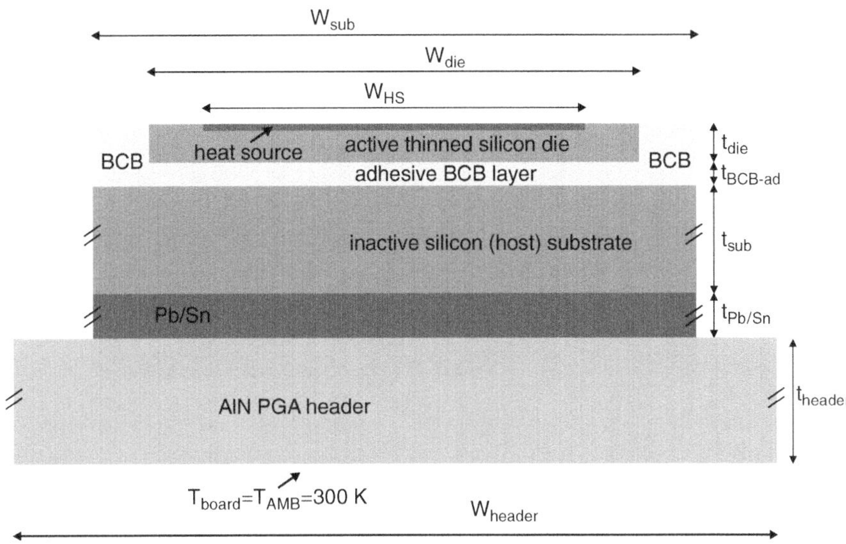

Fig. 23.1 2D schematic of the single-level UTCS module

Table 23.1 Geometrical parameter values of the reference module

Parameter	Value (μm)
$W_{\text{die}} = L_{\text{die}}$	5,620
$W_{\text{HS}} = L_{\text{HS}}$	4,240
$W_{\text{sub}} = L_{\text{sub}}$	6,200
$W_{\text{header}} = L_{\text{header}}$	15,000
t_{die}	10
$t_{\text{BCB-ad}}$	3
t_{sub}	500
$t_{\text{Pb/Sn}}$	50
t_{header}	760

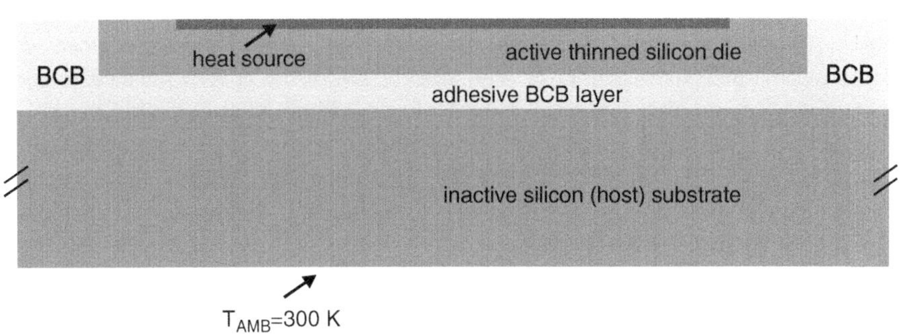

Fig. 23.2 2D representation of the reduced single-level UTCS structure

Table 23.2 Thermal conductivities at $T = 300$ K adopted for FEM simulations

Material	Thermal conductivity (W/mK)
Si	148
BCB	0.18
Pb/Sn	36
AlN	150

The self-heating thermal resistance R_{TH} is evaluated as the ratio between the temperature rise above ambient averaged over the whole heat source ΔT_{AV} and the dissipated power P_{D}, that is,

$$R_{TH} = \frac{\Delta T_{AV}}{P_D} = \frac{T_{AV} - T_{AMB}}{P_D}. \tag{23.1}$$

The AlN header bottom in contact with the board is considered isothermal at ambient temperature, while the top and lateral surfaces are assumed adiabatic. Due to the inherent system symmetries, only one quarter of the structure was simulated to reduce the CPU/memory requirements needed to achieve a high level of accuracy; adiabatic conditions were exploited in order to virtually restore the missing module portions.

The Comsol mesh – realised by resorting to smart refinement strategies available in the recent software releases – comprises about 5×10^5 elements and 7×10^5 degrees of freedom (see Fig. 23.3). If nonlinear thermal effects are disregarded, a steady state solution (i.e. the static temperature in any point) is evaluated in about 300 s via a workstation equipped with 2 hexacore 2.43 GHz CPUs and a 100 GB RAM. The self-heating thermal resistance of the reference single-die module was calculated to be 1.24 K/W, while the value corresponding to the reduced module version depicted in Fig. 23.2 was found to be 1.02 K/W. As a conclusion, the contribution of the layers beneath the silicon substrate amounts to about 17.5%. The R_{TH} values are rather low in spite of the low k_{BCB} value (k_{BCB} is about 800 times lower than k_{Si} [2]), due to the large dissipating areas. The adiabatic condition at top and lateral faces was justified by demonstrating that a natural convection with a heat transfer coefficient h amounting to 2 W/m^{2}K applied to the top and lateral surfaces [3] does not influence appreciably the results. In order to quantify the BCB heating-up action, a simulation was performed by replacing the adhesive BCB layer with silicon; the resulting R_{TH} reduced to about 0.39 K/W, i.e. a value less than a third of that of the BCB-equipped module in Fig. 23.1. Figure 23.4 reports the temperature rise over ambient normalised to dissipated power $\Delta T/P_D$ along a vertical cut crossing the centre of the dissipating region in the complete single-level module; the significant heating-up influence of the 3-μm thick adhesive BCB layer is apparent.

Similar to the analysis carried out in [2], the possibility of integrating circumscribed dissipating regions in the individual thin silicon dies was then investigated. Figure 23.5 illustrates the R_{TH} behaviour as a function of heat source area $A_{\mathrm{HS}} = W_{\mathrm{HS}} \times L_{\mathrm{HS}}$, which was varied by keeping all other parameters (including $A_{\mathrm{die}} = W_{\mathrm{die}} \times L_{\mathrm{die}}$) equal to the reference values, for both the complete and reduced

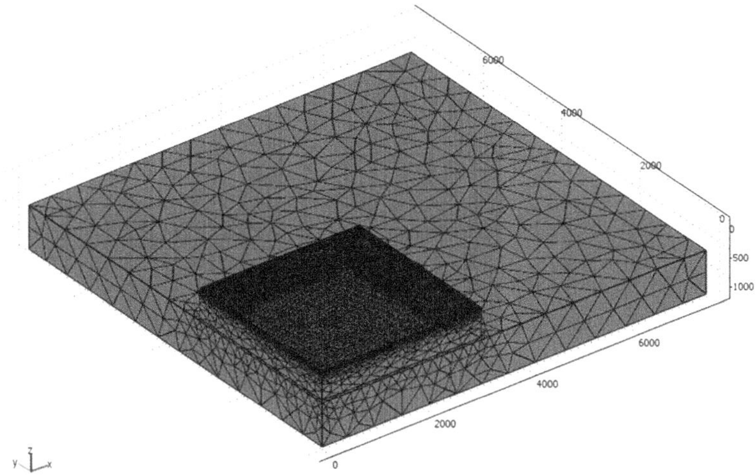

Fig. 23.3 Comsol mesh for the reference single-die UTCS module depicted in Fig. 23.1

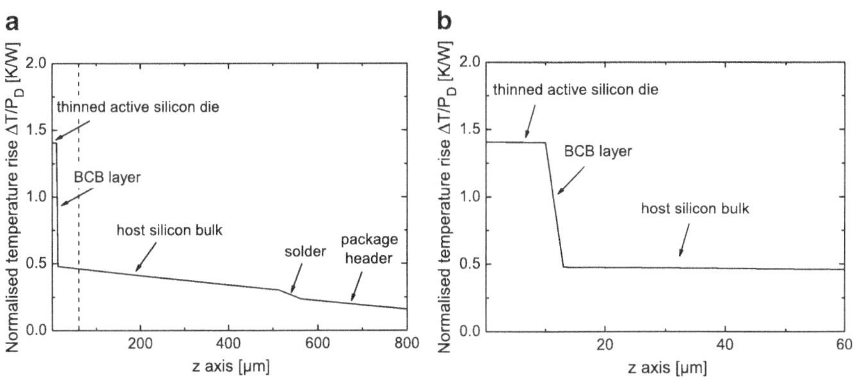

Fig. 23.4 Single-level module: (**a**) temperature rise over ambient normalised to dissipated power $\Delta T/P_D$ versus z as evaluated by a 3D FEM simulation along a vertical cut crossing the heat source centre; (**b**) magnification of the normalised temperature behaviour within the top module region delimited by the *dashed line* in (**a**)

modules. The influence of a thicker silicon die (with $t_{\mathrm{die}} = 20\ \mu\mathrm{m}$) was also studied. R_{TH} increases from 0.92 ($A_{\mathrm{HS}} = 2.5 \times 10^7\ \mu\mathrm{m}^2$) to 537.4 K/W ($A_{\mathrm{HS}} = 100\ \mu\mathrm{m}^2$) for the reference thickness $t_{\mathrm{die}} = 10\ \mu\mathrm{m}$, while the analogous values for $t_{\mathrm{die}} = 20\ \mu\mathrm{m}$ were found to be 0.93 K/W and 408.8 K/W, that is, the beneficial impact of the increased silicon die thickness (i.e. of the larger volume where the heat can easily spread) is higher for small dissipating areas; in particular, a maximum thermal resistance decrease of about 38% was detected. The influence of the soldering Pb/Sn grease and PGA header was estimated to be negligible for small heat source dimensions, while increasing up to 19% for $A_{\mathrm{HS}} = 2.5 \times 10^7\ \mu\mathrm{m}^2$.

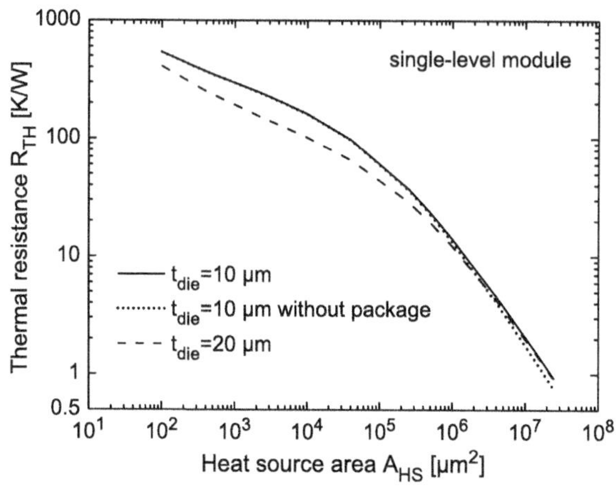

Fig. 23.5 Single-level module: self-heating thermal resistance R_{TH} as a function of heat source area A_{HS} for (*solid line*) t_{die} = 10 μm and (*dashed*) t_{die} = 20 μm; also shown is (*dotted*) the curve corresponding to the reduced domain with the host silicon substrate directly contacted to an ideal heat sink at T = 300 K

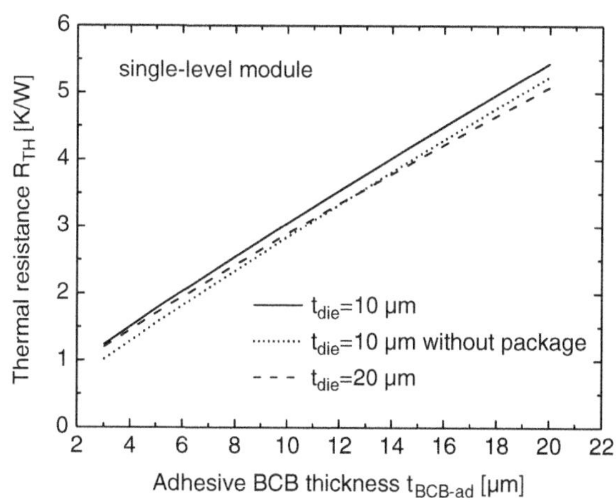

Fig. 23.6 Single-level module: self-heating thermal resistance R_{TH} as a function of adhesive BCB thickness $t_{BCB\text{-}ad}$ for two values of silicon die thickness t_{die}, i.e. (*solid line*) 10 μm and (*dashed*) 20 μm; also shown are (*dotted*) the results obtained by simulating the reduced structure

The impact of the thickness $t_{BCB\text{-}ad}$ of the adhesive BCB layer interposed between the thinned silicon die and the host substrate was also examined. In particular, $t_{BCB\text{-}ad}$ was varied from 3 (namely, the reference value) to 20 μm. The numerical results are presented in Fig. 23.6 for t_{die} equal 10 and 20 μm; all other parameters were kept equal to the reference values. It is shown that R_{TH} grows almost linearly with $t_{BCB\text{-}ad}$; in particular, R_{TH} increases from 1.24 to 5.44 K/W for

$t_{\text{die}} = 10$ μm. Thickening the silicon die up to 20 μm ensures a perceptible improvement in the thermal behaviour only for higher $t_{\text{BCB-ad}}$ values: the reduction in thermal resistance amounts to 6.6% for $t_{\text{BCB-ad}} = 20$ μm while being only 2.7% for $t_{\text{BCB-ad}} = 3$ μm; the t_{die} impact indeed decreases in the latter case due to the stronger cooling action of the silicon host substrate. Values spanning from 1.02 to 5.24 K/W were found for the reduced structure.

An analysis was carried out to quantify the influence of the thermal conductivity of the package header, which was assumed equal to 150 W/mK for the reference domain, where AlN is employed [2]. This parameter was varied within the range 50–250 W/mK. Results are given in Fig. 23.7 for the complete UTCS module; it is shown that the self-heating thermal resistance ranges from 1.51 to 1.17 K/W.

Lastly, a transient simulation was performed to identify the components of the vertical heat path; in particular, a power step was applied at $t = 0$ and the temperature over the heat source was monitored versus time until steady state conditions were reached. The thermal impedance Z_{TH} is defined as

$$Z_{TH} = \frac{\Delta T_{AV}(t)}{P_D} = \frac{T_{AV}(t) - T_{AMB}}{P_D}. \tag{23.2}$$

In Fig. 23.8, Z_{TH} is represented against time for various structures, namely, the complete single-level module (Fig. 23.1), the reduced package-free structure (Fig. 23.2), and the corresponding devices obtained by replacing the adhesive BCB layers with silicon. An inspection of the figure reveals that a longer transient behaviour is obtained for the package-equipped modules: steady state conditions are reached at about 5×10^{-2} s; conversely, reduced domains are characterised by a faster response: the impedance growth extinguishes at 7×10^{-3} s. It is noteworthy that for times shorter than 5×10^{-6} s the heat is still propagating along the 10-μm thick active silicon die, while at $t = 1 \times 10^{-3}$ s it is approaching the bottom of the

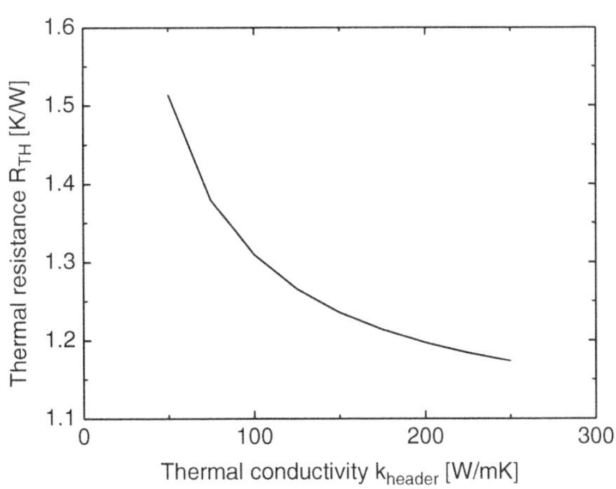

Fig. 23.7 Single-level module: self-heating thermal resistance R_{TH} as a function of thermal conductivity of the PGA header k_{header}

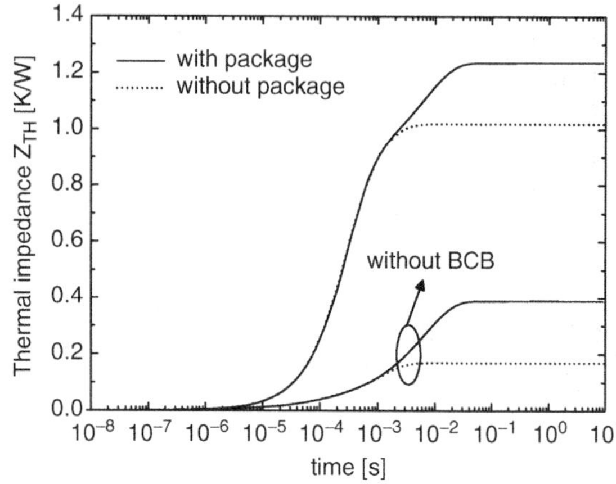

Fig. 23.8 Single-level module: thermal impedance Z_{TH} versus time after applying a power step at $t = 0$; (*solid lines*) the devices mounted on the board via AlN header are compared to (*dotted*) the package-free counterparts. For both cases, the adhesive BCB layer is either accounted for or removed and replaced by silicon

silicon substrate. The impact of the adhesive BCB layer on the efficiency of heat spreading within the structure is apparent, regardless of the presence of the AlN header. The values of specific heat and mass density employed for the FEM simulations can be found in [2].

23.2.2 *Two-Level Module*

In the following analysis, two identical silicon thin chips are vertically integrated in the BCB matrix over the host silicon substrate (Fig. 23.9) so as to obtain a stacked two-level module. The thickness t_{BCB} of the BCB layer separating the thinned silicon dies amounts to 3 μm in the reference structure. All other parameters were considered equal to those of the single-level module. A simplified module version is also defined, for which all layers beneath the inactive silicon substrate are removed and an isothermal boundary condition with $T = 300$ K is assumed for the silicon substrate bottom. Simulations were performed by alternately switching on the stacked silicon dies. The self-heating thermal resistance of the ith die R_{THii} and the mutual resistance between the ith and jth dies R_{THij} are defined as

$$R_{THii} = \frac{\Delta T_{AVi}}{P_{Di}} \quad R_{THij} = \frac{\Delta T_{AVi}}{P_{Dj}}, \tag{23.3}$$

where ΔT_{AVi} is the temperature rise above ambient averaged over the heat source of the ith thin silicon die and P_{Di}, P_{Dj} are the powers dissipated by the ith and jth dies, respectively.

The first study was carried out by activating the 1st-level silicon die (namely, the closest to the inactive host substrate) and evaluating R_{TH11} and R_{TH21} accordingly to (23.3). Figure 23.10 shows R_{TH11} and R_{TH21} for both the complete and reduced

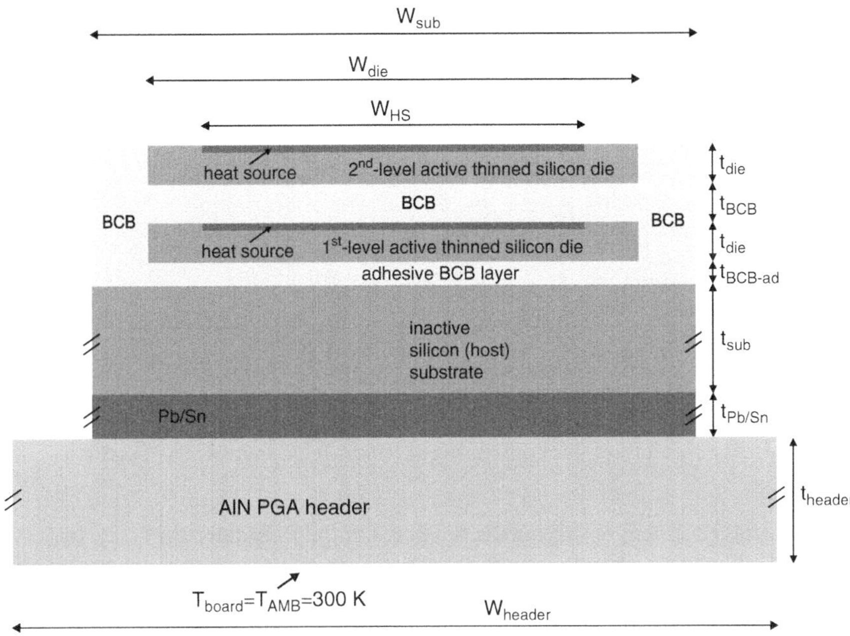

Fig. 23.9 2D scheme of the complete two-level UTCS module

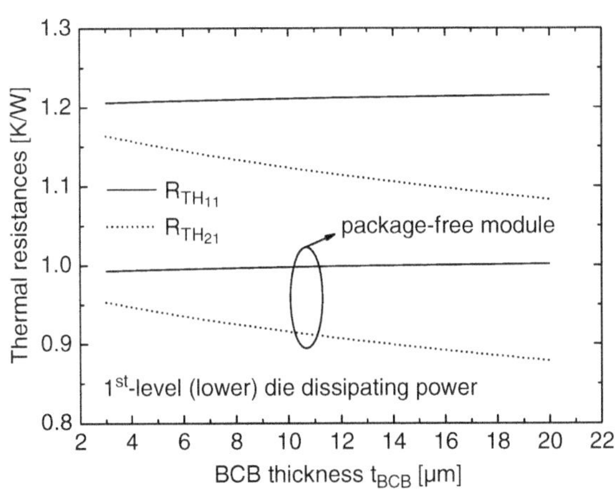

Fig. 23.10 Stacked two-level module in which only the lower thinned silicon chip dissipates power: (*solid lines*) self-heating thermal resistance R_{TH11} and (*dotted*) mutual thermal resistance R_{TH21} as a function of BCB thickness t_{BCB}. The thermal resistances of both the complete and the package-free modules are represented

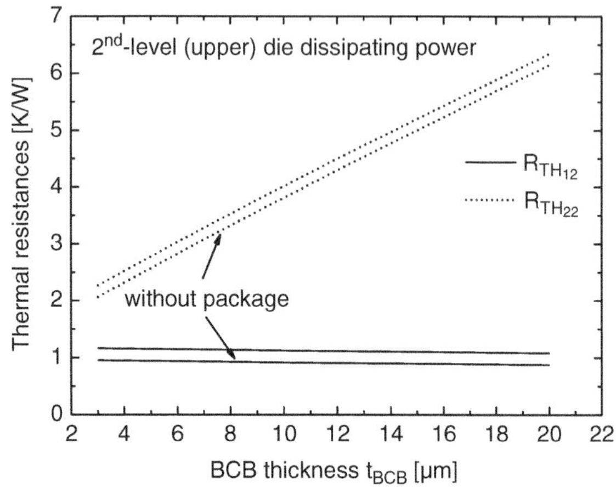

Fig. 23.11 Stacked two-level module in which only the upper thinned silicon die dissipates power: (*dotted lines*) self-heating thermal resistance R_{TH22} and (*solid*) mutual thermal resistance R_{TH12} as a function of BCB thickness t_{BCB}. Both the cases of module mounted on package and package-free structure are shown

structures as obtained by varying the thickness of the BCB interlayer t_{BCB} from 3 to 20 μm. All other parameters were kept constant and equal to those reported in Table 23.1; in particular, the silicon dies share the same value of t_{die}, namely, 10 μm, and the thickness of the adhesive BCB layer $t_{BCB\text{-}ad}$ is equal to 3 μm. As can be seen, R_{TH11} slightly increases due to the growing thickness of the BCB interlayer, which hampers the upward heat emersion from the active die, whereas R_{TH21} more perceptibly decreases (from 1.16 to 1.08 K/W for the complete two-die module and from 0.95 to 0.88 K/W for the package-free counterpart) due to the increasing reduction in thermal coupling between the active (1st) and inactive (2nd) thinned silicon dies.

Figure 23.11 depicts the case in which only the (upper) 2nd-level silicon die is active; both the self-heating (R_{TH22}) and mutual (R_{TH12}) thermal resistances are shown as a function of the BCB thickness t_{BCB}. It is found that the thermal behaviour of the 2nd-level die markedly degrades due to the reduction in the downward heat spreading from the heat source that is inhibited by the increasingly thicker BCB interlayer: R_{TH22} ranges from 2.27 (for $t_{BCB} = 3$ μm) to 6.34 K/W (for $t_{BCB} = 20$ μm) as far as the complete stacked module is concerned, whilst a growth from 2.06 to 6.15 K/W is detected for the reduced package-free structure. The mutual thermal resistance R_{TH12} slightly diminishes due to the lowering in thermal coupling between the 1st-level and the overhanging 2nd-level thinned chips.

23.2.3 Three-Level Module

The complete three-die module is represented in Fig. 23.12. As for the case of the stacked two-level structure, an analysis was performed by alternatively switching on the individual thinned dies in order to quantify the values of the thermal resistance matrix and their sensitivity to the BCB interlayer thickness t_{BCB}. Except

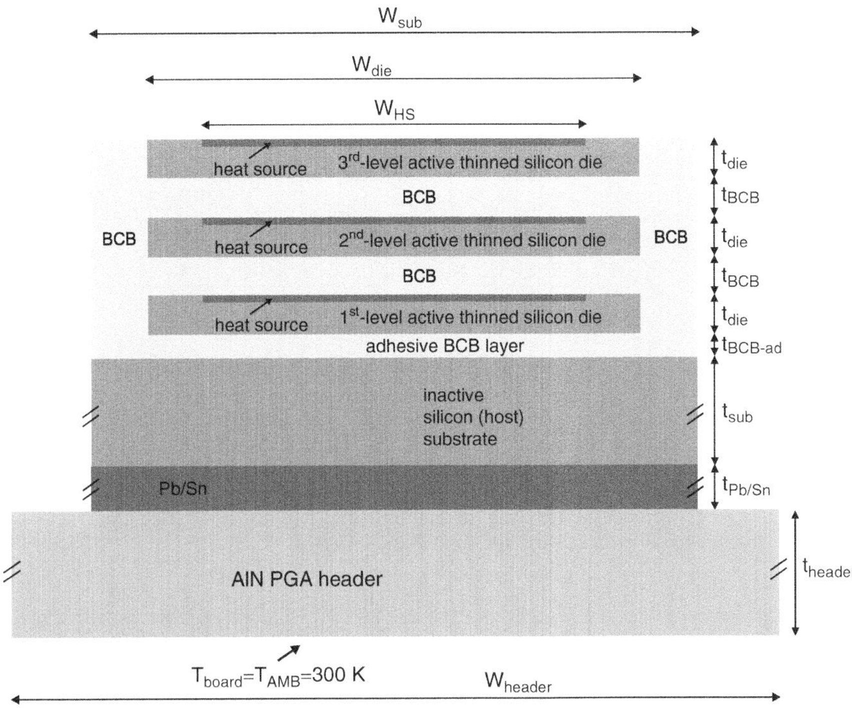

Fig. 23.12 2D representation of the complete three-level UTCS structure

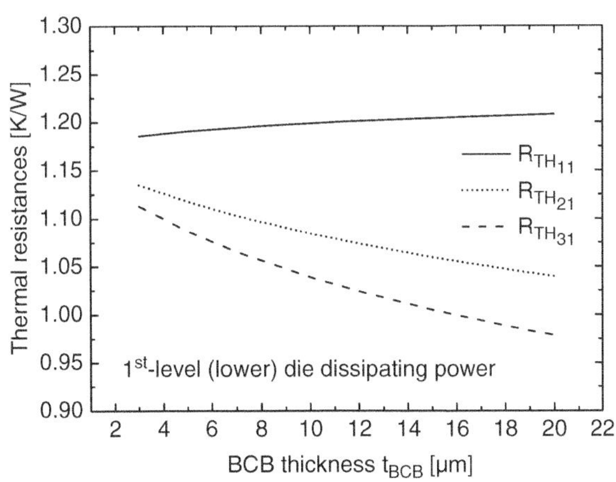

Fig. 23.13 Stacked three-chip module in which only the lower thinned silicon die is activated: (*solid line*) self-heating thermal resistance R_{TH11} and mutual thermal resistances (*dotted*) R_{TH21} and (*dashed*) R_{TH31} as a function of BCB thickness t_{BCB}

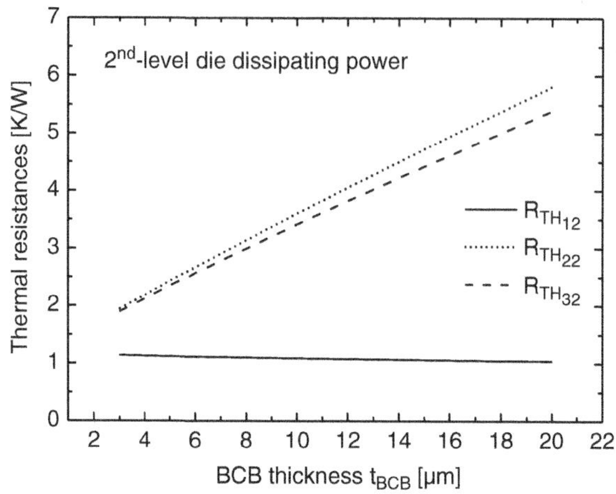

Fig. 23.14 Stacked three-chip structure in which only the intermediate 2nd-level thinned silicon die is activated: (*dotted line*) self-heating thermal resistance R_{TH22} and mutual thermal resistances (*dashed*) R_{TH32} and (*solid*) R_{TH12} against BCB thickness t_{BCB}

for t_{BCB}, all geometrical parameters were kept to the reference values (e.g. t_{die} = 10 μm for all dies, $t_{BCB\text{-}ad}$ = 3 μm, t_{sub} = 500 μm, A_{HS} = 4,240 × 4,240 μm^2).

As a first step, the 1st-level (lower) chip was switched on, while the 2nd- and the 3rd-level dies were assumed inactive. The thermal resistance variation versus t_{BCB} (the common thickness of the planarisation BCB layers interposed between the thinned silicon dies) is reported in Fig. 23.13. It is shown that R_{TH11} slowly increases due to the rising reduction in the upward heat flow (while the downward remains unaltered). Both thermal resistances R_{TH21} and R_{TH31} decrease since the thermal coupling is growingly inhibited: in particular, R_{TH21} lowers from 1.14 (t_{BCB} = 3 μm) to 1.04 K/W (t_{BCB} = 20 μm), i.e. an 8.4% reduction is observed; R_{TH31} diminishes up to 12.1% (namely, from 1.11 to 0.98 K/W) since two increasingly thicker BCB layers are located between the 1st- and the 3rd-level dies.

Subsequently, the 2nd-level thinned die is activated, while the others are kept switched off. The behaviour of the thermal resistances as t_{BCB} is varied is represented in Fig. 23.14; R_{TH22} swiftly increases from 1.94 to 5.81 K/W, since the heat is progressively more constrained within the dissipating die by the thicker overhanging and overhung BCB layers. The mutual resistance R_{TH32} between the upper die and the centre (dissipating) one linearly grows mainly due to the downward heat flow inhibition, which causes a temperature increase in the overall top region of the module; conversely, R_{TH12} weakly decreases.

Lastly, the 3rd-level silicon die is switched on, while the others are maintained inactive. Simulation results are shown in Fig. 23.15, which illustrates the behaviour of the thermal resistances as a function of t_{BCB}. As can be seen, the main effect of the t_{BCB} increase is the heat constriction within the upper device portion, which in turn causes a growth in both R_{TH33} (from 2.72 to 10.2 K/W) and R_{TH23} (from 1.89 to

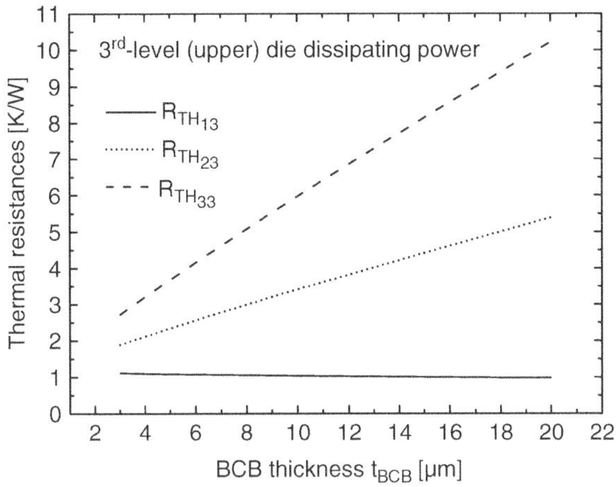

Fig. 23.15 Stacked three-chip module in which only the upper thinned silicon die is activated: (*dashed line*) self-heating thermal resistance R_{TH33} and mutual thermal resistances (*dotted*) R_{TH23} and (*solid*) R_{TH13} versus BCB thickness t_{BCB}

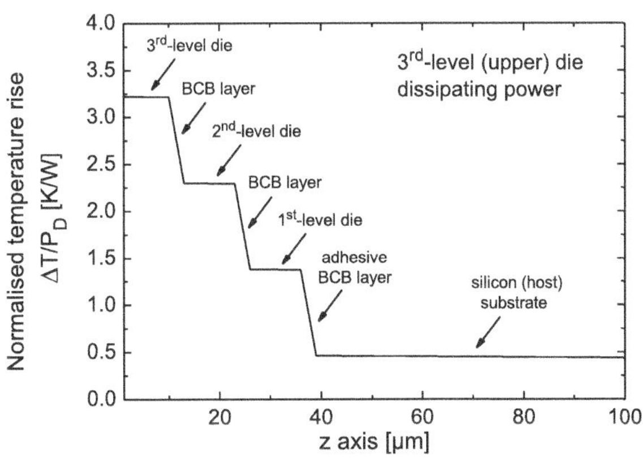

Fig. 23.16 Stacked three-layer module in which only the upper 3rd-level thinned silicon die is activated: temperature rise above ambient normalised to dissipated power $\Delta T(z)/P_D$ along a vertical cut crossing the heat source centre

5.39 K/W); on the other hand, a slight reduction is observed for R_{TH13} (the 1st-level chip is increasingly far from the top 3rd-level one).

Let us consider the latter case in which the upper die is switched on. In Fig. 23.16, the temperature rise over ambient normalised to dissipated power is represented, as evaluated through a 3D FEM simulation performed for $t_{BCB} = 3$ μm. It is worthwhile to note that the self-heating thermal resistance R_{TH33} amounts to 2.72 K/W, while the maximum normalised temperature rise is equal to 3.22 K/W.

Lastly, a transient analysis was performed by alternately activating the thinned silicon dies with a power step applied at time $t = 0$ and monitoring the temperature field over the regions of the thinned dies where the circuitry is located.

Figures 23.17–23.19 detail the simulation data corresponding to the cases in which 1st-, 2nd-, and 3rd-level dies are turned on, respectively. It is shown that the transient behaviours of the different configurations share a common feature: the temperature rise over ambient of the activated silicon die becomes perceptibly different from zero in correspondence of $t = 10^{-5}$ s, while that of the closest chip(s) starts growing at $t = 10^{-4}$ s, after the heat has propagated through the electrically (and thermally) insulating BCB layer.

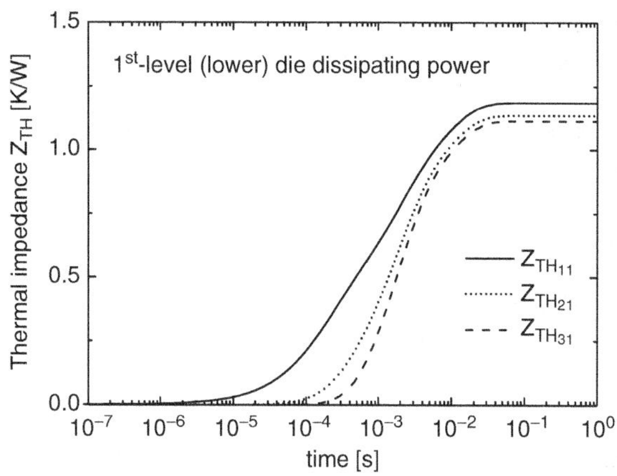

Fig. 23.17 Stacked three-layer structure in which only the 1st-level thinned silicon chip is turned on: (*solid line*) self-heating thermal impedance Z_{TH11} and mutual impedances (*dotted*) Z_{TH21} and (*dashed*) Z_{TH31} after the application of a power step at $t = 0$

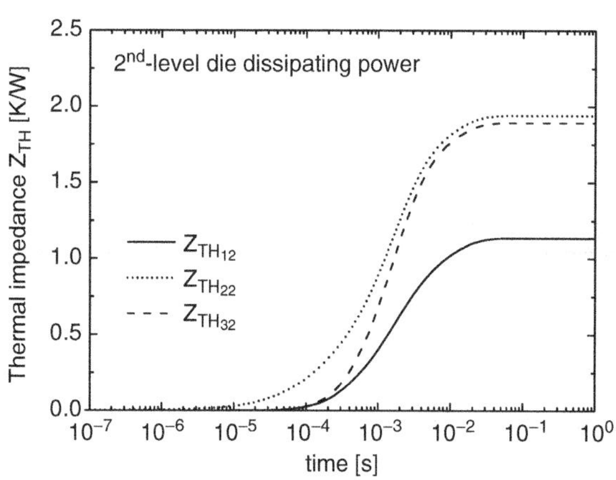

Fig. 23.18 Stacked three-chip module in which only the 2nd-level thinned silicon chip is activated: (*dotted line*) self-heating thermal impedance Z_{TH22} and mutual impedances (*solid*) Z_{TH12} and (*dashed*) Z_{TH32} after the application of a power step at $t = 0$

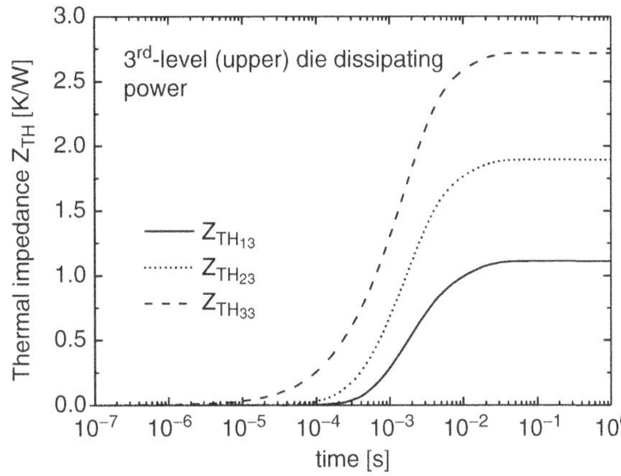

Fig. 23.19 Stacked three-level module in which only the 3rd-level silicon die is dissipating power: (*dashed line*) self-heating thermal impedance Z_{TH33} and mutual impedances (*solid*) Z_{TH13} and (*dotted*) Z_{TH23} after the application of a power step at $t = 0$

23.3 Thermal Modelling of UTCS Modules

23.3.1 Compact Thermal Models: A Perspective

As apparent from the previous section, a major issue in the numerical thermal analysis of UTCS modules is related to the complexity of the structure, in which heat flows through different regions showing a huge variation in size and material properties. This poses a considerable difficulty in mesh generation, and increases computation time and memory requirements. The investigation presented above could take advantage of the symmetry of the structure and dissipation pattern, so that only one quarter of the module could be considered. However, for nonsymmetric structures this simplification is not possible, and the numerical analysis becomes computationally demanding since the whole structure must be simulated. To simplify the numerical analysis of complex electronic systems one could resort to compact thermal models (CTM) [5–7]. The idea behind CTMs is to partition the whole system into sub-systems (parts), and develop a CTM for each part. For instance, the structure shown in Fig. 23.1 can be subdivided into two sub-systems A and B, as shown in Fig. 23.20. CTMs are developed by partitioning the contact surface between the two domains into surface elements S_i (referred to as the thermal nodes of the CTM), and by formulating relations between the (average) temperature and heat flux at the thermal nodes, which are typically linear equations involving parameters that can be determined by numerical simulation (or experimental characterisation) of each individual part. Thus, numerical thermal analysis of the entire system is not required, and simulations can be optimised for each part. In Fig. 23.20 it can be seen that the contact surface between parts A and B can be discretised into, e.g. two elements. Once the CTM parameters are extracted for all parts, the CTMs can be combined to solve the problem for the whole system. One common approach is to assume a uniform temperature distribution on the boundary surface elements S_i [5]. Since the temperature field is typically uneven, a sufficiently fine discretisation of the boundary surface is required. Hence, by a proper

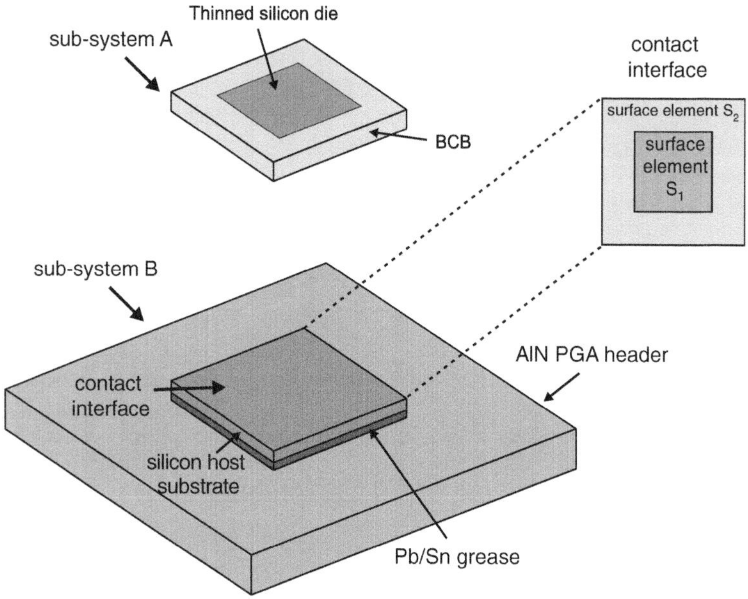

Fig. 23.20 CTM formulation for the stacked single-level module depicted in Fig. 23.1. The system is partitioned into two sub-systems denoted as A and B. The contact interface between sub-systems is discretised into 2 surface elements S_1 and S_2

boundary discretisation, the CTM allows reproducing complex temperature or heat flux distributions on the boundary interface, and is therefore said to be *boundary condition independent*. This implies that boundary condition independent CTMs retain their accuracy also if any part of the system, or environmental condition, is changed. It has been suggested that this approach can be improved by assuming a variable temperature or heat flux distribution on the boundary surface elements [6, 7]. In this case, the temperature distribution at the thermal nodes is assumed to have a prescribed behaviour given by *shape functions*. With a proper choice of the shape functions, the accuracy of the CTMs improves, that is, a lower number of thermal nodes is needed to achieve a prescribed accuracy.

23.3.2 Compact Thermal Model Formulation

In the following, we derive the CTM equations for the general case of variable shape functions. This CTM formulation is then applied to the single-die module shown in Fig. 23.1. This implies that we must derive CTM equations for both sub-systems A and B.

As a first step, we must find the solution of the heat conduction equation

$$\nabla^2 T + \frac{g(\mathbf{r})}{k} = 0, \tag{23.4}$$

where k represents the thermal conductivity (in general, position-dependent) and g is the power density (W/μm^3). In this analysis we consider the case in which only one silicon die dissipates power; a system with multiple dissipating sources can be treated by superposition starting from the single source case. In order to develop a CTM we need to approximate the power density as

$$g(\mathbf{r}) = P\frac{\sigma_g(\mathbf{r})}{V_g}, \tag{23.5}$$

where P is the dissipated power (W), V_g is the volume of the active region where the power is dissipated (μm^3) and $\sigma_g(\mathbf{r})$ is a suitable shape function that describes the position dependence of the power density. The shape function $\sigma_g(\mathbf{r})$ is zero outside the active region V_g. Since the volume integral of the power density g must be equal to the total dissipated power P, the shape function must satisfy the condition

$$\int\limits_{V_g} \sigma_g(\mathbf{r})dv = V_g. \tag{23.6}$$

Note that we assume the shape function to be *bias independent*, that is, σ_g does not depend on the current and voltage (or dissipated power). In many thermal studies [5] a constant shape function is assumed (i.e. a uniform power density is considered). In this case, the shape function is simply equal to unity in the active region, and zero outside.

Besides power generation, we must also model the heat exchange between each part and the outside environment (e.g. the heat exchange between adjacent parts). The heat exchange between sub-systems A and B occurs through the N surface elements S_i ($N = 2$ for the case shown in Fig. 23.20). Mathematically, we can specify the distribution of the temperature rise over ambient in a certain portion S_i of the boundary surface S as

$$T|_{\mathbf{r}_i \in S_i} = \phi_i(\mathbf{r}_i), \tag{23.7}$$

where the function ϕ_i defines the prescribed temperature on S_i. Similarly to (23.5), we can assume that the temperature distribution on S_i is described by a suitable shape function σ_i, i.e.

$$T|_{\mathbf{r}_i \in S_i} = \phi_i(\mathbf{r}_i) = T_i\sigma_i(\mathbf{r}_i). \tag{23.8}$$

The shape functions σ_i are assumed to satisfy the condition

$$\int\limits_{S_i} \sigma_i(\mathbf{r}_i)\mathrm{d}s_i = S_i \tag{23.9}$$

so that T_i can be seen as representing the *average temperature* on S_i

$$\frac{1}{S_i}\int\limits_{S_i} T(\mathbf{r}_i)\mathrm{d}s_i = \frac{T_i}{S_i}\int\limits_{S_i} \sigma_i(\mathbf{r}_i)\mathrm{d}s_i = T_i \tag{23.10}$$

The generalised CTM equations for a given sub-system are derived from the solution of the heat equation

$$\nabla^2 T + \frac{P\sigma_g(\mathbf{r})}{V_g k} = 0, \tag{23.11}$$

subject to the boundary conditions

$$T|_{\mathbf{r}_i \in S_i} = T_i \sigma_i(\mathbf{r}_i) \tag{23.12}$$

for all boundary surfaces S_i. For each boundary surface we specify a shape function σ_i subject to the constraint (23.9).

Using the Green's function method [8], the solution of the problem defined by (23.11) and (23.12) is given as

$$\begin{aligned} T(\mathbf{r}) = & T_g(\mathbf{r}) + \sum_{i=1}^{N} T_{BCi}(\mathbf{r}) = \frac{P}{kV_g}\int\limits_{V_g} \sigma_g(\mathbf{r}')G(\mathbf{r},\mathbf{r}')\mathrm{d}v' - \\ & \sum_{i=1}^{N} T_i \int\limits_{S_i} \sigma_i(\mathbf{r}'_i)\frac{\partial G(\mathbf{r},\mathbf{r}'_i)}{\partial \mathbf{n}'_i}\mathrm{d}s'_i \end{aligned} \tag{23.13}$$

where $G(\mathbf{r}, \mathbf{r}')$ is the steady Green's function (GF) for the problem at hand [8], and $\mathbf{n}_i$ is the outwardly pointing unit normal. The term T_g represents the solution of the problem accounting for power generation under homogeneous boundary conditions (i.e. $P \neq 0$ and $T_i = 0$). The term T_{BCi} is the solution for a zero dissipated power ($P = 0$), a nonzero excitation at the ith boundary surface S_i ($T_i \neq 0$) and zero excitation at all remaining boundaries ($T_j = 0$ for $j \neq i$). Therefore, all terms T_g and $T_{\mathrm{BC}i}$ can be easily determined from $N + 1$ numerical simulations, where only one forcing term is activated. In some cases, symmetry can be used to reduce the number of simulations.

A CTM provides two sets of equations:

1. An equation relating the (average) temperature T_0 at some location of interest and the forcing terms P, T_i.
2. A set of equations for the power flowing through the boundary surfaces and the forcing terms P, T_i.

In addition, the CTM must satisfy some constraints that derive from the heat flux conservation or by applying the properties of Green's functions. These equations are derived below.

23.3.2.1 Junction Temperature Equation

The ultimate goal of the thermal model is to calculate the temperature in a specific region of interest V_0, which is sometimes referred to as the 'junction' temperature. The junction temperature can be defined as the average temperature calculated in a volume V_0:

$$T_0 = \frac{1}{V_0} \int_{V_0} T(\mathbf{r})\mathrm{d}v. \tag{23.14}$$

Substituting (23.13) into (23.14), we obtain

$$T_0 = R_{TH0}P + \sum_{i=1}^{N} a_i T_i, \tag{23.15}$$

where coefficients R_{TH0} and a_i can be expressed in terms of GFs (see (23.13)). Note that R_{TH0} represents the thermal resistance related to the sole heat generation.

23.3.2.2 Boundary Heat Flux Equations

The heat flux (W) flowing out from the kth boundary surface is expressed as

$$P_k = -k \int_{S_k} \nabla T(\mathbf{r}_k) \cdot \mathbf{n}_k \mathrm{d}s_k. \tag{23.16}$$

Using (23.13), we obtain

$$P_k = q_k P + \sum_{i=1}^{N} \frac{T_i}{R_{ki}} \tag{23.17}$$

for $k = 1,\ldots,N$. These relations include N parameters q_k and N^2 parameters R_{ki} (thermal resistances between the thermal nodes). Also, these parameters can be expressed in terms of GFs.

23.3.2.3 Constraint Relations

Coefficients in (23.17) are subject to some constraints imposed by heat conservation. The conservation of heat flux implies that the power generated by the heat source equals that leaving the body. This yields [5]

$$\sum_{k=1}^{N} q_k = 1 \quad \sum_{i=1}^{N} \frac{1}{R_{ki}} = 0. \tag{23.18}$$

Additional constraints for the coefficients in (23.15) and (23.17) are obtained in the case of uniform boundary conditions. Assuming $\sigma_i(\mathbf{r}_i) = 1$, the following relations are derived [5]:

$$\sum_{i=1}^{N} a_i = 1 \quad R_{ki} = R_{ik} \tag{23.19}$$

for $k = 1,\ldots,N$. These relations can be used to reduce the number of simulations required to generate the parameters. However, in case of nonuniform shape functions, equations (23.19) do not hold.

The CTM formulation derived above can be straightforwardly applied to the problem of Fig. 23.20. First, we determine the parameters q_k^{A} and R_{ki}^{A} for sub-system A, and parameters R_{ki}^{B} for sub-system B (since $P = 0$ in part B, coefficients q_k^{B} do not apply). Then, we write equations (23.17) for parts A and B. This yields a system of 2 N equations for the 2 N unknowns T_k and P_k. Once the temperature values T_k are determined at the thermal nodes, (23.15) is used to calculate the junction temperature.

The calculation of CTM parameters a_k and R_{ki} is quite straightforward in the case of uniform shape functions, i.e. $\sigma_i(\mathbf{r}_i) = 1$. Instead, in the case of nonuniform temperature at thermal nodes, some choice for the shape functions $\sigma_i(\mathbf{r}_i)$ has to be made. Note that, strictly speaking, the shape functions depend on the whole thermal problem under investigation. That is to say, if any of the parts is modified (i.e. by

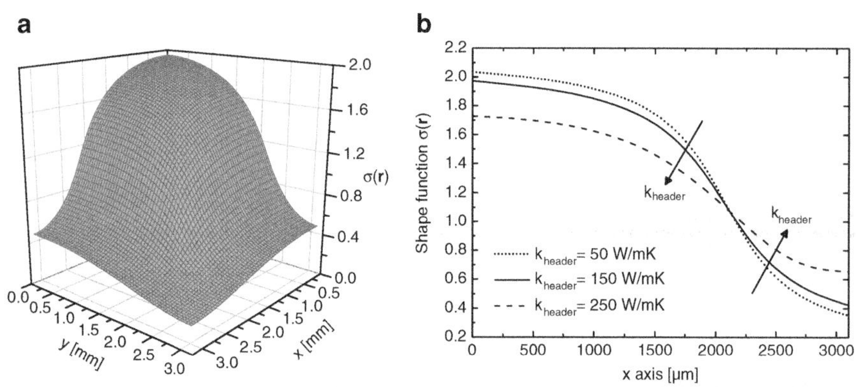

Fig. 23.21 (**a**) 3D and (**b**) 2D representations of the shape function σ(r) over the contact interface, as determined through a 3D thermal simulation of the single-level module illustrated in Fig. 23.1; the shape function shown in (**b**) is determined along a cut crossing the contact interface centre for three values of the thermal conductivity of the PGA header, namely (*dotted line*) 50, (*solid*) 150 (i.e. the reference value), and (*dashed*) 250 W/mK

changing the power dissipation pattern, the geometry and/or the thermal conductivities), then the shape functions should also change. Therefore, if the shape functions are determined or estimated for a reference case, then some error should be expected if any part is modified. Figure 23.21 shows the shape function across the contact interface between the adhesive BCB layer and the inactive host silicon substrate (i.e. the interface between sub-systems A and B), as determined from the thermal analysis of the complete single-level module in Fig. 23.1 for three values of the thermal conductivity of the package header. As can be seen, the shape function becomes more uniform as the thermal conductivity is increased. The accuracy of the CTM does not depend only on the shape function but also on the number of thermal nodes N. Similar to the uniform case, the accuracy can be improved by properly increasing N. With a judicious choice of the shape function, this error can be expected to be lower compared to the uniform shape function case ($\sigma_i = 1$) treated in [5].

We evaluated the thermal resistance of single-level modules characterised by PGA headers with different thermal conductivities by using the CTM approaches relying on either a uniform or the position-dependent shape function. In the latter case, the shape function used for the calculation of the CTM parameters was determined for the reference single-level module (i.e. that characterised by an AlN header with $k_{\mathrm{AlN}} = 150$ W/mK;). The calculation of the thermal resistance was repeated for $N = 1, 2, 3$ to quantify the accuracy improvement with increasing the discretisation level. CTM-based results are compared to numerical data for each thermal conductivity of the header in Fig. 23.22. Considering a single thermal node

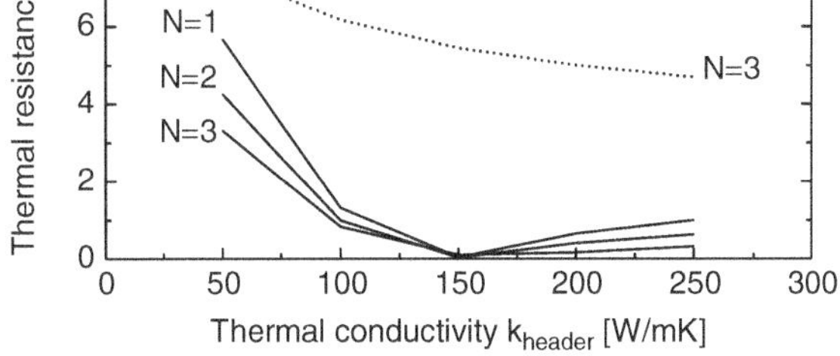

Fig. 23.22 Percentage error (in absolute value) in evaluating the thermal resistances by CTMs in comparison to FEM 3D results for the single-layer module shown in Fig. 23.1 as a function of thermal conductivity of the PGA header k_{header}. The CTM-based R_{TH} are calculated by (*dotted lines*) a uniform shape function $\sigma(\mathbf{r}) = 1$ [5] and (*solid*) the position-dependent $\sigma(\mathbf{r})$ determined for the reference domain with $k_{\mathrm{header}} = k_{\mathrm{AlN}} = 150$ W/mK illustrated in Fig. 23.21. Three discretisation levels ($N = 1, 2, 3$) are considered for both cases

($N = 1$), the error is above 10% for the $\sigma(\mathbf{r}) = 1$ case, and decreases with increasing k_{header}. Clearly, this is due to the fact that the temperature distribution at the contact surface becomes more uniform if k_{header} is higher (Fig. 23.21). In the case of the CTM based on a nonuniform shape function, the error is below 6% and vanishes for the reference value of $k_{\text{header}} = k_{\text{AlN}}$. The error can be lowered by means of a finer discretisation of the contact surface: increasing N from 1 to 3 leads to an error reduction of about 50% both for $\sigma(\mathbf{r}) = 1$ and for the reference $\sigma(\mathbf{r})$ as far as the k_{header} values farthest from k_{AlN} (i.e. 50 and 250 W/mK) are concerned.

References

1. Sung B (2006) Thermal enhancement of stacked dies using thermal vias. M.S. thesis, University of Texas at Arlington, Arlington
2. Pinel S, Marty A, Tasselli J, Bailbe J-P, Beyne E, Van Hoof R, Marco S, Morante JR, Vendier O, Huan M (2002) Thermal modeling and management in ultrathin chip stack technology. IEEE Trans Compon Packag Technol 25(2):244–253
3. Palacín J, Salleras M, Samitier J, Marco S (2005) Dynamic compact thermal models with multiple power sources: application to an ultrathin chip stacking technology. IEEE Trans Adv Packag 28(4):694–703
4. Comsol Multiphysics 3.5a (Dec. 2008), User's guide, Comsol AB
5. Gerstenmaier YC, Pape H, Wachutka G (2002) Rigorous model and network for static thermal problems. Microelectron J 33(9):711–718
6. Bosch EGT (2003) Thermal compact models: an alternative approach. IEEE Trans Compon Packag Technol 26(1):173–178
7. Sabry M-N (2003) Compact thermal models for electronic systems. IEEE Trans Compon Packag Technol 26(1):179–185
8. Beck JV, Cole KD, Haji-Sheikh A, Litkouhi B (1992) Heat conduction using Green's functions. Hemisphere Publishing Corporation, New York

Chapter 24
Optical Characterisation of Thin Silicon

Michael Reuter and Sebastian J. Eisele

Abstract This chapter presents the optical characteristics of thin silicon. It starts with the absorption of irradiation, discusses the efficiency potential for thin solar cells and introduces the quantum efficiency. The chapter further discusses the advantages of vertical versus lateral arrangement of the pn-junction and introduces laser processing of silicon, which allows, e.g. for low temperature contactless emitter formation.

24.1 Introduction

Crystalline silicon is an indirect semiconductor with a bandgap energy $E_g = 1.124$ eV at room temperature. Silicon absorbs irradiation fairly well, especially photon energies close to E_g. The efficiency of solar cells is defined as the quotient of electrical power output by the irradiated power input. The electrical power $P_{EL} = V_{OC}J_{SC}FF$ depends on the open circuit voltage V_{OC}, the short circuit current density J_{SC} and the fill factor FF. The FF describes the ratio of the output power $P = V_{mpp}J_{mpp}$ of a solar cell in the maximum power point to the optimum power output $P = V_{OC}J_{SC}$. In order to contribute to the J_{SC}, a photon has to be absorbed in the semiconductor and the light-generated carriers have to be collected by the metallic contacts. Thus, J_{SC} depends on the impinging photons, the amount of absorbed photons in the semiconductor and on the collection efficiency. The open circuit voltage $V_{OC} = V_{th}\ \ln(J_{SC}/J_0 + 1)$ depends on J_{SC}, the reverse saturation current density $J_0 = J_{0,rad} + J_{0,non\text{-}rad}$ due to radiative and nonradiative recombination and the thermal voltage $V_{th} = kT/q$ with the Boltzmann constant k, temperature T and elementary charge q. While the radiative recombination J_{0rad} is not thickness-dependent [1], the nonradiative recombination $J_{0,non\text{-}rad} = qnW/n_{bulk}$ [2] depends on

M. Reuter (✉)
Institut für Physikalische Elektronik (ipe), Universität Stuttgart, Pfaffenwaldring 47, Stuttgart, Germany
e-mail: michael.reuter@ipe.uni-stuttgart.de

J.N. Burghartz (ed.), *Ultra-thin Chip Technology and Applications*,
DOI 10.1007/978-1-4419-7276-7_24, © Springer Science+Business Media, LLC 2011

the injection level n, the bulk lifetime τ_{bulk} and the thickness W. Therefore, both J_{SC} and V_{OC} depend on the thickness of the solar cell: J_{SC} via absorption and carrier collection and V_{OC} via nonradiative recombination and the average carrier density in the bulk material [3].

24.2 Light Trapping

Complete absorption of irradiation with energies close to the bandgap energy requires a long light path through the silicon material, due to the low absorption coefficient of silicon in this energetic region. A long path either requires a thick solar cell material or good photon management, which prevents the photons from escaping out of the solar cell, a process which is the so-called 'light trapping'.

Figure 24.1a depicts a flat silicon wafer without light trapping. The irradiation enters perpendicularly, passes the wafer once and leaves the wafer on the rear side. This results in a light path $l = W$ equal to the wafer thickness W. Figure 24.1b presents a textured silicon wafer with a rear surface reflector. Here, refraction and rear surface reflection enlarge the light path. Figure 24.1c shows one type of Lambertian light trapping. A Lambertian surface randomises the light and allows for efficient light trapping. Yablonovitch [3] calculated an optimum light trapping with a light path length $l = n^2W$ where n is the refractive index of silicon. The solar cells discussed in this chapter, as well as in Chap. 27, feature a geometrical light trapping, while the simulations assume either no light trapping, as in Fig. 24.1a, and optimum light trapping according to Yablonovitch with $l = n^2W$.

24.3 Maximum Efficiency of Thin Crystalline Silicon Solar Cells

The maximum efficiency η_{max} for single junction solar cells amounts to $\eta_{max} \approx 33.5\%$ in the radiative limit for the bandgap energy of crystalline silicon under an insolation of AM1.5G [4], the so-called 'Shockley/Queisser limit' [1].

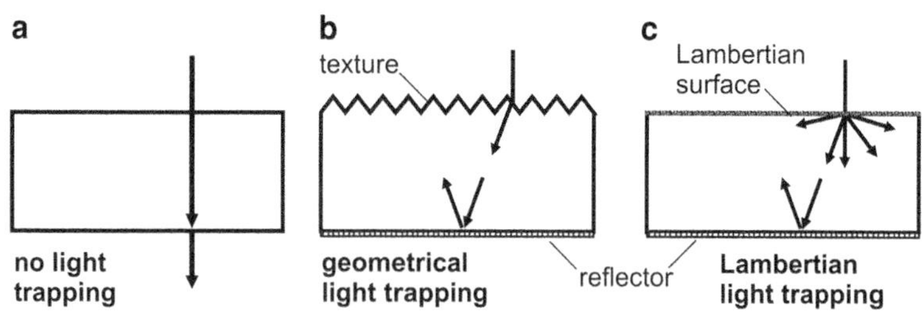

Fig. 24.1 (**a**) No light trapping (**b**) Geometrical light trapping with a back reflector. (**c**) Lambertian light trapping with a back reflector

The Shockley/Queisser limit forms an upper boundary for photovoltaic energy conversion, but neglects some material and thickness-dependent loss mechanisms: first, bulk recombination due to Auger [5] and Shockley–Read–Hall [6, 7] processes, and second, incomplete absorption due to the material properties of silicon.

Figure 24.2a depicts the simulated thickness dependence of the short circuit current density J_{SC}; Fig. 24.2b shows the efficiency η of single crystalline silicon solar cells without light trapping and for optimum light trapping according to Yablonovitch [3]. No light trapping equals one light pass through the solar cell, while optimum light trapping results in an average light path of 52 times through the solar cell. The simulation accounts for an insolation with AM1.5G [4], temperature $T = 25$ °C, radiative recombination and bulk lifetime $\tau_{bulk} = 200$ µs. Upon decreasing solar cell thickness, the efficiency of solar cells with optimum light trapping reaches a maximum efficiency $\eta \approx 24\%$ for a thickness range 3 µm $< W < 30$ µm [8]. Therefore, thin solar cells hold the potential of efficiencies $\eta > 20\%$. The current world record for single crystalline silicon solar cells under AM1.5G [9] insolation amounts to $\eta = 25.0\%$[1] [10, 11] on high quality float zone material with a thickness $W = 450$ µm.

The maximum efficiency depicts the upper limit for a solar cell bulk material showing a minority carrier lifetime $\tau_{bulk} = 200$ µs. The diffusion length L_{eff} is the distance the generated carriers travel before they recombine. A real solar cell shows further loss mechanisms: Auger recombination [5] especially in the highly n-type

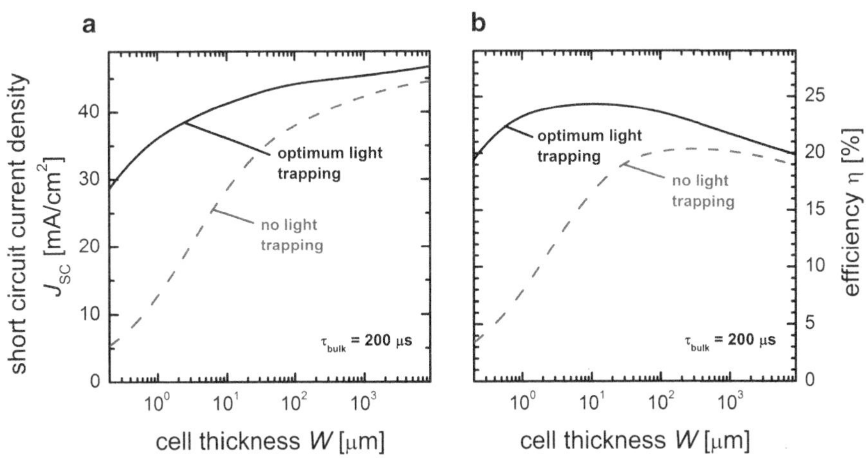

Fig. 24.2 Simulated solar cell (**a**) short circuit current density J_{SC} and (**b**) efficiency η depending on the cell thickness for optimum light trapping according to Yablonovitch [3] and no light trapping. The simulation accounts for radiative recombination and a bulk carrier lifetime $\tau_{bulk} = 200$ µs

[1]24.7% efficiency reported by Zhao et al. (see [10]) recalibrated by Green to new reference AM1.5G spectrum [9] (see [11])

doped emitter region, reflection losses, nonradiative recombination and incomplete carrier collection. Minority carriers are collected as soon as they reach the emitter region. The collection depends on L_{eff} and the distance the carriers need to travel in order to reach a contact. This distance depends on the solar cell thickness and the arrangement of the pn-junction, thus, on the design of the solar cell.

Two major designs are used in today's industrial crystalline silicon solar cell manufacturing. First, the vertical layout, where the pn-junction is distributed along a vertical plane with the emitter usually located at the front surface. The minority carriers are collected at the front surface, while the majority carriers are collected at the rear surface. Second, the lateral layout arranges both emitter and base regions on one surface, usually facing away from the sun. One example of a lateral solar cell layout is the interdigitated back contact solar cell from SunPower Inc., which exhibit maximum efficiencies η exceeding $\eta > 21\%$ [12].

24.4 Vertical and Lateral Solar Cell Designs

Today's solar cell industry uses two main cell designs: first, the conventional design, where the emitter is located on the solar cell front surface facing towards the sun. In such a layout, a metallic front contact grid shadows the solar cell to some extent and should be well designed to optimise between the shadowing losses and electrical losses [13]. The second design avoids this optimisation problem by locating both contacts on the solar cell backside. This design avoids all grid shadowing, but introduces another challenge: collection of the minority carriers at the backside of the cell. Therefore, the second layout requires high carrier lifetime in the bulk material, excellent frontside passivation and light trapping.

Figure 24.3a schematically shows the structure of the conventional (vertical) design with front and rear contacts, and Fig. 24.3b shows the back junction back contact (lateral) solar cell design. Both structures use amorphous silicon nitride (SiN_x) on the front side as passivation and antireflection coating (ARC) layer.

To provide a good understanding of the working principle of solar cells designed by both structures, we introduce a simple model for calculating cell efficiency based on the material and structural parameters. This model separates the optical response and the electronic response as follows: The incident radiation impinges onto the solar

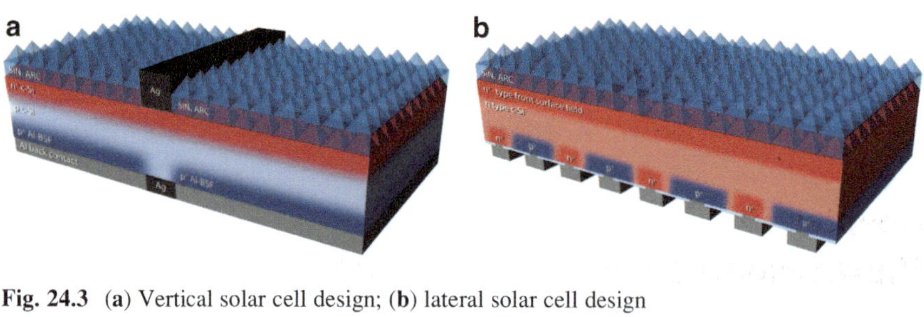

Fig. 24.3 (**a**) Vertical solar cell design; (**b**) lateral solar cell design

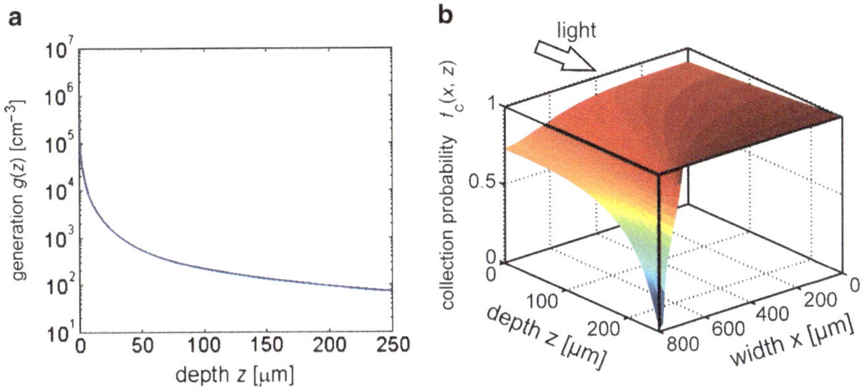

Fig. 24.4 (**a**) Generation profile $g(z)$, and (**b**) simulated collection probability f_c for a back junction back contact solar cell [14]. The drop in f_c ($x = 800$ μm) is caused by the base contact metallisation, which results in a high carrier recombination

cell and generates one electron hole pair for each absorbed photon. Figure 24.4a depicts the optical properties that are expressed by the generation profile:

$$g(\lambda, z) = \Theta[1 - R_f(\lambda)]\alpha(\lambda)e^{-\alpha(\lambda)z}, \tag{24.1}$$

which depends on the irradiance Θ, the front surface reflectivity $R_f(\lambda)$ as a function of the wavelength λ, the absorption coefficient $\alpha(\lambda)$ and the spatial coordinate z perpendicular to the cell surface. The electronic properties are defined by the collection probability $f_c(x,y,z)$ of the minority carriers. The collection probability is attained by solving the differential equation at a given minority carrier diffusion length L_{bulk} for electrons in the case of a p-type doped absorber. The diffusion length $L_{\text{bulk}} = (D\tau_{\text{bulk}})^{1/2}$ [15, 16] depends on the diffusivity D and the bulk lifetime τ_{bulk}. In the case of the conventional cell structure, f_c can be reduced to a one-dimensional problem and an analytical solution is simply found as [17, 18]

$$f_c(z) = f_{\text{SCR}}e^{-\frac{z}{L_{\text{bulk}}}}, \tag{24.2}$$

where f_{SCR} is the collection probability in the space charge region.

For lateral cell design, a two-dimensional numerical solution gives $f_c(x,z)$ and enables the calculation of the short circuit current density

$$J_{\text{SC}} = q\int_0^p\int_0^W\int_{E_g}^{\infty} g(z,\lambda)f_c(x,z)\,\mathrm{d}\lambda\,\mathrm{d}z\,\mathrm{d}x, \tag{24.3}$$

which depends on the local solar cell properties. The pitch p describes the solar cell dimension in the x-direction.

Figure 24.4b shows a simulated collection probability profile for a back junction back contact solar cell. The solar cell shows an emitter that covers 87.5% of the back surface, while the base region covers the remaining area. The base metallisation contacts an area with 20-μm width. The base metallisation results in an increased surface recombination velocity, and thus reduces the local collection probability dramatically. The low rear collection probability results in a reduced collection profile, eventually reducing the front surface collection, too. A high doping beneath the metallic contacts allows the reduction of the recombination by repelling the minority carriers, a so-called 'back surface field' (BSF). Unfortunately, a BSF usually requires an additional diffusion process, which results in higher production costs. In the following section, we present laser doping, which allows for low cost formation of highly p- and n-type doped areas.

24.5 Laser Doping of p- and n-Type Areas

Laser doping (LD) of semiconductors has been known since the 1960s [19] and saw a strong phase of research during the 1980s [20–22]. But, only in recent years has LD come to the attention of the photovoltaic industry, due to the rapid development of high performance and reliable solid state lasers. The main advantage of LD is local selectivity. The dopant incorporation only takes place at laser irradiated areas. The use of different types of precursors, like phosphorus or boron, results in n-type or p-type doped emitters [20, 21].

Figure 24.5 depicts two sets of laser-doped phosphorus emitters measured with secondary ion mass spectroscopy (SIMS). The increase of the laser pulse energy

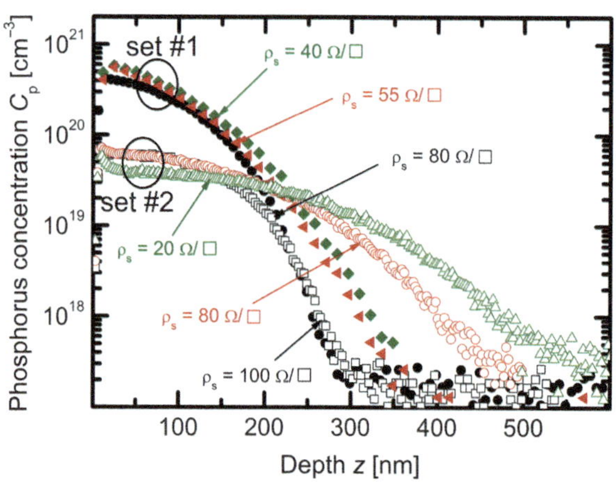

Fig. 24.5 Typical doping profiles of laser emitters, measured by SIMS. Filled symbols represent sample set #1, irradiated with low laser pulse duration $\tau_{p1} \approx 10$ ns at different pulse energy densities E_p. Open symbols of set #2 stand for longer $\tau_{p2} \approx 65$ ns. The maximum phosphorus concentration $C_{p,max}$ of set #1 is about one magnitude higher than set #2

density E_p results in longer and deeper melting of the silicon surface. Thus, the phosphorus atoms from the doping precursor have more time to diffuse deep into the melt, increasing the emitter depth and decreasing the emitter sheet resistance ρ_s. The variation of the LD parameters allows an accurate adjustment of the shape of the emitter profile and the emitter sheet resistance ρ_s in a wide range. Therefore, LD enables manifold applications like in situ selective emitter formation with high doped areas underneath contact fingers and lower doped areas in between. Particularly, for new solar cell concepts like back contacted solar cells, LD has great potential due to its selectivity and flexibility. The first step in the LD process is the deposition of the dopant precursor layer, followed by laser processing. Typically, a pulsed laser beam melts the wafer surface. During melting, dopant atoms from the precursor layer diffuse into the molten silicon. At the same time the solid precursor layer is evaporated or ablated. One part of the evaporated precursor layer recondenses on the wafer surface and serves as dopant source for subsequent laser pulses [23]. Unlike common high temperature furnace diffusion, which uses vapour phase diffusion, LD applies liquid phase diffusion. The more than five order of magnitudes higher phosphorus diffusion constant $D_p = 5.1 \times 10^{-4}$ cm^2/s in liquid silicon [24] enables the fast incorporation of phosphorus atoms from the precursor layer, up to 1 μm deep into the molten silicon within a few hundred nanoseconds. So far the record in full area laser-doped emitter solar cells is $\eta = 18.9\%$ [25]. As mentioned before, laser doping can also be used to create selective emitters. The selective emitter concept uses laterally different emitter doping: (1) high doping under the frontside metallisation for low contact resistance between contact metal and semiconductor interface; (2) lower doping between the contact fingers for a better short-wavelength response due to less Auger recombination [26] as well as improved emitter passivation [10, 27]. This process is easily implemented into existing lines for mass production of solar cells; through it, efficiency increases between 0.5% and 0.8% absolute [28].

24.6 Quantum Efficiency

The external quantum efficiency $Q_E = N_{collected}/N_{phot}$ is defined as the ratio of collected charge carriers $N_{collected}$ with respect to the number of impinging photons N_{phot}, and thus describes the electronic quality of a solar cell. Crystalline silicon shows high absorption for high energy photons, and low absorption for photons with energies close to the bandgap energy. Therefore, the wavelength-dependent analysis of quantum efficiency allows the analysis of different parts of the solar cell: The emitter properties influence the Q_E in the blue irradiation regime $\lambda < 400$ nm, while the bulk material quality affects the Q_E in the green to red irradiation regime 400 nm $< \lambda <$ 900 nm. The rear side quantities, as for example, the rear surface recombination velocity S_{rear} and the rear surface reflection R, dominate the Q_E in the infrared wavelength regime $\lambda > 900$ nm. The internal quantum efficiency

$$Q_I = \frac{Q_E}{1 - R - A} \tag{24.4}$$

is the Q_E corrected for optical losses like front surface reflection R and parasitic absorption A, e.g. in an antireflection coating. Therefore, the Q_I only shows the electronic properties of the solar cell.

Figure 24.6a shows the measured internal quantum efficiency Q_I for a 50-μm thin monocrystalline silicon solar cell and a simulated Q_I for three cases: a 50-μm thick solar cell with optimum and no light trapping and a 250-μm thick solar cell without light trapping. The 50-μm thin Q_I surpasses the Q_I of the 250-μm thick solar cell with the help of light trapping. In the long wavelength regime $\lambda > 1{,}000$ nm, the measured Q_I exceeds the simulation for no light trapping due to the geometrical light trapping depicted in Fig. 24.1b. While the measured Q_I suffers from front and rear surface recombination and insufficient rear reflection, the light trapping in the solar cell is better than the simulated case for no light trapping.

Figure 24.6b shows the inverse internal quantum efficiency

$$Q_I^{-1} = 1 + \cos(\varphi)\frac{L_\alpha}{L_{\text{eff}}}, \tag{24.5}$$

which depends linearly on absorption length L_α and the inverse effective diffusion length L_{eff}, according to Basore [31]. A light path enlargement due to refraction at a surface texture is accounted for by the refraction angle ϕ. The effective diffusion length L_{eff} describes the distance a generated minority carrier travels prior to recombination, and therefore forms a measure of the electronic quality of the solar cell. A rule of thumb states that highly efficient solar cells require an L_{eff}

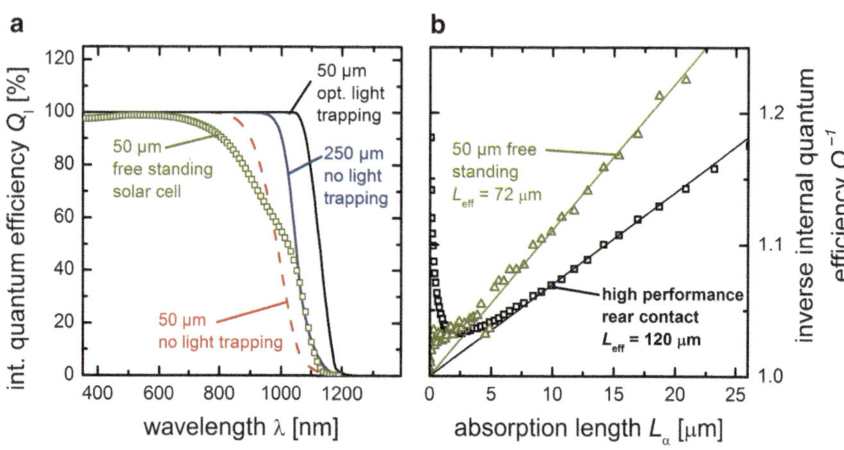

Fig. 24.6 (**a**) Internal quantum efficiency Q_I for a 50-μm thin monocrystalline silicon solar cell with a record efficiency $\eta = 17.0\%$ fabricated at *ipe* [29]. (**b**) Analysis of the inverse Q_I reveals the effective diffusion length $L_{\text{eff}} = 72$ μm for the freestanding solar cell [29] and $L_{\text{eff}} = 120$ μm for the solar cell featuring a high performance rear contact according to [30]

three times larger than the solar cell thickness W. The evaluated solar cells show $L_{eff} = 72$ μm and 120 μm, which mainly depend on the rear surface properties. Here, the high performance rear contact increases the L_{eff} due to a rear surface passivation with a hydrogenated amorphous silicon layer.

The effective diffusion length $L_{eff} = 120$ μm is more than twice the cell thickness for the solar cell featuring the high performance rear contact, thus enabling high efficiencies. Unfortunately, the solar cell shows an efficiency $\eta = 16.9\%$ [30] (independently confirmed by ISE CalLab), which is greatly reduced compared to the maximum efficiency $\eta \approx 24\%$ for optimum light trapping and well below $\eta \approx 19\%$ for no light trapping, according to Fig 24.2. Furthermore, the quantum efficiency in Fig. 24.6 reveals losses due to recombination in the bulk material and at the rear surface. The high performance rear contact allows for efficiencies $\eta = 20.5\%$ [32] on 350-μm thick FZ material. Therefore, the bulk material of the thin solar cells features a low minority carrier lifetime. Tobail et al. [33] ascribe this finding to a contamination with platinum during the electrochemical formation of the porous silicon in the layer transfer process.

Acknowledgments The authors would like to thank O. Tobail for fruitful discussions and help with the simulation. We thank J. Cichoszewski, T. Röder and P. Gedeon for carefully reading the manuscript.

References

1. Shockley W, Queisser HJ (1961) Detailed balance limit of efficiency of p-n junction solar cells. J Appl Phys 32:510–519
2. Brendel R, Queisser HJ (1993) On the thickness dependence of open circuit voltages of pn junction solar cells. Sol Energy Mater Sol Cells 29:397–401
3. Yablonovitch E (1982) Statistical ray optics. J Opt Soc Am 72:899–907
4. ISO (1992) International Organization for Standardization, Geneva, Switzerland, ASTMG173-03 from ISO 9845-1
5. Beattie AR, Landsberg PT (1959) Auger effect in semiconductors. Proc Royal Soc A 249:16–29
6. Shockley W, Read WT (1952) Statistics of the recombinations of holes and electrons. Phys Rev 87:835–842
7. Hall RN (1952) Electron-hole recombination in germanium. Phys Rev 87:387
8. Reuter M, Thin crystalline silicon solar cells, Dissertation, University of Stuttgart (to be published)
9. International Electrotechnical Commission (2008). International Standard IEC 60904-3, Edition 2, 2008. Photovoltaic devices-Part 3: Measurement principles for terrestrial photovoltaic (PV) solar devices with reference spectral irradiance data. International Electrotechnical Commission, ISBN 2-8318–9705-X
10. Zhao J, Wang A, Green MA (1999) 24.5% efficiency silicon PERT cells on MCZ substrates and 24.7% efficiency PERL cells on FZ substrates. Prog Photovolt Res Appl 7:471–474
11. Green MA (2009) The path to 25% silicon solar cell efficiency: history of silicon cell evolution. Prog Photovolt Res Appl 17:183–189
12. Mulligan WP, Rose DH, Cudzinovic MJ, De Ceuster DM, McIntosh KR, Smith DD, Swanson RM (2004) Manufacture of solar cells with 21% efficiency. In: Hoffmann W, Bal J-L, Ossenbrink H,

Palz W, Helm P (eds) Proceedings of the 19th European Photovoltaic Solar Energy Conference, pp 387–390
13. Cuevas A, Russell DA (2000) Co-optimisation of the emitter region and the metal grid of silicon solar cells. Prog Photovolt Res Appl 8:603–616
14. Eisele S (2006) Herstellung und Charakterisierung von Rückkontaktsolarzellen, Diploma thesis, University of Stuttgart, Stuttgart
15. Einstein A (1905) Eine neue Bestimmung der Moleküldimensionen. Ann Phys 322:549–560
16. von Smoluchowski M (1906) Zur kinetischen Theorie der brownschen Molekularbewegung und der Suspensionen. Ann Phys 326:756–780
17. Al-Omar A-AS, Ghannam MY (1996) Direct calculation of two-dimensional collection probability in pn junction solar cells, and study of grain-boundary recombination in polycrystalline silicon cells. J Appl Phys 79:2103
18. Rau U, Brendel R (1998) The detailed balance principle and the reciprocity theorem between photocarrier collection and dark carrier distribution in solar cells. J Appl Phys 84:6412
19. Fairfield JM, Schwuttke GH (1968) Silicon diodes made by laser irradiation. Solid State Electron 11:1175–1176
20. Fogarassy E, Stuck R, Grob JJ, Siffert P (1981) Silicon solar cells realized by laser induced diffusion of vacuum-deposited dopants. J Appl Phys 52:1076–1082
21. Sameshima T, Usuzi S, Sekiya M (1987) Laser-induced melting of predeposited impurity doping technique used to fabricate shallow junctions. J Appl Phys 62:711–713
22. Deutsch TF, Fan JCC, Turner GW, Chapman RL, Ehrlich DJ, Osgood RM (1981) Efficient Si solar cells by laser photochemical doping. Appl Phys Lett 38:144–146
23. Köhler JR, Eisele SJ (2010) Precursor layer ablation influences laser doping of silicon. Prog Photovolt Res Appl 18(5):335–339
24. Kodera H (1964) Diffusion coefficients of impurities in silicon melt. Jpn J Appl Phys 2:212–219
25. Eisele SJ, Röder TC, Köhler JR, Werner JH (2009) 18.9% efficient full area laser doped silicon solar cell. Appl Phys Lett 95:133501
26. Kerr MJ, Cuevas A (2002) General parameterization of Auger recombination in crystalline silicon. Appl Phys Lett 91:2473–2480
27. Kerr MJ, Schmidt J, Cuevas A, Bultman JH (2001) Surface recombination velocity of phosphorus-diffused silicon solar cell emitters passivated with plasma enhanced chemical vapor deposited silicon nitride and thermal silicon oxide. J Appl Phys 89:3821–3826
28. Röder TC, Eisele SJ, Grabitz P, Wagner C, Kulushich G, Köhler JR, Werner JH (2011) Add-on laser tailored selective emitter solar cells. Prog Photovolt Res Appl 18:505
29. Reuter M, Brendle W, Tobail O, Werner JH (2009) 50 μm thin solar cells with 17.0% efficiency. Sol Energy Mater Sol Cells 93:704–706
30. Brendle W (2007) Niedertemperaturrückseitenprozess für hocheffiziente Siliziumsolarzellen. Shaker Verlag, Aachen
31. Basore PA (1993) Extended spectral analysis of internal quantum efficiency. In: Proc. 23rd IEEE Photovoltaic Specialists Conference, IEEE, Piscataway, NY, pp 147–152
32. Brendle W, Nguyen VX, Grohe A, Schneiderlöchner E, Rau U, Palfinger G, Werner JH (2006) 20.5% efficient silicon solar cell with a low temperature rear side process using laser-fired contacts. Prog Photovolt Res Appl 14:653–662
33. Tobail O, Reuter M, Werner JH (2009) Origin of the open circuit voltage limit for transfer solar cells. In: Sinke WC, Ossenbrink HA, Helm P (eds) Proc. 24th European Photovoltaic Solar Energy Conference, WIP, Munich, Germany, pp 2593–2595

Part VI
Applications

As illustrated in Parts I through V ultra-thin chips provide several aspects that cannot be realized in conventional silicon technology: (i) ultra-thin form factor, (ii) special chip shape and adaptivity, (iii) 3D chip stacking, and (iv) mechanical flexibility (Fig. VI.1).

Chip thinning has for long been a technique used to reduce the thermal resistance between integrated devices and package. Novel power device structures require in addition that a thinned wafer be processable from the backside (Chap. 25). Locally

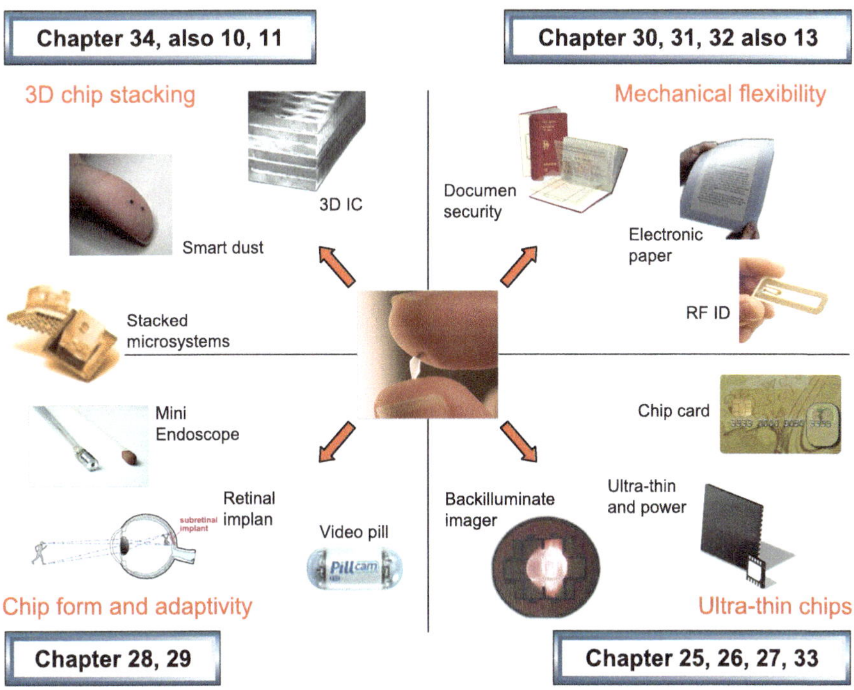

Fig. VI.1 Illustration of the application areas: 3D chip stacking, mechanical flexibility, ultra-thin chips, and chip form and adaptivity with the relevant chapters from Part VI and also from chapters of other parts, where applicable

thinned chips are of high interest for realizing back-illuminated imagers with maximum fill factor (Chap. 26). Globally thinned chips or wafers are the base substrate for cost-effective high-efficiency solar cells (Chap. 27) and for integrated radio-frequency (RF) systems (Chap. 33). Shape adaptivity, in addition to extremely small thickness, is a key requirement for sub-retinal implants and other biomedical devices (Chap. 28), as well as for certain sensor components (Chap. 29). Chip stacking leads to three-dimensional integrated circuits (3D ICs) and Microsystems (Chaps. 10 and 11) as well as to high-performance RF integrated circuits (Chap. 34). The excellent mechanical properties of ultra-thin silicon are exploited in flexible ICs in the emerging field of flexible electronics. There, high-performance compact flexible silicon chips are combined with thin-film or organic large area electronics to provide an overall optimum hybrid solution for RF ID tags in general (Chap. 30) and security applications (Chap. 31), for driver ICs in large flexible displays (Chap. 32), and for systems-in-foil (SiF).

The applications illustrated in the chapters of this part, and in some chapters of the other parts of this book (see Fig. VI.1), are just examples of what can be emerging in the near future.

Chapter 25
Thin Chips for Power Applications

Kurt Aigner, Oliver Häberlen, Herbert Hopfgartner, and Thomas Laska

Abstract Thin chips for power applications are the key enabler for energy efficiency. This chapter will view low voltage metal oxide semiconductor field effect transistors (MOSFETs), IGBTs (insulated gate bipolar transistors) and FWDs (free wheeling diodes). For these the trend to ultra-thin power transistor chips (20 μm thickness) and the resulting electrical and thermal benefits by chip thickness reduction is described. Further the influence of thin chips on product definition and application especially for automotive smart power devices is pointed out. Thin chips technology for power applications allow an increase in efficiency and new packaging processes of semiconductor devices. The application benefits include reduced area requirements for the same performance and new packaging processes.

25.1 Low Voltage MOSFETs

In the modern world the driving factors for power electronics are energy efficiency and high power density for space-saving designs. Here, low voltage power MOSFETs (defined as voltage classes between 20 and 250 V) play an important role in many applications like DC/DC converters, switched mode power supplies, motor drives and class D audio amplifiers [1, 2]. In recent decades there has been a tremendous development in silicon power technologies, especially in the voltage classes around 30 V, which have the largest portion of the low voltage MOSFET market.

The basic function of a power MOSFET is the switching of currents pretty similar to a relay's. One of the main optimisation criteria is, thus, the on-state resistance at a given blocking voltage. To achieve this power MOSFETs consist of a large number of paralleled MOSFET cells. To be able to switch high currents, so-called "vertical transistors" are typically used. In vertical transistors, the low-doped

K. Aigner (✉)
Infineon Technologies Austria, Villach, Austria
e-mail: Kurt.Aigner@infineon.com

J.N. Burghartz (ed.), *Ultra-thin Chip Technology and Applications*,
DOI 10.1007/978-1-4419-7276-7_25, © Springer Science+Business Media, LLC 2011

drift region to accommodate the space charge region in the blocking case extends into the device below the source and body regions, enabling higher packing density of the cells than what is possible for lateral MOSFETs [3]. A further advantage is that all parallel source terminals are arranged on the front-side of the device and all parallel drain terminals are connected to the backside. This basic construction principle can be seen in Fig. 25.1.

In the 1980s planar transistors were created with the MOS channel coplanar with the wafer surface. The need for reducing the specific on-state resistance ($R_{ON} * A$) led to the widespread usage of so called "trench transistors" in the 1990s, where the channel is oriented in the vertical direction along the sidewalls of a trenched structure (see Fig. 25.2). This increased channel density and eliminated the JFET region of the planar transistor.

In the last decade trench transistors have been further optimised to such an extent that in the chain of resistances from source to drain, shown in Fig. 25.2, the part of the highly doped wafer substrate (n^+ in the figures) could not be neglected any longer. Historically, this has been tackled by lowering the resistivities of substrate material from around 20 mΩ-cm for antimony doping to arsenic substrates down to 3 mΩ-cm to phosphorous-doped substrates, which, meanwhile, were approaching resistivities of 1 mΩ-cm. A further substantial reduction of the resistivity is not expected, so the only way now to reduce resistivity is to reduce the wafer thickness.

Shown in Fig. 25.3 are the last three generations of Infineons OptiMOS® low voltage power transistors in the 30 V class, in comparison with substrate resistance. The standard components operate with a wafer thickness of 175 μm, which was termed "thin wafers" just some years ago and still is quite thin for many companies.

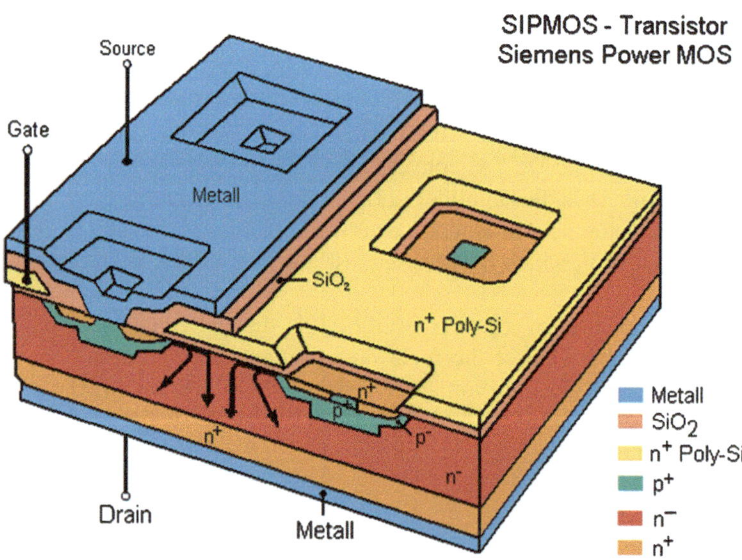

Fig. 25.1 Schematic drawing of a vertical power MOSFET

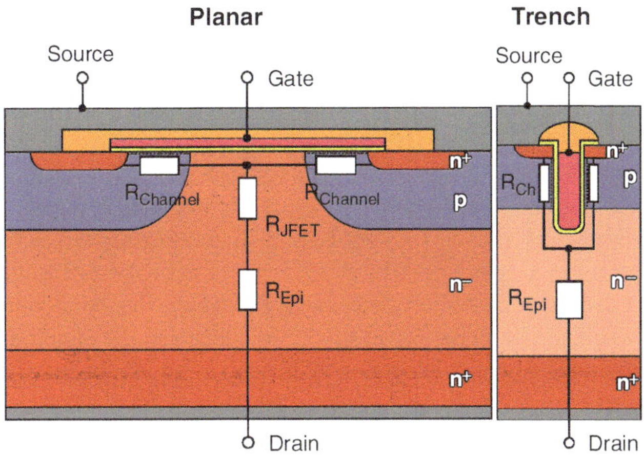

Fig. 25.2 Power transistor evolution from planar cell to trench cell (cross-section)

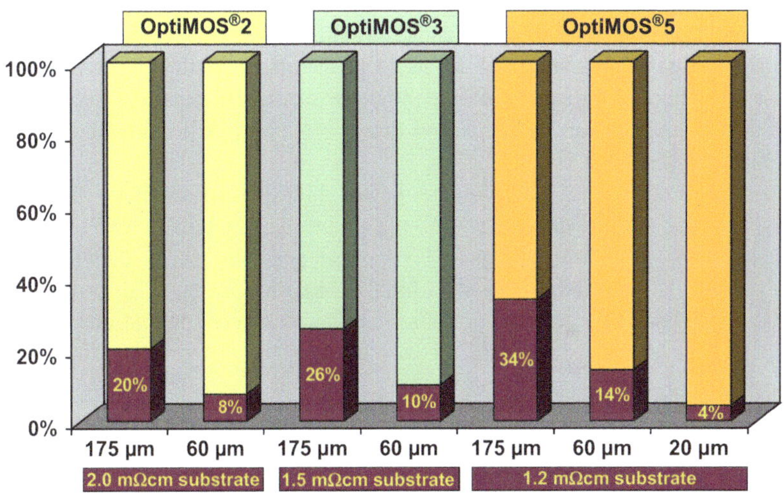

Fig. 25.3 Fraction of the substrate resistance of total transistor resistance (30 V class) for various wafer thicknesses

Retaining this wafer thickness would mean a shear 34% fraction of substrate resistance for the current released OptiMOS®5 generation, despite the ultra-low ohmic substrate with 1.2 mΩ-cm resistance. For this reason the high performance products in the SuperSO8 and S³O8 package platforms now operate with 60 μm wafer thickness, bringing down the substrate fraction to 14%. By reducing the wafer thickness even further, to 20 μm, substrate resistivity can be nearly neglected – a result that has been demonstrated in a feasibility study.

So far we have looked only at reducing parasitic electrical resistances of the wafer substrate by thin wafer technology. Of equal importance, however, is thermal resistivity, especially with increasing power densities due to shrinking die sizes. Here, thin wafers can bring substantial benefits if the designer considers, that the thermal path due to heat generation in the transistor cells, which travels through the silicon substrate to the leadframe, has to pass also through the solder joint. For example a typical bond line thickness of 50 μm for soft solders results in a thermal boundary resistance of 1.72×10^{-6} m² K/W, which is already larger than that of a 175 μm thick silicon substrate with 1.35×10^{-6} m² K/W (all values at room temperature). So it is absolutely necessary to reduce the bond line thickness, something which can be done, for example, by advanced die attach methods like diffusion soldering, wherein bond line thickness is on the order of only a few micrometres.

Figure 25.4 shows three different types of wafer technologies (220 μm thick wafers with thick and with thin solder; and 60 μm thin wafers with thin solder layer) subjected to constant heating for up to 1 ms. Changing the die attach already reduces peak temperature reached by about 60 K. Going to thin wafers in combination with thin solder brings down peak temperature by another 125 K. So in this case the thin chip with thin solder ends up with $T_j = 200°C$. Compare this to the thick chip with thick solder with $T_j = 385°C$ – a temperature that will bring the chip close to destruction! One thing to note, however, is that for duration <~60 μs there is no difference between the thick and thin wafer variants because the heat does not reach the wafer backside.

The first commercial low voltage power MOSFET technology that makes widespread use of ultra-thin wafers (60 μm) is Infineons OptiMOS™5 technology. This technology has been designed for highest possible energy conversion efficiency in DC/DC converters, which is highly important for the CPU power supply in PC motherboards, notebooks, graphics cards and server applications. Typically

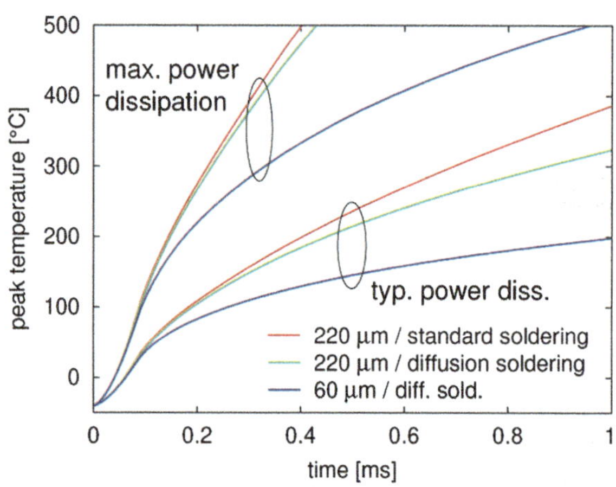

Fig. 25.4 Thermal benefits of thin wafers with different die attach methods

an input voltage of 12 V has to be converted to 1.2 V output voltage with currents ranging from 30 A up to 180 A for servers. The buck converters used typically operate at frequencies around 500 kHz, with a trend to even higher frequencies. This requires power MOSFETs with the lowest possible parasitic switching capacitances for a given resistance value [4]. The corresponding figure-of-merit (FOM) is the product of charge (= integrated capacitance Q) per area A and area-specific on-state resistance:

$$\text{FOM} = (Q/A)(R_{on}A) = QR_{on}$$

For DC/DC converters, the three relevant figures-of-merit are the gate charge (FOM_g), the gate-drain charge (FOM_{gd}) and the output charge (FOM_{oss}). Since switching capacitances mainly originate in the transistor cell's construction, a reduction of substrate resistance lowers on-state resistance without increasing the corresponding charges. Thus, thin wafer technology delivers a major contribution

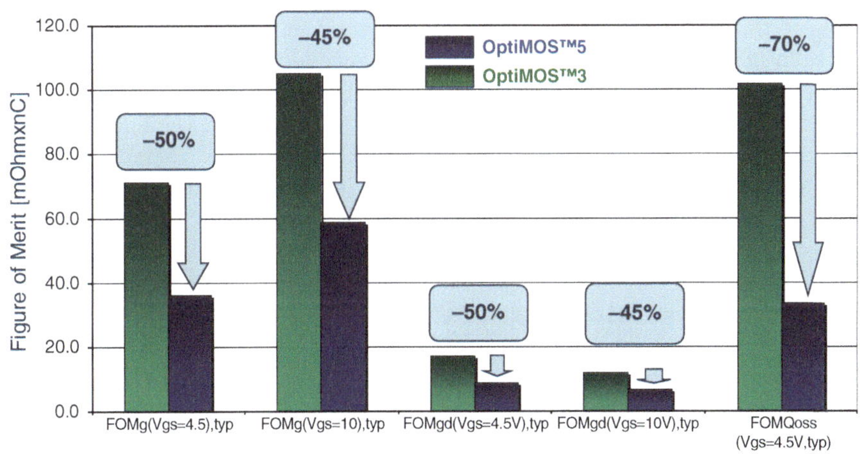

Fig. 25.5 Electrical benefits of thin wafers for low voltage power transistors (comparison for new OptiMOS™5 25 V versus previous OptiMOS™3)

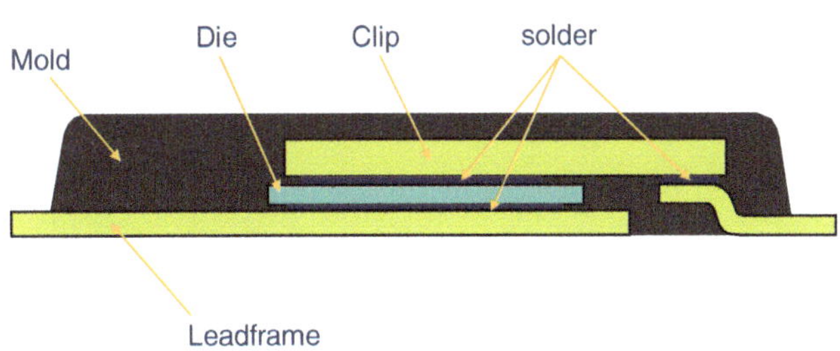

Fig. 25.6 Cross-section of a 60 μm thin trench power MOSFET in a SuperSO8 package with dual side soldering

to extreme improvement of the switching figures-of-merit of Infineons OptiMOS™5 technology shown in Fig. 25.5.

Ultra-thin wafer technology not only poses challenges for front end manufacturing technology but also for back end technology to convert ultra-thin wafers via ultra-thin chips to products that can be used by the customer. Examples include Infineons SuperSO8 and S³O8 packages, where the chip attach is made by solder interconnects on both sides to deliver the lowest possible resistivity and inductivity interconnects (Fig. 25.6).

25.1.1 IGBTs and FWDs

In the area of power electronics, we turn to *electric drives* for consumer, industrial and traction applications, as well as in hybrid and fully electrical vehicles. Key components of electric drives are the power semiconductor device insulated gate bipolar transistor (IGBT) in combination with a fast switching bipolar free-wheeling diode (FWD) [5, 6] (Fig. 25.7). In general, an electric drive performs the conversion of electrical energy to mechanical energy or vice versa. In the inverter part of the drive, a three-phase current/voltage with variable frequency and amplitude is generated out of a rectified DC link voltage. Thereby the IGBT acts as a switch, and the diode acts as a freewheeling path (or "valve"), typically in a three-phase configuration of six IGBTs with inverse connected diodes ("six pack") (Fig. 25.8).

Common switching frequencies for the IGBT and FWD inside the inverter are in the range of several kHz; typically, DC link voltages are between a few hundred volts (e.g., air conditioner, washing machine, hybrid vehicle) and several kV (e.g., fast trains). In this range of voltages and switching frequencies the IGBT + FWD is

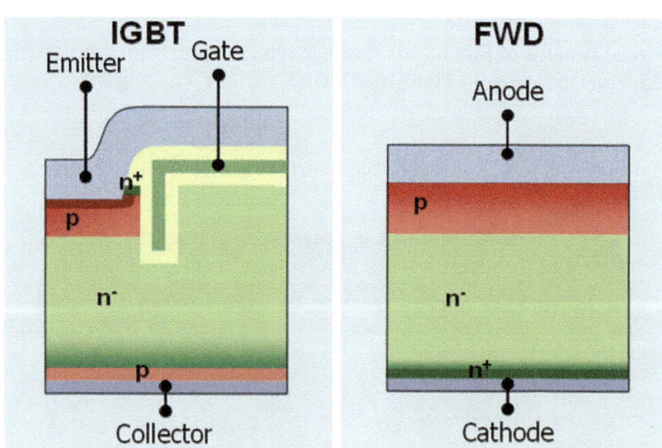

Fig. 25.7 Cross-section of IGBT and FWD

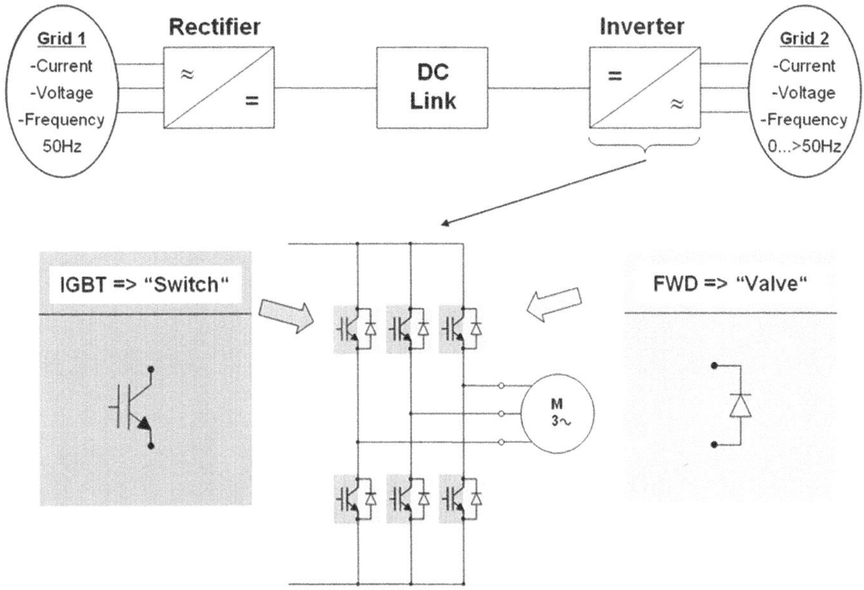

Fig. 25.8 IGBTs and FWDs as a part of the inverter in an electric drive

usually the best choice of power device compared to other power semiconductors like GTOs/thyristors, power MOSFETs or superjunction MOSFETs.

Key requirements for these quite simple (at a first glance) bipolar power devices are:

- Blocking voltage capability as required from the DC link and the additional inductive voltage overshoot pikes (e.g., 600 V rated blocking voltage for a DC link of 300–400 V or 1,200 V rated blocking for a DC link of 500–800 V).
- Lowest possible on state losses and lowest possible switching losses to achieve highest energy efficiency for the electric drive.
- Soft switching behaviour to ensure lowest EMI noise.
- Robustness against electrical malfunctions such as short circuit and/or over-current.

To ensure the required blocking capability both IGBT and FWD have a vertical pn-junction (thin p-doped layer on top of the device (p-body of the MOS part of the IGBT on the silicon top surface or anode layer on the top of the diode surface) and thick n-doped layer beneath (see also Fig. 25.7). If a reverse bias-VR is applied to this pn-junction, a depletion layer and a high electrical field is formed. The achievable blocking voltage depends on the thickness and the doping concentration of the n-doped layer (e.g., 600 V requires an n-doped Si layer of ~1e14 cm^{-3} doping and 60 μm thickness; 1,200 V requires ~5e13 cm^{-3} and 120 μm thickness) (Fig. 25.9).

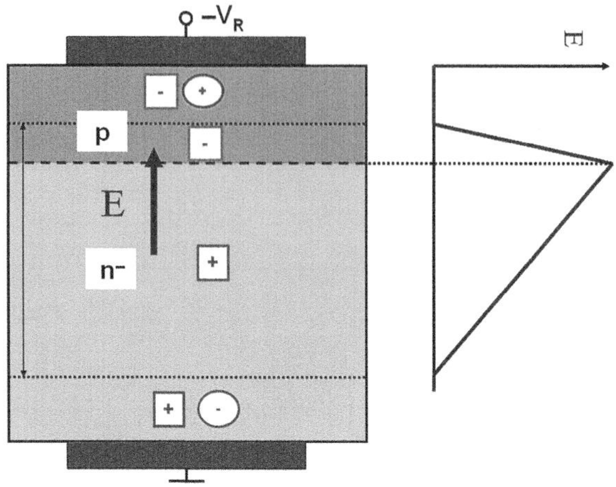

Fig. 25.9 One-dimensional pn-junction

Historically, this thick 60–120 μm n-layer was built by epitaxy on a Si substrate wafer (substrate doping with n+ in case of an FWD or p+ in case of an IGBT), but this technological approach has severe disadvantages in terms of costs and process complexity.

Another approach, by Siemens/Infineon, was to use, directly, a suitable doped silicon substrate wafer as the required n-layer [7–9]. In this case no expensive epitaxial processes were necessary. But on the other hand one must be able to handle and process very thin wafers of ~60 μm for a 600 V IGBT and FWD or 120 μm for 1,200 V in the wafer fab.

Usually all high temperature processes (diffusion, oxidation, deposition) are done on very stable thick wafers (500–700 μm) in a conventional, mechanical way, as are all ion implantation and photo lithography steps necessary to form the front-side power device structure. Then, at the process end, the wafers are ground down to the required minimal thickness to ensure, on the one hand the necessary blocking capability, and on the other hand the lowest possible on-state and switching losses [10]. Because IGBT and FWD are bipolar devices with an excess carrier concentration of electrons and holes (a factor of 10–100 higher than base doping concentration!) giving them a desired low on-state voltage, both on-state as well as switching losses are directly dependent on the total amount of stored carriers in the on state condition. This means the thinner the Si, the lower are both on-state losses and switching losses [11, 12].

Thus, it is evident that the key for modern high efficiency, low loss bipolar power devices is competent, available ultra-thin wafer technology.

Furthermore, the trend is to have as few process steps as necessary in the thin wafer condition to ensure high production yield at the lowest device thickness that is physically needed for achieving the required blocking voltage (on a long-term view 20 μm for 400 V devices, 40 μm for 600 V, 80 μm for 1,200 V and so on) at the lowest achievable power losses (Fig. 25.10).

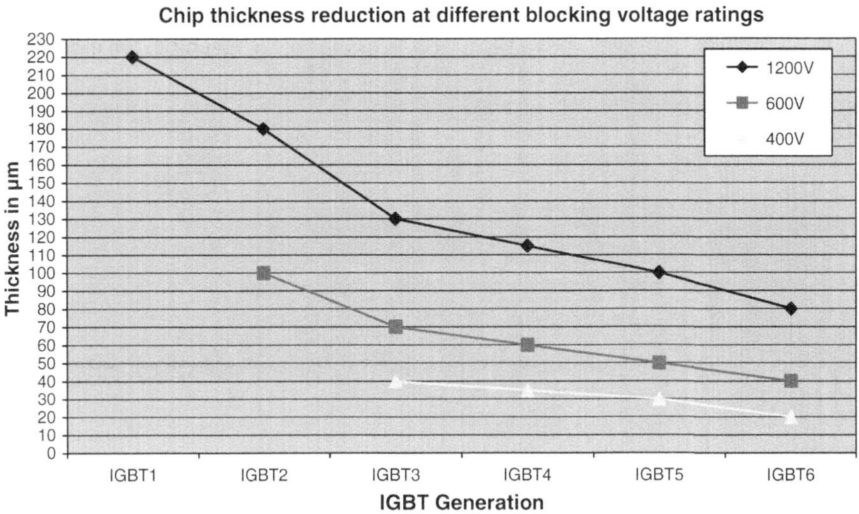

Fig. 25.10 Trend in IGBT chip thickness reduction

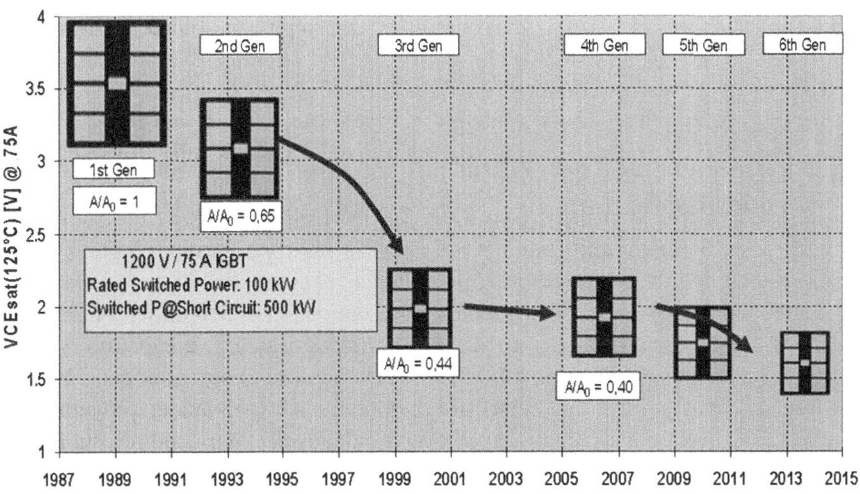

Fig. 25.11 On state voltage and chip size reduction, example of a Infineon 1,200 V IGBT

As a result it was possible up until now – and will be possible in the future – to continuously increase the output power of inverters by using shrunken silicon chips in smaller IGBT module packages or silicon chips and IGBT power modules with higher current capability [13] (Figs. 25.11 and 25.12).

Important enablers for today's as well as future ultra-thin wafer bipolar power technology are, in particular, low temperature, backside-doping techniques, e.g., by high energy, low-doping implantations that do not need diffusion or low energy,

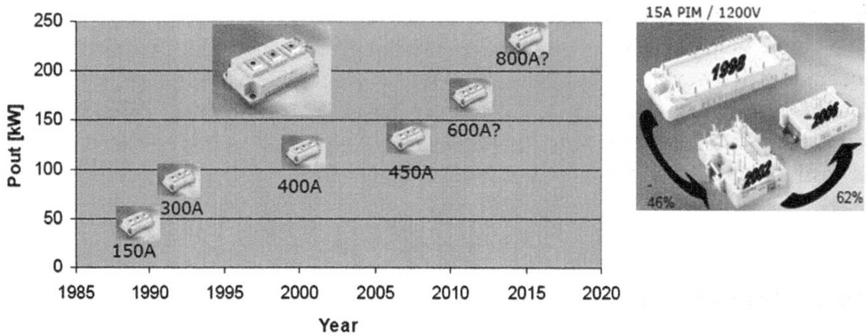

Fig. 25.12 Possible power output increase of an inverter (example of three 1,200 V Infineon half-bridge modules) or use of smaller module packages for same power (example of Infineon 1,200 V, 15-A Six Pack Power Integrated Module)

high-doping implantations with a subsequent backside laser annealing (melting of the Si backside while maintaining the wafer front side in a cold condition) [14].

To sum up: Ultra-thin Si wafer technology with all its tricks is the key to leading edge production of cost efficient, high performance power bipolar devices. This technology helps ensure increasingly power efficient electric drives, which becomes essential for worldwide energy conservation and reduction of CO_2 emissions.

25.2 Influence of Thin Wafers on Product Definition and Application

Semiconductor devices are designed to fulfill a required function, as defined by customer requirements or future trends. The semiconductor product definition is bound to the targeted functionality and on technical capabilities and feasibilities. Another limiting fact is the size of semiconductor devices. The device size impacts the required PCB (print circuit board) area, which is directly linked to the cost of the application – from consumer to automotive applications. Thin wafer technologies allow area requirements and costs to go down, impacting the product definition as well.

25.2.1 Thermal and Electrical Aspects

The most common manufacturing PCB processes within the automotive area use FR4 PCBs today. The semiconductor devices are soldered directly onto the PCB. The fully equipped and tested PCBs – known as modules – are put into housings consisting of plastic or metal. Metal housings are used in applications, where either

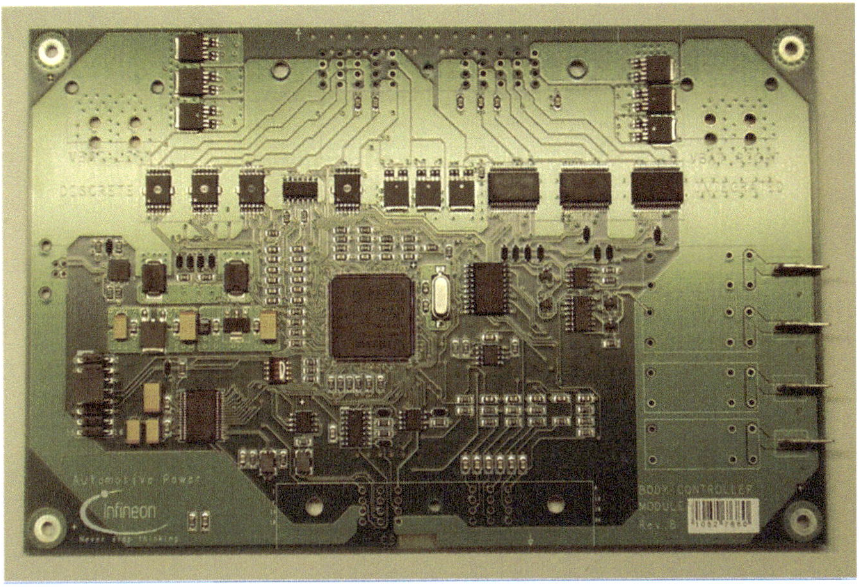

Fig. 25.13 Typical automotive module realized with Infineon semiconductor devices

the mechanical stability is mandatory, or thermal conductivity to the surrounding is needed. Automotive modules have to withstand harsh conditions for example, humidity or wide temperature ranges from −40°C to 125°C or higher. Furthermore, modules should be placed in the car's smallest areas. Usually the maximum module size and surrounding conditions are defined by the OEM (Original Equipment Manufacturer). According to this, the TIER1 designs the module to fit into the given size and to fulfil the functional requirements. In many cases it is challenging for the TIER1 to realize the full functionality within the given area because of two reasons: On the one hand, the semiconductors and the connections between them require a certain area on the PCB. On the other hand, all the power loss and heat generated inside the modules has to be transported to the outside of the module. Figure 25.13 shows an example, of how a typical module for automotive applications could look like realized with Infineon Smart Power Devices and Microcontroller.

For more complex applications, PCBs could be even more crowded, leading to strong area restrictions. As a result, semiconductor devices using less area are a clear benefit for automotive applications. The same is valid as well for consumer electronics. Unfortunately, there are some limitations for the minimum size of semiconductor devices. One limitation is the technical capability to manufacture extremely small structures within the silicon and mount the silicon into the package. Another limitation is the thermal heat transportation. Every semiconductor device has a certain power loss during its operation. This power loss results in a heating of the device. Considering a semiconductor device acting as a pure switch (e.g., Infineon ® PROFET devices) the power loss is defined as $P_{\text{Loss}} = \text{I}_{\text{Load}}^{2} {*} R_{\text{ON}}$,

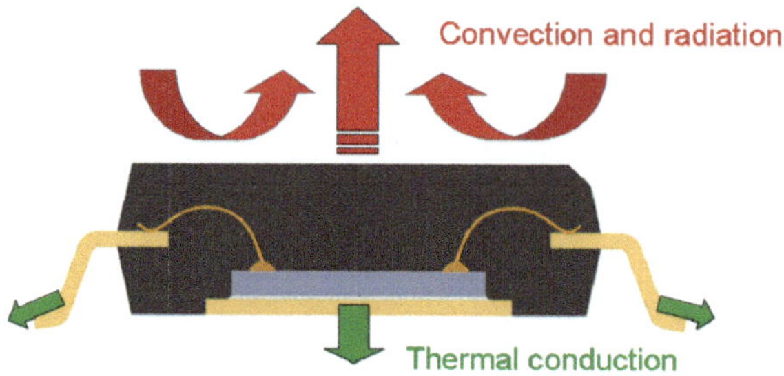

Fig. 25.14 Possible heat flows of a semiconductor device

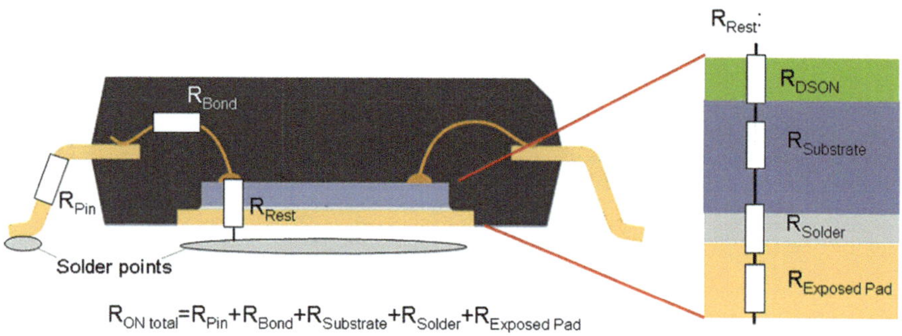

Fig. 25.15 Sum of resistances for R_{ON} of a semiconductor switch

with I_{Load} representing the load current and R_{ON} representing the electrical resistance of the semiconductor device. Switching losses are not considered in this example. The resulting heat needs to be transported from the silicon to the outside of the device. There are two ways of heat transportation: Convection, radiation and thermal conduction. Figure 25.14 shows the possible heat flow of a semiconductor device.

As a result of the generated heat, the heat is transported via convection, radiation and thermal conduction to the outside of the semiconductor device. This means the heat is conducted to the PCB and radiated or conducted to the surrounding air within the module. The major part of heat flow is the thermal conduction for semiconductor devices. Therefore, also the module has to transport the heat to the outside, which could mean for example the car's chassis or the air around the module. As far as the module's maximum dimensions are defined by the OEM the total power loss within the module has to be defined in an appropriate way. The total power loss is the sum of all devices within the module as $P_{Loss\ total} = \Sigma P_{Loss\ n}$. It has to be so low that for maximum ambient temperatures of the module, no overheating or degradation within the module occurs.

This leads to three key parameters of semiconductor devices within the automotive area:

- Low power loss (which means low ohmic resistance R_{ON} for the example of a semiconductor switch)
- Low area requirement
- Good thermal conductivity, i.e., low thermal resistance.

Unfortunately, low power loss, low area requirement and good thermal conductivity are contradicting. Low R_{ON} can be achieved mainly by using silicon area. A good thermal conductivity can be achieved mainly by big thermal conjunctions. Therefore, the benefits of thin wafer technologies improve the total situation a lot (Fig. 25.15).

Similar to the electrical resistance, overall thermal resistance is also reduced through thin wafer technologies. By using thin wafer technologies it is possible to achieve similar electrical and thermal behaviours as is possible with standard wafer technologies, doing so, however, with bigger silicon areas and packages.

25.2.2 Alternative Packaging Methods Enabled by Thin Wafer Technologies

On the one hand, thin wafers allow improvements on technical parameters of semiconductor devices. On the other hand, thin wafers also allow new manufacturing processes.

Infineon is well-known for the capability of chip-on-chip packaging for automotive semiconductor devices, where a silicon chip (top chip) is mounted onto another silicon chip (base chip). This is used, for example, for Infineon ® High Current PROFET devices. Figure 25.16 shows an example device using the chip-on-chip process. Another possibility to achieve the same functionality is the chip-by-chip

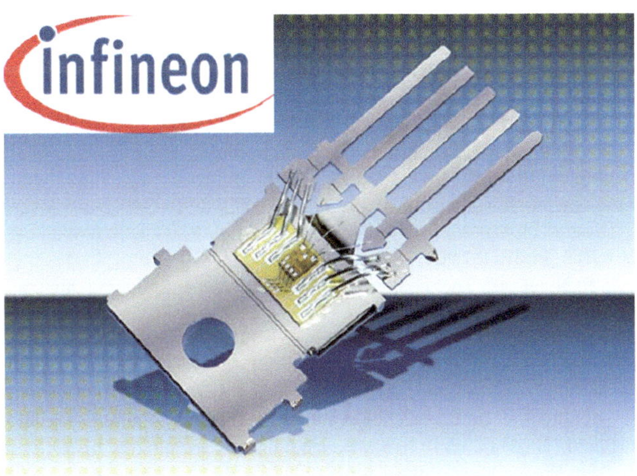

Fig. 25.16 Infineon® high current PROFET using chip on chip process

process, where the top chip is placed next to the base chip. A clear disadvantage of the chip-by-chip process is the increased package area.

In case of thin wafer technologies the total device height can be reduced for chip on chip packages, reducing the material costs. The amount of mould compound and bond wire length can be reduced.

Not only the automotive area benefits of thin wafers. As far as thin wafers are more flexible than regular wafers, a usage within other packaging technologies is possible, which are used in the consumer market. One example is chip cards, where certain flexibility is required and the chip is fully embedded in the card. Another packaging technology is the embedding of chips directly into PCBs, which is used today for mobile phones. The costs of packages can be avoided. The next step is to embed the chips into flexible foils, where the traces are also flexible. Well known applications for this are notebooks. The interface between the screen and the motherboard can be realized with these flexible foils.

Thin wafer technologies allow semiconductors to have increased efficiency and new packaging processes. As a result they impact product definition regarding the feature set, parameters and package design. The application benefits are the reduced area requirements at the same performance and new packaging processes to offer.

References

1. Siemieniec R, Hirler F, Schlögl A, Rösch M, Soufi-Amlashi N, Ropohl J, Hiller U (2006) A new and rugged 100 V power MOSFET. In: Proceedings of the 12th international EPE-PEMC 2006, Portoroz, Slovenia, pp 32–37
2. Siemieniec R, Häberlen O, Sánchez J (2009) Leistungs-MOSFETs im Niederspannungsbereich – Wege zu mehr Effizienz. Design & Elektronik 9:10–11
3. Stengl JP, Tihanyi J (1992) Leistungs-MOS-FET-Praxis. Pflaum Verlag, München, Germany
4. Görgens L, Siemieniec R, Sánchez J (2006) MOSFET technology as a key for high power density converters. In: Proceedings of the 12th EPE-PEMC 2006, Portoroz, Slovenia
5. Baliga BJ (1987) Modern power devices. Wiley, New York
6. Lutz J (2006) Halbleiter-Leistungsbauelemente. Springer, Berlin
7. Miller G, Sack J (1989) A new concept for a non punch through IGBT with MOSFET like switching characteristics. In: 20th annual IEEE PESC' 89 Record, Milwaukee, WI, pp 21–25
8. Laska T, Miller G (1990) A 2000 V non-punch-through-IGBT with dynamical properties like a 1000 V IGBT. In: IEDM 90, Abstract 32.6, pp. 807–810
9. Mauder A et al (2000) The field stop IGBT concept and Emcon high efficiency diode. In: Proceedings of the PCIM USA
10. Burns D et al (1996) NPT-IGBT-optimizing for manufacturability. In : Proceedings of the 8th ISPSD, Maui, HI, pp 331–334
11. Laska T et al (2000) The field stop IGBT (FS IGBT) – a new power device concept with a great improvement potential. In: Proceedings of the 12th ISPSD, Toulouse , France, pp 355–358
12. Rüthing H et al (2003) 600 V-IGBT3: trench field stop technology in 70 μm ultra-thin wafer technology. In: Proceedings of the 15th ISPSD, Cambridge, pp 66–69

13. Bayerer R (2008) Higher junction temperature in power modules – a demand from hybrid cars, a potential for the next step increase in power density for various variable speed drives. In: Proceedings of the PCIM Europe 2008, Nurnberg, Germany
14. Gutt T et al (2006) Laser thermal annealing for power field effect transistor by using deep melt activation. In : Proceedings of the 14th IEEE international conference on advanced thermal processing of semiconductors (RTP), Kyoto, pp 193–197

Chapter 26
Thinned Backside-Illuminated (BSI) Imagers

Koen De Munck, Kyriaki Minoglou, and Piet De Moor

Abstract Backside-illuminated (BSI) imagers are becoming popular due to their enhanced light sensitivity. During their fabrication advanced Si wafer thinning is used in combination with backside surface passivation techniques. In this section both technology for thin (hybrid) backside-illuminated imagers as well as the trade-offs they present between quantum efficiency and crosstalk are discussed and illustrated, with the use of imec imager results.

26.1 Introduction

26.1.1 Detecting Light

Today's digital cameras make use of silicon-based image sensors to convert incident light into an electrical signal. Roughly speaking, this conversion process consists of three stages. A first stage considers the coupling of incident light into the silicon substrate. In a second stage, the light reaching the substrate needs to be absorbed before it is able to travel through the substrate and is coupled out again. 'Absorption' in this context means the conversion of light photons into electron-hole pairs. Finally, the last stage considers the separation of generated electron-hole pairs using a suitable electric field and their collection in the correct imaging pixel before they recombine. This last stage is not as obvious as it may seem because in typical (not fully depleted) image sensors, the carriers diffuse in an isotropic way, i.e. not necessarily to the nearest collecting photodiode.

The efficiency by which these detection processes occur give rise to a few important imager performance metrics. Quantum efficiency (QE) is, e.g. the efficiency by which incident light photons are converted into collected charges.

P. De Moor (✉)
Interuniversity Microelectronics Centre (IMEC), Kapeldreef 75, 3001 Leuven, Belgium
e-mail: demoor@imec.be

J.N. Burghartz (ed.), *Ultra-thin Chip Technology and Applications*,
DOI 10.1007/978-1-4419-7276-7_26, © Springer Science+Business Media, LLC 2011

Crosstalk (XT) is the fraction of generated charges collected in the incorrect pixels during exposure in the centre of one pixel.

26.1.2 Limitations of Frontside Illumination

Traditionally, imagers are illuminated from the frontside of the device through the chip's backend, irrespective of whether they are CIS (CMOS image sensor) or CCD (charge-coupled device) chips. Disadvantages to this approach are twofold: Of the incident light, a part gets reflected on the metals in the backend. In other words the fill-factor (FF) is limited and therefore decreases QE. In addition, part of the light is reflected (and/or absorbed) on the backend dielectrics, as these are not optimised to limit reflections in terms of refractive index and thickness. Thus, backend dielectrics are not an optimal antireflective coating (ARC). In the end, without special measures the overall QE is less than 50% over a limited spectral range. A typical solution, implemented to overcome the QE loss in frontside illumination, is the use of microlenses, possibly in combination with light pipes, to guide the light around the metal interconnects. This goes a long way toward recovering QE loss (up to 80%), at least at specific wavelengths. Broadband response is still difficult as the microlens material is again not optimal in terms of ARC properties.

Frontside-illuminated CCD imagers have for a long time served the imager market. However, dedicated design effort and small process changes to the traditional CMOS electronics have enabled CMOS imagers. Although they have not as good optical performance (compared to CCDs), they have a number of advantages, namely, their low power consumption and the possibility they permit of incorporating smart electronics on the imager chip. Therefore, CMOS imagers are slowly replacing CCDs in the market. Another trend in consumer imager applications is ever improving resolution, i.e. the number of pixels recorded is increasing very fast (currently, up to several of megapixels in mobile phones). As, however, (for cost reasons) the total imager size has remained more or less equal, this resolution enhancement results in a smaller pixel size – down to about 1μm in 2010. As the number of photons per unit of area is more or less constant, it is obvious that sensitivity per pixel needs to increase for good performance. A natural way to overcome this limitation is illumination of the imager from the backside.

26.1.3 Backside Illumination

By illuminating the imager chip from the backside, i.e. directly to the Si, the imager chip's backend is moved out of the path of the incoming light. This solution thus offers maximal light coupling to the substrate as the fill factor is then 100%, and an optimised ARC can be used matched to the refractive index of the substrate. For

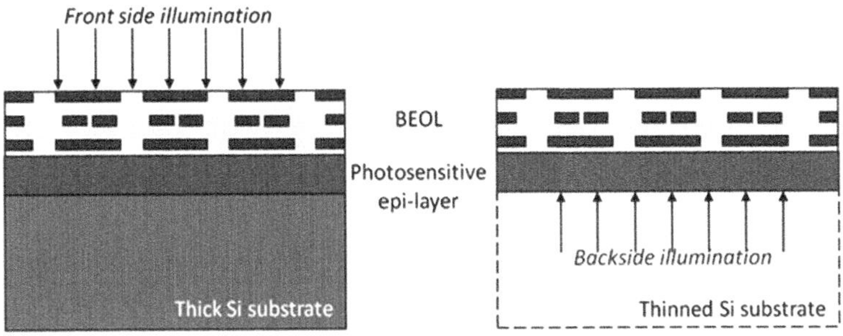

Fig. 26.1 Frontside- (*left*) and backside- (*right*) illuminated imager

backside illumination, the bulk of the non-sensitive bulk of the substrate faces the incoming light and so it needs to be removed. The substrate must be thinned to typical residual thicknesses of 30 μm or less. This requires wafer-thinning techniques with accurate thickness control and low amounts of subsurface damage. Moreover, the exposed backside will need to be treated appropriately in order to keep generated charges away from the defect-rich surface – where they recombine and are lost for collection in the photodiode. In spite of the additional cost related to these extra process steps, backside-illuminated imagers have begun to penetrate the imager market. For high-end imaging applications where optical requirements are important, both backside-illuminated CCDs and CMOS imagers are being used. However, they are also being put to use in consumer applications; thanks to their increased sensitivity, backside-illuminated CMOS cameras are becoming standard today (2010) (Fig. 26.1).

26.1.4 Relation Between QE, Crosstalk and Substrate Thickness

Optimal detection layer thickness of an imager is a trade-off between crosstalk and QE in the long wavelength range, and strongly depends on pixel size.

In typical frontside-illuminated imager applications, the sensitive layer over which absorption and collection occur is only a few microns thick. Charges from deeper absorbed light recombine in the low-resistive and defect-rich substrate long before they reach the collecting diode junctions. And even if those carriers did reach the sensitive layers, they would have first experienced significant lateral diffusion, resulting in large XT. For most applications, a few-micron substrate thickness is enough. On the other hand, as the absorption length of visible light in Si increases with the wavelength, light with a wavelength above 500–600 nm will pass through a few micron of Si without being absorbed. Therefore, the sensitivity of an imager with only a few micron of detection layer will sharply decrease, while in principle silicon can detect light of wavelengths up to ~1,100–1,200 nm.

An imager with a thicker detection layer thus has better performance, as long as charge recombination can be avoided and electrical crosstalk can be kept low. Charge recombination can be limited through the use of a high resistive detection layer. The purity of such layer is higher, which means there are less charge recombination centres. The crosstalk can be kept low by guiding the generated charges to the correct pixel either by generating a suitable electric field distribution or by somehow physically separating individual pixels.

26.1.5 Imager Types and Roadmap

There are two distinct types of BSI imager: monolithic and hybrid (Fig. 26.2a, b, respectively). In monolithic imagers both the light-sensitive detector array and the Read-Out Integrated Circuits (ROIC) are made in the same substrate and are thinned from the back. Several technology options are available for connecting the imager to the outside world. One such example is shown in Fig. 26.2, the option of using through-silicon cavities.

In hybrid imagers the sensitive detector array and the ROIC are made in a different substrate, allowing for both technologies to be optimised individually. The detector array is subsequently hybridised onto the ROIC using a pixel-wise bump connection. In this case, only the detector array needs to be thinned.

Each option has its own advantages and disadvantages. The hybrid imager has as its main advantage the freedom it offers for an independent design and optimisation of both the photo-detection layer and of the ROIC layer. For example, a different CMOS technology generation or an alternative substrate material can be used. For the photo-detection layer, which in the first instance can be merely a diode, this offers increased flexibility for the implementation of special features like pixel-separating trenches for crosstalk reduction. In addition, by stacking two layers on top of each other, the effective area of the pixel increases, giving more room for additional functionality.

While most current imaging needs have been covered by the monolithic imager technologies, a considerable performance improvement is still possible if we

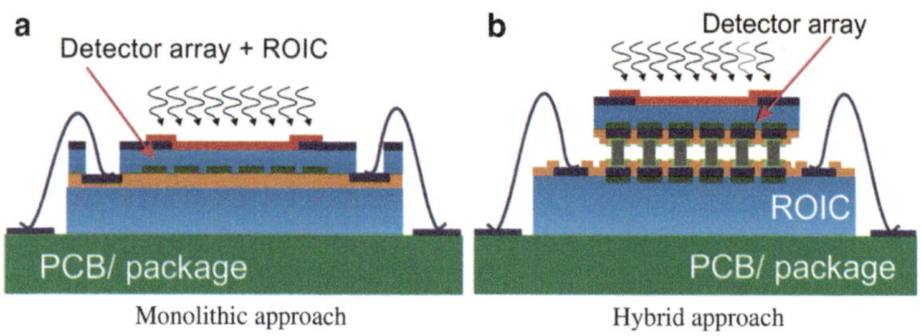

Fig. 26.2 Two ways of making backside-illuminated CMOS imager

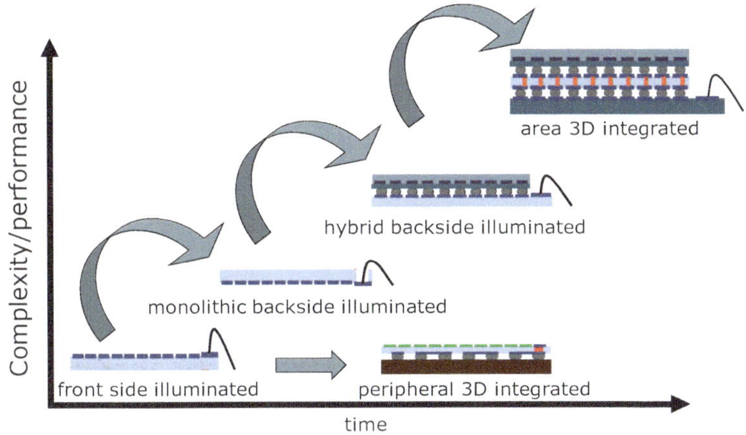

Fig. 26.3 Imager integration roadmap

expand from the traditional planar (single layer) imager architecture into the third dimension, i.e. using multiple active substrates in one imager system. In fact, the newest technologies such as through-silicon vias (TSVs) and high density bumping enable stacking of more than two layers. This new vertical interconnection technology has two types of conceptual application: peripheral and area 3D integrated imagers (see Fig. 26.3).

Peripheral 3D integration can be considered an advanced packaging technology. TSVs are used to vertically interconnect the bond pads of the imager die to the underlying printed circuit board. The main advantages of this approach are a highly miniaturised packaging and lower interconnect capacitance (as compared to the classical wire bonding), resulting in lower power consumption and/or high speed operation.

Area 3D integration uses TSVs at a higher density, e.g. per pixel or group of pixels. This enables a new 'vertical' pixel readout architecture: One can combine different CMOS (or other) technologies to realise a parallel pixel read-out chain. As a result, a wide variety of complex and performant read-out architectures will become possible.

26.2 Specific Hybrid BSI CMOS Imager Technologies

This section briefly describes several key technology steps used to realise (hybrid) backside-illuminated CMOS imagers. The ROIC is realised in a standard CMOS technology, produced at a foundry and only needs bump formation before it is ready for hybridisation. The detector array consists of customised backside-thinned diodes, enhanced to achieve the best possible performance.

Key process technology steps include substrate engineering, trench formation, wafer thinning, thin wafer handling, backside surface treatment, hybrid integration and antireflective coating.

26.2.1 Substrate Engineering

A very important parameter for the realisation of BSI imagers is the choice of detection layer. As mentioned earlier, the properties of the substrate (defect density and substrate thickness after thinning) have a direct impact on some of the most important imager performance metrics like QE and crosstalk. Three basic options can be discerned: EPI growth, high resistivity wafers and SOI substrates.

SOI substrates are interesting for several reasons. First of all, the buried oxide can be used as a stopping layer for the thinning process. And as the thickness and thickness variation of the top silicon layer can be very well controlled, response nonuniformities like interference fringes can be kept to a minimum. Second, the eventual BSI imager backside can be conditioned such that backside surface treatment can be avoided. Nevertheless, the flexibility in choice of SOI substrate is less than what could be obtained with custom EPI growth, especially in terms of dopant profile.

High resistivity substrates can be considered in case a large collection volume (in depth) is required. Most likely this will be in combination with advanced ROIC schemes like CTIA and backside biasing to achieve deep depletion and sufficient crosstalk-reducing electric fields. However, high resistivity substrates are not as easy to process. Moreover, depending on the required resistivity they are not commonly available commercially.

Finally, thick high quality EPI layers can be grown on standard silicon substrates. Tuning the growth conditions can let us obtain special EPI doping profiles. The exact doping concentration of the substrate is not important since afterwards the substrate itself will be thinned away from the back into the EPI layer. Using EPI layers with graded doping concentrations, one can obtain a built-in electric field which pulls generated carriers from the substrate to the diode at the surface, where they are collected in the photodiode. An example of a 50-μm thick graded EPI layer is shown in Fig. 26.4.

26.2.2 Trench Formation

Pixel-separating trench formation is an alternative solution to keeping XT low in case substrate engineering or full substrate depletion is not feasible. The concept is to etch deep and narrow trenches around each active pixel in the array and fill them with highly doped poly-silicon (Fig. 26.5b). By doing so, one creates a built-in electric field at the sides of the trench, repelling generated carriers to the bulk of the

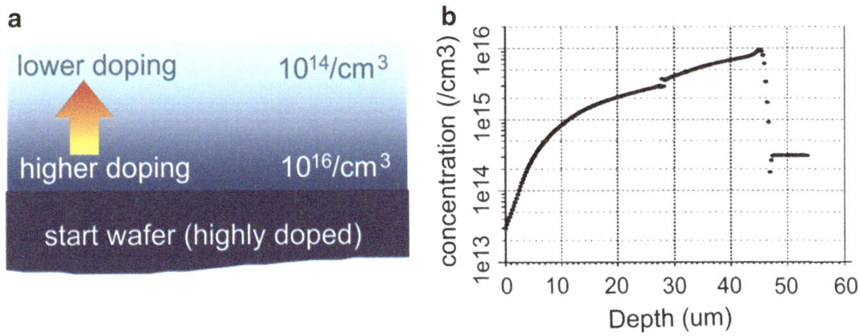

Continuous EPI profile

SRP depth profile of 50 μm continuous EPI

Fig. 26.4 Schematic representation of (**a**) a continuous EPI profile with (**b**) a measured scanning resistance probe (SRP) profile

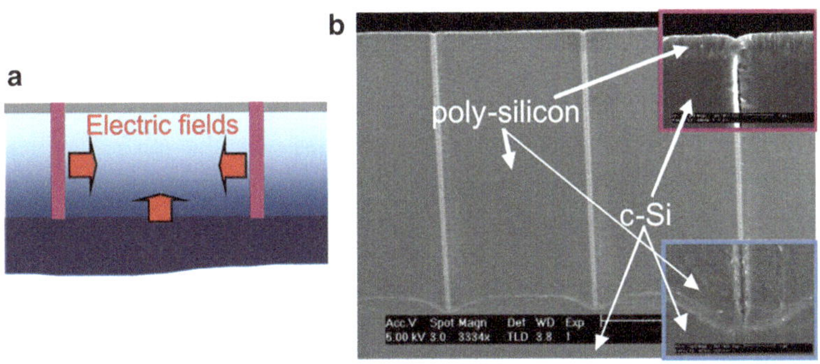

Fig. 26.5 Highly boron doped poly-silicon fill of the isolation trenches. (**a**) Schematic cross-section and (**b**) SEM pictures from a cleaved processed wafer

pixel, effectively creating a wall between the pixels and eliminating electrical crosstalk, though at the expense of fill factor.

26.2.3 Wafer Thinning and Handling

Though previous chapters discussed wafer thinning, backside-illuminated imagers have some specific issues in this respect that should be mentioned. As the entire remaining substrate is used for detection, it needs to be completely damage-free because damage to the Si crystal results in recombination centres and therefore in diminished collection efficiency. The applied thinning procedure should thus not leave damage present at the backside. Grinding, however, induces damage, as it is a mechanical, abrasive process. Although grinding damage with the latest technology

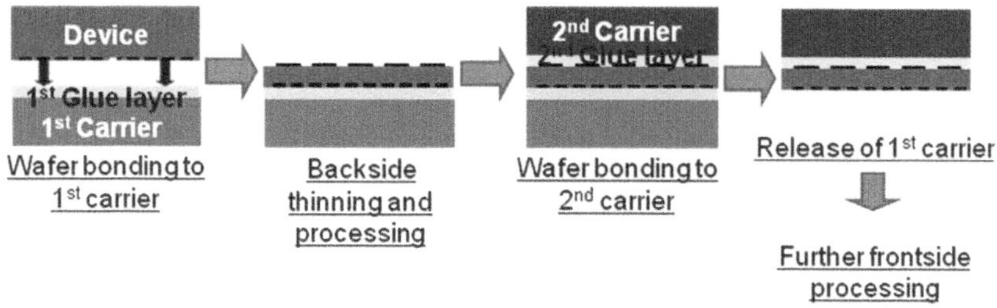

Fig. 26.6 Thin wafer flip process

can be at a submicron level, some additional damage removal is yet required. Therefore, grinding is typically stopped a few microns before the final thickness is reached and the remaining silicon is removed with wet etching, while any grinding damage also is removed.

Thinning to final thickness below 30 microns (and the optional further processing of the thinned wafer) requires the use of handling/support/carrier wafers. Both silicon and CTE- (coefficient of thermal expansion-) matched glass carrier can be used. Glass has the benefit of being transparent, which enables backside alignment for possible additional backside processing. On the other hand, infrared alignment allows silicon carriers to be used, as Si is transparent for infrared light, depending on thickness, substrate roughness and wavelength.

For monolithic BSI imagers, a permanent bonding to a Si substrate is typically used. For hybrid BSI imagers, we have to process both sides of a thinned wafer, and therefore a dedicated process has been developed, involving transfer between 2 carriers. After the backside thinning and processing, the thin device wafers are bonded to a secondary carrier. Then the first carrier can be removed, thereby exposing the frontside again. This ‘thin wafer flip’ allows processing to resume on the frontside, e.g. for bump formation (Fig. 26.6).

26.2.4 Backside Surface Treatment

In order to reduce the negative effect of backside surface traps on the spectral response of the detector diodes, a backside surface field is created. Typically, it consists of a shallow blanket boron implantation followed by laser annealing to activate the dopants (Fig. 26.7). Laser annealing is required as opposed to an oven anneal because of the limited temperature budget of the imager (CMOS) layer. Since the highly doped surface region (the so called ‘dead layer’) is not light-sensitive (as the generated carriers recombine very fast), then the shallower the (activated) dopant profile, the better. This is, in particular, important for UV detection, as the UV light has a very small penetration depth in Si.

26.2.5 Anti-reflective Coating

As mentioned earlier, contrary to frontside-illuminated imagers, antireflection coatings for backside illumination can be optimally matched to silicon to achieve maximum response for the wavelength range of interest. For example, use of a two-layer stack, composed of 56-nm ZnS and 110-nm MgF_2, lets us obtain a reflectivity over the entire optical range of less than 3%. As a result, the QE of an imager can be made larger than 80% – between 400 and 850 nm (depending on the silicon device thickness) (see Fig. 26.8). Without an ARC, a QE of maximally ~60% can be achieved.

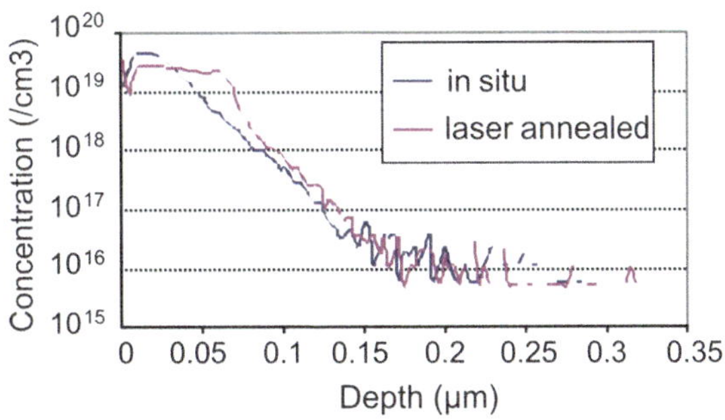

Fig. 26.7 SIMS measurements after shallow boron implantation and after laser annealing

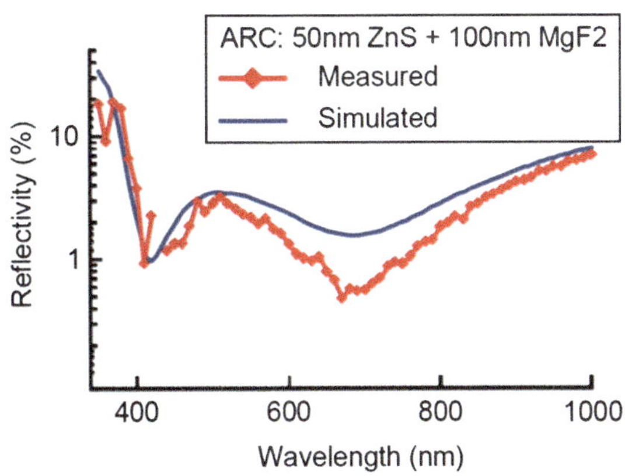

Fig. 26.8 Optimised ARC properties. Global reflective light loss is below 3% over the extended optical range of interest (400–850 nm)

26.2.6 Hybrid Integration

Both on the detector array and the ROIC, bumps need to be formed in order for interconnections to be created during flip chip. Typically, for indium bumps a lift-off process is used (Fig. 26.9). Alternatively, for SnCu bumps electroplating is used.

The major benefit of an indium bump process is ductility, which allows the take-up of CTE mismatch between different substrates or substrates at different temperature. As a result, indium can even be used at cryogenic temperatures.

Following bump formation, both the ROIC dies and detector arrays are ready for flip chip hybridisation using thermo compression bonding. This process is optimised to ensure good bump interconnect yield. Figure 26.10 shows an example of a flip chipped hybrid detector array.

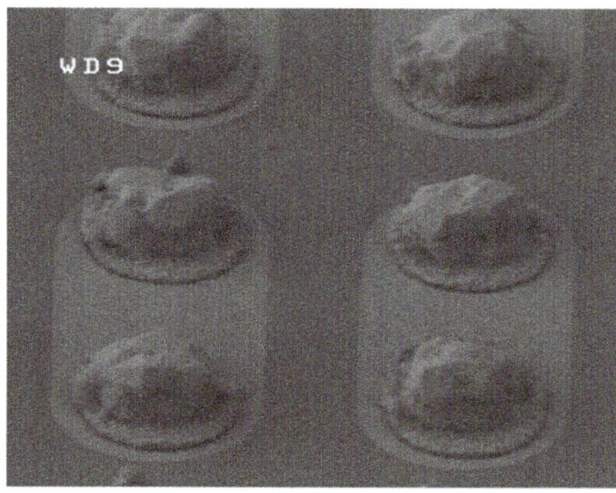

Fig. 26.9 Indium bumps with a pitch of 20 μm

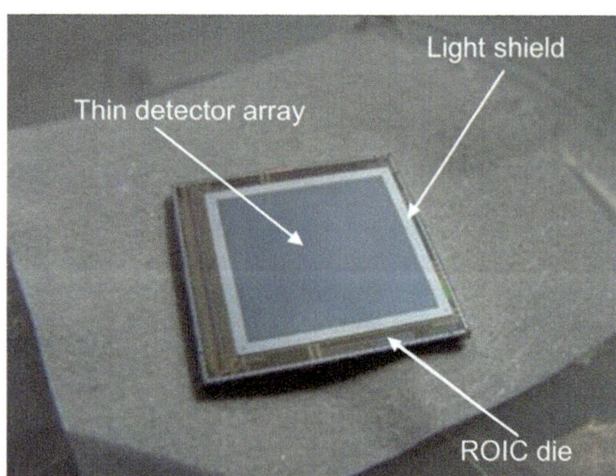

Fig. 26.10 Flip chipped 512 × 512 hybrid detector array on its ROIC

26.3 Results of IMEC-Thinned BSI Imagers

In this section, results typical of the most critical specifications from the characterisation of our thinned backside-illuminated imagers are presented.

26.3.1 Functionality and Yield

Fully packaged backside-illuminated prototypes of 512 × 512 and 1,024 × 1,024 array size have been tested. Initially, their functionality was checked. Especially for the case of hybrid imagers, functionality is directly related to the hybridisation yield. A pixel yield of 99.93% on a finished hybridised 1 × 1 k imager was obtained [3] (Fig. 26.11).

26.3.2 Dark Current

Dark current was initially measured on large test diodes and afterwards on the final BSI imagers. Depicted in Fig. 26.12. are the diode characteristics (absolute value) of thinned and nonthinned test diodes, measured both in (frontside) illumination and in dark. Dark current was found to be 300 pA/cm^2 at 25°C (9,500 e–/pixel·s) at 2-V bias voltage, independent of thinning. Also, an exponential decrease with decreasing temperature was observed, of which the slope (×2/10 K) points towards a mechanism of thermal generation in the depletion layer. The shift in the zero crossing is related to the diode measurement configuration and is dominant for nonthinned devices.

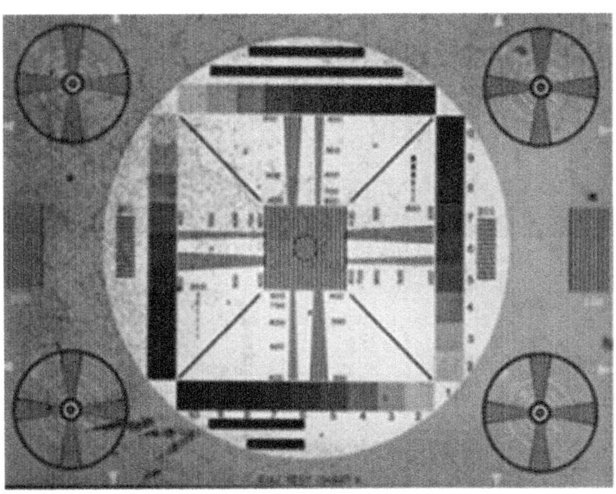

Fig. 26.11 Hybridised 1 × 1 k imager raw test image

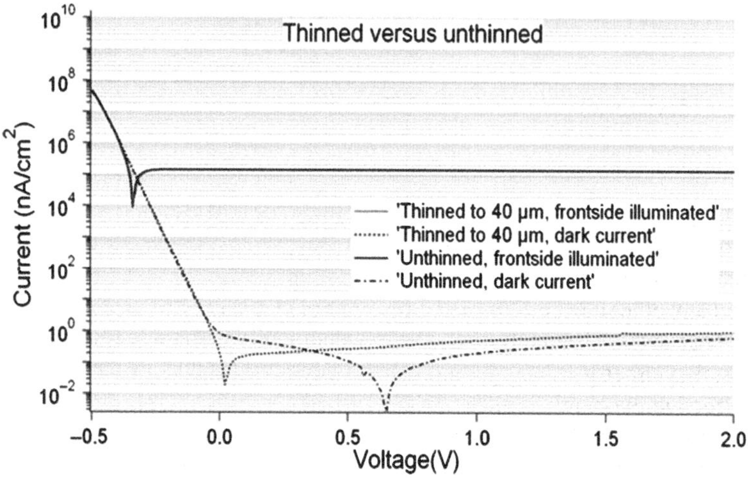

Fig. 26.12 Measured dependence of dark current with thinning at different bias

26.3.3 Quantum Efficiency

Shallow backside implantation and subsequent laser annealing are performed on the devices. The aim is to minimise the thickness of the dead backside layer, thus improving the QE, especially in the blue. Additionally, an antireflective coating, optimised for the broadband region, was deposited.

Quantum efficiency is typically measured with a spectral response bench. This consists of an illuminator with a Xe arc lamp, a 1/8 monochromator with motorised filter wheel and a manually moving stage (rail carrier) for adjusting the optical plane of the test imager and the calibrated reference. It is also possible to include an integrating sphere at the exit of the monochromator. The system illuminator-monochromator is optimised so that the illuminator output is focused and matched to the monochromator. The system is mounted on an optical table top that provides flatness and static rigidity. The illuminator coupling to the monochromator is fixed by mounting of both devices on a common base plate. This mounting kit includes the lightshield to enclose the beam path and which can be also used to mount other necessary hardware. In Fig. 26.13 the spectral response bench is pictured. Most of the key imager parameters like conversion gain, linearity, QE and photo response nonuniformity (PRNU) can be measured when we use this bench [1, 2].

Using the spectral responsivity test bench (uniform illumination with 5-nm wavelength resolution) we characterised both monolithic and hybrid devices. In Fig. 26.14 the expected results (from simulation) and the measurements are presented. The result (black line) is in agreement with a simulated device with ARC deposited and 15-nm dead layer thickness (green line). Also, the devices possess excellent broadband QE, over 80% from 400 to 800 nm. Similar results are obtained both from monolithic and hybrid devices. Our values are cited as

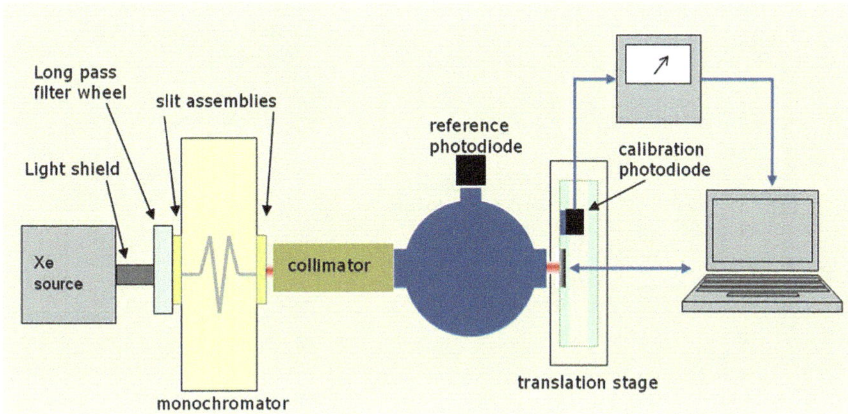

Fig. 26.13 Schematic representation of the spectral bench

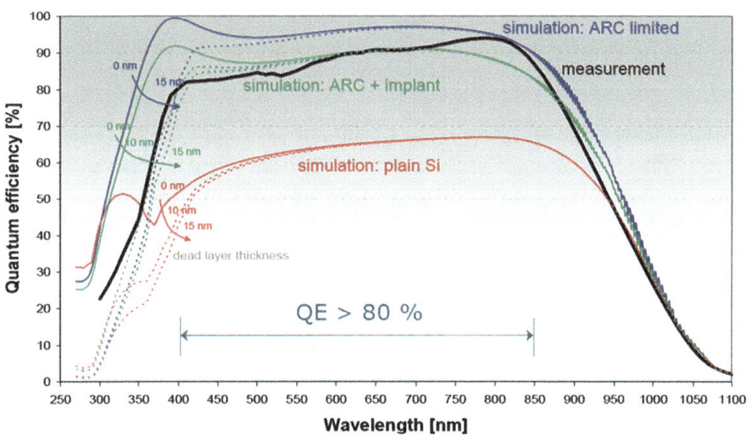

Fig. 26.14 Simulated and measured QE. Simulated structures: only Si (*red line*), Si with ARC (*blue line*), Si with ARC and implant (green line), all with various dead layer thickness

comparison with frontside-illuminated CMOS imagers [4], where the advantages of thinned backside-illuminated devices are evident (Fig. 26.15).

26.3.4 Crosstalk

The schematic of the crosstalk measurement using the single spot illumination technique is presented in Fig. 26.16. Stimulation of a single pixel of the detector is achieved using a HeNe laser source, providing a laser spot smaller than the pixel. The smaller the average spot size over the absorbed depth, the more accurate the

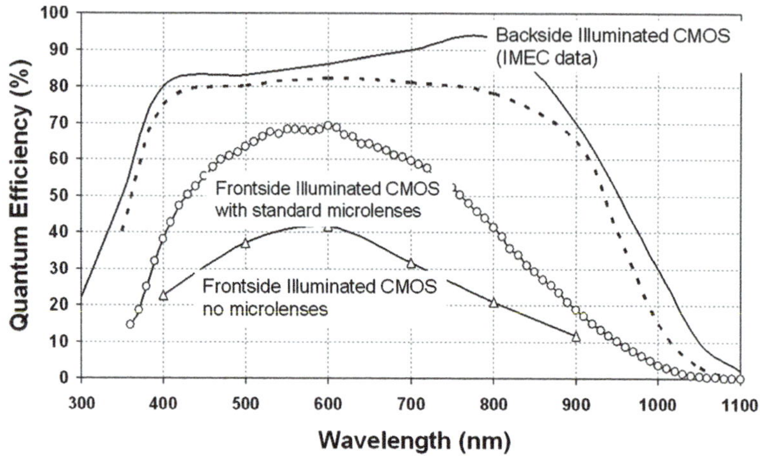

Fig. 26.15 Comparison of backside illuminated CMOS imagers (from imec) to frontside [4]

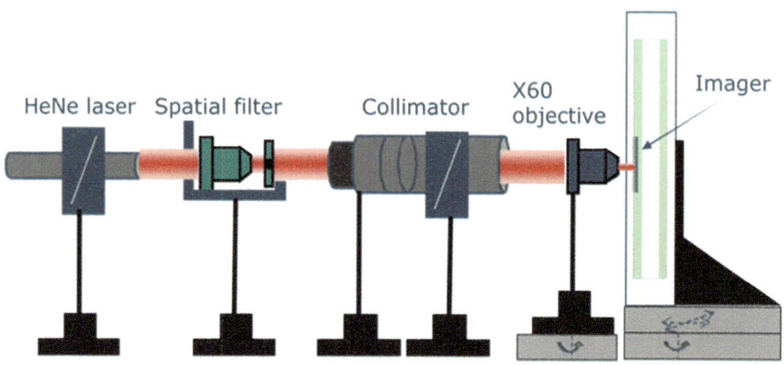

Fig. 26.16 Schematic representation of the crosstalk bench

measurement. It guarantees that surrounding pixels are not illumined. For that reason a focusing assembly, to adjust the spot size of the free-space laser beam, and an XYZ translation stage, for mounting the imager, are included in the setup. Effects of vibration are minimised by an optical table.

The crosstalk of different thinned BSI imagers is measured. We compare devices with and without pixel separating trenches, as explained in the previous section. Both devices have been fabricated using a graded epilayer structure, to partially control the diffusion mechanism. Figure 26.17. presents the signal per pixel of both trenched and nontrenched devices, acquired by single pixel illumination technique. Crosstalk is defined in this case by the ratio of the output signal of the neighbouring pixels to that of the centre pixel. As is clearly depicted, for the single pixel illumination of the trenched device, all the incident signal is collected by the central

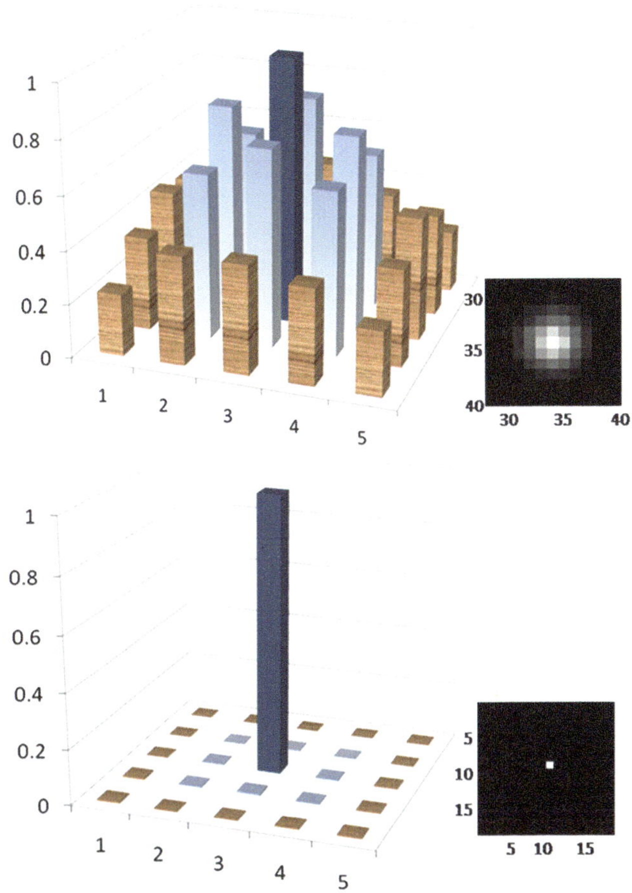

Fig. 26.17 Normalised pixel output comparison under spot illumination of nontrenched (*up*) and trenched (*down*) thinned BSI CMOS imagers

pixel. Quantitative measurements from the specific pixel illustrated in Fig. 26.17. give a maximum crosstalk value of 8.2×10^{-3} and average crosstalk of 5.2×10^{-3} and 3.9×10^{-3} from the 4 and 8 neighbouring pixels, respectively [5]. For the nontrenched, however, a significant portion of signal is collected by neighbouring pixels. The maximum crosstalk for the nontrenched device is measured to be 6.5×10^{-1} and average crosstalk is 6.2×10^{-1} and 5.0×10^{-1} from the 4 and 8 neighbouring pixels.

The obvious effect of the trenches on the crosstalk behaviour is a dramatic reduction by at least two orders of magnitude. For the trenched device, it should also be noted that the small signal in the neighbouring pixels is uniform in magnitude across the array. This leads to the conclusion that this signal is not crosstalk but mainly dark signal.

References

1. Hopkinson GR, Goodman TM, Prince SR (2004) A guide to the use and calibration of detector array equipment. SPIE Optical Engineering Press, Bellingham
2. ESA/SCC basic specification no. 25000 (July 1993) Electro-optical test methods for charge coupled devices, issue 1
3. De Munck K et al. (2006) High performance hybrid and monolithic backside thinned CMOS imagers realised using a new integration process. In: Proceedings of the IEEE international electron devices meeting, San Francisco, 2006, pp 139–142
4. Bai Y et al. (2008) Teledyne Imaging Sensors: Silicon CMOS imaging technologies for x-ray, UV, visible, and near infrared. Proc SPIE 7021:702102. doi:10.1117/12.792316 (2008)
5. Minoglou K et al. (2008) Reduction of electrical crosstalk in hybrid backside illuminated CMOS imagers using deep trench isolation. In: Proceedings of the IEEE international interconnect technology conference, San Francisco, June 2008

Chapter 27
Thin Solar Cells

Michael Reuter

Abstract This chapter outlines the advantages of thin crystalline silicon solar cells, gives an overview on the current status in research and development on solar cells with thicknesses below 50 μm and describes the current developments. The material utilisation of different wafer fabrication methods is discussed, and an overview on applications for thin and hence flexible solar cells is given.

27.1 Why Thin Solar Cells?

Thin monocrystalline silicon solar cells have four main advantages compared to thick solar cells:

1. *Less material consumption*: A thin crystalline silicon solar cell consumes less material than a thick solar cell, but absorbs less impinging irradiation, too. This drawback is addressed by introducing an advanced light trapping; therefore, a thin cell with light trapping generates a similar short circuit current density J_{SC} as a thick cell without light trapping [1].
2. *Less vulnerable to low bulk carrier lifetime*: A rule of thumb states that the bulk diffusion length L_{bulk} has to be on the order of the cell thickness W. Therefore, a thin cell tolerates a lower bulk diffusion length, and thus a lower material quality [2].
3. *Higher power conversion efficiency*: Thin cells allow for higher open circuit voltages V_{OC} for equal material quality due to less entropy generation [3] if the short circuit current density is kept constant.
4. *Flexible*: Thin monocrystalline silicon is flexible upon reaching thicknesses below $W = 100$ μm, which allows for new areas of applications.

M. Reuter (✉)
Institut für Physikalische Elektronik, Universität Stuttgart, Pfaffenwaldring 47, Stuttgart D-70569, Germany
e-mail: michael.reuter@ipe.uni-stuttgart.de

J.N. Burghartz (ed.), *Ultra-thin Chip Technology and Applications*,
DOI 10.1007/978-1-4419-7276-7_27,

In the following we will address the four main advantages of thin solar cells one by one, starting with the introduction of the term ‘silicon utility’. Silicon utility serves as a measure of the optimum power output of a solar cell weighted with the consumption of silicon feedstock material during solar cell production.

27.1.1 Less Material Consumption

One major argument for thin solar cells is always mentioned in before all others: reduced material consumption. The task of this section is to answer whether reduced material consumption affects the overall energy yield under air mass 1.5 global spectra (AM1.5 g) irradiation. If we consider the generated power per gram of silicon we see an increased power yield per gram. Thus, if we reduce the material thickness, then power generation per gram of silicon increases with decreasing thickness, as thickness decreases faster than efficiency, which was demonstrated in Chap. 24 (reuter and eisele). The assumption holds if we consider only the material present in the solar cells, but there is a considerable amount of silicon feedstock that is lost during silicon wafer fabrication.

In order to distinguish wafering technologies with respect to wafer thickness reduction, this section investigates silicon usage per generated watt of power output for today’s standard wafering by wire saw slicing from an ingot, direct wafering by string ribbon technology (SR) [4] and thin crystalline silicon by layer transfer technology. In this last, the cutting of thin crystalline silicon wafers from an ingot suffers from kerf loss per wafer of a minimum $w_{\mathrm{kerf}} \approx 120\ \mu\mathrm{m}$ [5], where ‘kerf loss’ describes the amount of material lost during cutting. Direct wafer growth by string ribbon or edge-defined film-fed growth (EFG) technology prevails with a maximum silicon output-to-input rate of up to 90%. The layer transfer process shows a major drawback as it relies on the epitaxial growth on a substrate wafer. I assume a substrate wafer thickness of 550 μm and a transfer repetition of 20 times. A kerf loss of 120 μm per substrate wafer is included, too. While all materials start with some sort of purified crystalline silicon, I do not account for losses during the transformation of trichlorosilane (HCl_3Si), i.e. during epitaxial growth or during polysilicon deposition.

Figure 27.1 shows the thickness-dependent generated power per gram silicon under AM1.5 g irradiation, called ‘silicon utility’, for three different solar cell materials: standard silicon wafers from ingots by wafer slicing, string ribbon material and thin single crystalline silicon from the layer transfer process. The silicon utility, first introduced by Tobail [6], is shown in Fig. 27.1 for two cases of light trapping as discussed in Chap. 24: (a) no light trapping with a light path length $l = W$ and (b) optimum light trapping according to Yablonovitch [7] with a path length $l = n^2W$, where n is refractive index of silicon and W is wafer thickness. Power output simulation accounts for radiative and nonradiative recombination resulting in a material-dependent minority carrier lifetime $\tau = 200\ \mu\mathrm{s}$. While the silicon utility of standard wafer material is limited due to kerf loss, silicon ribbon material shows an

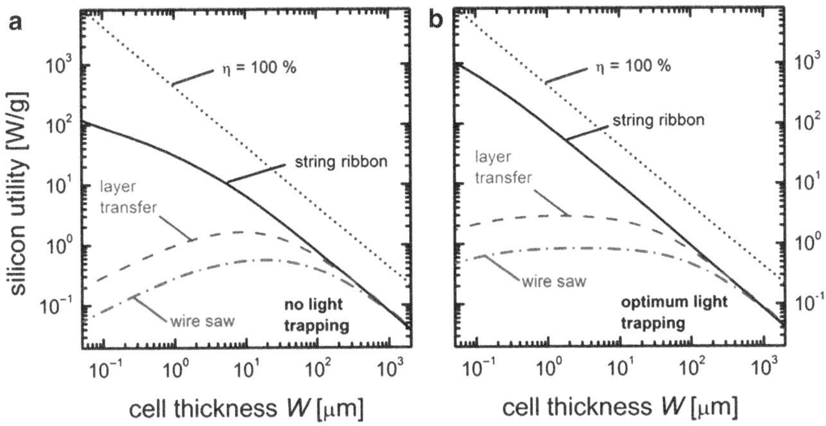

Fig. 27.1 Silicon utility for silicon wafers with thickness W and (**a**) no light trapping and (**b**) optimum light trapping for three wafering processes: Conventional wafer slicing, string ribbon growth and layer transfer process. Efficiency simulation accounts for radiative and nonradiative recombination resulting in a bulk minority carrier lifetime $\tau = 200$ μs

increased material usage for thinner wafers due to negligible material loss during wafer production. Silicon usage of the layer transfer process largely depends on the recycling rate and the thickness of the substrate wafer. For clarity, the overall insolation impinging on the surface covered by a silicon wafer with a weight of one gram is given, which corresponds to a conversion efficiency $\eta = 100\%$.

Solar cells fabricated on thin film crystalline silicon from the layer transfer process show a reasonable advantage compared to wafer slicing in terms of material usage. SR technology allows for even higher silicon utility, but results in multicrystalline wafers due to direct growth from the melted silicon. Multicrystalline silicon wafers show a reduced efficiency potential compared to monocrystalline silicon due to increased recombination in bulk material impurities and at grain boundaries. Grain boundaries result in higher vulnerability to mechanical stress. Lower efficiency potential and mechanical weakness outweigh the advantage in silicon utility.

27.1.2 Less Vulnerable to Low Bulk Carrier Lifetime

Thin solar cells are less sensitive to low bulk carrier lifetime. Highly efficient solar cells require a diffusion length in the bulk material of several times their material thickness. Thus, decreasing thickness lowers the requirements on bulk material quality. However, surface recombination velocity gains in importance. Effective diffusion length [8]

$$L_{\text{eff}} = L_{\text{bulk}} \frac{D + SL_{\text{bulk}} \tan h(W/L_{\text{bulk}})}{SL_{\text{bulk}} + D \tan h(W/L_{\text{bulk}})} \tag{27.1}$$

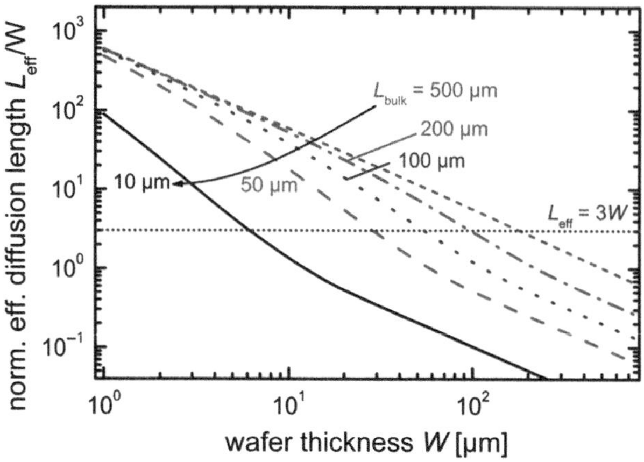

Fig. 27.2 Effective diffusion length L_{eff} normalised to the wafer thickness W calculated for varying wafer thickness W and different bulk diffusion length L_{bulk} = 10–500 µm. The simulation accounts for a bulk doping density $N_A = 7.2 \times 10^{15}$ cm^{-3} and a rear surface recombination velocity S = 500 cm/s. While thinner wafers tolerate a lower bulk diffusion length, the rear surface recombination limits effective diffusion length towards ultra-thin wafers

defines the electronic quality of the solar cell depending on bulk diffusion length L_{bulk}, diffusivity of minority charge carriers D and rear surface recombination velocity S. Especially, the collection of photo-generated charge carriers depends on L_{eff}, hence on the quantum efficiency, as discussed in detail in Chap. 24.

Figure 27.2 shows effective diffusion length normalised to thickness L_{eff}/W for varying wafer thickness and different material qualities characterised by bulk diffusion lengths L_{bulk} = 10–500 µm. Highly efficient solar cells require an effective diffusion length exceeding wafer thickness by a least a factor of three. Highly effective diffusion length $L_{eff} > W$ requires a high bulk diffusion length $L_{bulk} = (\tau_{bulk}D)^{1/2}$ [9, 10], and therefore a high bulk lifetime τ_{bulk}. In the numerical simulation of Fig. 27.2, rear surface recombination velocity S = 500 cm/s and bulk doping density $N_A = 7.2 \times 10^{15}$ cm^{-3} are assumed. Figure 27.2 demonstrates that thinner solar cells allow for lower bulk diffusion length, and therefore lower material quality compared to thick cells. The value for $L_{eff} = 3\ W$ is marked as a visual guide.

27.1.3 Higher Power Conversion Efficiency

Thin solar cells feature a higher efficiency potential compared to thick solar cells. The main reason for higher efficiencies lies in the higher open circuit potential due to a higher carrier concentration. As mentioned before, a thin cell absorbs an equal amount of impinging irradiation due to improved light trapping compared to a thick solar cell. The photo-generated carriers are distributed into a smaller volume

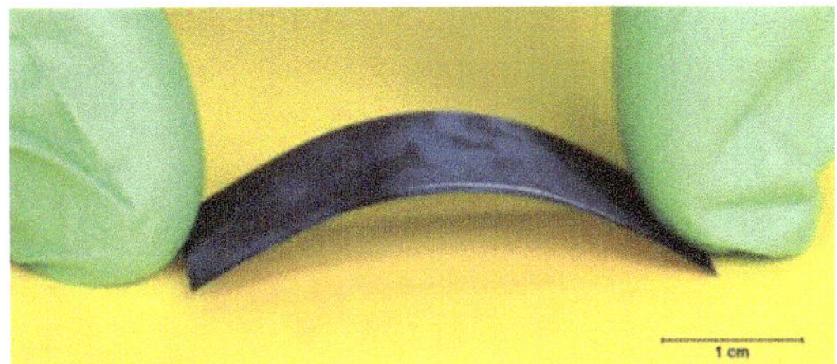

Fig. 27.3 Fifty micrometers thin, flexible monocrystalline silicon solar cell fabricated at Institut für Physikalische Elektronik, University of Stuttgart, *IPE*

resulting in a higher carrier concentration. The higher carrier concentration allows for reduced entropy generation per photon, thus allowing for a higher open circuit voltage [3]. Chapter 24 discusses the thickness dependence of the power conversion efficiency of crystalline silicon solar cells in general.

27.1.4 Flexible Solar Cells

Yield in solar cell production is of major interest for the PV market. A reduced yield might counteract an efficiency improvement caused by, e.g. thinner solar cells. While the force needed to break silicon wafers decreases linearly with thickness [11], their flexibility increases quadratically. Therefore, solar cell thicknesses in the range 100 μm $< W <$ 150 μm probably yield the highest fragility to mechanical stress, whereas even thinner wafers gain in stability due to the higher flexibility.

Figure 27.3 shows a bent monocrystalline silicon solar cell with a thickness $W = 50$ μm. This solar cell is much more flexible compared to a thick cell.

Considering the advantages of thin solar cells discussed above, we now present an overview of the current state of development of thin crystalline silicon solar cells. The overview focuses on the highest reported efficiencies rather than a complete summary.

27.2 Overview on Thin Crystalline Silicon Solar Cells

Thin crystalline silicon solar cells allow for higher conversion efficiencies compared to thick solar cells for similar material quality. The higher efficiency arises from an increased open circuit voltage in thin cells if the loss in the short circuit current density J_{SC} is diminished, e.g. by means of improved light trapping. Up to now, the highest efficiency for thin crystalline silicon solar cells is reported by Wang et al. [12] for a 47-μm thin single crystalline silicon solar cell with an efficiency $\eta = 21.5\%$. The group started with a thick wafer and reduced the thickness of their record solar cell by thinning the wafer from the backside. Wafer thinning is not viable for industrial production, but proves the efficiency potential of thin single crystalline silicon solar cells. In the following, the current state of thin crystalline silicon solar cells with thicknesses below 50 μm is discussed, starting with the highest efficient solar cells on small areas $A < 2\ cm^2$, followed by large area devices, and by back contact, back junction devices.

27.2.1 Record Small Area Solar Cells

This chapter presents the current world record for thin crystalline silicon solar cells without wafer thinning. The solar cells are fabricated on thin films prepared by the layer transfer process invented and implemented at the Institut für Physikalische Elektronik, University of Stuttgart (*IPE*) [13, 14]. The layer transfer process is also known as PSI-process [1, 15] and bases on the electrochemical anodisation of silicon in hydrofluoric acid and subsequent epitaxial silicon deposition from the gas phase. The production of thin single crystalline films by the layer transfer process is described in detail in Chap. 8. Monocrystalline silicon solar cells manufactured by the transfer process at Institut für Physikalische Elektronik, University of Stuttgart, *IPE* with thicknesses below 50 μm reach efficiencies as high as $\eta = 16.9\%$, attached to a glass superstrate [16] and $\eta = 17.0\%$ for freestanding solar cells [17].

Thin monocrystalline silicon solar cells are fabricated from thin monocrystalline silicon films by the layer transfer process. The solar cells presented in this chapter receive a front design that is capable of open circuit voltages $V_{OC} > 680$ mV and efficiencies $\eta > 21\%$ on monocrystalline FZ wafers at *IPE* [18]. Device fabrication starts with the layer system still attached to the wafer. The solar cell frontside is textured with random pyramids [19]. A thermal silicon oxide passivation and anti-reflection layer is formed on a lowly n-type doped emitter with a sheet resistance $\rho = 100\ \Omega$/sq. Photolithography defines a front contact with high efficiency features, consisting of an evaporated Ti/Pd/Ag layer stack. Up to this step, all solar cells receive equal processing. Now the thin films are separated from the substrate wafer. In the conventional transfer process, a glass superstrate is attached by an epoxy resin and mechanical force detaches the thin film. In the freestanding transfer process a vacuum pick-and-place tool [20] detaches the thin films without using a superstrate. The rear side of the device is now accessible for back contact formation. The stabilized solar cell yields superior handling properties due to the glass superstrate, which induces mechanical stability, but suffers from parasitic absorption in the glass and epoxy resin in the blue wavelength regime. The freestanding solar cell features an increased mechanical flexibility but is less robust, which induces a limitation to the processing of the rear contact. Thus, the free-standing solar cell obtains an easy, single step rear contact consisting of evaporated aluminium, which limits the performance due to rear surface recombination. The stabilised cell receives a high performance rear contact as proposed by Brendle [16]: The highly p-type doped silicon layer, which is a residue from the layer transfer process and results in high recombination, is removed by wet chemical etching. Then, a layer stack consisting of hydrogenated amorphous silicon for surface passivation and amorphous silicon nitride as transparent dielectric is deposited. An evaporated aluminium rear contact increases the reflection while laser-fired contacts [21] form the electrical contact to the base material. The superior performance of the sophisticated rear contact is proved by an extended analysis of the quantum efficiency, as shown in Chap. 24.

Both solar cells feature efficiencies $\eta \approx 17\%$ on an area $A = 2\ \text{cm}^2$, and thus are the maximum reported efficiencies for monocrystalline silicon solar cells with thicknesses below 50 μm, which are processed without the help of wafer thinning. The power per weight ratio R_{TL} for the freestanding solar cell is $R_{\text{TL}} = 1.46$ W/g, five times that of $R = 0.36$ W/g, in a decent performing, 17% efficient, 200-μm thick industrial monocrystalline silicon solar cell.

27.2.2 Thin Large Area Solar Cells

Thin single crystalline silicon solar cells on large area by means of the PSI-process were produced at the Institut für Solarenergieforschung (ISFH) by Brendel and coworkers. They reported a maximum efficiency $\eta = 14.1\%$ for a solar cell with an area $A = 95.5\ \text{cm}^2$ attached to a glass superstrate [22]. The major difference from the solar cells described in the previous section is that this approach avoids high costly processes such as photolithography and evaporated contacts at the price of decreased efficiency. Therefore, large area solar cells are closer to industrial production and still yield improvement potential.

27.2.3 Thin Back Junction Back Contact Solar Cells

The lateral arrangement of the pn-junction on the solar cell rear side features some advantages upon going to thinner wafers as discussed in Chap. 24. Here, Haase [23] showed a back contact back junction device on thin single crystalline silicon attached to a glass superstrate. The thin silicon substrate is produced after the PSI-process. A silicon dioxide passivation and antireflection layer covers the solar cell front surface. The thin silicon layer then attaches to a glass superstrate and now the back junction and back contacts are formed according to the process described in [23]. The main advantage lies in avoiding frontside metallisation, thus reducing reflection. Thin, back junction solar cells further allow for a better carrier collection as the collection path length for minority carriers decreases for thinner material. The front surface recombination has gained in importance, since most of the carriers are generated near the front surface but are collected at the rear surface.

Table 27.1 Reported I/V-parameters for monocrystalline silicon solar cells with thicknesses $W \leq 50\ \mu m$

	Method	η (%)	W (μm)	V_{OC} (mV)	J_{SC} (mA/cm²)	FF (%)	A (cm²)	Ref.	Year
UNSW	Wafer thinning	21.5[a]	47	698.5	37.9	81.1	4.0	[12]	1995
IPE	Layer transfer	17.0	47.6	634	36.0	74.6	2.0	[17]	2009
IPE	Layer transfer	16.9[a]	41.6	641	33.5	78.7	2.0	[16]	2007
ZAE	PSI	15.4[a]	25.5	623	32.7	75.5	3.88	[24]	2003
ISFH	PSI	14.1	25	616	29.0	78.8	95.5	[22]	2006
ISFH	PSI (back contact)	13.5	30	633	28.7	74	79.2	[23]	2009
ISE	Wafer thinning	20.7	37	677	37.5	81.6	4.0	[25]	2009

[a]Cell efficiency independently verified

27.2.4 Overview on Thin Monocrystalline Silicon Solar Cells

Table 27.1 presents an overview on reported performance for thin monocrystalline silicon solar cells with thicknesses $W \leq 50\ \mu m$. The maximum efficiency of solar cells fabricated without wafer thinning remains much lower than the 21.5% produced at University of New South Wales (UNSW), clearly pointing at an improvement potential for thin film single crystalline silicon solar cells. The simulation in Chap. 24 indicates efficiencies approaching 24% for bulk material lifetimes τ exceeding $\tau = 200\ \mu s$.

27.3 Thin Solar Cell Applications

Thin single crystalline silicon solar cells allow for new application areas, which demand that solar cells have a high power per weight ratio, e.g. in aeronautics or space application. The flexibility of thin c-Si films further allows for the application on curved surfaces. The application in consumer electronics, e.g. solar jackets [26] or mobile phones, seems possible, too, as these applications require flexible modules with a high power/weight ratio. Further applications include low weight irradiation sensors, flexible power supply on credit cards or portable outdoor power generators.

Figure 27.4 depicts an example for clothing integrated photovoltaics (CIPV). Here, amorphous silicon modules are integrated into a skiing jacket and empower consumer electronics via an USB port. The amorphous silicon modules are readily available, but show only low conversion efficiencies. Here, modules with thin, flexible mono-crystalline silicon solar cells with high power conversion efficiencies reduce the required module area for a similar power output.

Fig. 27.4 Clothing integrated photovoltaic, here carried out by flexible amorphous silicon modules [26]. Thin, highly efficient monocrystalline silicon solar cells allow for smaller module area or higher power output

References

1. Brendel R (1997) A novel process for ultrathin monocrystalline silicon solar cells on glass. In: Ossenbrink HA, Helm P, Ehrmann M (eds) Proceedings of the 14th European photovoltaic specialist conference. H.S. Stephens, Bedford, pp 1354–1357
2. Werner JH, Bergmann R, Brendel R (1994) The challenge of crystalline thin film silicon solar cells. Adv Solid State Phys 34:115–146
3. Brendel R, Queisser HJ (1993) On the thickness dependence of open circuit voltages on p-n junction solar cells. Sol Energy Mater Sol Cells 29:397–401
4. Sachs E, Ely D, Serdy J (1987) Edge stabilized ribbon (ESR) growth of silicon for low cost photovoltaics. J Cryst Growth 82:117–121
5. Möller HJ (2004) Basic mechanisms and models of multi-wire sawing. Adv Eng Mater 6:501–513
6. Tobail O (2008) Porous silicon for thin solar cell fabrication. Shaker, Aachen, p 96
7. Yablonovitch E (1982) Statistical ray optics. J Opt Soc Am 72:899–907
8. Green MA (1982) Solar cells: operating principles, technology, and system applications. Prentice-Hall, Englewood Cliffs
9. von Smoluchowski M (1906) Zur kinetischen Theorie der brownschen Molekularbewegung und der Suspensionen. Ann Phys 326:756–780
10. Einstein A (1905) Eine neue Bestimmung der Moleküldimensionen. Ann Phys 322:549–560
11. Jimeno JC, Rodriguez V, Gutierrez R, Recart F, Bueno G, Hernando F (2000) Very low thickness monocrystalline silicon solar cells. In: Scheer H, McNelis B, Palz W, Ossenbrink HA, Helm P (eds) Proceedings of the 16th European photovoltaic specialist conference. James & James, London, pp 1408–1411
12. Wang A, Zhao J, Wenham SR, Green MA (1996) 21.5% efficient thin silicon solar cell. Prog Photovoltaics Res Appl 4:55–58
13. Rinke TJ, Bergmann RB, Werner JH (1999) Quasi-monocrystalline silicon for thin-film devices. Appl Phys A 68:705–707

14. Bergmann RB, Berge C, Rinke TJ, Schmidt J, Werner JH (2002) Advances in monocrystalline Si thin film solar cells by layer transfer. Sol Energy Mater Sol Cells 743:213
15. Tayanaka H, Yamauchi K, Matssushita T (1998) Thin-film crystalline silicon solar cells obtained by separation of a porous silicon sacrificial layer. In: Schmid J, Ossenbrink HA, Helm P, Ehrmann H, Dunlop ED (eds) Proceedings of the second world conference on photovoltaic solar energy. IEEE, Piscataway, pp 1272–1277
16. Brendle W (2007) Niedertemperaturrückseitenprozess für hocheffiziente Siliziumsolarzellen. Shaker Verlag, Aachen
17. Reuter M, Brendle W, Tobail O, Werner JH (2009) 50 μm thin solar cells with 17.0% efficiency. Sol Energy Mater Sol Cells 93:704–706
18. Rostan PJ, Rau U, Nguyen VX, Kirchartz T, Schubert MB, Werner JH (2006) Low-temperature a-Si: H/ZnO/Al back contacts for high-efficiency silicon solar cells. Sol Energy Mater Sol Cells 90:1345–1352
19. King DL, Buck ME (1991) Experimental optimization of an anisotropic etching process for random texturization of silicon solar cells. In: Proceedings of the 22nd IEEE photovoltaic specialists conference. IEEE, Piscataway, pp 303–308
20. Werner JH (2006) Final report of project 0329818 A. Technical Report, BMU
21. Schneiderlöchner E, Preu R, Lüdemann R, Glunz SW (2002) Laser-fired rear contacts for crystalline silicon solar cells. Prog Photovoltaics Res Appl 10:29–34
22. Terheiden B, Horbelt R, Brendel R (2006) Thin-film solar cell and modules from the porous silicon processing using 6″ Si-substrates. In: Poortmans J, Ossenbrink H, Dunlop E, Helm P (eds) Proceedings of the 21st European photovoltaic solar energy conference. WIP-Renewable Energies, Munich, pp 742–745
23. Haase F, Horbelt R, Terheiden B, Plagwitz H, Brendel R (2009) Back contact monocrystalline thin-film silicon solar cells from the porous silicon process. In: Proceedings of the 34th IEEE photovoltaic specialists conference. IEEE, Piscataway, pp 244–246
24. Feldrapp K, Horbelt R, Auer R, Brendel R (2003) Thin-film (25.5 μm) solar cells from layer transfer using porous silicon with 32.7 mA/cm^2 short-circuit current density. Prog Photovoltaics Res Appl 11:105–112
25. Kray D, McIntosh KR (2009) Analysis of ultrathin high-efficiency silicon solar cells. Phys Status Solidi A 206:1647–1654
26. Schubert MB, Werner JH (2006) Flexible solar cells for clothing. Mater Today 9:42–50

Chapter 28
Bendable Electronics for Retinal Implants

Heinz-Gerd Graf

Abstract This chapter focuses on sub-retinal chip implants, a direct replacement technique of malfunctioning photoreceptors in the retina of the human eye. The physical space available requires the application of a thin chip with thickness of 30–50 μm. These chips are unpackaged and thus need to be passivated to warrant long-term stability and reliability. Both passive and active chip design concepts are discussed.

28.1 Introduction

28.1.1 Background

Television and press publications have highlighted the first successful treatments of vision-impaired persons through the use of electronic visual aids. The success of such aids makes it possible for the vision-impaired to see outlines, improve orientation and recognise light sources, shapes, letters and words in controlled test environments [1, 2]. Such experiences are sensational for people who lost the sense of sight completely, even many years previously, due to advanced degenerative illnesses of the retina such as retinitis pigmentosa or macular degeneration. The thousands of afflicted people – 17,000 cases per year in Germany alone – gain hope from such information.

Thus far, scientific publications mainly describe the results of electrical stimulation of the retinal tissue from in vitro and in vivo testing as well as medical results on the course of the illness, surgical techniques and results from case studies on animal and human testing. Also known are the follow-up technological approaches of various research teams in the United States, Europe, Asia and Australia.

H.-G. Graf
Institut für Mikroelektronik Stuttgart (IMS CHIPS), Stuttgart, Germany
e-mail: graf@ims-chips.de

J.N. Burghartz (ed.), *Ultra-thin Chip Technology and Applications*,
DOI 10.1007/978-1-4419-7276-7_28,

However, there have been relatively few publications on the technical challenges and impressive technological accomplishments that had to be solved in the later course of the technology's development, after implants had been placed. Lacking also in the literature are reports on the challenges to be met before these implants can be marketed in the long term.

28.1.2 A Look Back

Initial attempts to create technical visual aids by electrical stimulation of the retina go back to Tassiker [3] in 1956. In that time, technical capabilities for miniaturising semiconducting and stimulating electrodes did not exist. There is documentation, however, describing in detail electrically generated (evoked) stimulation of light impressions (phosphenes) in the retina through the use of a photo receptor current. Brindley [4] and Dobelle [5] researched generation of sight impressions in their work, which was effect by electrical stimulation of the visual cortex using arrays of surface electrodes [6] as early as the late 1960s. Then, in the 1990s, progress achieved in micro technology, as well as global interest in neuro technologies and neuro implants, resulted in the creation of several consortia and the detailing of different approaches used to stimulate the retina or optical nerve using microelectronic implants. There are currently close to 30 consortia or companies using different approaches to creating neuroelectronic visual aids.

28.2 Implant Concepts and Techniques

28.2.1 Basic Implant Concepts

Different approaches (Fig. 28.1) exist as to best anatomical location for stimulation electrodes and thus initiation of the cortical processing of the visual information. The process uses densely arranged photo receptors on the rear end of the retina (rods and cones) through the retinal processing levels (horizontal and bipolar cells) to the spiking ganglion cells on the front side of the retina. The nerve fibres attached to the ganglion cells concentrate themselves in the blind spot to the optical nerve that transmits the visual information of the eye to the visual cortex. Therefore, one differentiates between sub-retinal stimulation (done directly at the degenerated receptor), epi-retinal stimulation (done directly at the ganglion cells) [7–9] and surface nerve fibre and optical nerve stimulation, which uses a spiral electrode called the Cuff electrode, attached to the optical nerve outside the eyeball [10]. Transscleral implants lie on the outside of the surrounding sclera of the eyeball, whose aim is to generate sub-retinal stimulation of the retina through the sclera, using needle-type electrodes [11]. This approach, which is similar to optical nerve stimulation, allows a clinician to avoid extensive surgical procedures within the eyeball.

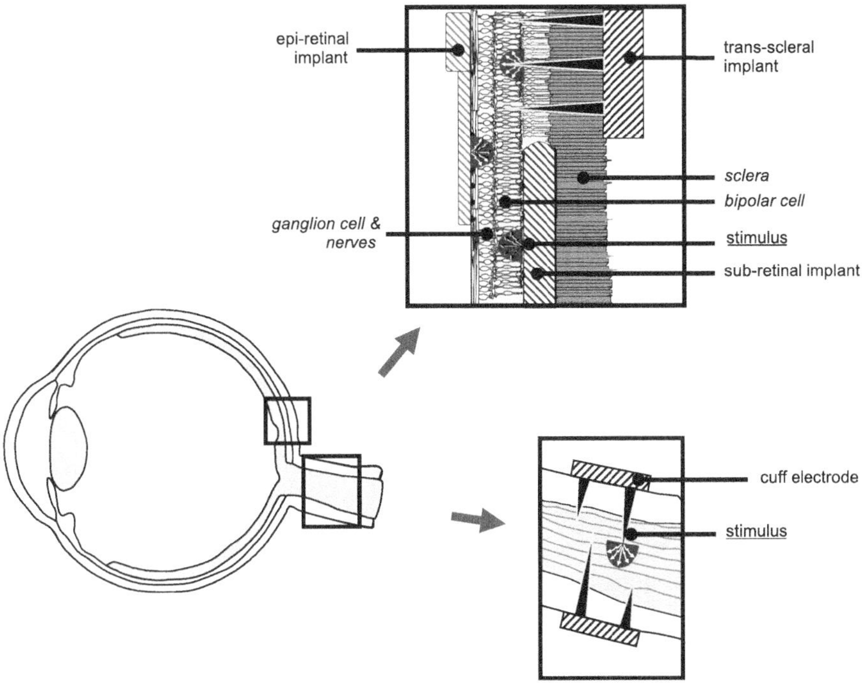

Fig. 28.1 Position and functionality of the various eye implants. Epi-retinal implants irritate the ganglion cell and nerve fibre. Sub-retinal and transscleral implants affect the bipolar cells of the retina. During optical nerve stimulation the stimulating electrodes affect the fibres of the nerve cord

Information derived from in vitro and in vivo tests (12–14) conducted so far makes it apparent that, with charge transfer on conductive electrodes within the range of 1–40 nC, electrically stimulated short light occurrences can be generated during sub-retinal and epi-retinal stimulation. The transmission of the charge between the electrode material and the tissue is carried out through the repolarisation of the H_2O molecules at the extremely large fractal surface of the electrode, without the voltage required for dissociation being surpassed. The signal transmission is therefore preferably capacitive, so toxic reaction products are avoided.

Thus, from a technical point of view there are several approaches to information transmission; the particular approach depends on the location inside the eye where the technique is applied. Epi-retinal and transscleral implants as well as optical nerve stimulation require processed image information by video cameras that serially transmit information via radio or optical coupling to the implant. The structure of the internal circuit therefore consists of a data/energy transmission, the stimulation signal generator (usually a biphasic amplitude-controlled electrical source) and a single or multilevel multiplexer serially forwarding the signals to the 16–200 stimulation electrodes. The cameras required for imaging are, preferably, attached to glasses enabling them to follow the movement of the head [15, 16]. Since the stimulation of the ganglion cells and the nerve cords cannot be assumed to be a location-conformal or stimulation-proportional correlation between the location of the stimulation electrode and the sensation in the visual field, extensive recoding is necessary in order for images to be illustrated through the perceived phosphenes. In the proposed concepts this task is conducted by a computer or neuro chip inserted in the data path between the camera and the transmitter. Sub-retinal implants, however, use the location-conform sensation of the remaining functioning neuronal connections of the bipolar cells in the retina. Therefore, the electronic generation of stimulation sensations is possible through a projected generated image done directly through the eye lens. Passive sub-retinal implants (Micro Photodiode Arrays or the ASR® Devices) generate the energy necessary to stimulate directly from incident optical radiant power [7, 8]. An increase in the radiation power using scanned laser, image intensifier or intensive LED display may be necessary. The more intricate sub-retinal implant here, however, amplifies the opto-electronic signal of the pixel and converts it into the required electronic signal. The energy supply and the transmission of the control signals are currently still executed transdermally via input leads. This way not only can the biggest amount of electrodes be put into use, but natural eye movement can also be incorporated into the generation of visual cognition.

Transscleral sub-retinal implants predominantly aim at reducing the necessity for surgical intervention at the eyeball or to reduce problems related to the applied material. But placement of the active electronics on the outside of the nontransparent sclera makes the use of the image incident of the eye lens no longer possible. The transmission of data therefore is conducted similar to what is done with epiretinal implants: transmitting images via external electronic cameras. The location of the electrodes determines the amount of stimulation of the sub-retinal tissue. The electrodes are pinned into the sclera [11]. Alternatively, they can be pushed underneath the retina using a foil [13].

Further designs of implants are based on chips with overgrown cells, chips to dispense neurotransmitters or on the use of photosensitive piezo-elements.

28.2.2 Electronic Design

Electronic designs mainly differentiate the implants with data transmission from sub-retinal implants with photosensitivity. The implants with data transmission (Fig. 28.2) [15, 16] usually consist of a rectifier, an RF unit, a processor, one or more digital programmable two-phase pulse generators and a multiplexer. The electronics are contained in a special compartment located in the lens space, vitreous space on the sclera or outside of the eye, depending on the workgroup. Up to 200 stimulation electrodes are mounted on flexible foil and connected to the chip via conducting paths. There are alternative solutions necessary for the growing requirements regarding the number of stimulation electrodes, since the amount of

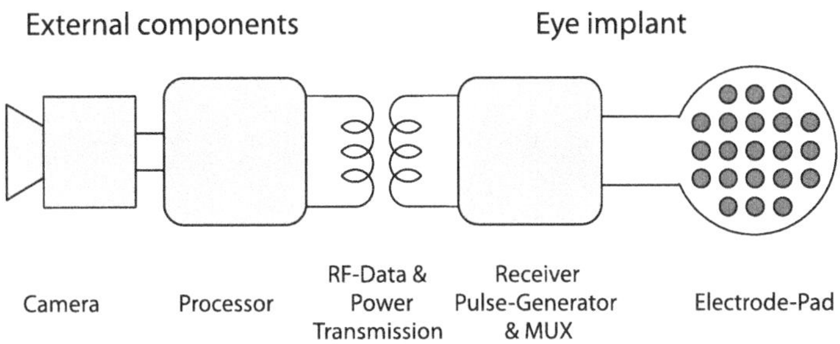

Fig. 28.2 Simplified diagram valid for epi-retinal implants and transscleral implants. These implants transmit the processed image of an electronic camera to an incorporated receiver. The signals are forwarded to electrodes with a multiplexer

conducting paths on the foil is limited. Solutions with several distributed electrode chips [20] connected to each other via a data-bus are currently being examined. However, the precondition for this approach is that the active electronics be located outside the save compartment.

With sub-retinal implants the active chip [17–19] is implanted directly into the retina. The incident light reaches the photodiodes through the eye lens and the vitreous, which forwards its signal directly or increases it to the stimulation electrodes.

Degenerated photoreceptors are also, in loco, replaced by the implant. For sub-retinal implants a high number of artificial stimulation cells are beneficial for replacing the light-sensitive human cells sufficiently. In many reports, researchers advocate passive photodiode arrays for electrical stimulation. After thorough investigation it appeared that generated energy from natural light was not strong enough to successfully stimulate the retina. Thus, active implant with analogue amplifiers for processing the photodiode signals was developed.

Similar to the micro photodiode arrays, active retinal implants utilise the eye lens image to create the electrical stimulation in the different image areas. The amplifier cells convert illumination into electrical voltages connected to stimulation electrodes placed at top of every amplifier. Also an amount of charge is supplied locally into the tissue by the electrodes' capacitance. The conversion of the brightness into a voltage is performed well by logarithmic cells working over a very wide dynamic range of more than 1 up to 10^6 for local and average illumination. This results in a logarithmic behaviour similar to the eyes' sensitivity to light. The average of the output for some of the logarithmic cells results in a very robust signal representing the average brightness of the actual scene. For the amplifier cells a differential configuration was chosen to get differences in lightness, independent of the absolute level of brightness (Fig. 28.3). Also the output of the amplifier is not dependent on the absolute voltage level in a very wide range of illumination and reacts only to the difference between the illuminations. Switched power lines and active discharging are used to create charge-balanced pulsed outputs.

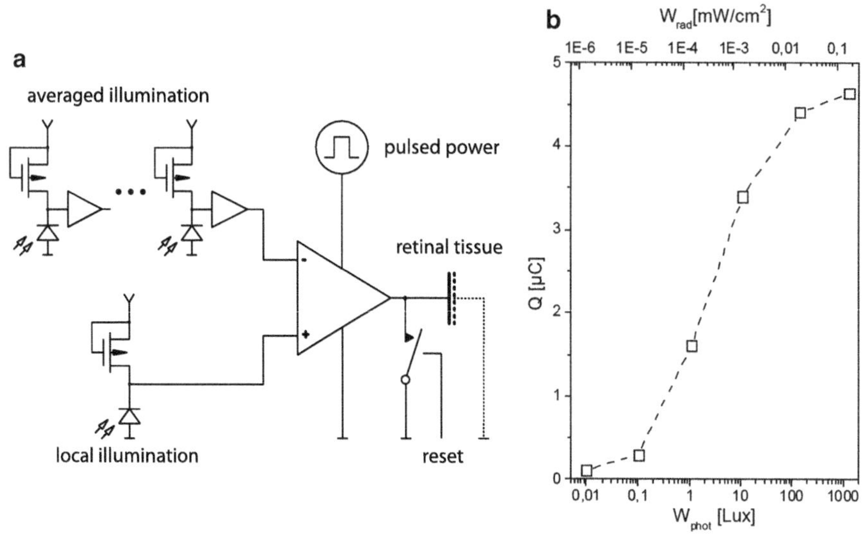

Fig. 28.3 Basic circuit of the active sub-retinal implant. The circuit (*left side*) forms constant electrical impulses from the difference between log (local illumination) and log (averaged illumination), applicable in a large range of illumination. These impulses represent the contrast of the illustrated image (*right side*)

28.3 Fabrication and Assembly of Retinal Implants

28.3.1 General Issues

The electronic and biological function of each retina implant is strongly connected to the implant location it will be exposed to for decades. Material is selected for stimulation electrodes for its long-term stability, as well as the bio stability; as such, the materials must fully comply with international safety regulations for medical products from an early phase on. Additional safety considerations with regard to the initial implant faults concern all predictable influences from every day life (such as positioning, eye rubbing, eye pressure) and the most common medical treatments (such as influence from electromagnetic fields, X-rays, NMR and so on).

Besides the selection of the material, the selected shape is also part of the bio compatibility that guarantees the function, suitability of implant technique as well as a nontraumatic positioning of the implant. Another issue is the cosmetic aspect for the implant-carrying person.

In most cases epi-retinal implants [15, 16] replace the eye lens or part of the vitreous with their active components. In addition, electrodes have to be attached to the retina with glue or pins since there is no natural adhesion guaranteeing a stable position of the electrodes. In common models (Fig. 28.4) the array is made of platinum or iridium electrodes fixed on silicon foils. The active electronic part and

Fig. 28.4 Epiretinal implant. Transmitter and circuit are designed as lens replacement. The electrodes are located at the end of the thread. The loops are for fixation at the sclera [15]

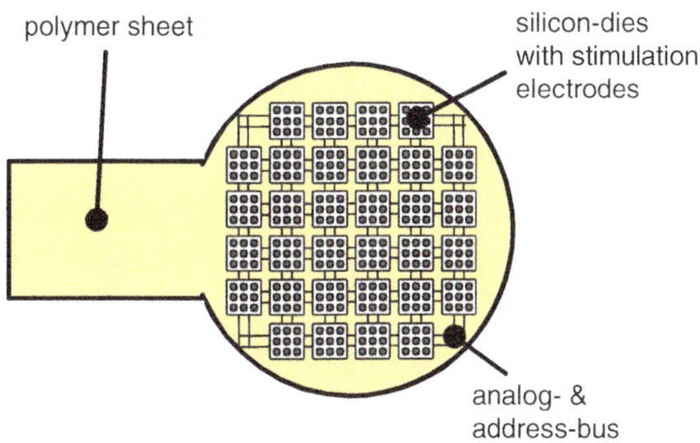

Fig. 28.5 Flexible large electrode field for epi-retinal and transscleral implants. The partial chips are connected to each other with a signalling line and an address bus [20]

the coil are arranged on thin flexible printed circuit board (PCB) or polyimide foils or connected to the electrode foil. The electronics and the coil are sealed for passivation afterwards. Integrated holes enable the fixation of the electronics to the sclera or the cornea with stitches or tags during implantation.

As an alternative to silicon the newer generation implants use titanium boxes as a compartment for the electronics [21]. The entire surface of the epi-retinal implants is therefore made of medically certified materials like silicon, titanium and iridium. The development of the material is focused primarily on the impermeability of the implant against diffusing ions, such as Na, Ka, H, OH and O.

A high degree of flexibility and low weight of all involved materials improve biological compatibility. This is also the case for the required integrated circuits. More electrodes cannot be realised with conventional connections because of the low number of conductor paths [20]. Satellite chips (Fig. 28.5) are being developed that contain several electrodes each, connected to one another via a digital bus system containing chip addresses and electrode addresses. The individual thin chips are connected to each other through flexible foil. As an alternative it is possible to integrate large-scale thin and flexible chips with the electrodes located on chip surface, similar to how it is done with sub-retinal implants.

Sub-retinal implants fixate themselves through the difference in pressure of the nutrition in the gap between the sclera and the retina. The fact that additional fixations are not necessary eases implantation and convalescence considerably. The position of the implant, however, requires an adequate mechanical electrode array.

The active optical function of the electrode array as image sensor and pulse generator additionally causes the direct integration of the almost bare, active Si circuits in the sub-retinal area.

Therefore, the active circuits need a post-processing by (1) addition of an optical transparent dense and bio-stable safety coating; (2) addition of and contacting of fractal electrodes having a large capacity made of TiN or Ir; (3) mechanical adjustment to the available space, i.e., low construction height (< 100 μm) and possible adjustment to the cranium shape of the eye; (4) integration of a thin flexible input lead with planar contacting technology for supply and the control signals; (5) integration of the rear side of the chip and the chip walls into the safety coating.

Another particular aspect is the handling of the implant during the operation. Trials with extremely thin and particularly flexible chips (< 20 μm Si thickness) resulted in a drastic loss of handling ability and spontaneous breakage [22]. Optimal results were achieved with 30 to 50-μm implants.

28.3.2 *Passive Sub-retinal Implants*

Passive sub-retinal implants (MPDAs or ASR™) (Fig. 28.6) [7], [8], [22], [23], gaining their energy supply directly from illumination, reduce themselves fully to the Si chip with adequate post-processing. The electrical function of these Si chips are the equivalent to an array of small insolated solar cells with various spectral sensitivities and a

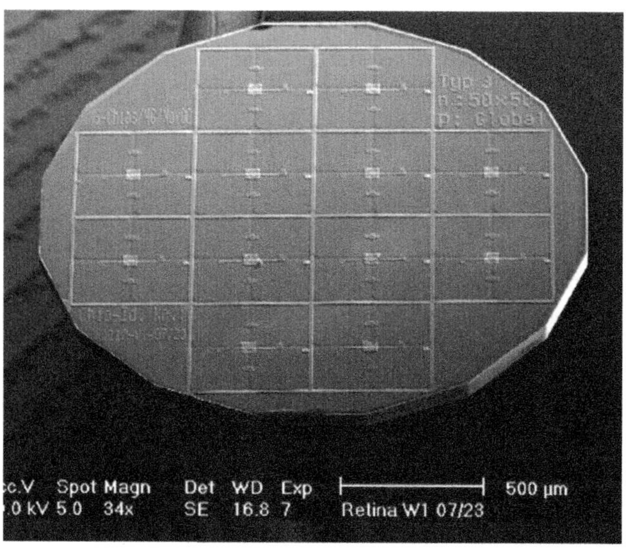

Fig. 28.6 Passive sub-retinal implant. In this type, four dielectric insolated photo elements are connected in a row in order to increase output voltage. Mechanically seen, the implant merely consists of an easily integrable Si plate with a diameter of 2 mm and a thickness of 50 μm. The outer shape is defined by deep trench etching prior to scaling down

density of up to 1,000 cells/mm^2. The individual solar cells are equipped with electrodes made of gold, platinum, iridium or titanium nitride. A passivation coating with Si_3N_4 or other inorganic or organics dielectric coatings serve as seal passivation. Implants for higher voltage and higher currents require a more complex dielectrical insolation of each cell.

During production process the individual MPDAs are cut out of wafers that have been scaled down through a special grind and polish technique. In a well-proven process the outer shape of the implant is generated by an anisotropic Si etching process (KOH or trench/STI process), creating a trench down to proposed final thickness of 50 μm. The generated bare trenches are now edge-passivated in the following process steps: The chips are glued in groups on 5×5-mm^2 dies on grinding fixtures and scaled back on a Struers DAP-V preparation grinder with semiautomatic polishing holders using a grinding technique (1,200/2,400 granularity) until the trenches become visible at a remaining thickness of about 7 μm. The final polishing with a granularity of 4,000 separates the dies afterwards. The different local grinding pressure thus generates local variations of the final thickness between the centre and the edge of the implant (with a diameter of 2 mm) of around 30%. The silicon-on-insulator (SOI) wafer used in another alternative process makes acquiring the correct thickness easier. The desired final thickness is now determined by the thickness of an epitaxial layer. The trench etching that is done down to the buried oxide layer defines the shape of the final chips, which are later separated following a combination of a grind and etch process. The SOI buried oxide layer forms an etching stop temporarily and, finally, the passivation of the rear side of the implant. This process is, however, limited by the costs of the SOI wafers and the thick epitaxial layer.

Furthermore, adjustment of the spherical shape of the eye with a diameter of 25 mm was researched in [22]. The adjustment of internal tension of a layer system with various expansion coefficients was used. These coefficients are generated during the cooling phase, after a coating in higher temperatures. The materials SiO_2 and Si_3N_4 commonly used in the semiconductor technology display a different behaviour here. Si_3N_4 on the surface or SiO_2 on the rear side results in the desired bending direction. The Si_3N_4 surface has to be coated prior to the contact and electrode lithography phase and therefore requires a handling of the tensioned wafer. A rear side coating with SiO_2 is possible with separated dies without lithographic processing. With a TEOS oxide thickness of 10-μm, die bends of 7 μm were measured during testing. In order to achieve the desired 40-μm bend, thinner implants with an approximate silicon thickness of 20 μm is required, which will result to a total thickness of 30 μm during the process. Another positive effect of these coatings is the rounding of all edges of the passive implant.

28.3.3 Active Sub-retinal Implants

The active sub-retinal implants (Fig. 28.7) require the complete CMOS technology for the manufacture of the analogue circuits. The additional optical function

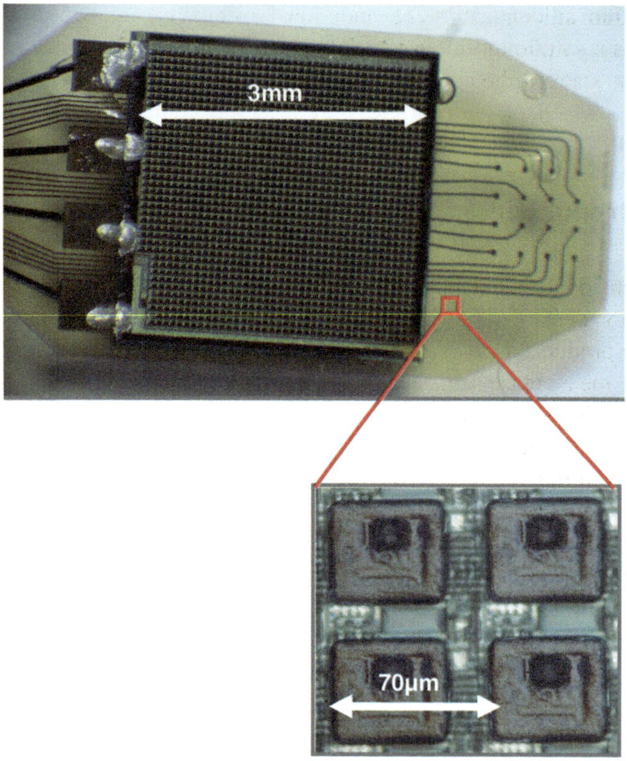

Fig. 28.7 Active sub-retinal implant. The part displayed at the top is located below the retina. The section displays the photo diodes and circuits below and the TiN electrodes on the polymer passivation layer

presents high demands on the pn-junctions of the photo diodes and the crystalline quality of the substrate. The chip's constant and pulsed supply voltage and individual control signals provide the connection to a thin and flexible PCB made of bio-compatible materials [17, 19]. In order to test the stimulation ability of the retinal tissue there are additional stimulation electrodes mounted on the top of the PCB. During the eye implantation the top of this PCB, including the integrated circuit, is pushed below the retina using a transscleral access. The extension of the connector foil extends below the skin to the plug behind the ear. Textile flaps attached to the connector enable the fixation of the implant with stitching or pins.

The total 80–100-µm thickness of the implant contains the PCB, glue and the 50-µm CMOS chip with the passivation layer deposited onto the chip. This 7-µm polymer passivation and also the Ti/TiN electrodes are structured on the CMOS wafer using photo lithography and etching. The silicon grouting of the chip edges insolates the bonds and protects the chip side walls.

Using 1,400 active electrodes arrayed at a distance of 70 µm to each other on a $3{\times}3\text{-mm}^2$ chip the active sub-retinal implant delivers the highest density of

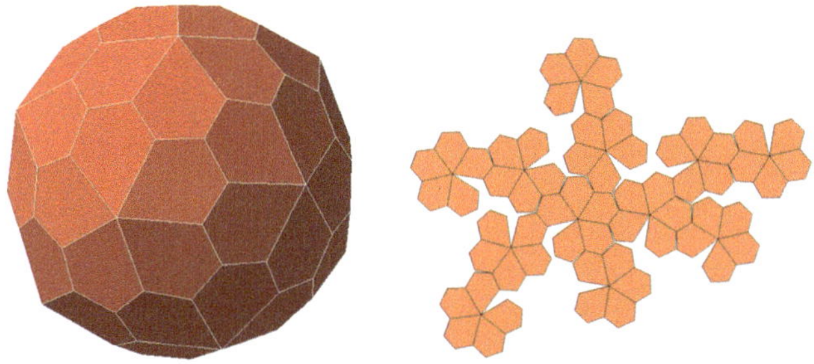

Fig. 28.8 Hexecontaedro pentagonal shape and flat projection [27]

electrical information in comparison to other approaches. With its chip measurement the implant, however, covers merely 7°, which corresponds to little more than 2° of the visual angle in the area of focused vision.

Reports on the clinical trials talk about recognition and localisation of individual unknown objects, such as the recognition of letters and words [23–25]. Larger chips with more pixels and extended visual fields are more likely to enable a better orientation in the visual field; however, they require a significantly higher adjustment to the shape of the eyeball. The approach to adjusting the cranial shape of the eyeball in passive sub-retinal implants can be applied also in active implants. The multilayer design of the entire implant, with all its internal tension deriving from processing and mounting, requires a comprehensive view of the entire system.

Extensions of the visual field to 30–60° requires a chip diameter of 12 mm and a cranium-shaped bend of more than 130 μm. The adjustment of the spherical shape with this chip diameter will result in a change of the lattice distance of the crystal tangentially by more than 4%, and therefore it is outside the area of plastic deformation, which reaches about 1% [26]. A solution may be in free definable chip shapes with radial cutouts or divisions into several single chip segments that are connected to each other. As far as the geometrical point of view is concerned, there are 20 radial cutouts with a width of 0–90 μm necessary for a visual field of 30°. A chip's division into segments that approximate a hexecontaedro pentagonal (Fig. 28.8) [27], allows a visual angle of more than 30°. This requires 15 interconnected pentagonally shaped individual chips with a 5-mm diameter.

28.4 Summary

Electronic circuits can reproduce for the vision-impaired several shapes of lost rudimentary visual impressions, doing so by reclaiming for the individual the orientation and the recognition of large objects. The technical challenge is to design the implants with regard to bio compatibility and adjusted to their location. Thinned flexible chips can add to more stimulation electrodes and increase the usable visual field in sub-retinal and epi-retina implants.

Acknowledgements Retina Implant AG und EpiRet GmbH are acknowledged for helpful discussions and for providing some illustrations used in this chapter. Astrid Hamala arranged the English translation of the text.

References

1. Artificial retina restores sight to the blind. http://www.vcstar.com /news/2010/feb/24/artificial-retina-restores-sight-blind/
2. Der Spiegel Digitales Wiedersehen 52:126–128, 2009
3. Tassicker GE (1956) Retinal stimulator. US Patent # 2,760,483
4. Brindley GS, Lewi WS (1968) The sensations produced by electrical stimulation of the visual cortex. J Physiol 196:479–93
5. Dobelle WH (2000) Artificial vision for the blind by connecting a television camera to the visual cortex. ASAIO J 46:1–7
6. Rousche PJ, Normann RA (1999) Chronic intracortical microstimulation (ICMS) of cat sensory cortex using the Utah intracortical electrode array. IEEE Trans Rehab Eng 7:56–68
7. Chow AY (1993) Electrical stimulation of the rabbit retina with subretinal electrodes and high density microphotodiode array implants. Invest Ophthalmol Vis Sci 34:835 (ARVO abstracts)
8. Zrenner E (2002) Will retinal implants restore vision? Science 295(5557):1022–1025
9. Weiland J, Fink W, Humayun M, Liu W, Rodger D, Tai YC, Tarbell M (2005) Progress towards a high-resolution retinal prosthesis. Conf Proc IEEE Eng Med Biol Soc 7:7373–7375
10. Brélen ME, Duret F, Gérard B, Delbeke J, Veraart C (2005) Creating a meaningful visual perception in blind volunteers by optic nerve stimulation. J Neural Eng 2:522–528
11. Gerding H (2008) Entwicklung eines minimal-invasiven Retinaimplantats. Ophthalmologe:1–10
12. Stett A, Barth W, Weis S, Haemmerle H, Zrenner E (2000) Electrical multisite stimulation of the isolated chicken retina. Vis Res 40(13):1785–1795
13. Rizzo JF, Wyatt J, Loewenstein J, Kelly S, Shire D (2003) Perceptual efficacy of electrical stimulation of human retina with a microelectrode array during short-term surgical trials. Investig Ophthalmol Vis Sci 44:5362–5369
14. Fujikado T, Morimoto T, Kanda H, Kusaka S, Nakauchi K, Ozawa M, Matsushita K, Sakaguchi H, Ikuno Y, Kamei M, Tano Y (2007) Evaluation of phosphenes elicited by extraocular stimulation in normals and by suprachoroidal-transretinal stimulation in patients with retinitis pigmentosa. Graefes Arch Clin Exp Ophthalmol 245(10):1411–1419
15. Schwarz M, Hostica BJ, Hauschild R, Mokwa W, Scholles M, Trieu HK (1996) Hardware architecture of a neural net based retina implant for patients suffering from retinitis pigmentosa. In: Proceedings of IEEE international conference. neural networks, Washington DC, pp.653–658, June 1996
16. Ortmanns M, Rocke A, Gehrke M, Tiedtke HJ (2007) A 232-channel epiretinal stimulator ASIC. IEEE J Solid-State Circuits 42(12):2946–2959
17. Graf HG, Harendt C, Engelhardt T, Scherjon C, Warkentin K, Richter H, Burghartz JN (2009) High dynamic range CMOS imager technologies for biomedical applications. IEEE J Solid-State Circuits 44(1):281–290, ISSN: 0018-9200
18. Dollberg A, Graf HG, Höfflinger B, Nisch W, Schulze Spuentrup JD, Schumacher K (2003) A fully testable retinal implant. In: Proceedings of IASTED conference on Biomed. Eng., Salzburg, Austria, pp. 255–259, 2003
19. Rothermel A, Wieczorek V, Liu L, Stett A, Gerhardt M, Harscher A, Kibbel S (2008) A 1600-pixel subretinal chip with DC-free terminals and ±2 V supply optimised for longlifetime and high stimulation efficiency. In: Proceedings of IEEE international solid-state circuits conference, San Francisco CA, pp. 144–145, 2008

20. Ohta J, Takuda T, Sasagawa K, Noda T (2009) Implantable CMOS biomedical devices. Sensors 9:9073–9093, ISSN1424-8220
21. DOI/10.1007/s00347-007-1631-9 April 14, 2010 http.//web.mit.edu/newsoffice/2009/microchip-blind-092309.html
22. Graf M (2002) Technologie des passiven Retina-Implantats, Fortschritt- Berichte VDI, Reihe 17, Nr. 223 VDI Verlag Düsseldorf (2002), ISBN 3-18-322317-1
23. Zrenner E, Gekeler F, Gabel VP, Graf HG, Graf M, Guenther E, Haemmerle H, Hoefflinger B, Kobuch K, Kohler K, Nisch W, Sachs H, Schlosshauer B, Schubert M, Schwahn H, Stelzle M, Stett A, Troeger B, Weiss S (2001) Subretinal microphotodiode arrays to replace degenerated photoreceptors? Ophthalmologe 98:357–363
24. Zrenner E, Wilke R, Zabel T, Sachs H, Bartz-Schmidt K, Gekeler F, Wilhelm B, Greppmaier U, Stett A (2007) Psychometric analysis of visual sensations mediated by subretinal microelectrode arrays implanted into blind retinitis pigmentose patients, 2007. ARVO Annual Meeting, 659/B283, May 2007
25. Zrenner E, Wilke R, Sachs H, Bartz-Schmidt KU, Gekeler F, Besch D, Benav H, Bruckmann A, Greppmaier U, Harscher A, Kibbel S, Kusnyerik A, Peters T, Porubská K, Stett A, Wilhelm B, Wrobel W (2009) Subretinal microelectrode arrays implanted into blind retinitis pigmentosa patients allow recognition of letters and direction of thin stripes. IFMBE 25/9 pp. 444–447
26. Wang L, Bartek M, Polyakov A, Jansen KMB, Ernst LJ (2005) Flexibility studies on ultra-thin silicon substrates. In: Krautschneider W (ed) Proceedings of the STW annual workshop on semiconductor advances for future electronics and sensors (SAFE 2005). Veldhoven, The Netherlands, November 17–18, 2005, Technology Stichting STW, Utrecht, 2005, pp 175–180
27. DOI/10.1007/s00347-007-1631-9 April 14, 2010 http://commons.wikimedia.org/wiki/Category:Catalan_solids?uselang=de

Chapter 29
Thin Chip Flow Sensors for Nonplanar Assembly

Hannes Sturm and Walter Lang

Abstract A thermoelectric mass flow rate sensor on a 10-μm thick polyimide foil for nonplanar assembly has been developed. For principal operation of thin film sensors a few hundred micrometres thick substrate (usually a 525-μm or 380-μm thick silicon wafer) is not necessary and might be detrimental for system integration, e.g., it could cause turbulences due to step in flow channel. A fabrication has been developed by removing the functional layers from the silicon substrate and transferring them onto a 10-μm thick polyimide foil. The sensor foil has been integrated on different materials by means of flip chip mounting and tested on its electrical and mechanical characteristics with air as flowing medium. Here, it was shown that thermoelectric flow sensors on polymer foils have comparable characteristics to flow sensors on silicon substrates. The outcome of this development is a highly integrated flow sensor on a polymer foil, which can be placed on planar, nonplanar as well as flexible surfaces and still has a comparable performance to thermoelectric flow sensors on silicon substrate.

29.1 Introduction to Thermal Flow Sensors

The measurement principle of the flow sensor shown here can be classified as calorimetric (alternative names: thermal dilution, thermotransfer) [1]. The flow sensor consists of at least one heater and a downstream temperature sensor, where the temperature sensor measures the quantity of heat released by the heater and its shift due to forced convection. Heater and downstream temperature sensor are integrated on a thin membrane for maximum thermal insulation. A second temperature sensor is located upstream. This gives the sensor a differential output signal so that conduction effects through the membrane are removed. Furthermore,

H. Sturm (✉)
Institute for Microsensors, -Actuators and -Systems (IMSAS), Microsystems Center Bremen (MCB), University of Bremen, Bremen, Germany
e-mail: hsturm@imsas.uni-bremen.de

J.N. Burghartz (ed.), *Ultra-thin Chip Technology and Applications*,
DOI 10.1007/978-1-4419-7276-7_29,

the second temperature sensor leads to another advantage: The flow sensor measures the mass flow rate now in a bidirectional way.

In operational mode the working principle of the flow sensor is as follows [2, 3]: The heater is operated in a constant temperature difference mode (CTD), where the temperature is controlled by continuously measuring the heater's resistance. Other operating modes such as constant current (CC) or constant power (CP) are also possible but have lower measurement ranges. The induced heat leads to a symmetric temperature field on top of the membrane, as shown in Fig. 29.1. A flowing medium changes the temperature distribution, which then becomes asymmetric, due to forced convection. The resulting temperature field is measured by temperature sensors and can be interpreted as mass flow rate of the fluid/medium.

As it can be seen in Fig. 29.1, heater and thermometer are located on a membrane. The membrane's thickness should be as thin as possible for maximum thermal insulation and lie in the range of some hundred of nanometres to tens of micrometres for MEMS technology [1, 4, 5]. Typical membrane materials to be used here are silicon nitride, silicon dioxide or just silicon [1, 6, 7].

The thermometers are based either on resistive principles [8] or on the thermoelectric effect (alternative name: Seebeck effect [9]). When the thermoelectric effect is exploited, a temperature difference between two points is directly converted into an electrical voltage. The conversion is effected by thermocouples – a combination of two electrically connected materials with different Seebeck coefficients. A series connection of two or more thermocouples is known as thermopile [10]. For thermopiles in flow sensors a combination of polysilicon and metal is used most often, due to the comparably high Seebeck coefficient of polysilicon [11]. Here, one contact area is located on the membrane and the other area is located on

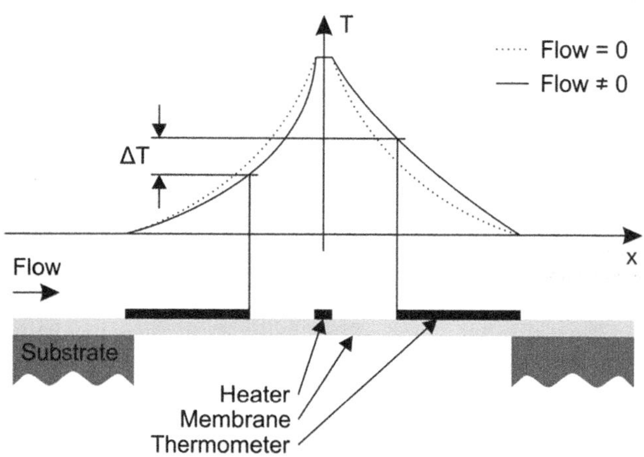

Fig. 29.1 Cross-sectional view of a calorimetric flow sensor operated in constant temperature difference mode and its temperature distribution over the membrane. The temperature field becomes asymmetric due to forced convection and leads to a temperature difference ΔT, which can be interpreted as mass flow rate

the substrate for thermal decoupling. Thus, all temperature measurements on the membrane are referred to the substrate temperature.

Figure 29.2 shows a thermoelectric flow sensor with heater and thermopiles fabricated by MEMS technology on silicon substrate. Silicon is best known as substrate material in MEMS micromachining and also mainly used for MEMS flow sensors. But silicon has disadvantages; especially in terms of system integration, a thickness of some hundred micrometres of the substrate material could be detrimental. Here, flush mounting of the sensor – which is expensive and might damage the flow channel or overflowed surface – is required to reduce turbulences in order to get more accurate output signals.

The problem of turbulences and the resulting flush mounting of the sensor can be avoided through use of thin flexible substrate materials. Furthermore, another advantage emerges: Flow sensors on flexible substrates can easily be integrated on nonplanar surfaces as tubes or even wings. Figure 29.3 shows a thermoelectric flow sensor on silicon substrate compared to a flow sensor on flexible substrate. Due to its marginal thickness the flexible sensor can be easily connected through its substrate by means of flip chip bonding. In addition to its height, the flow sensor on silicon substrate is usually connected by bond wires rising into the flow channel, influencing the flow profile.

A first approach of a thermal flow sensor on flexible substrates has been done in [12]. Here a hot-wire anemometer has been realized by placing either one or two

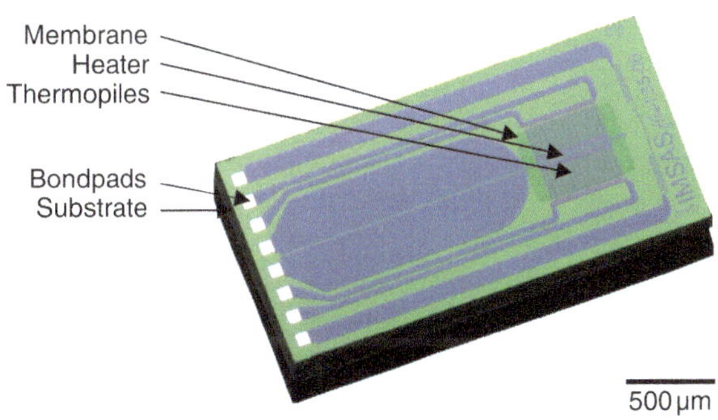

Fig. 29.2 Top view of thermoelectric flow sensor on nonflexible silicon substrate fabricated at IMSAS [3]

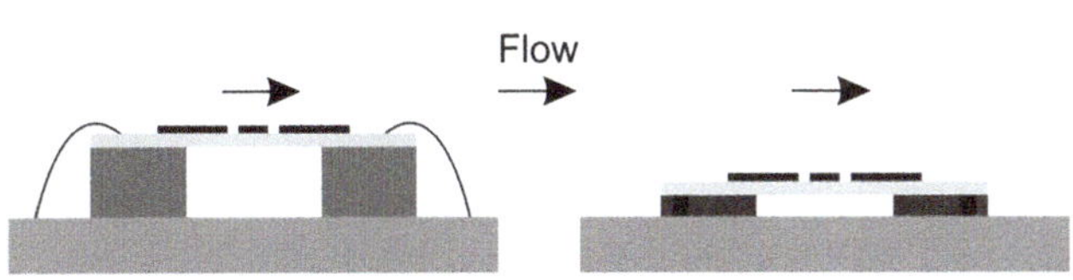

Fig. 29.3 Thermoelectric flow sensor on silicon substrate (*left*) and flow sensor on flexible substrate (*right*). The flow sensor on silicon substrate is usually connected with bond wires while flexible flow sensors presented here can be connected by means of flip chip bonding

freestanding nickel wires on a 125-μm thick polyimide foil. Flow measurements have been done by using the anemometric principle: A heated element (here: nickel wires) is cooled by convective heat transfer. The resulting temperature can be measured by its resistance and interpreted as flow rate.

A polyimide foil is also used as substrate material in [13]. Here, an Au/Cr heater is deposited on a 25-μm thick polyimide film. The film is mounted in an inner wall surface of a straight tube, also exploiting the anemometric measurement principle. There are no references about thermoelectric flow sensors on flexible substrates so far, especially using materials with high thermoelectric effect like polysilicon [9].

This leads one to the following question: How can a flexible substrate be realized by means of MEMS technology without losing a basic amount of sensitivity in comparison with silicon substrate?

29.2 Technological Approach

The main idea behind the fabrication process is to start with conventional manufacturing on silicon [3, 14] and transfer of the functional layers onto a flexible substrate. The transfer must be done because of the thermal load applied to the wafer during processing. Noteworthy are low pressure chemical vapour deposition (LPCVD) processes for deposition of polysilicon (560°C) and silicon nitride (up to 800°C) [15]. A schematic view of the whole fabrication process is shown in Fig. 29.4.

For processing, a 380-μm thick double-polished silicon wafer is used. In the first step the substrate is thermally oxidized to an oxide thickness of 500 nm. The thermal oxide acts as an etch stop layer for the subsequent sensor release. On top of the oxide layer a silicon-rich low-stress LPCVD silicon nitride of 300-nm thickness is deposited. LPCVD silicon nitride has a very high thermal and chemical stability and acts as a protective layer for the functional structures. The thermopiles are made of an in situ p-doped polysilicon and an alloy of 90% tungsten and 10% titanium. For the heater, tungsten titanium is used as well.

A diffusion barrier of titanium nitride is deposited between the polysilicon and tungsten titanium of the thermopiles to increase thermal stability, especially for the subsequent second deposition of 300-nm LPCVD silicon nitride [16]. For electrical contacting of heater and thermopiles the silicon nitride is structured.

In the next step, a photodefinable polyimide, HD-4100 from HD Microsystems™ [17], is deposited via spin coating; this will later become the flexible substrate. The polyimide has a thickness of approximately 20 μm and can be structured by means of photolithography. A final curing process gives the polyimide its high thermal and chemical stability and reduces its thickness to 10 μm. As it can be seen, the membrane area is also coated with polyimide. This will reduce the sensor's sensitivity but is necessary for mechanical stability during the packaging process (flip chip bonding).

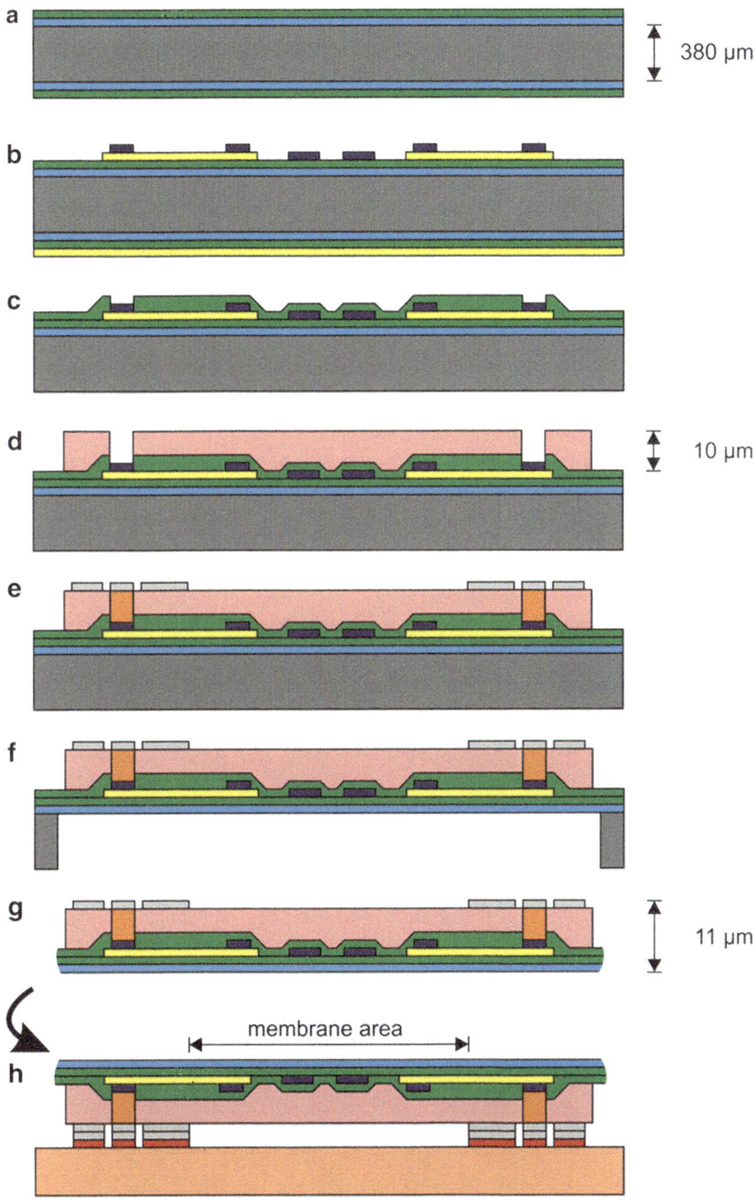

Fig. 29.4 Schematic view of the sensor fabrication process: (**a**) deposition of silicon dioxide and silicon nitride; (**b**) realisation of thermopile and heater structures made of polysilicon and tungsten titanium; (**c**) deposition and structuring of silicon nitride passivation layer; (**d**) deposition and structuring of polyimide; (**e**) electroplating of bumps and deposition of tin layer for flip chip bonding; (**f**) release of sensor structure using DRIE; (**g**) separation of flexible sensor dies; (**h**) flip chip bonding on flexible printed circuit board

To realize an electrical connection gold bumps are electroplated and coated with a 2-μm layer of tin for flip chip bonding. Further tin layers are placed all over the flexible substrate to ensure a mechanical mounting beside the electrical connection.

The silicon below the polyimide is removed using a deep reactive ion etching (DRIE) process with the thermal oxide as stop layer. The polyimide substrate is slightly smaller than the etched cavity. Thus, the whole flexible sensor structure is held by a thin rim of silicon nitride/thermal oxide only and can easily be removed. A SEM picture of the sensor is shown in Fig. 29.5. To obtain an internal stress compensation between the polyimide/silicon nitride/thermal oxide stack, the thermal oxide thickness can be reduced by wet etching. This minimizes the curvature of the sensor and improves its handling.

The packaging, including mounting and electrical connection, is done by means of flip chip bonding. Therefore, the sensor is picked up using a vacuum head and placed onto a flexible printed circuit board (PCB). The flexible PCB has copper tracks with a 2-μm tin layer on top corresponding to the structures on the flexible sensor. A short heating of the sensor and flexible PCB melts the tin and bonds both layers with high electrical and mechanical quality. An image of a flexible flow sensor flip chip bonded on a PCB is shown in Fig. 29.6.

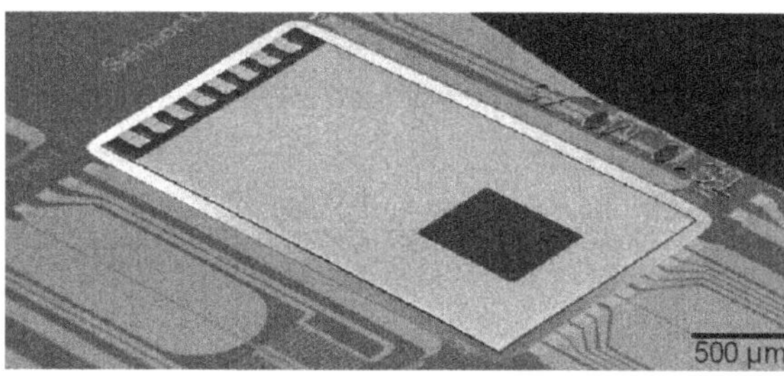

Fig. 29.5 SEM picture of flexible thermoelectric flow sensor on wafer held by a thin rim of silicon nitride/thermal oxide only. Bond pads and membrane area can easily be distinguished due to the interrupted tin layer

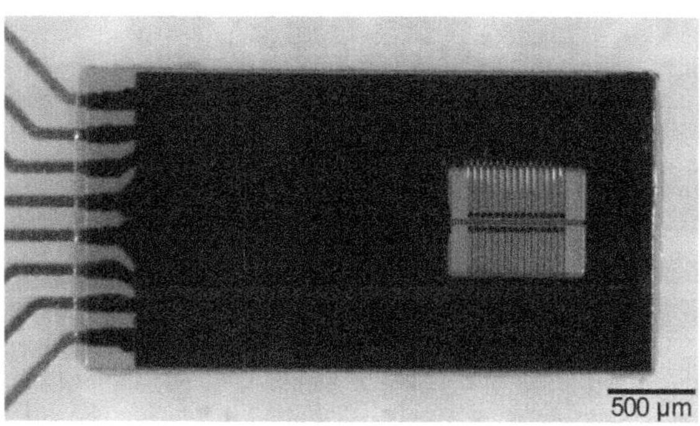

Fig. 29.6 Thermoelectric flow sensor on 10-μm polyimide substrate mounted on PCB. Rests of the holding rim can still be seen on edge of the polyimide substrate

As it can be seen in Fig. 29.6, most areas of the sensor are coated with a thin layer of tin for flip chip bonding. The tin layer was removed via lift-off around the bond pads for electrical isolation and in the membrane area for thermal insulation. A close contact to the PCB via the thin layer ensures a maximum decoupling of the thermopile contact area, which is far from the heater.

29.3 Sensor Characteristics

After having shown the technological approach of building thermoelectric flow sensors on 10-μm polyimide substrate, we now give an overview of the measured characteristics. Measurements have been made in terms of sensitivity, time response, power consumption and flow metering characteristics.

29.3.1 Electrical Characteristics

Sensitivity measurements for thermopiles have been made by application of a certain power to the heater and measurement of the output voltage. The heater temperature was controlled by its resistance since the used tungsten titanium alloy has a temperature coefficient of resistance of $2.6 \times 10^{-4}\ K^{-1}$.

The measured thermopiles show a linear behavior within heater temperatures of 0–100 K above ambient temperature. A sensitivity of 1.8 $mV\text{-}K^{-1}$ related to heater temperature could be measured for the thermopiles consisting of 15 thermocouples each. The measured value is much lower than that which is measured for the same sensor design on silicon substrate and a silicon nitride membrane only. Here, a sensitivity of 4.3 $mV\text{-}K^{-1}$ related to heater temperature could be observed [14]. The lower sensitivity is caused by the additional 10-μm polyimide layer on the 600-nm silicon nitride membrane. An HD-4100 polyimide has a thermal conductivity of 0.25 $W\text{-}m^{-1}\text{-}K^{-1}$ [17], and the used silicon nitride could be determined to 3.11 $W\text{-}m^{-1}\text{-}K^{-1}$–4.04 $W\text{-}m^{-1}\text{-}K^{-1}$ [18–20]. The membrane is freestanding. However, the comparably thick additional polyimide layer causes an additional heat flux, resulting in a more effective cooling of the membrane and a lower output signal. Furthermore, the thermopile end that is far from the heater is not entirely thermally decoupled as on silicon substrates, which have a much higher thermal conductivity of 150 $W\text{-}m^{-1}\text{-}K^{-1}$ [21]. This also leads to a reduction of the temperature difference between both thermopile contact areas and to a lower output signal.

The additional heat flux through the polyimide layer can also be observed by measuring the heat dissipation for different heater temperatures. During operation, flexible flow sensors have a power consumption[1] of 0.25 $mW\text{-}K^{-1}$, flow sensors

[1]Air has been used as medium

Table 29.1 Characteristics of flexible flow sensor compared with results for flow sensors on silicon substrate [14]

	Conventional flow sensor	Flexible flow sensor
Substrate material	Silicon	Polyimide
Membrane material	Silicon nitride	Silicon nitride + polyimide
Sensitivity of thermopiles[a] (mV-K^{-1})	4.3	1.8
Power consumption (mW-K^{-1})	0.13	0.25
Time constant (ms)	3–5	12–13

[a]Related to heater temperature

with the same design on silicon substrates achieve values of 0.13 mW-K^{-1} for heater temperatures of 0–100 K above ambient temperature.

Further measurements have been made on time response of the thermopile output signals. Therefore, a voltage pulse has been given on the heater and the time response of the thermopile signals has been recorded. For flexible flow sensors, a time constant[2] of 12–13 ms could be measured, flow sensors with the same design on silicon substrates reach values of 3–5 ms. The difference can again be explained by the additional polyimide layer on the membrane, which increases the thermal capacity and therefore the time constant.

Summing up, flexible flow sensors are not competitive with equivalent flow sensors on silicon substrates in terms of electrical characteristics. This is caused by the additional polyimide layer acting as substrate material and also being located on the membrane area for mechanical stability. A summary of electrical characteristics is shown in Table 29.1. In the next part, flow measurements are presented.

29.3.2 *Flow Measurements*

The sensor has been mounted on a PCB. A flow channel with a cross-sectional area of 1 × 1.5 mm has been attached on top in order to create defined fluid flow conditions. Measurements have been made using air as medium. The temperature of the heater has been chosen to 30 K above fluid temperature.

Figure 29.7 shows the voltage difference between the downstream and upstream thermopile in mV for different flow velocities in m-s^{-1}. As it can be seen, the output signal of the flexible flow sensor is lower than the signal of the flow sensor on silicon substrate. The lower output signal can be explained by the additional polyimide layer on the membrane. The layer increases the heat loss due to heat conduction to both thermopiles with the effect that heat transport with respect to forced convection through the flowing medium gets comparatively smaller. This makes a thicker membrane more insensitive to heating by the fluid above. An advantage of this effect can be seen at higher flow rates. A saturation of the sensors

[2] Time after 63% of final value has been reached.

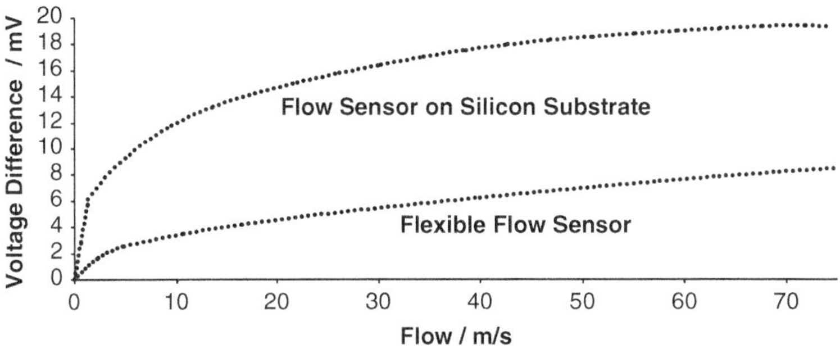

Fig. 29.7 Characteristic line (voltage difference between downstream and upstream thermopile) of flexible flow sensor and flow sensor on silicon substrate in air. The heater temperature is 30 K above fluid temperature. A distance of 10 μm between heater and thermopiles has been chosen

output signal appears at higher flow velocities with the result that differences in flow velocity can be detected better.

29.4 Conclusions and Outlook on Applications

A thermoelectric flow sensor on a 10-μm thick polyimide foil for nonplanar assembly has been presented. The functional layers have been fabricated on a standard silicon wafer and then transferred onto a polyimide substrate. For packaging, the flexible sensor system has been placed onto a PCB by means of flip chip bonding with electrical and mechanical connection.

Measurements have been made for sensitivity, power consumption, time constant and sensor outputs at different flow velocities. The resulting characteristics show lower performances than equivalent sensors on silicon substrate. This can be attributed to an additional polyimide layer, which is used as substrate material and also located on the membrane due to mechanical stability. This polyimide layer increases the heat conduction between heater and substrate with the effect of higher power consumption and lower output signals for different flow velocities. Furthermore, the thermal capacity is increased leading to higher time constants.

Summing up, flexible flow sensors do not have the same characteristics as thermoelectric flow sensors on silicon substrate. Their advantage lies in the higher integration due to their low thickness, which makes applications on nonplanar surfaces feasible.

The need for flow measurements on nonplanar surfaces can appear in many challenges of engineering, but integration of flexible flow sensors only makes sense in some cases. It could be the avoidance of turbulences, which can be observed on silicon sensors due to their height, or the need for structural health of the flow channel.

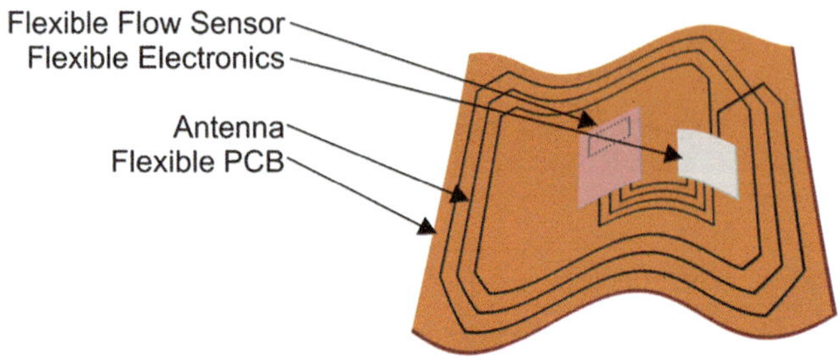

Fig. 29.8 Schematic view of highly integrated "stamp" flow measurement system to be developed. Sensor and electronics are integrated on flexible PCB. Energy and data transfer is realized by wireless only

A comparably thin flexible flow sensor system can avoid destructions of the flow channel due to integration that might happen with silicon substrate flow sensors. However, it has to be taken into consideration that an electrical connection of the sensor system is necessary. To avoid wires, wireless communication could be an interesting solution here. Therefore, a flexible flow sensor – together with flexible electronics [22–24] and antennas – is placed on a flexible PCB, which could easily be integrated in most nonconductive tubes.

Figure 29.8 shows a schematic view of a flexible sensor system to be developed. The working principle is as follows: A reader outside the tube induces an electromagnetic field into the antenna resulting in an electrical voltage. The electrical voltage is rectified by the flexible electronics and given to the flexible sensor for measurement. After measurement, the thermopile signal of the sensor is digitized and sent back by short-circuiting the antenna. The change of the electromagnetic field can be detected by the reader and interpreted as a flow measurement signal.

The use of a wireless communicating flexible flow sensor system opens other fields of applications (e.g., stall measurements in wind tunnels), which lead to more accurate results in flow measurements and a better understanding of processes in thermal and velocity boundary layers.

References

1. Haasl S, Stemme G (2008) Flow sensors. In: Gianchandani Y, Tabata O, Zappe H (eds) Comprehensive microsystems. Elsevier, Amsterdam, The Netherlands
2. Ashauer M, Glosch H, Hedrich F (1998) Thermal flow sensor for liquids and gases based on combinations of two principles. Sensor Actuat A-Phys 73:7–13
3. Sosna C, Buchner R, Lang W (2010) A temperature compensation circuit for thermal flow sensors operated in constant temperature difference mode. In: IEEE transactions on instrumentation and measurement, pp 1715–1721
4. Buchner R, Sosna C, Maiwald M et al (2006) A high-temperature thermopile fabrication process for thermal flow sensors. Sensor Actuat A-Phys 130–131:262–266
5. Kim S, Nam T, Park S (2004) Measurement of flow direction and velocity using a micromachined flow sensor. Sensor Actuat A-Phys 114:312–318

6. Ernst H, Jachimowicz A, Urban GA (2002) High resolution flow characterization in Bio-MEMS. Sensor Actuat A-Phys 100:54–62
7. Nguyen NT, Kiehscherf R (1995) Low-cost silicon sensors for mass flow measurement of liquids and gases. Sensor Actuat A-Phys 49:17–20
8. Fürjes P, Legradi G, Dücso C et al (2004) Thermal characterisation of a direction dependent flow sensor. Sensor Actuat A-Phys 115:417–423
9. Rowe DM (2006) General principles and basic considerations. In: Rowe DM (ed) Thermoelectrics handbook: macro to nano. CRC Press, Boca Raton, FL
10. Sarro PM, van Herwaarden AW, van der Vlist W (1994) A silicon-silicon nitride membrane fabrication process for smart thermal sensors. Sensor Actuat A-Phys 42–42:666–671
11. Nguyen NT, Meng AM, Black J et al (2000) Integrated flow sensor for in situ measurement and control of acoustic streaming in flexural plate wave micropumps. Sensor Actuat A-Phys 79:115–121
12. Buder U, Petz R, Kittel M et al (2008) AeroMEMS polyimide based wall double hot-wire sensors for flow separation detection. Sensor Actuat A-Phys 142:130–137
13. Tan Z, Shikida M, Hirota M et al (2007) Characteristics of on-wall in-tube flexible thermal flow sensor under radially asymmetric flow condition. Sensor Actuat A-Phys 138:87–96
14. Buchner R, Froehner K, Sosna C et al (2008) Toward flexible thermoelectric flow sensors: a new technological approach. J Microelectromech Syst 17:1114–1119
15. Ohring M (2002) Materials science of thin films. Academic Press, San Diego, CA
16. Buchner R, Sosna C, Lang W (2009) Temperature stability improvement of thin-film thermopiles by implementation of a diffusion barrier of TiN. In: IEEE sensors, Christchurch, New Zealand, pp 483–486
17. Hitachi DuPont Microsystems (2010) HD Microsystems^{TM} technical information library. http://hdmicrosystems.com/HDMicroSystems/en_US/tech_info/tech_info2.html. Accessed 28 February 2010
18. Völklein F, Kessler E (1984) A method for the measurement of thermal conductivity, thermal diffusivity, and other transport coefficients of thin films. Solid State Phys 81:585–596
19. Völklein F (1990) Thermal conductivity and diffusivity of a thin film SiO_2-Si_3N_4 sandwich system. Thin Solid Films 188:27–33
20. Sun R, White MA (2004) Thermal properties and applications of emerging materials. In: Tritt TM (ed) Thermal conductivity – theory, properties and applications. Springer, New York
21. Yang B, Chen G (2004) Experimental studies on thermal conductivity of thin films and supperlattices. In: Tritt TM (ed) Thermal conductivity – theory, properties and applications. Springer, New York
22. Zimmermann M, Burghartz J N, Appel W et al (2006) A seamless ultra-thin chip fabrication and assembly process. In: Technical digest of IEEE international electron devices meeting, San Francisco, CA, pp 1–3
23. Burghartz JN, Harendt C, Tu Hoang et al (2009) Ultra-thin chip fabrication for next-generation silicon processes. In: IEEE bipolar/BiCMOS circuits and technology meeting, Capri, pp 131–137
24. Richter H, Rempp HD, Hassan MU et al (2009) Technology and design aspects of ultra-thin silicon chips for bendable electronics. In: IEEE international conference on IC design and technology, Austin, TX, pp 149–154

Chapter 30
RFID Transponders

Cor Scherjon

Abstract Today, radiofrequency identification (RFID) is omnipresent and used in a vast amount of applications, ranging from pet registration to manufacturing logistics. As each application has its own requirements, a lot of different RFID systems have been developed in the past. The only similarity in these systems is that they consist of two basic elements, which are an interrogator and a so-called 'transponder'. This chapter will first provide a brief overview of RFIDs and shows typical applications and their special requirements. Finally, the chapter focuses on existing ultra-thin RFID transponders and new developments that will yield small, thin RFID transponders.

30.1 Overview

Radio frequency identification is a means of identifying entities through the air interface without physically contacting them [1, 2]. One of the very first applications of RFID was an active transmitter used in airplanes for identifying either friend or foe during World War II. In his paper [3] in the Proceedings of I.R.E., Harry Stockman describes reflected power communication that involves an interrogator and a target, which modulates the received signal that can be detected at the interrogator. This basic concept is still being used in all RFID systems in more or less sophisticated ways. RFID systems can be divided into different categories depending on their typical characteristics. These can be the number of bits used for storing the identity, the operating range of the system, the operating frequency or the level of intelligence at the transponder. This section discusses the RFID systems from the latter point of view.

The simplest system uses a 1-bit transponder, which is, for example, used in electronic article surveillance in order to prevent theft of objects. In this case, the

C. Scherjon (✉)
Institut für Mikroelektronik Stuttgart (IMS CHIPS), Stuttgart, Germany
e-mail: scherjon@ims-chips.de

J.N. Burghartz (ed.), *Ultra-thin Chip Technology and Applications*,
DOI 10.1007/978-1-4419-7276-7_30,

transponder consists of a circuit containing an LC resonant circuit, which absorbs the alternating magnetic field, generated by the interrogator. This absorption can be detected by a receiver. By applying a strong alternating magnetic field, a large current will be induced in the transponder, which leads to its destruction. The operating frequency of this kind of system lies in the range of 8 MHz.

Other systems operate in the microwave frequency range 860–960 MHz, 2.45 or 5.6 GHz, in which the transponder consists of a dipole antenna and a small capacitance diode. If the transponder is within the interrogator's range, the induced current in the capacitance diode will cause the generation of an harmonic signal of two or three times the base frequency due to the nonlinear behaviour of the component. The transmitted higher harmonic signals can then be detected at the interrogator.

In long wave range at approximately 125 kHz, a typical 1-bit transponder contains a coil with a parallel capacitance adjusted in such a way that the resulting LC resonant circuit is tuned to the carrier frequency. This causes the induced voltage in the system to be as high as possible, which is needed for supplying a microchip that divides the received carrier frequency into a subharmonic signal. The latter is used to control a load element connected to the receiver coil. In this way, the transponder absorbs the received energy in a varying way, which can be detected at the interrogator.

For the majority of applications the number of objects that need to be identified is very large, so transponders with an identification code of more than a few bits are needed. In typical transponders a guaranteed unique 64-bit code is stored in a read-only memory for each transponder. The transmission of these data can be done in a number of different configurations, depending on the communication protocol and operating frequency. Surface acoustic wave devices, for example, are operated at microwave frequencies and act as a reflector of the impinging electromagnetic waves. A number of reflective strips on the piezoelectric substrate will generate a unique pulse sequence that can be detected by the interrogator.

Another very basic RFID system that can read only one transponder at a time generates a carrier frequency and the transponder will start transmitting the 64-bit code immediately and repetitively as long as enough power is available for operation. It is obvious, that this system – although very easy to implement – has its drawbacks; for example, if two transponders are within the range of the interrogator, both transponders will send their identification code simultaneously, thus resulting in code collisions. A solution to this problem on transponder side is the implementation of a simple ALOHA protocol, where a code collision is detected by the transponder and a retransmission is started after a random interval. Solutions on interrogator side require the transmission of data from interrogator to transponder and a protocol for searching and addressing the transponders.

In a number of applications it is not only necessary to identify objects, but also to store additional data into the transponder. This information will then be used as the tagged objects are being tracked. The requirements for this kind of transponder is that a memory is integrated that stores data throughout the entire process. So the data will be either stored into a programmable nonvolatile memory like EEPROM or FLASH memory or into RAM memory if the transponder is supplied with power continuously. The memory on the transponder is accessible from the interrogator, so data can be read, changed and written depending on the needs of the application. It is obvious that some applications need a way of controlling access to this memory, as the data might contain confidential information or as data should not be read or changed by a third party. Transponders addressing this kind of application mostly contain cryptographic features that handle access control and/or provide encrypted data transmission in order to deny access to intruders or to avoid manipulation of sensitive data stored in the transponder.

It is obvious from the above descriptions that a large diversity of transponders exists that fulfil different requirements for numerous applications. Depending on the complexity involved, more or less intelligence is needed in the transponder, which can be characterised by the computing power needed for interaction with the interrogator. The power needed in a transponder is a crucial point, as this puts constraints on power transmission, carrier frequency, antenna dimensions, used chip technology, chip dimensions and usage of an extra external power source.

So-called 'passive transponders' that don't use an external power supply are supplied via the air interface only. Two different techniques for wireless power transmission can be used; one using inductive coupling and one using the energy from electromagnetic waves (backscattering).

Inductive coupling is used for RFID systems in which the transponder is located in the near field (see Fig. 30.1) of the interrogator, which means within a radius of the wavelength of the carrier signal divided by 2π. The received power depends on the distance of the transponder to the interrogator, as the coupling factor of this loosely coupled transformer decreases inverse proportionally to this distance by a power of three. Normally, the carrier frequency used in this kind of systems equals 13.56 MHz and the reading distance of commercially available systems lies in a range of 1.3 m, depending on the implementation of the interrogator regarding radiation and interrogator sensitivity.

The backscattering technique uses the energy of the electromagnetic field in the far field (see Fig. 30.1) of the interrogator and automatically allows longer distances between transponder and interrogator, as the available energy decreased by the free

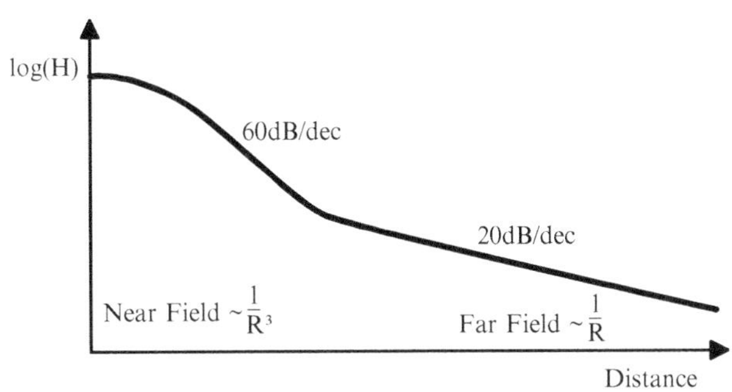

Fig. 30.1 Magnetic field strength as a function of distance between transponder and interrogator

space path losses drops in inverse proportion to the distance. The reading distance in this kind of RFID system is limited by the carrier frequency and maximum power radiated by the interrogator, which is a limitation that is based on international regulations. Using the equation for free space path losses

$$10 \cdot \log(FSPL) = 10 \cdot \log\left[\left(\frac{4\pi df}{c}\right)^2\right] = 20 \cdot \log(d) + 20 \cdot \log(f) - 147.56$$

it can be easily seen that the losses at higher frequencies increase. To illustrate this, the path losses for two RFID frequency bands as a function of the distance between transponder and interrogator are shown. If the transponder needs 10 μW for correct operation and the radiated power by the interrogator equals 1 W, the free space path losses should not exceed 50 dB; so, for an RFID system operating at 868 MHz, the maximum distance between interrogator and transponder is 9 m, whereas a system operating at 2.45 GHz allows a maximum reading distance of 3 m, see Fig. 30.2.

30.2 Applications

As described in the last section, a vast amount of RFID systems and transponders exist that fulfil the special requirements for individual applications. This section will provide an overview over the different applications and their requirements.

One of the biggest applications for RFID systems is tracking of products in warehouse management systems that allow web-based monitoring of goods. In this way, manufacturers and distribution centres can efficiently handle transportation of

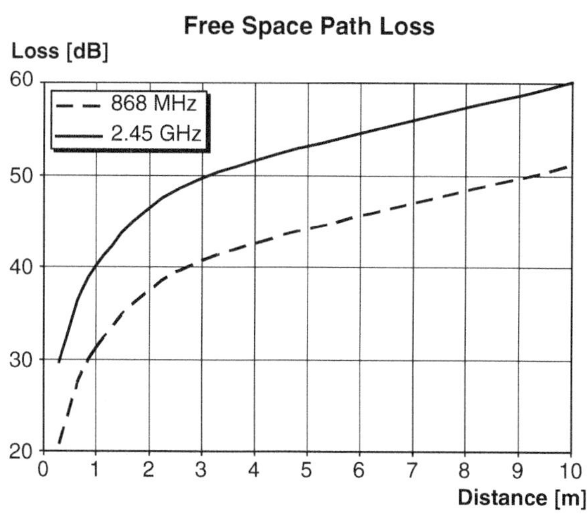

Fig. 30.2 Free space path losses at different carrier frequencies

goods and reduce storage costs. Challenges for RFID transponders lie in the area of material composition of the items to which these are attached. Most packaging materials like cloth, paper and plastics are quite transparent to electromagnetic waves, but other materials like polycarbonate and metal absorb or reflect incident radiation, resulting in poor performance of transponders. Tracking performance will also be reduced if the contents of the transported items are water-based. The influence of material on the performance of the transponder depends on the operating frequency of the RFID system. In metallic environments frequencies around 125 kHz are preferred, whereas frequencies at 13.56 MHz and 860–960 MHz can be used in 'watery' environments. In order to keep transponders inexpensive and small, transponders at higher frequencies, 860–960 MHz or 2.45 GHz are being used. These transponders are also able to communicate at a higher bit rate, which reduces the interaction time between interrogator and transponder.

The combination of RFID transponders and sensors allows a completely new field of applications [4, 5]. Such sensing transponders can be used to monitor goods during transportation or storage. Environmental parameters such as temperature or humidity can be logged for perishable foods. The continuous monitoring by these sensors needs for the integration of an ultra-thin battery into the transponder.

Another application field for RFID transponders is prevention of counterfeits and to guarantee authenticity of products or items. This latter might be of critical, life-or-death importance in, for example, pharmaceutical applications. In this case, cryptographic algorithms need to be implemented into the transponder for guaranteeing authenticity. Other applications that require authentication are the tagging of documents or the use of eSeals [6]. Additional requirements for this kind of application are ultra-thin RFID transponders, whereas the reading distance can be in the millimetre range.

30.3 Ultra-thin RFID Transponders

Applications like eSeals require ultra-thin RFID transponders, as they need to be integrated into thin foils or paper documents. This section provides a small selection of currently available ultra-thin transponder modules. A selection of available transponder modules are discussed in descending order of computational power.

In May 2008 Texas Instruments announced an ultra-thin transponder module [7], which contained the industry's thinnest contactless payment chip. The background of said thinnest module is the production of smartcards that comply with the ISO standards, a requirement that also put limitations on the thickness of such cards. As the card's thickness should be between 680 and 840 μm, card manufacturers are facing yield problems during assembly, especially when cards with more colourful artwork are produced, as these require thicker card substrates. The thin transponder module containing a 150-μm thick chip has a total thickness of 280 μm, which allows card manufacturers to use thicker card substrates, thus increasing production yield. The transponder offers all security features needed for payment transactions and offers fast transactions between interrogator and transponder, which is a key feature for the acceptance of contactless payment by end users.

Cooperation between Semiconductor Energy Laboratory Co., Ltd., and TDK Corporation led to the development of an ultra-thin flexible RFID transponder as thick as 30 μm [8, 9]. These flexible substrates can be formed as thin inlets and can easily be integrated into high quality paper, which has a thickness of 100 μm. The RFID integrated circuit, consisting of 71,000 transistors, operates at a frequency of 13.56 MHz and was manufactured in a high performance, thin-film transistor technology. The thin-film technology was applied to a plastic film substrate and the transistor dimensions used are approximately 1 μm. The transponder that implements parts of the ISO/IEC 15693 standard contains a CPU that can perform basic encryption.

In September 2003, Hitachi, Ltd. announced the development of a new version of their RFID transponder chip called μ-Chip [10]. Revolutionary in this product is the embedded on-chip antenna in combination with the ultra-small chip dimensions of only 400 × 400 μm. The predecessor chip's dimensions are 75 × 75 μm at a thickness of only 7.5 μm, and it is realised in an SOI process [11]. The combination of on-chip antenna and small chip dimensions allows applications that entail very close proximity requirements, like integration in banknotes. The 128-bit identification number of the chips is read-only, and these numbers are assigned individually during the manufacturing process using electron beam technology. The on-chip integration of the antenna eliminates the need for expensive assembly techniques at the cost of a reduced reading distance and that it offers only basic functionality.

Though larger than the μ-Chip, the Maxell Coil-on-Chip RFID [12] transponder also incorporates an on-chip antenna. The chip dimensions are 2.5 × 2.5 mm; it operates at a frequency of 13.56 MHz in very close proximity applications; the distance between interrogator and transponder should be less than 3 mm. The chip can store up to 2 kilobytes of user data. The antenna is mounted on top of the RFID chip using Maxell's proprietary electro-fine forming technology, which uses lithography and electroplating techniques. A wafer-level scale package method for on-chip antenna integration is proposed by Abe et al. [13]. As opposed to the Maxell RFID chip, this transponder is operated at frequencies above 2 GHz, using backscattering transmission and not via inductive coupling, which increases the communication range of RFID transponders by a factor 100.

Another interesting emerging technology for thin RFID applications is the use of organic electronics. In February 2009, Holst Centre announced a breakthrough in organic electronics presenting a 128-bit RFID transponder with high data rate and implementation of a basic anticollision protocol [14, 15]. These functions were realised using 1,286 organic p-type transistors with channel length of 5 μm. The design is operated at a carrier frequency of 13.56 MHz and occupies an area of 10.6 × 9.5 mm; transponder thickness is 25 μm. The high data rate that equals 2 kbps exceeds the transmission speeds of other comparable organic RFID transponders; however, data rates of more than 26 kbps needed to comply with the EPC specifications are not reached.

30.4 RFID Transponders with On-Chip Antenna

In order to overcome the assembly costs for transponders when the RFID chip is connected with an external antenna, the realisation of on-chip antennas will become increasingly important. The total costs for manufacturing an external antenna and assembling it to the RFID chip are greater than the costs for the RFID chip [16], so savings of around 50% for hardware costs can be achieved. The integration of an on-chip antenna will also positively influence the robustness of transponders, as the connection between antenna and chip experiences mechanical stress during bending of ultra-thin transponders. This stress might eventually lead to damage of the transponder. The μ-Chip mentioned in the last section is equipped with an on-chip antenna; however, only basic RFID functionality is given. This section elaborates on current research, carried out in the field of on-chip antennas, aiming at more complex RFID transponders.

The fully integrated transponder proposed by Shamali et al. [17] is designed in a 0.18-μm CMOS process in which the on-chip antenna is constructed by 5 coils realised in metal layers 2–6, each containing two loop turns. This large number of turns is needed for inducing the largest possible voltage in the close-coupled system between transponder and interrogator. This induced voltage is a function of the carrier frequency, the number of turns in transponder and interrogator coils, the radii of both coils and the distance between the coils. The area of the coils covers 550 × 550 μm, and the carrier frequency is 900 MHz. The induced voltage is about 0.3 V, and by a charge-pump as a power harvesting technique, a power supply of 1 V with load current of 5 μA is achieved. The transponder does not incorporate an identification number; it only generates a 208-kHz signal, which modulates the amplitude of the carrier frequency using load modulation, thus realising a 1-bit transponder. The operating range of the system is approximately 3 mm.

Chen et al. [18] describe a transponder with an on-chip antenna of 1 × 0.5 mm that operates at a frequency of 2.45 GHz. The transponder that has been designed in an 0.18-μm CMOS technology contains a 128-bit nonvolatile memory that can be read and written wirelessly. The challenges in the design lie in generating enough power and in storing this energy on chip for writing the memory, which requires an antenna with high quality factor. Due to eddy current losses in the substrate, plus the parasitic capacitances between coil windings and substrate, losses in the on-chip antenna would be high, so the on-chip antenna was realised in a post-process on top of the CMOS chip. An undoped silicon glass of 15 μm was used as an insolation layer between the chip's top level metal and copper antenna coil, whereas the interconnect was made with copper vias. The production costs for the post-processing is claimed to be 80% less compared to external antenna assembly [19].

Another RFID transponder operating at a carrier frequency of 900 MHz is reported by Xi Jingtian et al. [20]. The transponder with on-chip antenna is compatible with EPC Class 1 Generation 2 and includes a 640-bit EEPROM memory. The chip has been designed in 0.18-μm CMOS technology, and the antenna is processed during the normal CMOS process, so no post-processing for the on-chip antenna is needed. The optimised antenna design takes the substrate losses which can be modelled as an RC network into account and is operated at resonance. The on-chip antenna is connected via a matching network to the rectifier, which uses self-bias and threshold compensation techniques to generate a suitable power supply from the weak input signal. Combined with low power design methods for the digital base band, the transponder can be operated up to a distance of 10 mm from the interrogator.

Whereas transponders of most research groups and currently available products with on-chip antennas are operated using inductive coupling techniques, Radiom et al. [21] present an on-chip antenna that uses the backscattering technique. In their publication they describe an on-chip dipole antenna with rectifier and charge pump developed in a 0.13-μm CMOS process. The chip stores the harvested energy at a distance of 60 cm from the interrogator into a capacitor. Though they did not implement an RFID transponder, it is clearly visible that if the voltage across the capacitor is large enough, the circuit that is connected to the capacitor can be powered on and data can be sent to the interrogator. The on-chip dipole antenna is constructed as a meandering structure that measures 4.4 × 1.5 mm and is operated at a frequency of 1.45 GHz. The biggest challenge in designing the on-chip antenna is finding ways to overcome the losses in the substrate that are substantial; so, if substrate losses can be reduced, efficiency of the on-chip antenna will increase.

30.5 Conclusion

Radio frequency identification systems cover a large field of applications, ranging from electronic product codes to authentication of documents using secure eSeals. Latter applications require high computational effort needed by cryptographic algorithms implemented in the RFID transponder. Mechanical dimensions of the transponder should be as small as possible; the thickness must be less than 1/3 of the paper thickness, so transponders as thin as 30 μm need to be realised. Attaching an external antenna to the ultra-thin RFID chip and realising this desired thickness is a big challenge. Another way to create ultra-thin transponders is to utilise an on-chip antenna, which also increases the mechanical stability of the transponder, as no fragile interconnect is needed between antenna and chip. A number of promising RFID transponder architectures with on-chip antennas operated at high frequencies can be found in literature and have been discussed in this chapter. Most architectures use the inductive coupling method, which requires a careful design of the coil and selection of operating frequency. When operating frequency increases, induced voltage in the transponder increases. However, parasitic substrate losses simultaneously decrease the coil's quality factor which reduces the induced voltage. These substrate losses are also present when the backscattering technique is used; thus, if substrate losses can be reduced, the reading distance between transponder and interrogator could be increased, which would provide opportunities for new applications.

References

1. Finkenzeller K (2003) RFID handbook: fundamentals and applications in contactless smart cards and identification. Wiley, New York
2. Paret D (2005) RFID and contactless smart card applications. Wiley, New York
3. Stockman H (1948) Communication by means of reflected power. Proc IRE 36(10): 1196–1204
4. Ferrer-Vidal A, Rida A, Basat S, Yang L, Tentzeris MM (2006) Integration of sensors and RFIDs on ltra-low-cost paper-based substrates for wireless sensor networks applications, Wireless Mesh Networks, 2006. WiMesh 2006. 2nd IEEE Workshop, pp. 126–128, 25–28 Sept 2006
5. Blue Spark Technologies, Press release sealed air corporation deploys blue spark thin printed batteries to power its new rfid time and temperature data logging sensors. http://bluesparktechnologies.com/press_2008.09.08.cfm. publicated on 8 Sept 2008.
6. http://www.flexsecure.de/de/loesungen/flexsecure-e-seal/
7. Texas Instruments, Press release Texas instruments' ultra-thin chip module enables production of high-quality graphics-rich branded contactless cards. http://www.ti.com/rfid/shtml/news-releases-05-02-08.shtml. publicated on 5 May 2008
8. TDK Corporation, Press release New ultra-thin flexible rfid tag developed. http://www.tdk.co.jp/teaah01/aah24300.htm. publicated on 25 Sept 2007
9. Kurokawa Y, Ikeda T, Endo M, Dembo H, Kawae D, Inoue T, Kozuma M, Ohgarane D, Saito S, Dairiki K, Takahashi H, Shionoiri Y, Atsumi T, Osada T, Takahashi K, Matsuzaki T, Takashina H, Yamashita Y, Yamazaki S (2008) UHF RFCPUs on flexible and glass substrates for secure rfid systems. IEEE J Solid-State Circuits 43(1):292–299
10. Hitachi, Ltd., Press release Hitachi develops a new RFID with embedded antenna μ-chip, http://www.hitachi.com/New/cnews/030902.html. publicated on 2 Sept 2003
11. Noda H, Usami M (2008) 0.075×0.075 mm^2 ultra-small 7.5 μm ultra-thin RFID-chip mounting technology. Electronic components and technology conference 2008. ECTC 2008. 58th vol, pp 366–370, 27–30 May 2008
12. Maxell Seiki co., Ltd. Website Coil-on-chip^{TM} RFID systems. http://www.maxei.co.jp/products/coc/eng-index.html, accessed March 2010
13. Abe H, Sato M, Itoi K, Kawai S, Tanaka T, Hayashi T, Saitoh Y, Ito T. Microwave operation of on-chip antenna embedded in WL-CSP [RFID applications], Antenna technology: small antennas and novel metamaterials 2005. IWAT 2005. IEEE International Workshop, pp 147–150, 7–9 March 2005
14. Holst Centre, Press release Holst Centre reports major step towards organic RFID. http://www.holstcentre.com/en/NewsPress/PressList/Holst%20Centre%20reports%20major%20step%20towards%20organic%20RFID.aspx. publicated on 9 Feb 2009
15. Myny K, Beenhakkers MJ, van Aerle, NAJM, Gelinck GH, Genoe J, Dehaene W, Heremans P (2009) A 128b organic RFID transponder chip, including Manchester encoding and ALOHA anti-collision protocol, operating with a data rate of 1529b/s, solid-state circuits conference – digest of technical papers, 2009. ISSCC 2009. IEEE international, pp 206–207, 8–12 Feb 2009
16. Vojtech L (2008) RFID transponder in X-band and its feasibility, sensor technologies and applications, 2008. SENSORCOMM '08. Second international conference, pp 99–102, 25–31 Aug 2008
17. Shameli A, Safarian A, Rofougaran A, Rofougaran M, Castaneda J, De Flaviis F (2008) A UHF near-field rfid system with fully integrated transponder. IEEE Trans Microw Theory Tech 56(5):1267–1277
18. Chen X, Yeoh WG, Choi YB, Li H, Singh R (2008) A 2.45-GHz near-field RFID system with passive on-chip antenna tags. IEEE Trans Microw Theory Tech 56(6):1397–1404

19. Guo LH, Popov AP, Li HY, Wang YH, Bliznetsov V, Lo GQ, Balasubramanian N, Kwong D-L (2006) A small OCA on a 1 × 0.5-mm2 2.45-GHz RFID Tag-design and integration based on a CMOS-compatible manufacturing technology. IEEE Electron Device Lett 27(2):96–98
20. Jingtian X, Na Y, Wenyi C, Conghui X, Ciao W, Yuqing Y, Hongyan J, Hau M (2009) Low-cost low-power UHF RFID tag with on-chip antenna. J Semicond 30(7):1–6
21. Radiom S, De Roover C, Vandenbosch G, Steyaert M, Gielen G (2009) A Fully integrated pinless long-range power supply with on-chip antenna for scavenging-based rfid tag powering, silicon monolithic integrated circuits in RF systems, 2009. SiRF '09. IEEE Topical Meeting, pp 1–4, 19–21 Jan 2009

Chapter 31
Thin Chips for Document Security

Thomas Suwald

Abstract This article underscores the importance of thin silicon for the development of new RFID security products. Quality requirements are presented that show how the utilisation of thin silicon enables new business opportunities. A presentation of some typical RFID applications, based on thin silicon, gives an impression of thin silicon integration into RFID security products.

Keywords Wafer thinning · Chip side rounding · Robustness · Flexible substrates · Printable electronics · Mechanicale stress tests · Die breakage · Printed antenna · Hot-wire antenna · Printed interconnections · Electronic passport · Electronic ID card · Display card · ePaper display · Chip in paper · Hybid systems

31.1 Ultra-Thin Chips – Enabler for Robust and Flexible RFID Applications

Despite the economic situation in 2009, the radiofrequency identification (RFID) market has expanded in 2010. According to ABI research, 'RFID orders surged in 2009; experts predict the technology will continue to grow through the coming year and beyond. Despite continued challenges regarding RFID adoption, new applications multifunctional chipcards for authentication applications emerged in the course of 2010 in the commercial and industrial sectors'. Over the next 5 years the global RFID market will grow from \$5.35 billion in 2010 to \$8.25 billion by 2014 at an expected compound annual growth rate of 14%. New application domains will grow faster than traditional applications like car immobilisers or access control [1].

T. Suwald (✉)
NXP Semiconductors Germany GmbH, Business Unit Identification, Georg-Heyken-Str. 1, 21147 Hamburg, Germany
e-mail: thomas.suwald@nxp.com

J.N. Burghartz (ed.), *Ultra-thin Chip Technology and Applications*,
DOI 10.1007/978-1-4419-7276-7_31, © Springer Science+Business Media, LLC 2011

Growth is driven by cost improvement programs in supply chain management, public administration (via electronic government – eGovernment – documents) and payment transactions. Manual processes for item or individual tracking are replaced by automated processes, where the exchange of authentication information between tagged items and a database application is implemented by RFID. The increase of miniaturisation and the significant improvements in mechanical robustness are prerequisites to the opening up of new application domains. An analysis of potential new RFID applications (Fig. 31.1) unveils mechanical flexibility as a key enabler. This is exactly where thin silicon enters the scene.

The eGovernment domain is a good example how RFID component properties influence the development of new application areas. Among the expected growth areas are electronic identification cards (eID), citizen cards, electronic passports (ePP), electronic vehicle registration and electronic driver licenses. Document security is a key requirement for those applications and is achieved by a combination of printed and electronic security layers. As the form factors of the documents are fixed, an increase of layers automatically results in reduced layer thicknesses. The remaining electronic layer thickness will reduce further to levels well below 200 μm – for chip-in-paper applications well below 25 μm.

Figure 31.2 indicates the relation between chip document applications and die thicknesses. New applications like ultra-thin electronic layers of eGovernment documents and chip-in-paper applications are not feasible without ultra-thin silicon components.

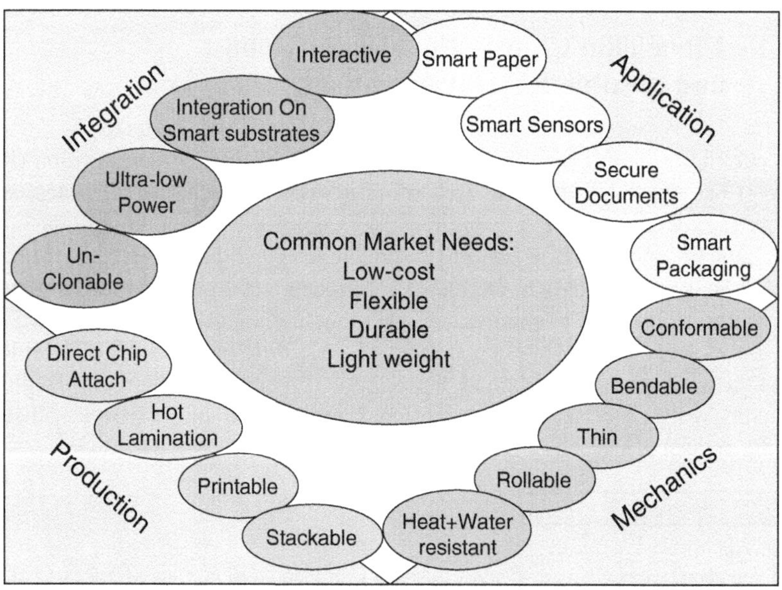

Fig. 31.1 RFID application properties

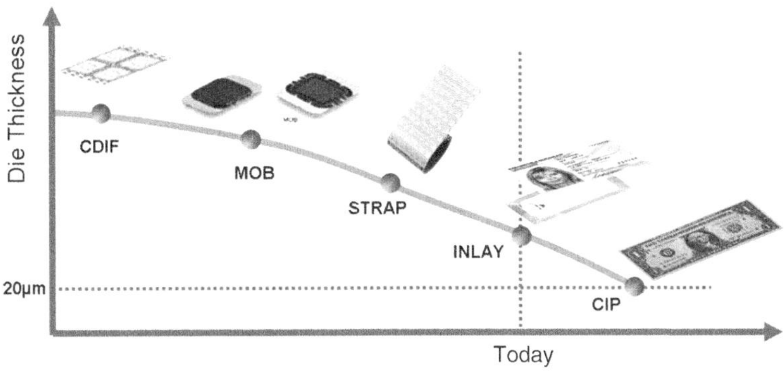

Fig. 31.2 Trend in chip thickness reduction, stimulated by applications

Thin silicon components introduce some positive side effects: Due to the increase of mechanical flexibility more efficient and lower-cost assembly methods can be applied, resulting in an overall cost reduction of RFID transponders. The reduction in die thickness is also an enabler for document compatible 3D-stacks, which are combinations of die functions stacked on each other.

31.2 RFID Quality Requirements

Still today most RFID applications are based on single-chip implementations. Adding user interface components to electronic documents like buttons, displays and biometrical sensors results in more complex systems. Display card systems for authentication purposes combine functions like a security controller, memory chips and sensor applications, including biometrical sensors. With the introduction of more capable systems, the utilised silicon area is bound to grow. Switching to new process technologies hardly compensates the increase in die area, as a lot of area-consuming blocks are analogue and power management blocks. An increase of power results in larger capacity areas for the voltage converters that cannot be compensated by denser processes. An increase in die surface area requires special measures to make the chips resistant against mechanical stress like bending and torsion. Increasing the mechanical flexibility by thinning is a good countermeasure against the mechanical tortures that a high percentage of the RFID applications have to endure today.

The main factors that influence a die's mechanical robustness are wafer thickness, die surface uniformity and chip side and chip edge (Fig. 31.3) smoothness. Wafer thicknesses of 50 µm and below, in combination with chip side healing and a chemical mechanical backside polishing (CMP), will make the chips robust against the common stress situations applied to the host application like bending, torsion and wrapping. Conformity with these mechanical requirements is

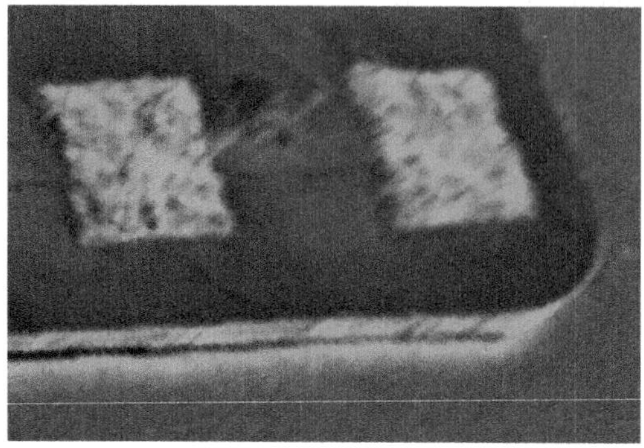

Fig. 31.3 Rounded chip corners

Table 31.1 SmartCard ISO/IEC 10373 mechanical tests

No	Abbr.	Test description	Method	Condition	Requirement (R/extended)
1	BEND	Bending test	ISO/IEC 10373-1/5.8	10/20 mm deflection	**1,000**×
				Test: cold RST, warm ATR	2,000× 3,000× 4,000× 5,000× 6,000×
2	TORS	Torsion test	ISO/IEC 10373-1/5.9	±15°	**1,000**× 2,000× 3,000×
				Test: cold RST, warm RST	4,000× 5,000× 6,000×
3	CERE	Chemical resistance	ISO/IEC 10373-1/5.4 short term contamination	20–25°C, 1 min Test: cold RST, warm RST, ATR	
			According to ISO 7810 long term contamination	20–25°C, 24 h	
			According to ISO 7810 MIL-STD 883 Meth. 1009	Test: cold RST, warm RST, ATR	
4	WRAP	Wrapping test, mandatory	CQM Rel:1.9E Date: 2007-02-09 Chap. 10.3.23	$R = 20$ mm	**10 × FS** **10 × BS**

verified by a couple of critical tests, such as what is defined by the ISO/IEC 10373 standard. Typical quality requirements of a smartcard host application are listed in Table 31.1.

The mechanical tests as defined by the ISO/IEC 10373 standard simulate typical mechanical stress situations as applied to an RFID component inside a host application during its lifecycle. Typical setups for the bending test (both axes) and for the torsion test are displayed in Fig. 31.4.

The wrapping test, as indicated by Fig. 31.5, supplements the mechanical stress tests: It simulates stress that is typically applied during RFID card personalization and mailing. The card under the test is bent over a metal cylinder with a radius of 20 mm. The test is performed bending over frontside and rearside. The FEM simulation results [2] on the right indicate a smart card with an encapsulated chip-module bending over the card's backside. The mechanical stress approaches the highest tension level of 390 MPa in the die centre.

The wrapping test is quite critical in case of more complex smartcard systems because next to the die the components involved in the chip-attach process (bumps, adhesives) are also stressed. Figure 31.6 indicates a negative wrapping test result. A 150-μm silicon die embedded in a multichip ID1-smartcard was bent over a 20-mm metal radius. The crack started from the rear side of the chip.

The radius of 20 mm represents the typical transportation rolls with a diameter of 40 mm. Further reduction of the bending radius below 20 mm results in dramatically increased failure as Table 31.2 underscores.

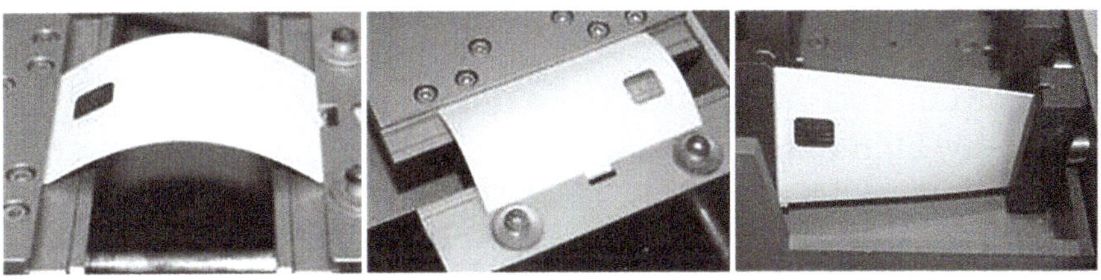

Fig. 31.4 ISO bending and torsion tests

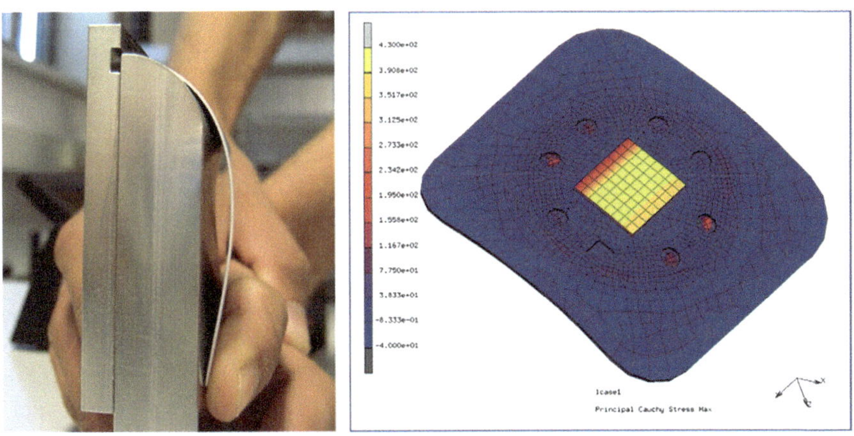

Fig. 31.5 Wrapping test setup and simulation results [2]

Fig. 31.6 Die breakage during wrapping test [2]

Table 31.2 Radius impact on wrapping rest

Wrapping radius (mm)	Failure rate (fail/total)
20	0/20
18	2/20
16	18/20

31.3 Manufacturing Cost Reduction Stimulated by Thin Silicon

The request for chip thickness reduction is not solely defined by the host application's mechanical stability requirements. A strong pull for thinner silicon results from an assembly cost reduction demand. So far the manufacturing of RFID systems still involves serial processes like antenna wiring and pick-and-place. Parallelisation can here only be achieved by implementing a sufficient amount of process lines in parallel. One example is the manufacturing process for ePP antenna inlays involving the hot-wire technology (SmarTrac, Fig. 31.7): The antenna substrate is locally heated up by an ultrasonic beam, and the thin wire tracks are pushed into the softened material by a wire router.

This technology is straightforward but it is a serial process. As printing processes are already applied during the document finishing and personalisation phase, it is a logical response that one utilise available printing equipment for the antenna manufacturing process as well. Printing the antenna directly on the flexible document layers (Fig. 31.8) with conductive ink makes unnecessary the purchase of a margin-sensitive antenna pre-product – a first step towards product cost reduction.

In this case, existing printing machines need only be modified in for major investment for new process equipment to be avoided. This is an important factor for smaller manufacturers. The assumed yearly production volume of the German eID cards as an example is at a level of 8 million units per year – a modest volume that permits sharing of existing equipment. At production capability of 10,000 units per hour for a typical pick-and-place machine a volume of 8 million units can, theoretically, be processed within 32 days. A single printing machine is hence capable of printing inlays and running the other document printing processes.

A logical next step in assembly cost reduction will be the replacement of the traditional interconnection process by a process based on existing printing equipment. Interconnection based on printing technology (Fig. 31.9) is highly parallel and will result in a visible manufacturing cost reduction, as long as the silicon components can be made resistant against the expected mechanical stress.

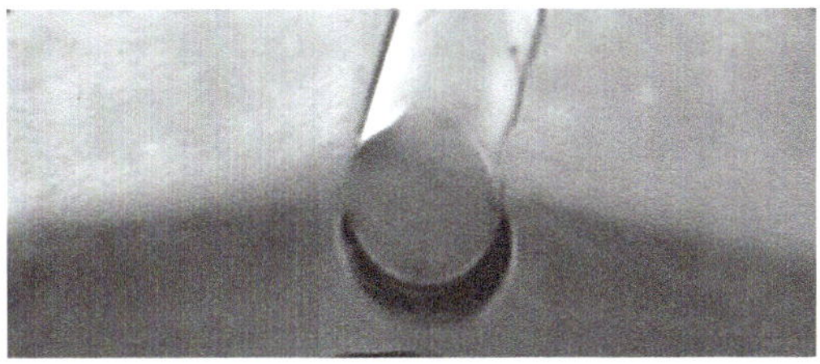

Fig. 31.7 Hot-wire antenna manufacturing [3]

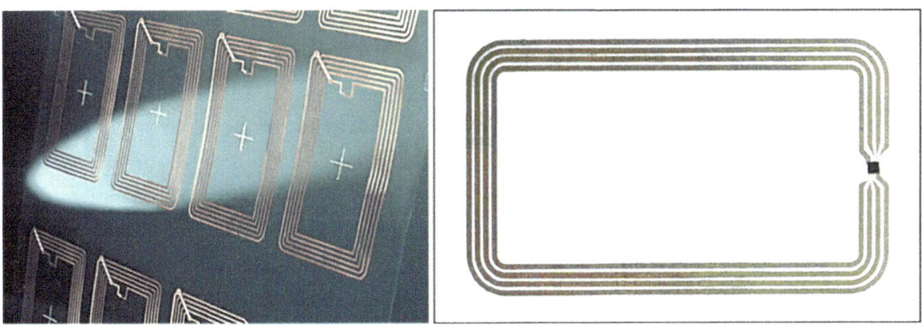

Fig. 31.8 Printed silver ink antennae and assembled inlay with security controller

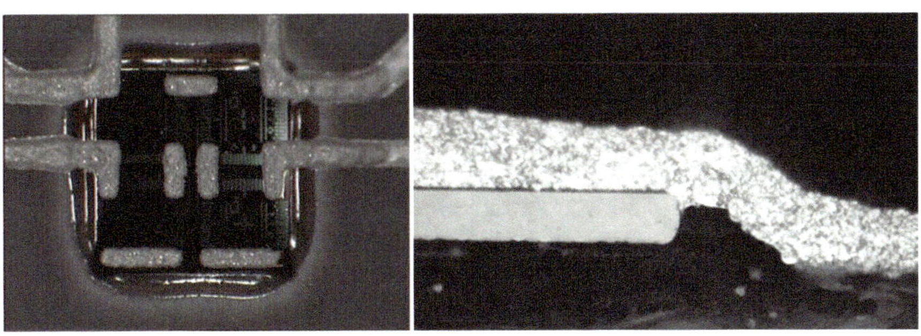

Fig. 31.9 Silver-ink printed IC interconnection [4]

Mechanically robust silicon components are a prerequisite for the introduction of low cost and fast printing interconnection processes and hence constitute an enabler for new, price-sensitive volume applications. Next to manufacturing cost (cost per unit and investment/depreciation) substrate material is a key cost contributor. Cheap substrate materials like paper will become feasible if the silicon components can get close enough to the flexibility properties of the substrate. Substrate materials like paper will enable disposable volume products. Paper, compared to, i.e., PET, polyimide and polycarbonate, is from an ecological perspective very attractive, as it is fully recyclable and provides a good CO_2 balance.

Among the new applications for thin silicon components are security seals, banknotes and secure paper document applications. These applications demand, beside low height levels, excellent mechanical stability of the components involved. Typical thicknesses of document papers and banknotes are in the order of 80–100 μm. The remaining height level for the chip layer in these applications is $\leq$25 μm including the required bumps and interconnect layers.

31.4 Thin Silicon Security Application Examples

31.4.1 Multifunctional Electronic Passports and ID Card

Electronic government (eGovernment) documents have been compared to other RFID applications for the extremely high quality requirements of their silicon components. An eGovernment document has to withstand harsh treatments by its owner. Often stored in a wallet somewhere in a back pocket, it is subjected to all kinds of mechanical stress to its inner components. The majority of the eGovernment documents have a projected expiration period of up to 10 years; some of them, like driver licenses (at this writing), never expire. These constraints make the design of eGovernment systems a challenge. Nevertheless, electronic passports (ePP) and electronic identification cards (eID) are well introduced to the current document landscape.

An example of an electronic passport that utilises a multilayer design is pictured in Fig. 31.10. The electronic inlay is based on a hot-wired antenna with a mounted security controller module; the security chip inside is enclosed by an encapsulation that protects it against stress.

While the ePP is typically stored at a safe place most of the time, an eID is an everyday item. The eID (Fig. 31.11, new German eID) has to withstand by far more mechanical stress than an ePP. Due to steadily increasing security requirements the remaining height for the electronic layer reduces constantly. The thickness of a standard ID1 card is $\pm$800 μm. Traditional thickness of the electronic layer was $\pm$400 μm. Due to additional optical anti-counterfeiting features implemented in the printing layers, the available height for the electronic layer reduces to less than 200 μm. Target thicknesses demanded by eID manufacturers are close to 100 um. A thickness budget of 100 um leaves just 40 μm for the silicon components. 50 μm

Fig. 31.10 The German electronic passport is a multifunctional document [5]

Fig. 31.11 New German eID [5]

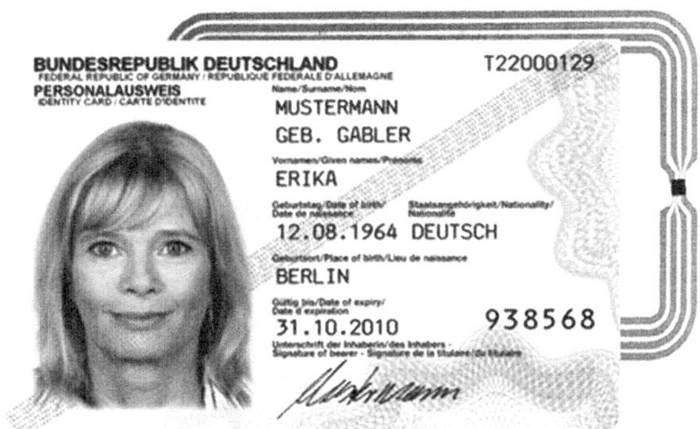

will be used for a substrate and 10 μm for contact bumps and conductive adhesive or conductive film for the chip attach.

Latest designs of the electronic inlay are based on a printed inlay with the security controller flip chipped directly to the antenna. The chip is placed across the antenna loops and makes an additional antenna bridge obsolete. The security controller for eIDs has been tested at thicknesses of 30–75 μm.

31.4.2 Display Card – A Multi-Chip SmartCard Application

Display cards (Fig. 31.12) are good examples of higher integrated systems with increased silicon area. As a rule of thumb die sizes beyond 3 mm^2 require special

Fig. 31.12 Dual-power display token card

means to achieve the required mechanical stability; otherwise, the risk of die breakage is unacceptably high. Thinning the wafers to a thickness ≤ 50 μm in combination with laser-dicing and chip-side healing (rounding the die edges) guarantees the necessary mechanical properties. The inlay design of this display card is a combination of two substrates: an antenna substrate (FR4) and a display module substrate (PI). An optional battery makes this design a dual power system. The display is an E-Ink electrophoretic electronic paper type.

The integration of electronic paper (E-Ink) display drivers becomes a special challenge: pin-counts of 100 and above are common due to the single segment drive scheme. Attaching the die via 100 connections to the substrate creates a sandwich construction that significantly reduces the flexibility of the overall display module. Thinning the chips to ≤ 50 μm helps in this case to compensate the flexibility loss. Mechanical FEM simulations are required to identify card areas with lower stress areas (corners).

31.4.3 Integration on Printed Inlays

Figure 31.13 gives an example of a flip chip-bonded display module. The two chips in the middle are 100-pin display segment drivers assembled by an anisotropic conductive adhesive (ACA) approach. The substrate material is polyimide.

Figure 31.14 displays an X-ray of the display module of Fig. 31.13 mounted to a printed inlay. An ACA flip chip assembly process is used to assemble the module to the antenna.

31.4.4 Security Tags and Labels

Security tags for documents represent small and flexible RFID transponders that can be attached to secure documents for tagging or as protection against counterfeiting, e.g., for pharmaceutical products (Fig. 31.15). Due to the small antenna area the reading range of these transponders is limited to a few millimetres.

Fig. 31.13 Flexible ePaper display module

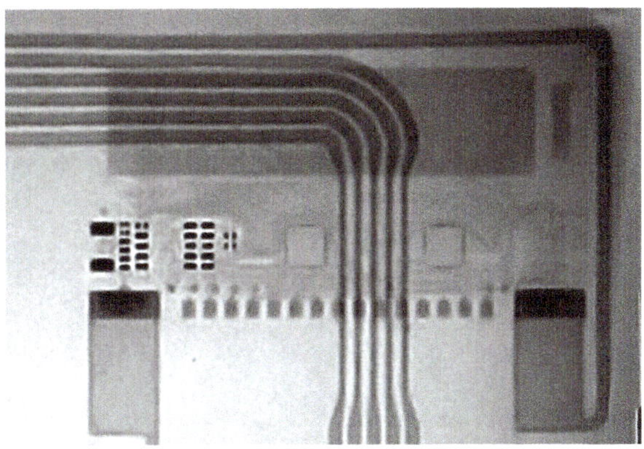

Fig. 31.14 Display module mounted on a printed antenna

Fig. 31.15 Security tag [6] and UCODE chip with integrated RF antenna

In combination with optical security features (holograms, bar codes) a handy tag can be created that is compatible with existing optical scanners for bar codes, hologram scanners and RFID readers [6].

31.4.5 Chip-in-Paper

Adding security features to paper documents or banknotes requires silicon thicknesses of less than 25 μm to match the flexibility properties of the host material. Thickness of paper is in the range of 80–100 μm. A conductive antenna layer has a thickness of 20 μm. Depending on the assembly method an additional bump layer of 10 μm might be required. Considering a bottom and top paper layer of 30 μm leaves less than 15 μm for the silicon chip. First samples of a 7.5-micron thick chip have been presented by Hitachi during the 2006 IEEE International Solid-State Circuits Conference (ISSCC) in San Francisco [8]. Target applications for paper integration are intelligent watermarks and electronic seals.

31.4.6 Thin Components in Hybrid Systems

The introduction of flexible polymer electronics created excitement about the technology's potential and how it predicted a low cost alternative for silicon electronics in RFID applications. But it became obvious rather soon that polymer electronics will take a while before it is able to succeed silicon as the main integration medium of RFID solutions. Systems with high processing speed requirements currently cannot be integrated in polymer electronics – performance and cost targets cannot be achieved yet. A key benefit of polymer electronics is its

Fig. 31.16 RFID label with polymer-IC (PolyIC [7])

mechanical flexibility and compatibility with printing technologies. A feasible way to integrate RFID systems is a combination of polymer and silicon components. The silicon components in such hybrid systems have to match the flexibility properties of the polymer electronics in order to maintain the flexibility benefit of the overall system. Silicon components with thicknesses of ≤ 50 μm and with carefully improved chip sides provide the required material properties. An example of a polymer RFID design is given in Fig. 31.16.

Acknowledgements The information related to multifunctional smartcards presented in this article is partially based on publically funded innovation activities. The author would like to thank the BSI (Bundesministerium für Sicherheit in der Informationstechnologie) for funding the InSiTo project and the BMBF (Bundesministerium für Bildung und Forschung) for funding the SECUDIS project.

References

1. Leybovich I (2010) RFID market projected to grow in 2010. 11 Mar 2010, http://news.thomasnet.com/IMT/archives/2010/03/radio-frequency-identification-rfid-market-projected-to-grow-in-2010-beyond.html
2. Kasemset B, Zenz C, Eiper E, van Driel WD, Yang DG (2009) Reliability modeling of IC package in smart card applications. Business Line Identification/Operations BE Innovation. NXP Semiconductors Austria GmbH, 10 Mar 2009
3. SMARTRAC Recognizing the world (2009) SmarTrac Company Presentation. p 11, http://www.smartrac-group.com/en/download_2/20071030_SMARTRAC_Brochure.pdf
4. Feil M, Landesberger C, Bock K (2010) The challenge of ultra thin chip assembly. Fraunhofer Institute for Reliability and Microintegration (IZM), Hansastr. 27d, Germany. Downloaded 25 Apr 2010, www.izm.fraunhofer.de/Images/114_feil_tcm357-92246.pdf
5. Bundes Druckerei, www.bundesdruckerei.de
6. Tesa-Scribos, www.tesa-scribos.com
7. PolyIC, www.polyic.com
8. Presentation 7.5-micron-thick chip, IEEE International Solid-State Circuits Conference (ISSCC) in San Francisco. 5 Feb 2006

Chapter 32
Flexible Display Driver Chips

Ali Asif and Harald Richter

Abstract Rapid advancements in the field of flexible display have lifted expectations of the availability of full-flexible display products in near future. A full-flexible, paperlike display requires ultra-thin driver chips embedded in the same flexible substrate. This reduces external connections and cost of product, and it increases system reliability. Organic materials, hydrogenated amorphous silicon (a-Si:H), poly-silicon and single-crystal silicon are potential candidates for ultra-thin chips, each with its pros and cons. High mobility transistors are indispensible for video streamings. Threshold instability and self-heating issues demand proper addressing. Different placement schemes of driver chips around TFT matrix and various driving waveforms can be employed to drive heavy loads, especially in case of large-size displays. Technical obstacles must be eliminated in order to realise mass productions of full-flexible display systems.

Keywords Flexible display · Ultra-thin driver · Thin-film transistors · Single-crystal silicon · Matrix addressing

32.1 Introduction

Viable and dynamic flexible display technology has enticed the electronic display industry for the last few years. Flexible displays are supposed to be thin and sturdy, and they are designed to have modifiable shapes and sizes. Based upon degree of freedom of flexibility, flexible displays can be broadly categorised as one-time

A. Asif (✉)
Institut für Mikroelektronik Stuttgart (IMS CHIPS), Allmandring 30a, 70569 Stuttgart, Germany
e-mail: aliasif@ims-chips.de

J.N. Burghartz (ed.), *Ultra-thin Chip Technology and Applications*,
DOI 10.1007/978-1-4419-7276-7_32,

flexible, semi-flexible and full-flexible. One-time flexible displays are curved or conformed at the time of manufacturing and retain their shape afterwards. Semi-flexible displays can only be partially bent, rolled or folded. Full-flexible displays can be rolled, folded, bent and/or stretched from any direction, as many times as possible without breakage [1–3].

A typical flat-panel flexible display comprises a flexible substrate, a layer (the 'back plane') of either conducting stripes or a matrix of thin-film transistors (TFTs), a layer of functional display material and a layer (the 'front plane') for encapsulation. The front plane also serves as an electrode (Fig. 32.1). Additional coatings of thin films and sealings are used to enhance barrier properties and reliability, as well as other features of the display [4].

Material selection for flexible substrates depends on a number of factors: flexibility, robustness, thermal, optical and barrier properties, compatibility with other layers and feasibility for roll-to-roll manufacturing, among others. Flexible glass and metal sheets, thin sheets of polymers like polyimide, polyethylene terephthalate (PET) and polyethersulphone (PES) have shown good results in this regard, but none of them can yet be considered colloquial. A functional display material reflects or transmits light upon application of proper voltage across it. Different technologies, e.g., electrophoretic, cholesteric LCD, electrowetting, electrofluidic and electrochromic, have been adopted by various vendors to get better display properties. New technologies like photonic crystal technology etc. are also showing their presence in industry, but they are still in their early/initial development stages. Thin films of transparent conducting oxides (TCO), e.g., indium-tin oxide (ITO), are used to form a front encapsulation layer. Beside adequate optical properties TCOs also show good electrical characteristics, which enable them to be used as electrodes in displays.

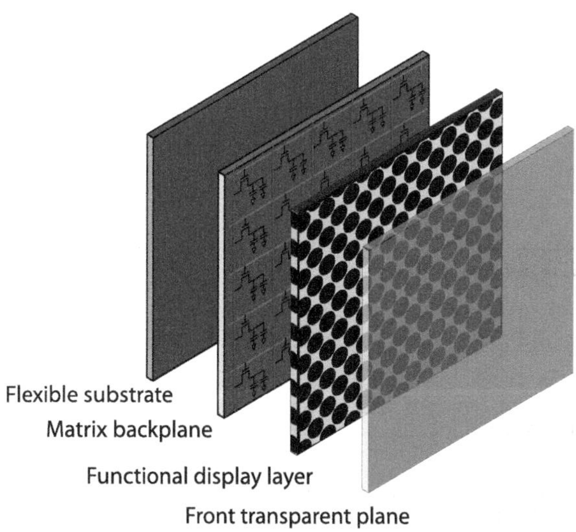

Fig. 32.1 Layer arrangement in a typical flexible flat panel display

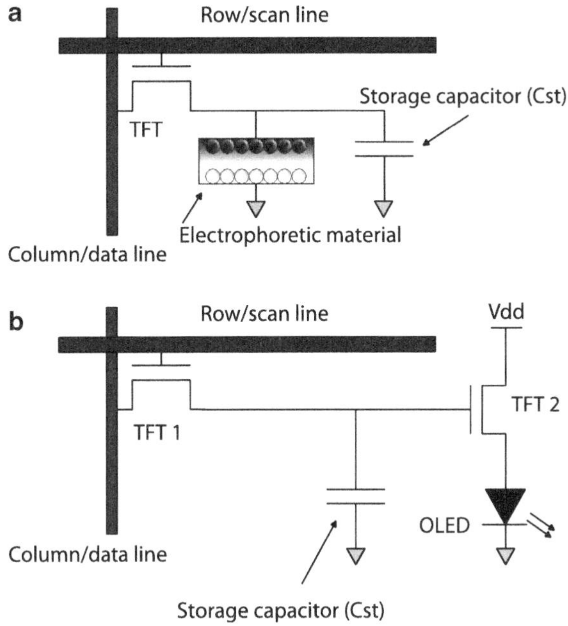

Fig. 32.2 A simple pixel circuit for (**a**) electrophoretic and (**b**) OLED display

The functional display layer can change its optical properties upon application of suitable voltage across it. Such voltage is provided by employing either a passive matrix or an active matrix technique. In the case of the passive matrix, conducting stripes are patterned on both front and back planes. They are patterned perpendicular to each other in order to form a matrix. The intersection points of the stripes define pixels in the passive matrix. In the active matrix, TFT-based pixel circuits (TFT switches) provide appropriate voltages. The pixels circuits are patterned in matrix form on the back plane, whereas in this case, the front plane serves as a common electrode. The number and density of pixels is determined by the number and density of TFT switches in the backplane. The number of TFTS in pixel circuits can vary. For example, a 1-TFT pixel circuit is sufficient for electrophoretic displays, but in OLED displays at least a 2-TFT pixel circuit is required (Fig. 32.2). A storage capacitor is also used in every pixel circuit. It helps in maintaining voltage level at the TFT gate terminal and in reducing effects due to leakage current.

Generally, addressing with active matrix is preferred over a passive matrix scheme because of the former's higher refreshing capabilities, its better control of applied voltage, reduced crosstalk and fewer flickering problems. On the other hand, due to its simplicity and low cost, passive matrix schemes still find application in systems where large displays, higher resolutions and fast refreshing rates are not involved.

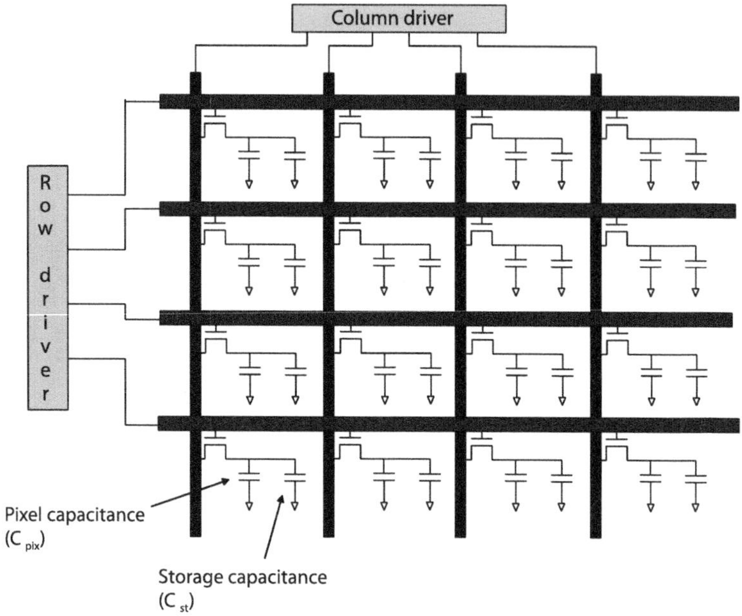

Fig. 32.3 4 × 4 active matrix backplane with row and column drivers

32.2 Display Drivers

A matrix back plane scheme requires row and column drivers for providing precise voltage signals to each pixel through pixel circuits (Fig. 32.3). Packed in tape carrier packages, these drivers are placed outside the display and connected to matrix back planes via tape-automated bonding (TAB).

Row (or scan) drivers are shift registers that generate sequential voltage pulses for addressing rows one-by-one. For a frame with N rows, a row driver with N buffers is required. The driver generates N pulses with frequency $1/T_{\mathrm{f}}$, where T_{f} is the frame-refreshing time period [5]. The clock frequency required by the row driver is determined by the product of number of rows and refresh rate [6]. It lies in the range of a few tens of kilohertz. The time available for pixels in a row (T_{row}) to activate, accept respective data, settle and deactivate is given by

$$T_{\mathrm{row}} = \frac{1}{\text{Frame frequency}} \times \frac{1}{\text{Number of rows}} \quad (32.1)$$

The level shifter in row drivers shifts the voltage levels from 0 and 3.3 V to –5 and 20–25 V, respectively [6]. The values of these voltage levels are set compatible with the ON and OFF requirements of the TFT gates in matrix back planes. In contrast to row drivers, a column driver has to provide relatively high analogue voltages (several tens of volts) at the source terminals of TFTs in the

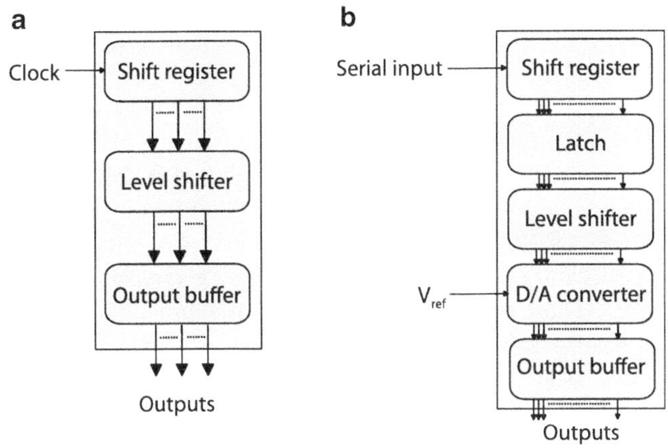

Fig. 32.4 Simplified form of typical architecture of (**a**) row and (**b**) column driver for monochrome display

matrix back plane. This makes the design of a column driver more complex compared to a row diver. A typical column driver accepts serial low voltage digital input data, shifts the voltages to higher levels, converts the data into parallel analogue signals and applies it to all the pixels selected row simultaneously. Figure 32.4 depicts the internal structure of typical row and column drivers for a monochrome display. For a full colour display, the signals become threefold in order to represent RGB colours. The minimum clock frequency limit for shift registers in a column driver is determined by the product of pixel count and refresh rate [6]. The value of the required frequency also depends on colour depth and number of drivers.

In a high definition colour display (1920 × 3 × 1080), the back plane matrix can be driven using 4 row-drivers, each having 270 channels and 8 column-drivers with 720 channels each. For 24-bit colours and at a 60-Hz refresh rate, the column drivers have to operate at a frequency greater than 373 MHz. The high frequency requirement can be reduced by increasing the number of drivers, adopting techniques like dual scan driving, grouping of columns, simultaneous addressing and so on.

32.3 Ultra-Thin Driver Chips

Realisation of a full-flexible, paperlike display demands embedding of ultra-thin driver chips on a flexible substrate. Integration of such chips on flexible substrates reduces not only external connections but also the product cost. It also helps in increasing reliability of the system, especially when it is in a rugged environment.

Development of ultra-thin driver chips, compatible with flexible substrates, is a cumbersome task. Plastic substrates cannot withstand high processing temperatures. They also possess lower dimensional stability compared to rigid substrates. Therefore, modifications in conventional fabrication processes, employed for devices on rigid substrates, are required. Two approaches have been widely adopted by researchers to deal with this issue. One is the fabrication of low-temperature-processed TFTs directly on plastic substrates. The other is transfer of prefabricated devices on plastic substrates. The former approach targets amorphous silicon, low temperature polysilicon (LTPS) and organic semiconductors (OTFT) whereas the latter deals with high temperature processed polysilicon or single crystal silicon TFTs. Decrease in processing temperatures results in low carrier mobility. Therefore, the direct fabrication approach is unsuitable for producing high performance circuits on plastic substrates. On the other hand, high mobility transistors are indispensible especially in driver circuits for meeting the high refreshing rates required in video streaming. Also, significant degradation occurs in TFT on flexible substrates due to threshold voltage variation, self-heating and piezoresistive effects.

Hydrogenated amorphous silicon (a-Si:H) TFTs require low processing temperatures, which make them suitable for direct fabrication on flexible substrates. The a-Si:H TFT-based drivers on flexible substrates [7, 8] perform well at lower refreshing rates (up to a few tens of hertz [8]) or for occasional image refreshing applications. However, because of low mobility values (~0.1–1 $cm^2/V\text{-}s$) they are unable to cope with video signal refreshing requirements (Table 32.1). Integration of both source and row drivers reduces the total number of external interconnects significantly, as in [7], where for a QVGA display the total number was reduced from 560 to 76, i.e., reduced by a factor of 7.4.

Hydrogenated amorphous silicon (a-Si:H)-based circuits face severe degradation in performance due to their threshold voltage instability. This instability is attributed to charge injection or charge trapping in the gate insulator and defect formation in the channel region in amorphous silicon [9]. The contribution of the two mechanisms is a function of applied voltage stress, where defect formation in a-Si dominates at low applied voltage, and charge trapping in the gate insulator dominates at high applied voltage stress. The shift in threshold voltage affects column voltages, where a 30-V signal can be decreased by 10% within only 3 h [10]. The effects of threshold voltage shift can be minimised, if not eliminated, by accurate modelling of instability mechanisms and their impacts on device

Table 32.1 Grain sizes and mobility value range for different silicon types

		Mobility in transistor channel	
Silicon type	Grain size	Holes ($cm^2/V\text{-}s$)	Electrons ($cm^2/V\text{-}s$)
Amorphous	<1 nm	~0.01	~0.1–1
Nanocrystalline	~10 nm	~0.2	~40
Microcrystalline	~0.1–100 μm	~50	~300
Large-grained poly	~0.1–100 mm	~200	~350
Single crystal	–	~250	~600

performances. Also, improved circuit designs may compensate undesirable effects due to threshold voltage shift.

Polycrystalline silicon TFTs show mobility values much higher than amorphous silicon TFTs. High temperature (>600°C) processed polysilicon TFTs show better electrical performances than low temperature processed polysilicon (LTPS) TFTs, but such high processing temperatures make them unsuitable for direct fabrication on flexible plastic substrates. However, they can be fabricated on flexible metal sheets, which can withstand temperatures as high as 1000°C. On flexible metal substrates these devices can provide mobility values upto 300 cm^2/V-s for n-channel MOS (NMOS0) and 150 cm^2/V-s for p-channel MOS (PMOS) TFT [11]. With such high field-effect mobility, poly-Si TFT drivers can easily drive video signals. For producing driver circuits on plastic substrates, preprocessed low temperature polysilicon TFTs can be transferred onto a flexible plastic substrate without showing significant performance degradation [12]. On plastic substrates, the achieved mobility values are, however, much lower than claimed in [11] on metal substrates. With continuous research aimed at decreasing maximum processing temperatures, low temperature polysilicon TFTs with mobility ~50 cm^2/V-s can now be fabricated directly to plastic substrates with relatively high glass transition temperature, e.g., polyimide (T_g > 350°C) [13].

Organic TFTs offer the advantage of low processing temperatures and compatibility with roll-to-roll printing technique. These properties permit direct fabrication of organic TFTs on all flexible substrates. Design of row driver with organic TFTs has already been reported in [14]. The performance is, however, limited to low resolution and small display sizes. This is mainly due to very low mobility values of organic transistors, which lie in the range of about 1 cm^2/V-s for p-type and about 0.2 cm^2/V-s for n-type TFT. Recently, mobility values up to 5 cm^2/V-s in solution-processed organic TFTs have been reported [15]. This mobility value exceeds that of amorphous silicon TFTs, and provides ease in mass production by printing technique. The mobility values of organic semiconductors are a function of internal ordering, defects and interface properties. In the case of pentacene, various techniques yield different mobility values that range from 1.8 cm^2/V-s for solution processed pentacene to 40 cm^2/V-s for single crystal pentacene [16].

With respect to circuit design, organic TFTs have to face another challenge due to the performance of n-type organic TFTs. The mobility values of n-type TFTs are far less than p-type TFTs, a situation that is opposite to the case of inorganic TFTs. Furthermore, in ambient environment, n-type organic materials are very much unstable. These two factors suggest a need of modification in design techniques of organic TFT-based circuits. A PMOS-only or pseudo-PMOS approach can eliminate the limitations set by n-type TFTs. It has been shown that employing a PMOS-only approach can improve overall circuit performance by a factor of 2–4 compared to CMOS arrangement of organic TFTs [17]. However, in integrated source drivers the CMOS method dissipates much less power than n-type or p-type only TFTs [18]. Another approach for using a CMOS scheme with organic TFTS is the hybrid CMOS device approach [18–20]. The hybrid CMOS technique makes use of p-type organic TFTs and n-type inorganic TFTs to form a complementary scheme. The hybrid CMOS approach has so far shown satisfactory performances and stability on flexible plastic substrates, but at the cost of enormous processing effort.

The well-established single-crystal silicon technology was previously thought to be incompatible with plastic substrates because of high processing temperatures. With the development of the techniques of transferring prefabricated single-crystal silicon devices on flexible substrates [21, 22], single-crystal silicon technology has now found its scope in the field of flexible displays. For this, ultra-thin silicon chips are required to maintain the degree of flexibility of product after embedment. For reducing the thickness to $\leq$20 μm, conventional backside grinding technique is not suitable as it can produce nonuniform thicknesses and defects; furthermore, handling of ultra-thin wafer is not easy. However, techniques like Chipfilm™ [23] can be used to solve these problems.

With the help of single-crystal silicon technology, high speed transistors with mobility >500 cm^2/V-s and on/off ratios $>10^5$ on flexible substrates can be obtained [24]. For column drivers, in level shifter stage, high voltage transistors are required to shift the voltage level up to high supply voltage. The conventional high voltage designs for LDMOS transistors rely on deep wells along with RES-URF technique to sustain high voltages. The depth of well, usually a few tens of microns, puts a limit on the reduction of chip thickness. This suggests the necessity of modifications in design of the LDMOS transistor in order to make it ultra-thin so that it can be embedded on flexible substrates without any compromise in flexibility of the product. A few attempts have been made in this regard so far. In one such an attempt an ultra-thin (35 μm) high-voltage transistor design was reported [25]. The breakdown voltage of the transistor was >110 V [26]. At high voltages significant reduction in the current occurs because of the self-heating effect. However, for switching purpose this reduction in current does not produce any effect. In another design of an ultra-thin (~10 μm) LDMOS transistor based on the Chipfilm™ technique was proposed [27]. The transistor can withstand a high voltage of about 100 V. Performance of a single-crystal silicon transistor in bent form has also been studied by [26, 28]. Results showed that under bent state, drain current changes due to piezoresistive effects. For switching purposes, these effects do not affect the device performance significantly. However, in the design of analogue circuits these effects must be taken into consideration.

The driver chips are meant to provide several tens of volts with current values sufficient to help quick charging and discharging of load capacitances. The heat produced as a result must be extracted from the chips in order to minimise performance degradations due to thermal effects. The temperature distribution for a given dissipated power depends largely on substrate thermal conductivity [29]. In the case of external driver chips, metal substrates behave as a heat sink and provide enough space for heat dissipation. In a flexible display system the driver chips have to be embedded between two layers of flexible material, which in most cases (except flexible metal sheet) are thermal insulators. This arrangement significantly reduces the amount of heat dissipated into environment. As a result, the self-heating deteriorates device performances. Reducing on-resistance of transistors through improved designs can reduce power dissipation in driver circuits. Besides, schemes like charge recycling and selective updates [30] can be adopted to further reduce power dissipation.

For high resolution and large size flexible displays, handling of insufficient charging time of pixels and signal delay in long rows becomes a challenge. A pixel can be represented as a capacitance that needs to be charged and discharged to desired voltage levels according to the input data signal. The time required for charging and discharging depends largely upon the response time of display material. With the increase in resolution, available row-line time is reduced. Shorter row-line times can lead to insufficient charging of pixels, which affects picture quality and uniformity. For large display sizes, the row-lines behave like a distributed resistance-capacitance (RC) network. Due to this RC network, a signal delay time is involved between pixels of same row which reduces the available charging time of pixels. Methods like line time extension (LiTEX) [31] can be applied to increase effective row-line time for a given resolution. For signal delay compensation on the row-line, techniques like horizontal line delay compensation (H-LDC) [32] and adaptive charging [33] can be adopted.

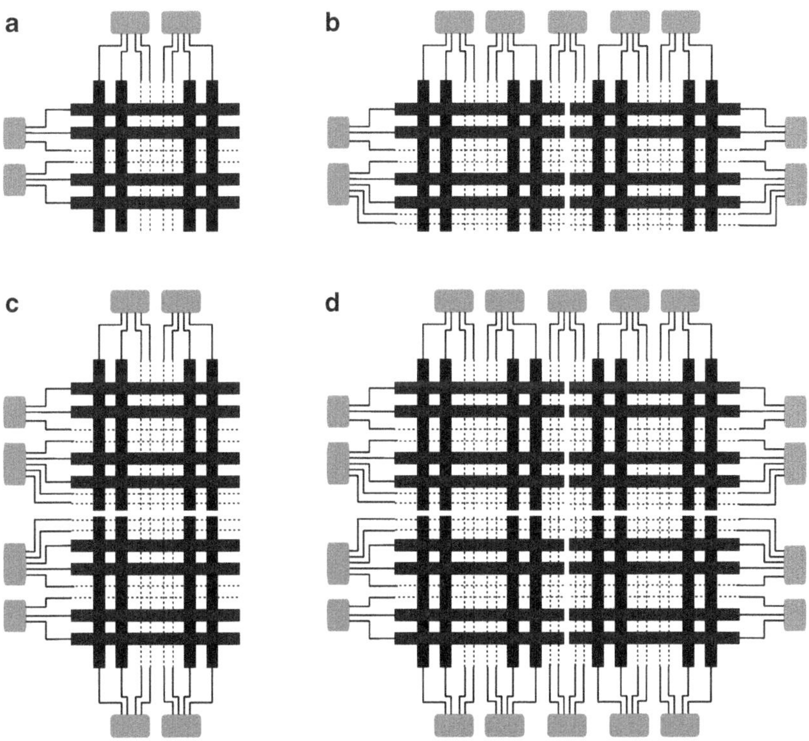

Fig. 32.5 Placement of drivers around matrix back plane. (**a**) Typical placement scheme for normal resolution and small display sizes. (**b**, **c**) and (**d**) Various possibilities of dual scan method for high resolution and large display sizes

Dividing the frame into two halves vertically or horizontally and driving the lines from both sides can significantly increase the row-line time and reduce signal delays. However, the cost of this approach is the increase in price of the product due to increased number of drivers. Figure 32.5 shows the various options of the placement of driver chips around matrix back plane.

Advancement in flexible display technology has opened doors to many new applications that were once either unimaginable or thought to be unachievable. Small size flexible displays have already been released and more are announced to be released in near future. Success in developing flexible display as large as A3-sized newspaper has also been claimed recently [34]. Besides making displays only for viewing purpose, the concept of smart windows and smart textile has also appeared in literature. In addition to electronic shading and utilization as a large screen, the purpose of smart windows is to block and transmit the infrared portion of sunlight according to need [35]. Similarly, smart textile is meant to display, communicate and perform thermal management. All this and many other such applications demand an autonomous and multipurpose flexible system, which can be used not only for display purposes but also be able to transmit, receive, store and process data. All this can be achieved by having thin and flexible batteries, memories, transceivers, logical and driver circuitry on the same flexible substrates. There exists, however, a number of technical [36] and technological [37] obstacles to surmount before the realisation of mass produced full-flexible and autonomous display systems.

References

1. Allen KJ (2005) Reel to real: prospects for flexible displays. Proc IEEE 93(8):1394–1399
2. Crawford GP (2005) Flexible flat panel displays. Willey, West Sussex
3. Sekitani T, Nakajima H et al (2009) Stretchable active-matrix organic light-emitting diode display using printable elastic conductors. Nat Mater 8:494–499
4. Collins L (2003) Roll-up displays: fact or fiction. IEE Rev 49(2):42–45
5. Lueder E (2004) Liquid crystal displays – addressing schemes and electro-optical effects, Wiley-SID series in display technology. John Wiley & Sons Ltd, West Sussex
6. Den Boer W (2005) Active matrix liquid crystal displays – fundamentals and applications. Elsevier (Newnes), Oxford
7. Venugopal SM, Allee DR (2007) Integrated a-Si:H source drivers for 4// QVGA electrophoretic display on flexible stainless steel substrate. J Display Technol 3(1):57–63
8. Venugopal SM, Shrigarpure R et al (2008) Integrated a-Si:H source drivers for electrophoretic displays on plastic substrates. IEEE. doi: 10.1109/FEDC.2008.4483875
9. Powell MJ (1989) The physics of amorphous silicon thin-film transistors. IEEE Trans Electron Devices 36(12):2753–2763
10. Allee DR, Clark LT et al (2009) Circuit-level impact of a-Si:H thin-film transistor degradation effects. IEEE Trans Electron Devices 56(6):1166–1176
11. Troccoli MN, Roudbari AJ et al (2006) Polysilicon TFT circuits on flexible stainless steel foils. Solid State Electron 50:1080–1087

12. Inoue S, Utsunomiya S et al (2002) Surface-Free technology by Laser Annealing (SUFTLA) and its application to poly-Si TFT-LCDs on plastic films with integrated drivers. IEEE Trans Electron Devices 49(8):1353–1360
13. Pecora A, Maiolo L et al (2008) Low-temperature polysilicon thin film transistors on polyimide substrates for electronics on plastic. Solid State Electron 52:348–352
14. Gelinck GH, Edzer H, Huitema A et al (2004) Flexible active-matrix displays and shift registers based on solution-processed organic transistors. Nat Mater 3:106–110
15. Takafumi Uemura, Yuri Hirose et al (2009) Very high mobility in solution-processed organic thin-film transistors of highly ordered [1]Benzothieno[3,2-b]benzothiophene derivates. Appl Phys Express 2:111501-1–111501-3
16. De Leeuw DM, Cantatore E (2008) Organic electronics: materials, technology and circuit design developments enabling new applications. Mater Sci Semiconductor Process 11:199–204
17. Wu Q, Zhang J et al (2006) Design consideration for digital circuits using organic thin-film transistors on a flexible substrate. In: IEEE circuits and system symposium, Island of Kos, Greece, 2006, pp 1267–1270
18. Gowrisanker S, Quevedo-Lopez MA et al (2009) A novel low temperature integration of hybrid CMOS devices on flexible substrates. Org Electron 10:1217–1222
19. Dodabalapur A, Baumbach J et al (1996) Hybrid organic/inorganic complementary circuits. Appl Phys Lett 68(16):2246–2248
20. Min Suk Oh, Wonjun Choi et al (2008) Flexible high gain complementary inverter using n-ZnO and p-pentacene channels on polyethersulfone substrate. Appl Phys Lett 93:033510-1–035510-3
21. Menard E, Nuzzo RG, Rogers JA (2005) Bendable single crystal silicon thin film transistors formed by printing on plastic substrates. Appl Phys Lett 86:093507-1–093507-3
22. Li HY, Guo LH et al (2006) Bendability of single-crystal Si MOSFETs investigated on flexible substrate. IEEE Electron Device Lett 27(7):538–541
23. Burghartz JN, Appel W et al (2009) A new fabrication assembly process for ultrathin chips. IEEE Trans Electron Devices 56(2):321–327
24. Ahn J-H, Kim H-S et al (2006) High-speed mechanically flexible single crystal silicon thin-film transistors on plastic substrates. IEEE Electron Device Lett 27(6):460–462
25. Sakuri R, Hattori R et al (2007) Ultra-thin and flexible LSI driver mounted electronic paper display using Quick-Response Liquid-powder technology. SID Digest 07:1462–1465
26. Asakawa M, Nakashima T et al (2008) Electrical characteristics of driver LSI with 35 μm thickness for flexible display. IEEE Trans Electron E91-C(10):1570–1575
27. Asif A, Richter H et al (2009) High-voltage (100V) chipfilm™ single-crystal silicon LDMOS transistor for integrated driver circuits in flexible displays. Adv Radio Sci 7:1–6
28. Richter H, Rempp HD et al (2009) Technology and design aspects of ultra-thin silicon chips for bendable electronics. In: IEEE international conference on IC design and technology (ICICDT '09), Austin, 2009
29. Maiolo L, Cuscuna M (2009) Analysis of self-heating related instability in n-channel poly-silicon thin film transistor fabricated on polyimide. Thin Solid Films 517:6371–6374
30. Monte A, Bauwens P et al (2008) New driving scheme for intelligent power-efficient high-voltage display drivers. J SID 16(11):1171–1180
31. Choi BD, Kwon OK (2004) Line time extension driving method for a-Si TFT LCDs and its application to high definition televisions. IEEE Trans Consum Electron 50(1):33–38
32. Kim SH, Kim GB et al (2004) A new driving method to compensate for row line signal propagation delays in an AMLCD. SID 04 Digest 35(1):280–283
33. Chen C-H, Kiang Jean-Fu (2008) Active and adaptive charging method on data lines for delay compensation. J Display Technol 4(2):198–203
34. Yu Y (2010) LG display unveils newspaper-size flexible e-paper. http://www.digitimes.com/news/a20100115PR201.html

35. Heikenfeld J (2010) Lite, brite displays. IEEE Spectr 3(10):23–28
36. Miyasaka M, Hara H et al (2008) Technical obstacles to thin film transistor circuits on plastic. Jpn J Appl Phys 47(6):4430–4435
37. Van den Brand J, de Baets J et al (2008) Systems-in-foil – devices, fabrication processes and reliability issues. Microelectron Reliab 48:1123–1128

Chapter 33
Microwave Passive Components in Thin Film Technology

Behzad Rejaei

Abstract This chapter is dedicated to a theoretical investigation of the electrical behaviour of inductors and transmission lines integrated on thin silicon wafers. Using electromagnetic simulations, it is shown that thinning a standard silicon substrate to ~20–70 μm yields a ~20–30% increase in the maximum quality factor and resonance frequency of typical integrated spiral inductors. This improvement is due to a higher spreading substrate resistance between the inductor and the physical ground. However, further improvement of the quality factor requires substrates thinned to several microns in order to reduce the effect of stray electric fields induced between different points on the conductor, e.g., between adjacent windings. In the case of coplanar waveguide and microstrip transmission lines, significant improvement in device quality is only observed when silicon thickness reaches values below 10 μm. On highly conductive silicon, the behaviour of inductors and transmission lines as a function of substrate thickness becomes strongly dependent on silicon resistivity. In particular, thinning substrates with a resistivity in the 0.01–2 Ω-cm range may even have an adverse effect on device performance.

Keywords Integrated spiral inductor · Integrated transmission line · Thin flim inductors · Passive microwave components · Thin flim technology · *Q* factor · Equivalent circuit model · Substrate loss

33.1 Introduction

The technological and social revolution brought about by modern communication systems owes its existence to monolithic microwave integrated circuits (ICs), where all electric and electronic components are built on a single underlying substrate. Without the development of fast, small and cheap microwave ICs, the mass

B. Rejaei (✉)
Department of Electrical Engineering, Sharif University of Technology,
11365 Tehran, Iran
e-mail: rejaei@sharif.ir

J.N. Burghartz (ed.), *Ultra-thin Chip Technology and Applications*,
DOI 10.1007/978-1-4419-7276-7_33,

production of mobile phones, wireless local area networks and GPS navigators would have been unimaginable. Yet, to those of us accustomed to the notion of perpetual progress of technology, it may come as a surprise to know how often this development was and is hindered by poor performance of mundane electrical elements that date from before the twentieth century!

Digital ICs, such as memory and central processing units, use CMOS transistors as their main building block. These components, which essentially work as electronic switches, have been continually downsized in past three decades, with the gate length of modern CMOS transistors now reaching a few tens of nanometres. However, unlike digital ICs microwave ICs have to contend with analogue microwave signals. This requires, besides transistors, components that can store (capacitors, inductors) or transfer (transmission lines) the microwave energy without significant loss of microwave power.

Energy-storing lumped passive elements have been known for centuries. The earliest example of a capacitor, one called the 'Leyden jar', can still be seen in many technical museums around the world. It is basically a glass jar with inner and outer walls coated by two metal layers forming the two electrodes of the device. A modern capacitor in an IC looks much more refined, but works in the same manner. Opposite electric charges are stored on two (often parallel) metallic electrodes, which are separated by an insulating layer such as silicon dioxide (which is, by the way, also glass). The electric field induced by a charged parallel-plate capacitor is almost entirely confined to the dielectric region between its two electrodes. It does not spread in space, interfering with other devices on the substrate.

It is an unfortunate law of nature (at least for microwave engineers) that isolated magnetic charges do not exist. One cannot build a magnetic capacitor by storing opposite magnetic charges on two separate electrodes. Instead, the magnetic field necessary for storing the magnetic energy has to be induced by passing an electric current through a conductor. The solenoid, first invented in nineteenth century, is

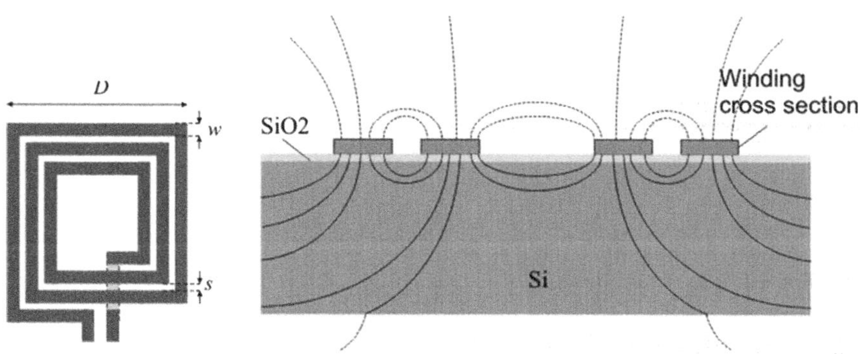

Fig. 33.1 *Left*: Integrated spiral inductor. *Right*: Schematic view of the electric field lines surrounding the device on its vertical cross-section

such a device. It consists of a wire that is wound many times around a magnetic or nonmagnetic core. Integrated inductors are seldom built as solenoids. Rather, they are usually implemented as planar spirals (Fig. 33.1) due to ease of fabrication [1]. Compared to a capacitor, the inductor is much less benign when integrated on a chip. Unlike a parallel-plate capacitor, the electromagnetic field surrounding a spiral inductor is not confined to its immediate vicinity. When an alternating, high frequency current passes through the device, the electromotive force induced along the inductor wire results in accumulation of negative and positive charges on different points on the wire. These local charges cause the electric field lines to start from a point on the wire, traverse the surroundings (air or the substrate) and terminate on another point on the wire or stretch all the way to the infinity (Fig. 33.1). This may sound harmless, but note that the majority of consumer ICs are silicon-based. Silicon (Si) technology is cheap, mature and highly suitable for active devices like CMOS transistors. However, Si is conductive, so that when the electric field enters the substrate it induces electric currents. Part of the microwave energy stored in the magnetic field is thus dissipated in the lossy substrate, significantly worsening the inductor performance [2–4].

Besides lumped passives, distributed components such as transmission lines are also employed in microwave ICs, especially at frequencies higher than 30 GHz. In its simplest form a transmission line consists of two parallel conductors in a medium whose properties do not vary along the line. Unlike lumped elements, which are actually quasi-static concepts, a transmission line is a true microwave device whose operation is based on the propagation of voltage (or, equivalently, current) waves along the line. It is also quite versatile. It can be used to physically transfer the microwave power in a controllable way, without the risk of radiation loss, or to implement different functions such as filtering and impedance matching. Yet again, on conventional Si substrates, transmission lines become almost totally useless due to substrate loss caused by stray electric fields [5, 6].

The account given above of hurdles of Si-based integrated passives may prompt the following questions: Can one do away with Si material? Is it possible to replace it with other substrates or modify it? All these solutions have in fact been tried. Si material below the passive device has been removed by various techniques, including wet etching [7, 8]. It has been oxidised [9] and even bombarded with protons to make it less conductive [10]. Thick dielectric spacers have been used to vertically separate the device from the Si substrate [11]. Alternative substrates like high-resistivity Si and glass have been suggested and experimented upon [12, 13]. And, while they all work, these techniques have never been commercially adopted for the simple reason that they entail the use of new materials or diversion from the mainstream silicon IC processing techniques.

In recent years we have witnessed the emergence of thin-chip technology, which was mainly developed to enable the fabrication of three-dimensional ICs [14, 15] as well as mechanically bendable and flexible chips in applications such as ID tags and wearable electronics [16]. These properties are achieved by thinning standard Si wafers of 500–800-μm thickness using grinding techniques or by exploiting pre-fabricated buried cavities [17]. Already functional devices and circuits on wafers with a thickness of ~20 μm and below have been demonstrated [17–19]. The presence of less Si material beneath passive elements readily hints at the potential of thin-chip technology to resolve (or at least alleviate) the substrate loss problem. But the extent to which this happens has not yet been systematically investigated. In particular, it is unknown exactly how the quality factor of spiral inductors and transmission lines fabricated on thin Si films behave as a function of substrate thickness. In what follows we shall try to answer these question by exploiting simple electric models as well as numerical electromagnetic simulations.

33.2 Spiral Inductors on Thin Chips

33.2.1 Electrical Parameters and the Equivalent Circuit Model

The inductor, being a two-port network[1], offers several ways for its impedance to be defined. But, it is common to consider one port (say port 2) grounded. The inductor impedance (Z), inductance (L) and quality factor (Q) are then defined as

$$Z = \frac{1}{Y_{11}}, L = \frac{\mathrm{Im}(Z)}{j\omega}, Q = \frac{\mathrm{Im}(Z)}{\mathrm{Re}(Z)}, \tag{33.1}$$

where Y_{ij} ($i,j = 1,2$), denotes the admittance matrix of the device and ω is the radial frequency. Besides these parameters, the (lowest) resonance frequency (f_{res}) of the device is also of practical importance as it limits the range of its applicability. f_{res} is the frequency at which L crosses zero.

The above parameters, once known, might be enough for a circuit designer to know in order to proceed. But to study their dependence on the device layout and substrate properties, the designer would need empirical or physical models. A widely used model is the lumped-element model of Fig. 33.2 [3, 4, 20, 21]. Here L_s, R_s and C_p are, respectively, series inductance, series resistance and self-capacitance of the inductor, and $C_{\mathrm{ox},1}$ and $C_{\mathrm{ox},2}$ represent the capacitance formed between the inductor metal and the surface of the conducting substrate through the insulating dielectric layer (usually silicon dioxide). The dielectric and ohmic response of the substrate itself is modelled by a simple network consisting of the resistors $R_{\mathrm{sub},k}$ ($k = 1,2,3$) and capacitors $C_{\mathrm{sub},k}$ [22]. Finally, the mutual and secondary inductances M and L_{sub} and the resistance R_{ed} account for eddy currents induced in the conductive substrate by the high frequency magnetic field of the

[1] When an inductor is separately characterised, a metallic patch is placed near the device where the ground tips of the microwave probes are landed. But, in a circuit, such a ground patch may not be present. Even if present, there is no guarantee that it can effectively be connected to a common ground for all devices. Physically, the best choice is then to consider a point far away (infinity) as the true ground. This is what we adopt in our analysis.

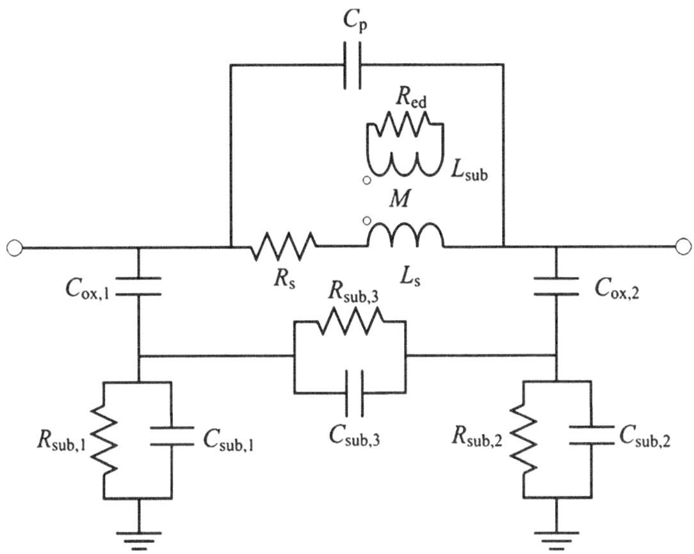

Fig. 33.2 Equivalent lumped-element model of an integrated inductor

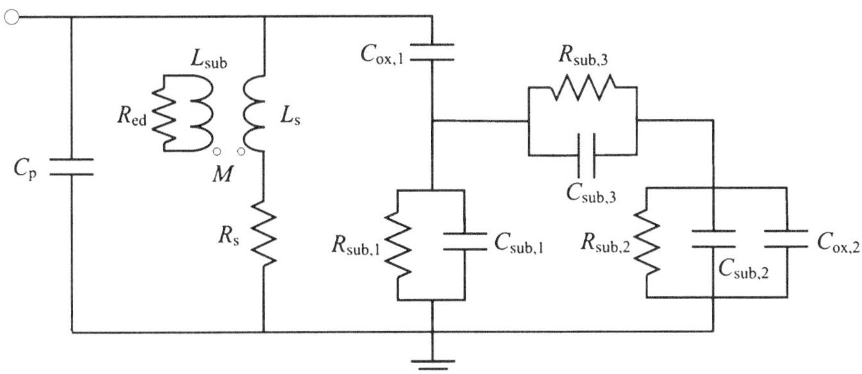

Fig. 33.3 Equivalent lumped-element model of an integrated inductor with port 2 grounded. Effects of substrate eddy currents are neglected

inductor and the resulting ohmic loss [4].[2] (The equivalent-circuit model with a grounded port 2 is shown in Fig. 33.3.)

How does substrate thinning affect the value of the components of the substrate network of Fig. 33.2? While a precise answer to this question is impossible to give,

[2] This is a semi-heuristic model which neglects the true distributed nature of the device. Nonetheless, such models have been proven functional, at least at low microwave frequencies, where inductors are often used.

one may attempt to get a qualitative picture by considering the electric field lines surrounding the inductor. As Fig. 33.1 shows, one expects some field lines to emerge from the conducting coil, cross the thin oxide layer, spread into the substrate and eventually stretch to the infinity.[3] In the quasi-static language, these fields are due to potential difference between the coil and the physical ground at infinity and their effect is taken into account by $R_{sub.1,2}$ (and $C_{sub.1,2}$), which link the inductor ports to the ground like in a spreading resistance problem. On the other hand, the potential difference between different points on the coil causes some field lines to terminate on the coil metal itself, entering and leaving the Si substrate on the way. These fields are effectively accounted for by $R_{sub,3}$ and $C_{sub,3}$.

If the substrate were infinitely thick, the distribution of electric field and current would have had a 3D nature, mainly determined by the size and shape of the coil and its windings. But, with a thin substrate, the electric field lines have a narrower low-impedance region to spread. This, in turn, implies less volume in which substrate currents may flow, leading to larger substrate resistances. Quasi-static calculations show the effect of substrate thickness on the distribution of the electric field lines between the coil and infinity to become significant when $t_{sub} < D/\pi$, where t_{sub} is substrate thickness and D is overall diameter of the inductor. The influence of t_{sub} on field lines terminating on the coil metal itself is much harder to analyse, but it is expected to depend on inductor spacing (s) and line width (w).

Note that the resistor R_{ed} also contributes to the device loss, although its effect becomes remarkable only on highly conductive substrates (<0.2 Ω-cm), where R_{ed} becomes sufficiently small to allow the flow of substantial eddy currents [4]. This not only increases substrate loss, but also leads to a reduction of overall inductance, due to mutual inductance M. This situation resembles the case in which a conductive ground plane is placed under a current-carrying structure: Induced eddy currents in the ground plane flow in a direction opposite to current in the wire, reducing the total magnetic field and flux, thus causing a lower inductance. Nevertheless, one would expect the influence of R_{ed} to become less important on substrates whose thickness is less than the skin-depth in Si, which is given by $\delta_{Si} = (2\rho_{Si}/\mu_0\omega)^{1/2}$ with ρ_{Si} the Si resistivity and μ_0 the permeability of vacuum.

33.2.2 Numerical Experiments

While the equivalent circuit model of Figs. 33.2 and 33.3 is useful for imparting a qualitative understanding of inductor behaviour, detailed analysis of substrate thinning effect calls for accurate electromagnetic simulation. To that end, we have selected three square spiral inductors with inductances of ~2.2, 5.7 and 10.5 nH. The geometrical parameters of the coils are listed in Table 33.1. The substrate

[3] Since the Si substrate has a high relative dielectric constant (11.9) with respect to the environment, and is conductive, it provides a low-impedance path for the field lines. Thus, electric field lines tend to stay in the substrate.

Table 33.1 Geometrical parameters of three simulated spiral inductors

Inductor	N	D (μm)	w (μm)	s (μm)	L_0 (nH)
L1	3	200	15	5	2.2
L2	5	280	15	5	5.7
L3	6.5	340	15	5	10.5

N number of turns, D outer diameter, w line width, s spacing, L_0 denotes inductance of the coils at low frequencies

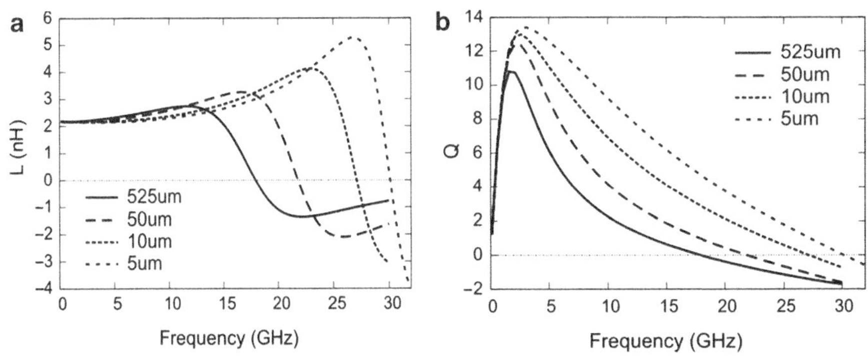

Fig. 33.4 Simulated inductance and quality factor of the coil L1 (Table 33.1) as function of frequency for Si substrates with the thicknesses 525, 50, 10 and 5 μm

consists of 8 Ω-cm Si covered by a 2 μm-thick SiO_2 layer. The coil metal is assumed to be 3 μm-thick aluminium. The simulations were carried out using Agilent's ADS.

Figure 33.4 shows the inductance and quality factor of the 2.2 nH coil as function of frequency for Si substrates of various thicknesses. While substrate thinning does not affect L at low frequencies, it strongly increases the resonance frequency f_{res}. In this particular case f_{res} is 17.9 GHz on a conventional Si substrate (525-μm thick). When the substrate is eventually thinned down to 5 μm, f_{res} is increased to ~30 GHz. The effect on Q may seem less drastic with the maximum quality factor Q_{max} showing just a 25% increase. But one should not forget that Q also includes the effect of conductor loss, which is indifferent to substrate thinning.

More insight can be gained by plotting f_{res} and Q_{max} as functions of substrate thickness (t_{sub}) for the three inductors (Fig. 33.5). In particular, the behaviour of Q_{max} with decreasing t_{sub} is rather surprising. Starting from a full wafer (525 μm) reduction of t_{sub} initially causes a rise in Q_{max}. However, below 50 μm no significant increase in Q_{max} is observed until t_{sub} reaches a lower threshold value of ~7 μm. Further thinning of the substrate then leads to a rapid rise in Q_{max}. The initial rise in Q_{max} can be attributed to an increase in the spreading resistances $R_{sub,1}$

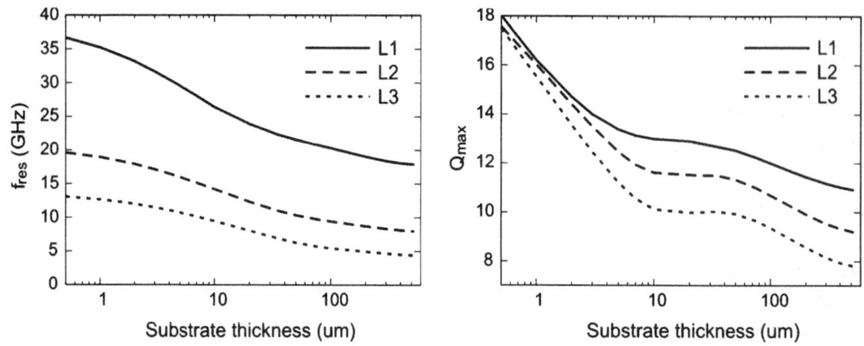

Fig. 33.5 Resonance frequency and maximum quality factor of the inductors L1, L2, and L3 (Table 33.1) as functions of Si thickness. A logarithmic scale has been used so the horizontal axis may better depict device behaviour at small thicknesses

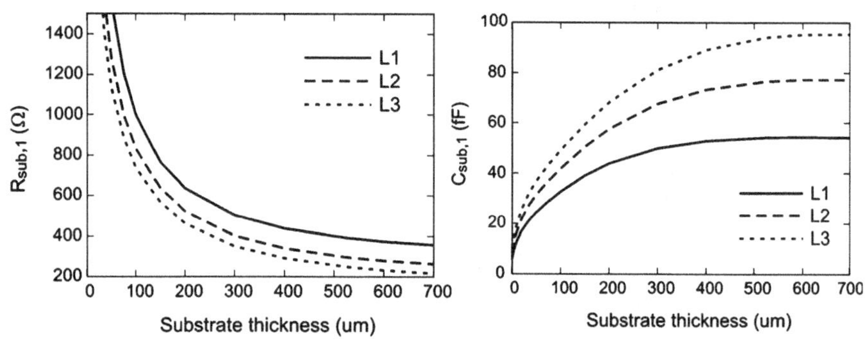

Fig. 33.6 Substrate resistance (*left*) and capacitance (*right*) of inductors L1, L2, and L3 as function of Si thickness. Since the values of the components of the equivalent model are, in general, frequency-dependent, we have chosen a frequency of 3 GHz for the extraction

and $R_{sub,2}$ (Fig. 33.6)[4], which hinders current flow through the $C_{ox,1}/R_{sub,1}/C_{sub,1}$ path, resulting in less substrate loss (Fig. 33.3). Moreover, note that through this path the oxide capacitance $C_{ox,1}$ acts in parallel with C_p, causing the LC resonance to occur at lower frequencies compared to what happens with an isolated inductor. As an increasing $R_{sub,1}$ renders this path less effective, f_{res} is shifted to higher frequencies. However, when t_{sub} reaches a value roughly corresponding to D/π a sharp rise in $R_{sub,1}$ occurs, which almost completely inactivates the $C_{ox,1}/R_{sub,1}/C_{sub,1}$ path. With $R_{sub,1}$ ceasing to contribute to substrate loss, further reduction of t_{sub} has no effect on Q_{max}, and the curve becomes almost flat.

[4] At frequencies where Q reaches its maximum value (less than 5 GHz in our examples), $C_{sub,1,2}$ is practically bypassed by $R_{sub,1,2}$, as the former's reactance value is much higher than the latter's. For that reason its effect on inductor behaviour is negligible.

But, then, what causes the sharp rise in Q_{max} for very thin substrates (<7 μm)? The answer lies in the behaviour of the $C_{ox,1}/R_{sub,3}/C_{sub,3}/C_{ox,2}$ path, which accounts for the electric field lines emanating from a point on the coil, crossing the thin oxide layer and moving into the substrate, just to cross the oxide again and return to another point on the coil. These field lines traverse relatively short paths, e.g., in between adjacent windings, and do not go deep into the substrate. Consequently, their distribution (reflected in the values of $R_{sub,3}$ and $C_{sub,3}$) is only affected when the Si substrate becomes very thin. When this happens the increase in the impedance of the $C_{ox,1}/R_{sub,3}/C_{sub,3}/C_{ox,2}$ path reduces both the current flow through $R_{sub,3}$ and the associated substrate loss, yielding an increase in Q_{max}. The same effect leads to a relatively rapid increase in f_{res} since, with the $R_{sub,1}/C_{sub,1}$ path already inactive, the flow of current through $C_{ox,1}$ now becomes almost totally blocked, leaving C_p alone as the only capacitor in parallel with L_s (Fig. 33.3).

The results discussed so far were obtained assuming an 8 Ω-cm Si substrate. This value is fairly standard, as typical Si wafers have a resistivity in the 2–20 Ω-cm range. Nonetheless, in many practical situations one may have to deal with highly doped or almost intrinsic substrates. It is thus instructive to see how changing the substrate resistivity (ρ_{Si}) affects the behaviour of inductors built on a thin chip.

On a standard Si wafer one may distinguish three modes of operation, depending on ρ_{Si}. These modes, termed 'inductor', 'resonator' and 'eddy current' modes in [4], are analogous to the quasi-TEM, slow wave and skin effect modes of Si-based microstrip transmission lines [23]. In the inductor mode ($\rho_{Si} > 10$ Ω-cm) the substrate essentially behaves as a (lossy) dielectric as $\omega R_{sub,k}C_{sub,k} \sim \omega\rho_{Si}\varepsilon_{Si} \gg 1$ with ε_{Si} the dielectric constant of Si. The substrate resistances are almost bypassed by substrate capacitances, which themselves are typically much smaller than the oxide capacitances $C_{ox,1}$ and $C_{ox,2}$. The inductor resonance frequency is thus mainly determined by C_p, $C_{sub,1}$ and $C_{sub,3}$ (Figs. 33.3 and 33.7a). This mode of operation is characterised by a relatively high, almost constant f_{res} whose value depends on substrate thickness. Note also that increasing ρ_{Si} yields a lower dielectric loss-tangent $1/\omega\rho_{Si}\varepsilon_{Si}$ and, therefore, a higher Q_{max}.

Below $\rho_{Si} = 10$ Ω-cm, where $\omega R_{sub,k}C_{sub,k} \sim \omega\rho_{Si}\varepsilon_{Si} < 1$, Si starts to behave as a semiconductor with $R_{sub,k}$ now bypassing $C_{sub,k}$. If ρ_{Si} is sufficiently lowered such that also $\omega R_{sub,1}C_{ox,1}$, $\omega R_{sub,2}C_{ox,2} \ll 1$ the inductor resonance occurs through the relatively large oxide capacitance $C_{ox,1}$ and the 'small' effective resistance $R_{sub,1} \parallel (R_{sub,2} + R_{sub,3})$. The onset of this mode (the resonator mode) is thus indicated by a drop in f_{res}. Besides, lowering ρ_{Si} now causes Q_{max} to increase as the quality factor of the capacitive path involving $C_{ox,1}$ increases with decreasing substrate resistances (Fig. 33.7c). This trend is disturbed by substrate eddy currents (skin effect) if ρ_{Si} is lowered so much that the skin depth $\delta_{Si} = (2\rho_{Si}/\mu_0\omega)^{1/2}$ becomes less than the penetration depth of the magnetic field inside the substrate.[5] The appearance of

[5] From quasi-static arguments it follows that this depth roughly equals $\min\{t_{Si}, (D_o - D_i)/2\pi\}$, where D_o and D_i denote the outer and inner radius of the inductor, respectively. On a sufficiently thin substrate, the depth simply equals the substrate thickness.

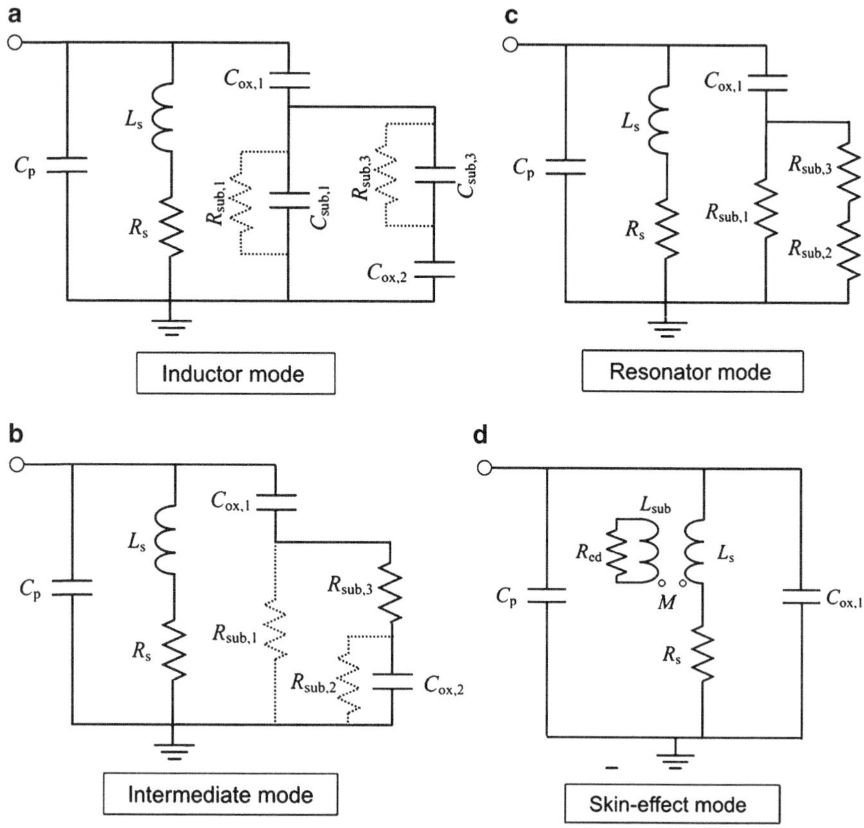

Fig. 33.7 Modes of operation of an integrated inductor

eddy currents causes inductance L and thus Q_{max}, to decrease again, while f_{res} increases due to the smaller L (Fig. 33.7d).

To check the validity of this picture on thin substrates, in Fig. 33.8 we have plotted f_{res} and Q_{max} of the 2.2-nH coil as a function of substrate resistivity for different values of t_{sub}. The onset of the inductor mode, heralded by a relatively high and flat f_{res} above 10 Ω-cm, is visible on all substrates. The same can be said of the onset of the skin effect mode, which is manifested by the rise in f_{res} for very low values of ρ_{Si}, albeit it now depends on substrate thickness. Moreover, a resonator mode region, characterised by a low, flat f_{res} of 7.9 GHz, can be observed for all values of t_{sub} (this frequency is determined by $C_{ox,1}$ which is independent of t_{sub}). However, the nature of the transition region between the inductor and resonator modes depends very much on the substrate thickness. On a thin substrate (5, 10 μm in Fig. 33.8), this transition occurs in two steps: a drop in f_{res} from its maximum, inductor mode value to ~14 GHz is followed by another drop from ~14 to 7.9 GHz. No such transition is observed on a thick substrate.

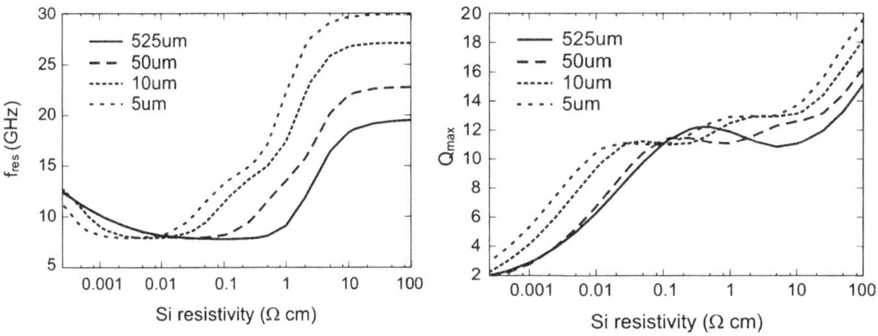

Fig. 33.8 Maximum quality factor of the inductor L1 as function of substrate thickness for different values of substrate resistivity (0.1, 8 and 100 Ω-cm)

To understand this behaviour, note that thinning the substrate will at the first instance lead to a huge increase in $R_{sub,1}$ and $R_{sub,2}$ (Fig. 33.6). The resonator mode condition, i.e., $\omega R_{sub,1}C_{ox,1}$, $\omega R_{sub,2}C_{ox,2} \ll 1$, can then only be satisfied at the expense of a much lower ρ_{Si} compared to a standard substrate. This is why the resonator mode region is shifted to lower values of ρ_{Si} for thinner substrates, as shown in Fig. 33.8. However, as argued in previous sections, $R_{sub,3}$ is less susceptible to substrate thinning. Thus, resonance might still occur through $C_{ox,1}$, $R_{sub,3}$ and $C_{ox,2}$ if $\omega R_{sub,3}C_{ox,1}C_{ox,2}/(C_{ox,1} + C_{ox,2}) \ll 1$, leading to a higher f_{res}, as the two oxide capacitors are now effectively connected in series (Figs. 33.3 and 33.7b). That which we observe as the second drop from 13 to 7.9 GHz is, in fact, the transition region between this intermediate mode and the conventional resonator mode of the inductor.

The behaviour of Q_{max} as function of substrate resistivity is complicated on thin substrates. Compared to a standard wafer (525 μm in Fig. 33.8), Q_{max} shows an extra dip in the intermediate region mentioned above. This can be attributed to the competition between the $C_{ox,1}/R_{sub,3}/C_{ox,2}$ and $C_{ox,1}/R_{sub,1} \parallel (R_{sub,2} + R_{sub,3})$ branches, the latter being dominant in the resonator mode. The quality factor of each branch (the ratio between the capacitive reactance of the path to its resistance) rises as ρ_{Si} is lowered. But the transition from the intermediate (Fig 33.8b) to the resonator mode (Fig. 33.7c) causes the path reactance to decrease as path capacitance becomes higher. This causes Q_{max} to fall slightly and, then, to rise again as ρ_{Si} is further lowered. Note also that for inductors operating in the resonator or intermediate modes (the ~0.1–2 Ω-cm range in Fig. 33.8), thinning the substrate may even lead to a lower Q_{max}, as it will increase substrate resistances and, thus, lower the quality factor of the two paths mentioned above. Therefore, a thinned Si substrate will only guarantee better performance in inductor and skin effect regions. However, this may not always be an issue as conventional Si substrates usually have a resistivity in the 2–20 Ω-cm range. Moreover, even in the intermediate and resonator regions, the effect of substrate thinning on Q_{max} is relatively small, as Fig. 33.8 shows.

33.3 Thin Chip Transmission Lines

33.3.1 Electrical Parameters and the Distributed Element Model

The primary quantities that determine the electric behaviour of a transmission line are the complex propagation constant, $\gamma = \alpha + j\beta$, and the characteristic impedance, Z_0. Here, β and α denote, respectively, the propagation and attenuation constant of the line, where $\beta = 2\pi/\lambda$, with λ the wavelength and α is a measure of per-unit-length power loss as the signal propagates along the line. Instead of β, it is customary to use the effective dielectric constant $\varepsilon_{\mathrm{eff}} = (\lambda_0/\lambda)^2$ where λ_0 is the wavelength of an electromagnetic wave propagating in vacuum at the same frequency. Moreover, it is common to use the transmission line quality factor $Q = \beta/2\alpha = \pi/\lambda\alpha$, which is a measure of line loss per wavelength instead of per unit length.

As it is for inductors, it is instructive to describe the behaviour of integrated transmission lines by means of lumped, equivalent circuit models. However such models can only be applied to a small line segment, with the transmission line, then viewed as the cascaded connection of (infinitely) many equivalent networks. Such a model is shown in Fig. 33.9. It is similar to the inductor model of Fig. 33.2 except for the fact that the capacitive-resistive substrate network is now considerably simpler. This is because the electric field surrounding the transmission line has (almost) no longitudinal component. The field lines lie in vertical, cross-sectional planes, emerging from points on one conductor (signal) and terminating on the other conductor (ground). Note also that, since the actual device consists of the cascaded connection of networks, as shown in Fig. 33.9, it suffices to introduce a single substrate branch. The combination $C_{\mathrm{ox}}/R_{\mathrm{sub}}/C_{\mathrm{sub}}$ then accounts for the electric field lines that cross the oxide layer and the Si substrate. Moreover, the capacitance C_{p} accounts for the electric field lines travelling through air (required for transmission lines such as coplanar waveguides). As is true for integrated

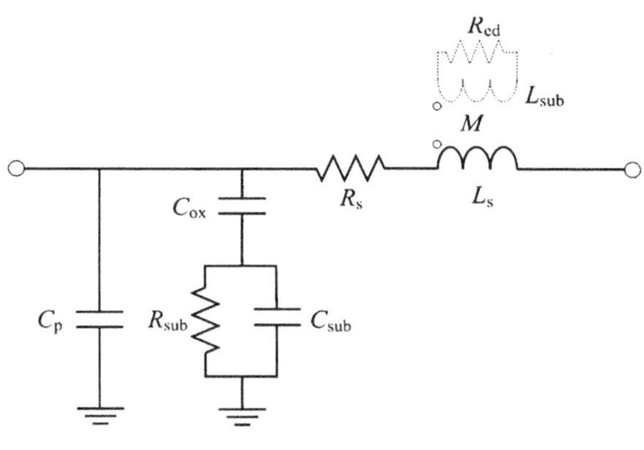

Fig. 33.9 Lumped element model of a segment of an integrated transmission line

inductors, the mutual inductance M and the elements L_{sub} and R_{ed} are included to describe the substrate eddy current effects.[6]

33.3.2 *Numerical Results*

For our numerical experiments we limit our scope to the two most widely used transmission line types: the coplanar waveguide (CPW) and the microstrip (MS). These designs are not of equal standing when it comes to implementation on standard Si wafers. The integration of MS lines on thick Si substrates is hindered by the difficulty of accessing the ground conductor on the wafer backside. By contrast, the conductors that make up a CPW line are all built within a single metal layer. This distinction, however, becomes less crucial on thin substrates where through-wafer vias (TSVs) can be more easily implemented [15]. Therefore, we shall also pay attention to Si-based MS lines.

33.3.2.1 CPW Transmission Lines

We consider three different CPW lines (Fig. 33.10) built using the technology described in Sect. 33.2.2 (on 8 Ω-cm Si). The CPW lines have the same signal line width of $W_{CPW} = 30$ μm, but they differ in signal-to-ground spacing, which is given by $S_{CPW} = 10, 30, 50$ μm for the lines T1, T2 and T3, respectively. Figure 33.11 shows the effective dielectric constant (ε_{eff}) and quality factor (Q) of the line T1 as a function of frequency for different values of Si thickness. Note that, in all cases, ε_{eff}

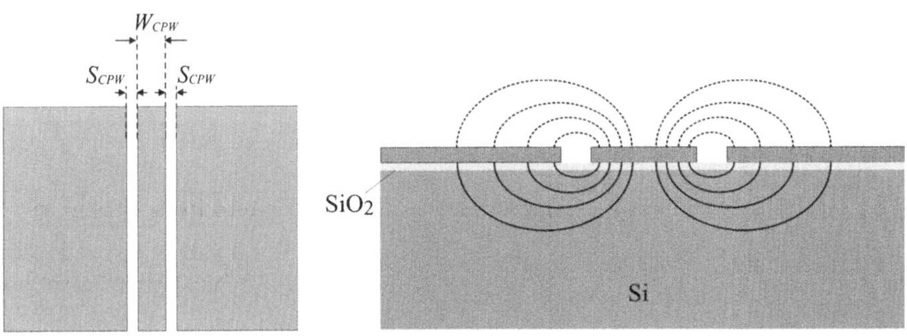

Fig. 33.10 *Left*: A CPW transmission line consisting of a central signal conductor and two ground stripes (the strips operate as a single conductor). *Right*: The electric field distribution around a CPW line

[6] If the overall impedance of the network connecting the two ports in Fig. 33.9 is Z_s and the overall admittance of the network connecting the left port to the ground is Y_g, then the complex propagation constant and characteristic impedance of the line follow from $\gamma l = (Z_s Y_g)^{1/2}$ and $Z_0 = (Z_s/Y_g)^{1/2}$, respectively, where l is the segment length.

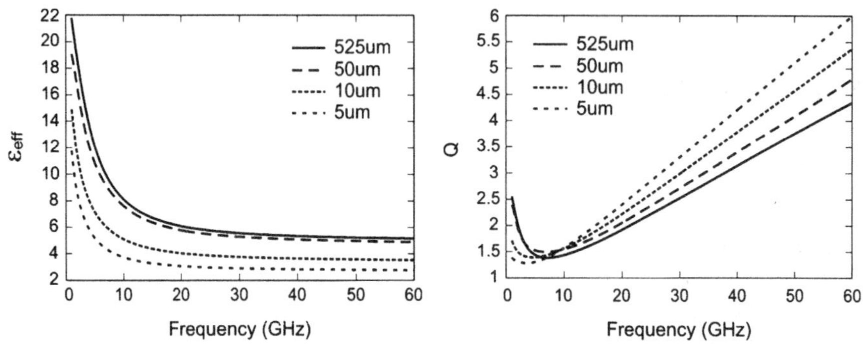

Fig. 33.11 Simulated effective dielectric constant and quality factor of the CPW line T1 as functions of frequency for Si thicknesses of 525, 50, 10 and 5 μm

is large at frequencies below 10 GHz after which it drops and becomes almost flat. This is due to the Si substrate acting as a conventional conductor at low frequencies. In the language of the model of Fig. 33.9, C_{sub} has a high impedance at low frequencies and is practically bypassed by R_{sub}. The line capacitance is thus mainly determined by C_{ox}, which is large due to the small thickness of the insulating oxide layer, leading to a large β and ε_{eff}. This effect wears off with increasing frequency, as the Si substrate starts to behave like a normal dielectric. Now R_{sub} becomes less relevant and the line capacitance is lowered, as it now consists of the series connection of C_{sub} and C_{ox}. This results in a smaller β and, therefore, a drop in ε_{eff}. The same effect, i.e., a decrease in the line capacitance, is responsible for the increase in (the real part of) the characteristic impedance Z_0 with frequency (Fig. 33.12).

Thinning the substrate will reduce ε_{eff}, because less Si material is present in the environment, which, in turn, causes an increase in the characteristic impedance. It also yields a higher Q as substrate loss is diminished. To get a more detailed picture we fix the operation frequency at 50 GHz and plot ε_{eff} and Q as functions of the Si thickness for the three CPW lines (Fig. 33.13). Practically no effect on ε_{eff} is seen until the Si substrate is thinned down to 40 μm. (In the case of Q the threshold is reached at 10 μm.) This is in contrast with inductors where an (initial) increase in Q was observed as the wafer was thinned from 525 to ~100 μm. But, note that the latter effect was due to an increase in $R_{sub,1,2}$, which accounts for conduction currents induced by electric fields spread far away from the device (to infinity). In the case of a CPW line, however, practically all electric field lines originating from the central signal line terminate on the two ground conductors and do not go deep into the underlying Si substrate. Thus, the behaviour of R_{sub} resembles that of $R_{sub,3}$ in Fig. 33.2, i.e., it is only affected when Si becomes quite thin.

Finally, let us briefly address the influence of substrate resistivity on the electrical parameters of CPW lines built on thin substrates. Figure 33.14 shows ε_{eff} and Q of T1 as functions of ρ_{Si} for different values of substrate thickness. The three modes of operation of a microstrip line [4] can also be distinguished for the

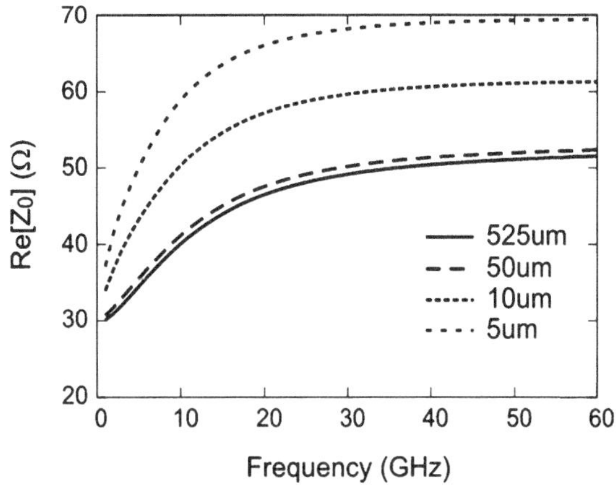

Fig. 33.12 Characteristic impedance (real part) of the CPW line T1 as a function of frequency for Si thicknesses of 525, 50, 10 and 5 μm

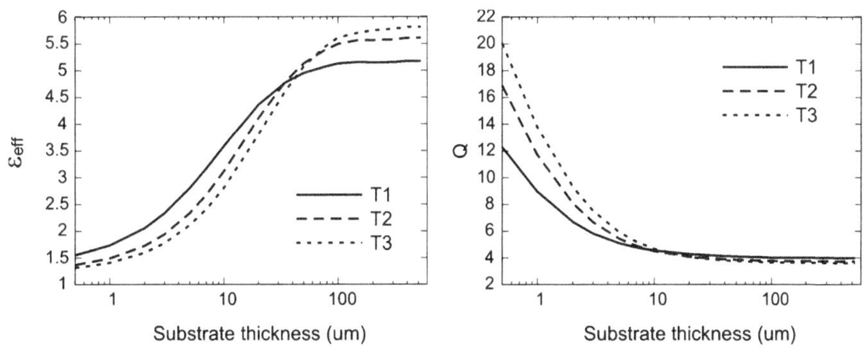

Fig. 33.13 Effective dielectric constant (*left*) and quality factor (*right*) of the CPW transmission lines T1, T2 and T3 at 50 GHz as functions of Si thickness. The substrate resistivity is 8 Ω-cm

CPW line. In the quasi-TEM mode where $\omega R_{\mathrm{sub}} C_{\mathrm{sub}} \sim \omega \rho_{\mathrm{Si}} \varepsilon_{\mathrm{Si}} \gg 1$, Si just behaves like a dielectric with a tangent loss inversely proportional to ρ_{Si}. The substrate resistance is almost bypassed by C_{sub}, which, being much smaller than C_{ox}, practically determines the line capacitance and, with it, the effective dielectric constant (Fig. 33.9). In this regime ($\rho_{\mathrm{Si}} > 10$ Ω-cm), $\varepsilon_{\mathrm{eff}}$ becomes almost independent of substrate resistivity but depends on substrate thickness, whereas Q increases with ρ_{Si}. The slow wave mode (analogous to the resonator mode of an inductor) sets in when $\omega R_{\mathrm{sub}} C_{\mathrm{sub}} \sim \omega \rho_{\mathrm{Si}} \varepsilon_{\mathrm{Si}} < 1$ and $\omega R_{\mathrm{sub}} C_{\mathrm{ox}} \ll 1$. Now, the relatively small C_{sub} is bypassed by R_{sub} and resonance takes place through the large oxide capacitance C_{ox}. The result is a high effective dielectric constant and a rising Q with decreasing ρ_{Si}. This region is visible for all thicknesses, even though it is shifted to lower values of ρ_{Si} for thinner substrates. Finally, for very low values of ρ_{Si} the

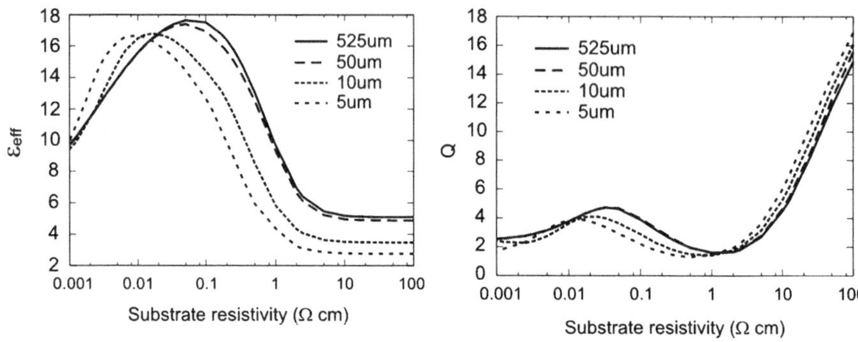

Fig. 33.14 Effective dielectric constant (*left*) and quality factor (*right*) of the CPW transmission lines T1 at 50 GHz as functions of Si resistivity for different substrate thicknesses

eddy currents developed in the substrate become strong enough to reduce the inductance per unit length, yielding a smaller β and, therefore, ε_{eff} (skin effect mode). Note that, as with inductors, substrate thinning will not ensure a higher quality factor for all values of ρ_{Si}. In particular, in the slow wave region a thinner substrate leads to a lower Q for reasons similar to those discussed for inductors.

33.3.2.2 Microstrip Transmission Lines

The effective dielectric constant and quality factor of three microstrip transmission lines (Fig. 33.15) with line widths of $W_{MS} = 30$, 50 and 70 μm are shown in Fig. 33.16 as function of the Si thickness (t_{sub}). The substrate resistivity is assumed to be 8 Ω-cm. The decrease in ε_{eff} with decreasing t_{sub} can be easily understood in terms of the equivalent circuit model of Fig. 33.9. For an integrated MS line, the effective line capacitance consists of the series connection of C_{ox} and C_{sub} (unlike a CPW, there is no electric field line directly connecting the signal and ground conductors through air, hence no C_p). On a thick wafer, the electric properties of the line are determined by the lossy Si material as $C_{ox} \gg C_{sub}$. The opposite situation occurs when t_{sub} becomes small enough so that $C_{ox} \ll C_{sub}$. The oxide layer, with its relatively low permittivity of ~4, will now be dominant. Obviously, for the same reason, thinning the substrate will lead to less substrate loss, causing the quality factor to rise for thicknesses less than 10 μm.[7]

[7] The per-unit-length value of C_{ox} can be estimated by the parallel-plate capacitor formula $C_{ox} = \varepsilon_{ox} W_{MS}/t_{ox}$ with ε_{ox} the dielectric constant and t_{ox} thickness of the oxide layer. This is due to t_{ox}/W_{MS} being very small in our examples (<0.06). However, the Si thickness is generally not too small compared to W_{MS}. As a result, the fringing field is not negligible, and C_{sub} has to be approximated by the well-known formulas for microstrip lines built on a single dielectric layer.

Fig. 33.15 Cross section of a microstrip line on a Si substrate and its electric field distribution

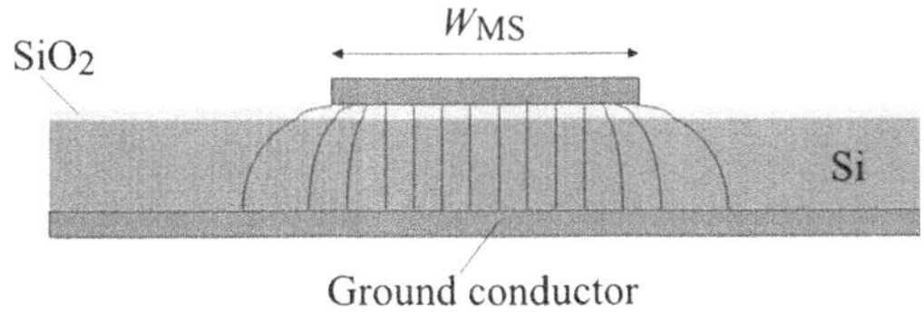

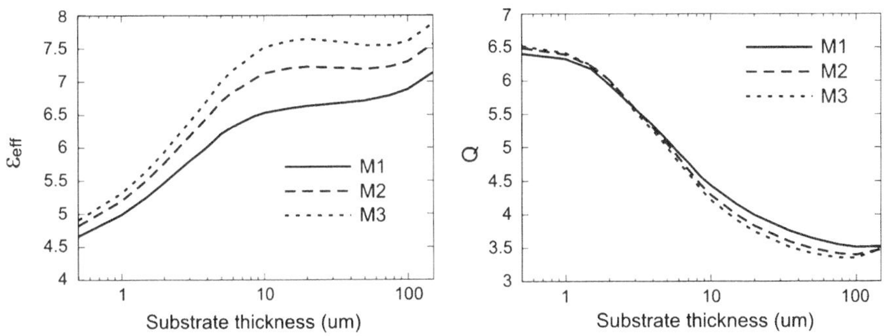

Fig. 33.16 Effective dielectric constant (*left*) and quality factor (*right*) of MS transmission lines with the width of 30 (M1), 50 (M2) and 70 μm (M3) at 50 GHz as functions of Si thickness. The substrate resistivity is 8 Ω-cm

Comparison of Figs. 33.13 and 33.16 shows the MS line to have a higher ε_{eff} than a CPW, especially on thin substrates. This is due to the fact that the electric field of a MS line is mostly confined within the Si (and oxide) layer whereas for a CPW almost half of the field lines travel through air. Thus, from the point of view of device miniaturisation, the integration of MS lines may seem more beneficial. On 10-μm thick Si, for instance, the simulated MS lines have an almost twice higher ε_{eff} compared to the CPW lines, leading to a 40% shorter propagation wavelength. However, the same field confinement causes the simulated MS lines to exhibit a somewhat lower quality factor due to the lossy nature of Si material.

What is more interesting, however, is the behaviour of these lines as function of substrate resistivity, shown in Fig. 33.17. The three operation modes of the device are clearly visible. Like it is for a CPW, in the quasi-TEM mode ($\rho_{Si} > 10$ Ω-cm) the effective dielectric constant is relatively low and the quality factor decreases with decreasing ρ_{Si}. In the slow wave region, ε_{eff} reaches high values while lowering ρ_{Si} leads to a higher Q. Finally, in the skin effect mode both ε_{eff} and Q drop with decreasing ρ_{Si}. But unlike a CPW line, reducing t_{sub} in the slow wave region actually yields a significantly higher quality factor. For example, at $\rho_{Si} = 0.05$ Ω-cm, reducing the substrate thickness from 50 to 10 μm results in an almost fourfold increase in Q. This is because in this region the quality factor of the dominant C_{ox}/R_{sub} path is inversely proportional to R_{sub}, which is reduced by lowering t_{sub}. At the

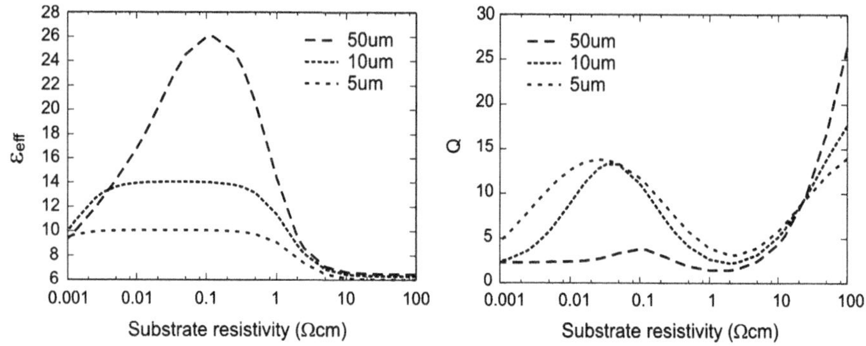

Fig. 33.17 Effective dielectric constant (left) and quality factor (right) of a 30 μm-wide MS transmission with at 50 GHz as function of Si resistivity for different substrate thicknesses

same time the slow wave MS lines with thin Si dielectric exhibit a relatively high ε_{eff} [8], which makes them quite attractive for practical applications.

Finally, note that MS transmission lines also can be fully implemented within the multi-metal/dielectric stack of a conventional IC process without any need for Si. As is true for an integrated capacitor, such devices just need two metal layers separated by an oxide dielectric. The latter is nearly lossless, so that a relatively high Q can be achieved compared to what is possible for Si-based devices. With 10 μm of oxide, for instance, a 30-μm wide MS line has a Q of ~21 at 50 GHz, which is nearly five times higher than that of the same line built on 8 Ω-cm Si. However, oxide-based lines have a low ε_{eff} because of the low dielectric constant of the oxide material compared to Si (3.9 vs 11.9).[9]

33.4 Conclusion

Conventional substrate grinding techniques can provide us with thin Si wafers with a thickness down the 20 μm. The presented calculations show the gain in inductor quality factor and resonance frequency to be less than 30%, in comparison with

[8] In contrast to CPW transmission lines the effective dielectric constant of MS lines in the slow wave region strongly depends on substrate thickness (Fig. 33.17). This has nothing to do with C_{ox}, which determines the line capacitance in this mode of operation and is insensitive to t_{sub}. Rather, the increase in ε_{eff} with substrate thickness is caused by a higher line inductance (L_s). When the vertical separation between the microstrip and the ground conductor becomes larger, the negative effect of (always present) ground currents on inductance is reduced. The resulting higher L_s leads to a shorter propagation wavelength, which is interpreted as a higher ε_{eff}.

[9] The main reason for the limited use of oxide-based MS lines is that, until recently, the maximum thickness of the oxide stack in a standard IC process did not exceed several microns. The resulting small vertical separation between the microstrip and ground conductors led to a low characteristic impedance limiting the applicability of such lines. Today's processes allow the overall thickness of the dielectric to reach 10 μm, which alleviates this problem. Nonetheless, to achieve high characteristic impedances (>50 Ω) one is still better served by the CPW design.

devices built on a thick, standard Si wafer. Higher gains can only be achieved if the technology is pushed to extreme limits, offering Si layers that are just several microns thick. However, if one adheres to the conventional grinding methods then, from an inductor-performance point of view, it may not make much sense to use wafers much thinner than 70 μm as the gain in quality factor is small. Moreover, the gain in quality factor depends on the resistivity of the substrate used. On highly doped substrates with a resistivity in the ~0.1–2 Ω-cm range, thinning of the Si wafer may even have an adverse effect on device quality.

Unlike spiral inductors, CPW transmission lines only benefit from a thin Si substrate for thicknesses below 10 μm. One should, however, note that substrate thinning leads to a lower effective dielectric constant, hence larger propagation wavelength. This is not a nuisance if the line is solely used for the transfer of microwave energy. But, quite often, half- or quarter-wavelength transmission line segments are used to build components such as resonators and impedance transformers, which will then require a larger chip area. In comparison with CPWs, microstrip transmission lines with a thin Si dielectric can offer a higher effective dielectric constant, but at the expense of a lower quality factor. One exception is the slow wave MS line on thin, highly doped Si, which offers both a high effective dielectric constant and quality factor.

In conclusion, the true potential of thin chip technology – improving the quality of spiral inductors and transmission lines – comes to light if the Si wafer is thinned down to below 10 μm. This is not a trivial feat as even wafers that are ground to 20 μm are prone to thickness nonuniformity and generation of crystalline damage [24, 25]. Nonetheless, the advent of new chip thinning techniques [17] may offer new prospects for reaching several-micron thick Si layers, which can greatly improve the performance of integrated microwave passives.

References

1. Ulrich RK, Schaper LW (2003) Integrated passive component technology. IEEE and Wiley Interscience, New York
2. Burghartz JN, Edelstein DC, Soyuer M et al (1998) RF circuit design aspects of spiral inductors on silicon. IEEE J Solid-State Circuits 33:2028–2034
3. Long JR, Copeland MA (1997) The modeling, characterization and design of monolithic inductors for silicon RFICs. J Solid-State Circuits 32:357–369
4. Burghartz JN, Rejaei B (2003) On the design of RF spiral inductors on silicon. IEEE Trans Electron Devices 50:718–729
5. Milanovic V, Ozgur M, DeGroot DC et al (1998) Characterization of broad-band transmission for coplanar waveguides on CMOS silicon substrates. IEEE Trans Microwave Theory Tech 46:632–640
6. Zheng J, Hahm Y-C, Tripathi VK et al (2000) CAD-oriented equivalent-circuit modeling of on-chip interconnects on lossy silicon substrate. IEEE Trans Microwave Theory Tech 48:1443–1451
7. Chang JYC, Abidi AA, Gaitan M (1993) Large suspended inductors on silicon and their use in a 2-μm CMOS RF amplifier. IEEE Electron Device Lett 14:246–248

8. Jiang H, Wang HY, Yeh J-LA et al (2000) On-chip spiral inductors suspended over deep copper-lined cavities. IEEE Trans Microwave Theory Tech 48:2415–2423
9. Miao J, Sun J (2004) Integrated RFMEMS inductors on thick silicon oxide layers fabricated using SIDeox process. In: Proceedings of the international conference on solid state and integrated circuits technology, Beijing, 2004, pp 1683–1686
10. Chan KT KT, Huang CH, Chin A et al (2003) Large Q-factor improvement for spiral inductors on silicon using proton implantation. IEEE Microwave Wireless Compon Lett 13:460–462
11. Carchon GJ, De Raedtand W, Beyne E (2004) Wafer-level packaging technology for high-Q on-chip inductors and transmission lines. IEEE Trans Microwave Theory Tech 52:1244–1251
12. Reyes AC, El-Ghazaly SM, Dorn SJ et al (1995) Coplanar waveguides and microwave inductors on silicon substrates. IEEE Trans Microwave Theory Tech 43:2016–2022
13. Rong B, Burghartz JN, Nanver LK et al (2004) Surface-passivated high-resistivity silicon substrates for RFICs. IEEE Electron Device Lett 25:176–178
14. Patti RS (2006) Three-dimensional integrated circuits and the future of system-on-chip designs. Proc IEEE 94:1214–1224
15. Swinnen B, Ruythooren W, De Moor P et al (2006) 3D integration by Cu-Cu thermo-compression bonding of extremely thinned bulk-Si die containing 10 μm pitch through-Si vias. In: Proceedings of the IEEE international electron device meeting (IEDM) 2006, San Francisco, 2006, pp 371–380
16. Jung E., Neumann A, Wojakowski D et al (2002) Ultra thin chips for miniaturized products. Proceedings of the electronics components and materials conference (ECTC) 2002, San Diego, 2002, pp 1110–1113
17. Zimmermann M, Burghartz JN, Appel W et al (2006) A seamless ultra-thin chip fabrication and assembly process. Proceedings of the IEEE international electron device meeting (IEDM) 2006, San Francisco, 2006, pp 1–3
18. Takyu S, Sagara J, Kurosawa T (2008) A study of chip thinning process for ultra-thin memory chips. In: Proceedings of the electronics components and technology conference (ECTC) 2008, Lake Buena Vista, 2006, pp 1511–1516
19. De Munck K, Chiarella T, De Moor P et al (2008) Influence of extreme thinning on 130-nm standard CMOS devices for 3-D integration. IEEE Electron Device Lett 29:322–324
20. Yue CP, Wong SS (2000) Physical modeling of spiral inductors on silicon. IEEE Trans Electron Devices 47:560–568
21. Niknejad A, Meyer R (1998) Analysis, design and optimization of spiral inductors and transformers for Si RF ICs. IEEE J Solid-State Circuits 33:1470–1481
22. Gilj J, Shin H (2003) A simple wide-band on-chip inductor model for silicon-based RF ICs. IEEE Trans Microwave Theory Tech 51:2023–2028
23. Hasegawa H, Furukawa M, Yanai H (1971) Properties of microstrip lines on Si-SiO_2 systems. IEEE Trans Microwave Theory Tech 19:869–881
24. Wee CY, San TB (2002) Study of interaction between the function of grid size and residual damage of an ultra thin wafer. In: Proceedings of the international conference on semiconductor electronics 2002, Malaysia, 2002, pp 163–168
25. Kröninger W, Mariani F (2006) Thinning and singulation of silicon: root causes of the damage in thin chips. In: Proceedings of the electronic components and technology conference (ECTC) 2006, San Diego, 2006, pp 1317–1322

Chapter 34
Through-Silicon via Technology for RF and Millimetre-Wave Applications

Alvin Joseph, Cheun-Wen Huang, Anthony Stamper, Wayne Woods, Dan Vanslette, Mete Erturk, Jim Dunn, Mike McPartlin, and Mark Doherty

Abstract This chapter focuses on some of the radiofrequency (RF) and millimetre-wave (MMW) applications that a through-silicon via (TSV) technology can enable. We will first discuss a grounded TSV application for a RF wireless communication power amplifier that is in mass production. A grounded TSV is essentially a metal connection that takes the active device interconnect to a package ground without needing to go through wire bond. A tungsten-filled grounded TSV delivers over 75% reduction in inductance compared to a traditional wire bond, thus enabling higher frequency applications in silicon technology. Such a grounded TSV is seamlessly integrated into a 350-nm SiGe BiCMOS.

Next we show feasibility on insulated TSV that is utilised in a digital technology for high speed I/O functions, such as, in an advanced microprocessor. The insulated TSV sits in a 3D-IC carrier silicon chip that is made compatible with standard 90-nm high performance CMOS processing. The TSV arrays are designed to provide minimal resistance and inductance with excellent yield and breakdown voltage characteristics. A commercial application of this concept is for the emerging 100-Gb Ethernet that requires almost 1 Tbps data transmission between ICs.

This concept of insulated via in a 3D-IC environment is then extended to a 130-nm MMW SiGe BiCMOS technology. We introduce the concept of a Wilkinson power divider that can be utilised in a 60-GHz phased array radar system.

34.1 Grounded-TSV for RF Power Amplifiers

The ground through-via is extensively used for both improved gain and thermal behaviour in a GaAs technology, the dominant RF power amplifier (PA) IC technology. In order for a silicon PA to compete in the marketplace a low

A. Joseph (✉)
IBM Semiconductor Research and Development Center, 1000 River Rd, Essex Junction, VT, USA
e-mail: josepha@us.ibm.com

J.N. Burghartz (ed.), *Ultra-thin Chip Technology and Applications*,
DOI 10.1007/978-1-4419-7276-7_34,

inductance approach needs to be introduced. The first commercial introduction of TSV in IBM technology is based on a 0.35-μm SiGe BiCMOS process [1].

In a 350-nm SiGe BiCMOS process flow, the TSV module is inserted as a modular feature at the contact module after the dielectric is deposited and the contact is opened (see Fig. 34.1). Masks are inserted after the contact module to define and etch the TSV areas. Tungsten is deposited and polished to simultaneously fill both the regular contact and the TSVs. The standard 0.35-μm Al metallisation flow is continued thereafter. Finally, the wafer is backside ground to expose the TSV and a backside metallisation completes the process. The wafer can then proceed for testing, dicing and sorting for packaging with only slight modification to the existing standard manufacturing flows. Since the ICs have to be packaged in a standard low cost standard QFN plastic package, with a die thickness on the order of 100–150 um, the TSV need to meet the required electrical performance of $R < 20$ mΩ/via, and inductance < 30 pH/via for such grounded TSVs.

TSV contacts require a special deep silicon etch tool designed to utilise the BOSCH etch process [2]. This process has excellent selectivity to resist and enables the creation of very high aspect ratio features at a reasonable etch rate within the existing litho processes. The TSV etch must be carefully controlled to create a contact with the proper shape to make metal fill easier. Care is taken to design the TSV feature such that proper metal fill is attained within this tungsten film thickness.

We have utilised this feature to design a state-of-the-art WLAN (wireless local area network) PA [3]. Figure 34.2 shows fully integrated dual band WLAN PA with a competitive die size of 1.7×1.6 mm^2. The PA integrates input and output matching network, out-of-band spur rejection filters, harmonic rejection filters, voltage regulator and bias circuits, temperature-compensated power detector and CMOS compatible enable circuitry. In addition to the integration, the dual-band PA

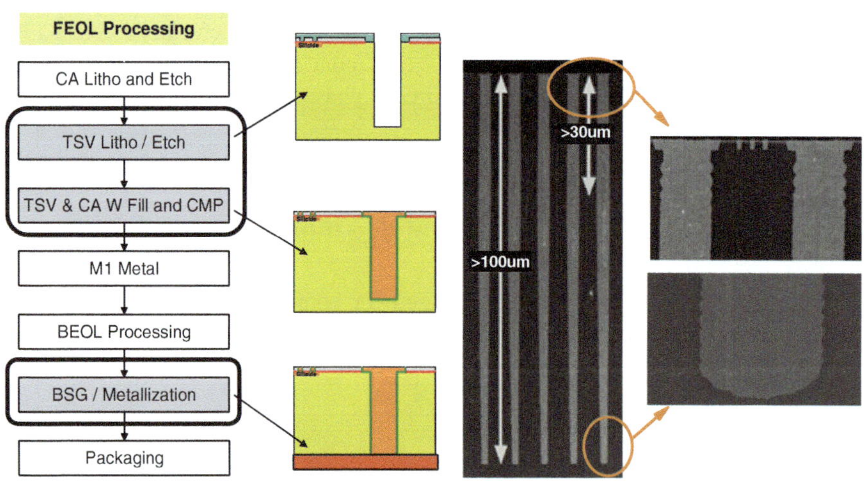

Fig. 34.1 Modular process flow of a TSV in the 0.35-μm SiGe BiCMOS technology. The SEM shows the details of the tungsten-filled grounded TSV

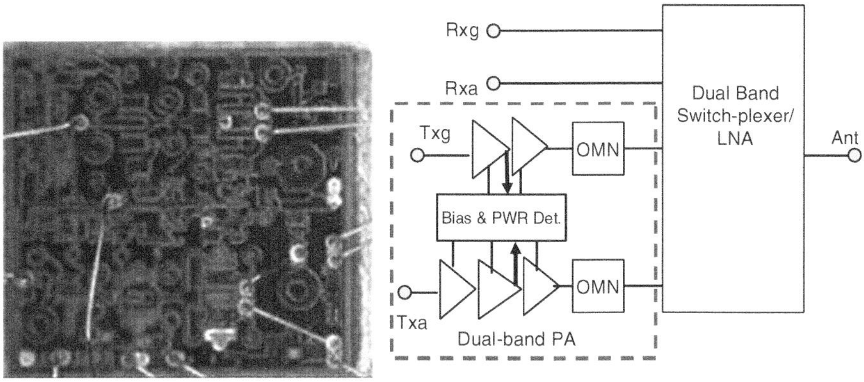

Fig. 34.2 A novel dual-band WLAN FEM architecture for WLAN a/b/g/n radios. Die photo of the dual-band WLAN amplifier based on SiGe BiCMOS technology

also demonstrates unparalleled RF performance from 4.9 to 5.9 GHz, exceeding previously published designs, thanks to the grounded TSV feature.

For 2.4–2.5 GHz, the b/g band[1] PA delivers 28 dB gain and 19.5-dBm linear power at 54 Mbps with error vector magnitude (EVM) $< 3\%$ and total current consumption < 185 mA. For 4.9–5.9 GHz, the a-band PA delivers 29 dB gain and > 19 dBm at 54 Mbps with EVM $< 3\%$ and total current consumption < 220 mA.

A GaAs PA presented in [4] has insufficient a-band bandwidth (4.9–5.9 GHz) and low integration due to its dependence on external surface mount component for matching networks. In addition, SiGe PA presented in [5] and CMOS PA presented in [6] both have high EVM levels, which will result in high packet error rate and requires more transmit time or current consumption. Moreover, [5] has noticeable gain slope in a-band PA. These comparisons confirm the view that a highly integrated SiGe-based amplifier serving both the b/g- and a-band delivers unprecedented performance and advantages. These advantages encompass both the traditional RF figures-of-merit (such as dynamic linearity) and other control features that are highly useful to the system integrator in a variety of WLAN applications.

34.2 3D-IC Isolated TSV for High-Speed Applications

The 3D-IC technology is clearly the next step in Moore's law scaling for obtaining increased bandwidth, at lower power and improved form factor [7]. The 3D-IC stacked systems create enormous opportunities for circuit designers to develop circuit elements that were previously unavailable to them. Many of these stacked chip-to-chip designs are targeted for digital applications, but mixed signal and analogue applications can also realise advantages by moving to stacked chip-to-chip systems.

[1]Radio frequency band: a-band: < 250 MHz; b-band: 250-500 MHz; g-band: 4-6 GHz

Typically, a silicon carrier is used in this 3D technology, which employs an insulated TSV to connect flip chips to the underlying substrate. The insulated TSV technology employs processes similar to the grounded TSV technology, especially the deep silicon etch process. The additional challenges of insulated TSV involves electrically isolating them from the silicon substrate and integrating them into the base-CMOS without impacting active and passive device parameters. An application of this technology is for a high speed 100-Gb Ethernet transmit-receive module that needs to interface high speed links between the DSP and ADC chips. In this case, one can envision the DSP chip being implemented in advanced RFCMOS node and the ADC chip in an advanced SiGe BiCMOS. However, the high-speed interface between these chips is built by using a silicon carrier chip with insulated TSV for optimized system performance.

We have utilised the qualified 200-mm RF CMOS at 90-nm node to demonstrate the insulated TSV silicon carrier technology [8]. The fusion of the high performance digital and low power RF capability in this technology node allows silicon carriers their optimal performance; the carriers require the high capacitance density (~40 $fF/\mu m^2$) of deep trench capacitors, high performance of low threshold voltage, thin gate oxide transistors, low parasitic capacitance of the tight pitch BEOL (back-end-of-line) wires and the low sheet resistance of the relaxed pitch super thick analogue wiring.

The 100-μm deep TSVs are insulated with a high temperature (>800°C) thermal oxide and are metallised with a CVD TiN/tungsten process similar to the one employed for grounded TSV technologies. A furnace-grown thermal oxide is used for the TSVs instead of a lower temperature CVD oxide to improve the dielectric integrity between the TSV and the silicon substrate. After wafer frontside processing is complete the wafers are attached to a temporary wafer handler, backside-ground to expose the TSVs, solder bumps are formed on the exposed TSVs on the wafer backside, chips are diced and mounted to a laminate package and one or more chips are flip-chip mounted to the active silicon carrier frontside. Figure 34.3 shows a simplified cross-sectional drawing and SEM of the laminate package. The silicon carrier chip has solder bumps on the backside, with flip chips mounted to the surface of the silicon carrier.

The TSV array and thinned wafer backside solder bump pitch is 185 μm, which translates into about 25,000 backside solder bumps per chip. The TSV bars are placed in parallel ~60×60-μm^2 arrays, with seven TSVs per array. The insulated TSVs are bar-shaped and are placed in farms of 7 bars to minimise inductance and resistance, and to enhance reliability by providing redundant conduction paths. The bars are ~5 μm wide at the top, ~3 μm wide at the bottom, ~60 μm long and 100 μm deep; the surrounding oxide insulator is ~1 μm thick. The insulated TSV farm resistance has been measured for TSVs with direct contact to M1 at ~4 mΩ.

Figures 34.4 and 34.5 show SEMs of the insulated TSV and the silicon wafer backside after solder bump formation. In order to attach the solder bumps the silicon carrier wafer is attached to a glass handle wafer, and the wafer backside silicon is removed to expose the bottom of the insulated TSV. Then the wafer backside is covered with CVD SiO_2. After which the Pb-free solder bumps are

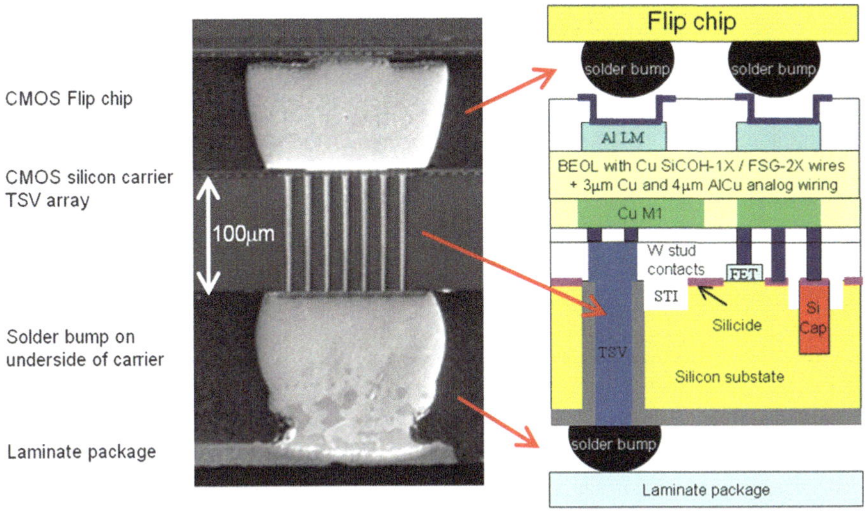

Fig. 34.3 Simplified drawing and SEM [2] cross-section of silicon carrier technology

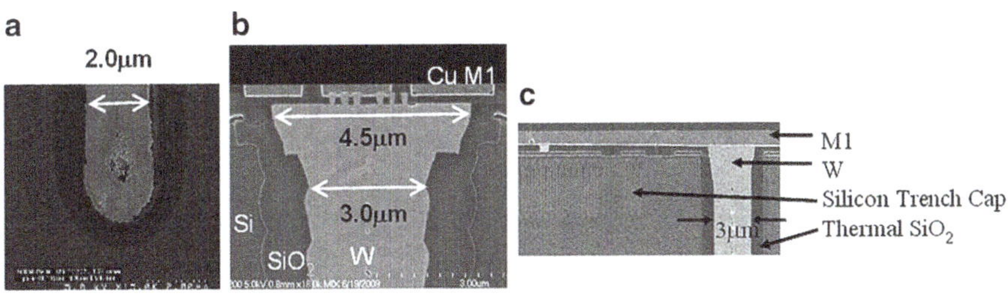

Fig. 34.4 (**a**) TSV bottom, (**b**) TSV-C1-M1 top and (**c**) details of TSV and 40 fF/μm^2 silicon deep trench capacitor

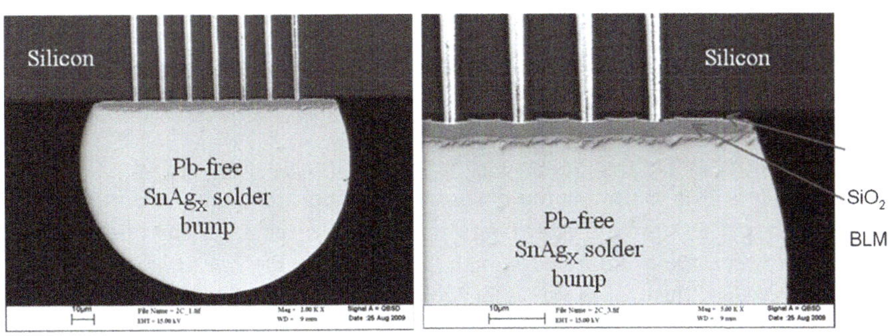

Fig. 34.5 SEM cross-sectional images of solder bumps formed on thinned silicon carrier wafer backside connected to an insulated TSV array

formed with standard processing under the exposed TSVs. At this point the silicon carrier chips are diced, the glass handle is detached, one or more flip chips are attached to the silicon carrier frontside and the solder bumps on the silicon carrier backside are attached to a laminate package.

The addition of insulated TSVs did not impact the electrical properties of the active 90-nm generation devices; the isolation between the TSVs and silicon substrate is excellent. We measure the breakdown voltage of the 7-bar TSV arrays to be over 400 V, using a 1-μA fail criteria by ramping the voltage between the TSV conductor and silicided silicon adjacent to the TSV. Overall, the integration show a feasibility of isolated TSVs built on a silicon carrier wafer with a method of mating the active device IC and providing good electrical connection.

34.3 MMW Wilkinson Power Divider

The silicon substrate is inherently lossy for RF and MMW applications; however, the usage of isolated vias with a combination of 3D-IC stacking can lead to new concepts for the next generation RF and MMW ICs. Wilkinson power dividers are needed in MMW circuit applications such as phased array antenna systems. Reducing the size of MMW Wilkinson power dividers reduces overall system cost. There is a limit to the minimum size that can be achieved for adequate power divider performance at a given frequency of operation in a conventional metal-dielectric BEOL stack. We describe a novel MMW Wilkinson power divider that utilises TSVs to reduce the area while realising good power divider performance at 60 GHz [9].

By utilising TSVs in a Wilkinson power divider one can make use of the third dimension extending into the silicon substrate to transmit signals. A novel TSV Wilkinson power divider presented here is designed in a 130-nm IBM SiGe BiCMOS technology. Ansoft's HFSS was used to simulate the performance of the TSV Wilkinson power divider. The design studied here requires two patterned metal layers on the backside of the wafer, having the same thickness of roughly 4 μm. This configuration allows 50-Ω microstrip transmission lines to be created on both the front- and backsides of the chip.

The general structure of a Wilkinson power divider is shown in Fig. 34.6. Its function is to split the input power evenly between two output lines while also providing high isolation between the two output paths. The two 'arms' of the Wilkinson power divider are $\lambda/4$ long. In the MMW Wilkinson power divider embodiment presented here, the two arms are constructed using TSVs.

The wafer thickness used in our design was 300 μm. Intra-substrate microstrip-like TSV signal and ground paths were used to form the $\lambda/4$ arms of the power divider. Between the ground and signal TSVs used in the $\lambda/4$ arms, oxide-filled TSVs were placed to allow higher impedance by decreasing the effective dielectric between the signal and ground TSV paths. The frequency of operation of the proposed TSV Wilkinson power divider is determined by the thickness of the silicon substrate. For the 300-μm thick silicon substrate and the Wilkinson power divider

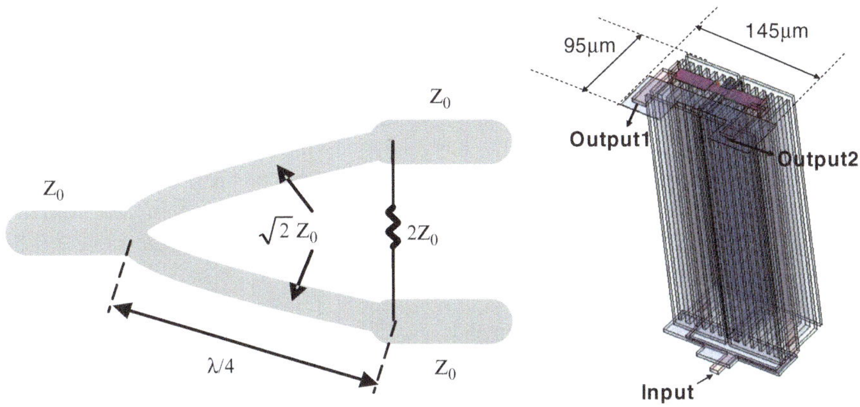

Fig. 34.6 Wilkinson power divider and its TSV embodiment

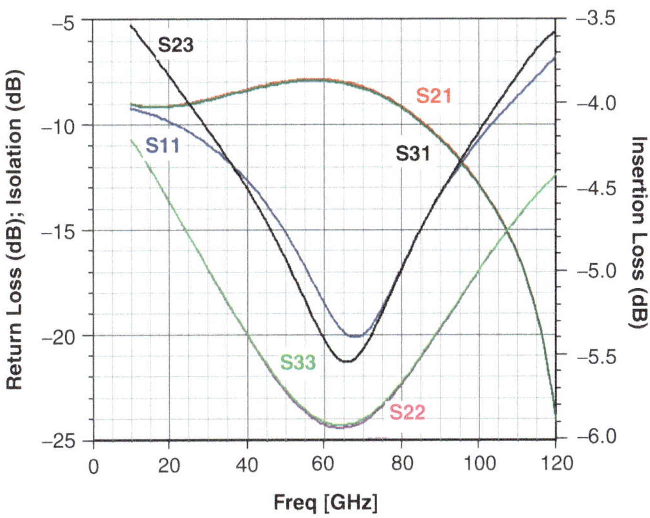

Fig. 34.7 HFSS simulated return loss, isolation and insertion loss of 60 GHz TSV Wilkinson power divider

geometry studied here, an operating frequency around 60 GHz was targeted. The silicon footprint of the structure on the frontside of the chip is 95 μm × 145 μm.

There are many TSVs utilised in the conceptual structure in Fig. 34.6. However, only four of the TSV structures are tungsten-filled conductors; the rest are silicon-dioxide filled TSVs. These oxide-filled holes reduce the overall footprint of the structure by reducing the effective dielectric constant between the signal TSVs and the ground TSVs.

HFSS simulations of the TSV Wilkinson power divider shown in Fig. 34.7 revealed insertion loss per λ/4 arm of 0.9 dB at 60 GHz. Both return loss and isolation of the power divider were better than 18 dB at 60 GHz with good matching in both phase and amplitude. The S23 plot is the simulated isolation in dB between the two outputs: the "Output1" and "Output2" microstrip lines on the frontside of the silicon chip. It is seen that the S23 isolation between the output paths of the power divider reaches best performance around 65 GHz of around 21 dB. The S11 return loss at the input of the device has a best-case performance of 20 dB around 67 GHz. The S22 and S33 curves almost exactly overlap one another, and they both show best-case performance of 24 dB around 65 GHz. The best-case insertion loss performance from the input to the two outputs, S21 and S31, is seen to be 3.9 dB around 57 GHz. These results are very encouraging and show a path for highly integrated MMW passives in a silicon technology. Silicon substrate thickness ranging from 300 μm to less than 100 μm are possible in this technology, allowing circuits with operating frequencies much higher than 60 GHz to make use the proposed power divider design.

34.4 Conclusions

The promise of TSV and 3D-IC stacking has, at last, arrived. As a breakthrough technology for silicon PA, we have successfully introduced a grounded TSV in a 350-nm SiGe BiCMOS technology, now in mass production. The grounded TSV has shown to meet the RF performance and reliability, thus expanding the competitiveness of silicon based PA technology. As an example of a state-of-the-art, we have demonstrated a $1.7 \times 1.6\text{-mm}^2$ SiGe BiCMOS dual-band WLAN PA. The utilisation of the grounded TSV feature in this PA design has enabled the performance to surpass previously published WLAN dual-band PA designs in linearity, gain flatness, out-of-band rejection and die size compactness.

For a 3D-IC stacking we have developed an isolated TSV feature that is fully compatible with the 90-nm CMOS technology with a 200-mm wafer silicon carrier technology. To minimise TSV resistance and inductance and to maximise yield and reliability, the TSVs are placed in redundant arrays. An extension of the grounded TSV along with the 3D-IC stacking has been explored for usage in a MMW phased array application. A novel on-chip MMW TSV Wilkinson power divider has been explored through HFSS simulations. A 60-GHz Wilkinson power divider presented here provides adequate performance and significantly reduces the total silicon footprint. Future direction will focus on exploring the fabrication and demonstration of such TSV integrations to provide critical system advantages of size reduction and potential performance gains in and RF and MMW ICs.

Acknowledgement The success of TSV and 3D-IC integration for this work was the countless efforts from many people spread amongst the various IBM and other groups. For their efforts we are grateful.

References

1. Joseph A.J et al (2008) Through-silicon vias enable next-generation SiGe power amplifiers for wireless communications. IBM Journal of Research and Development, Vol. 52, No. 6, pp. 635–648, August 2008.
2. Laermer F et al (2005) Technical digest. *13th international conference on solid-state sensors, actuators and microsystems*, vol 2, pp 1118–1121.
3. Huang C-W.P et al (2010) "A highly integrated dual band SiGe BiCMOS power amplifier that simplifies dual-band WLAN and MIMO front-end circuit designs," Microwave symposium digest (MTT), 2010. *IEEE MTT-S international*, pp 256–259.
4. Chien-Cheng Lin et al (2005) "Single-chip dual-band WLAN power amplifier using InGaP/GaAs HBT", *IEEE Gallium Arsenide and Other Semiconductor Application Symposium*, 2005, pp 489–492.
5. Liao H-H et al (2008) A fully integrated 2 × 2 power amplifier for dual band MIMO 802.11n WLAN application using SiGe HBT technology. *2008 IEEE RFIC symp dig*, June 2008.
6. Afsahi A et al (2009) Fully integrated dual-band power amplifiers with on-chip Baluns in 65-nm CMOS for an 802.11n MIMO WLAN SoC. *2009 IEEE RFIC symp dig*, June 2009.
7. Shapiro M et al (2009) *Proceedings of the international interconnection technology conference (IITC)*, p 63, Sapporo, Japan, 2009.
8. Dang B et al (2008) IBM J Res Dev 52(6):599–609
9. Woods W et al (2010) Novel on-chip through-silicon via MMW Wilkinson power divider. *IEEE electronic components and technology conference 2010*, June 2010.

Index

J.N. Burghartz (ed.), *Ultra-thin Chip Technology and Applications*,
DOI 10.1007/978-1-4419-7276-7,

D

GPSR Compliance

The European Union's (EU) General Product Safety Regulation (GPSR) is a set of rules that requires consumer products to be safe and our obligations to ensure this.

If you have any concerns about our products, you can contact us on ProductSafety@springernature.com

In case Publisher is established outside the EU, the EU authorized representative is:

Springer Nature Customer Service Center GmbH
Europaplatz 3
69115 Heidelberg, Germany

Batch number: 10614938

Printed by Printforce, the Netherlands